MW01032017

Throughout the twentieth century, from the furor over Percival Lowell's claim of canals on Mars to the sophisticated Search for Extraterrestrial Intelligence (SETI), otherworldly life has often intrigued and occasionally consumed science and the public. Is our species biologically unique – in the sense of mind and intelligence – in the universe? Are there other histories, religions, and philosophies outside of those on Earth? Do extraterrestrial minds ponder the mysteries of the universe? The attempts to answer these often asked questions form one of the most interesting chapters in the history of science and culture, and *The Biological Universe* is the first book to provide a rich and colorful history of those attempts during the twentieth century.

Covering a broad range of topics, including the search for life in the solar system, SETI, the origins of life, UFO's, and aliens in science fiction, Steven J. Dick's study shows how the concept of extraterrestrial intelligence is a worldview of its own, a "biophysical cosmology" that seeks confirmation no less than do physical views of the universe. No matter how much we learn about the varied life forms of Earth and the physical nature of the universe of which we are part, the question of biological uniqueness is central to the quest for who we are and what our role in nature may be, questions as much a part of religion and philosophy as of science. This history is not only important for an understanding of the nature of science, but is also central to any forward-looking concept of religion, philosophy, and other areas of human endeavor.

The Biological Universe

"THERE ARE CERTAIN FEATURES IN WHICH THEY ARE LIKELY TO RESEMBLE US. AND AS LIKELY AS NOT THEY WILL BE COVERED WITH FEATHERS OR FUR. IT IS NO LESS REASONABLE TO SUPPOSE, INSTEAD OF A HAND, A GROUP OF TENTACLES OR PROBOSCIS LIKE ORGANS"

Drawing by William R. Leigh from H. G. Wells, "The Things That Live on Mars," a nonfiction article that appeared in *Cosmopolitan* magazine for March 1908 at the height of the Martian canals furor.

The Biological Universe

The Twentieth-Century Extraterrestrial Life Debate and the Limits of Science

STEVEN J. DICK

Published by the Press Syndicate of the University of Cambridge
The Pitt Building, Trumpington Street, Cambridge CB2 1RP
40 West 20th Street, New York, NY 10011-4211, USA
10 Stamford Road, Oakleigh, Melbourne 3166, Australia

© Cambridge University Press 1996

First published 1996

Printed in the United States of America

Library of Congress Cataloging-in-Publication Data
Dick, Steven J.
The biological universe : the twentieth-century extraterrestrial
life debate and the limits of science / Steven J. Dick.
p. cm.
Includes bibliographical references and index.
ISBN 0 521 34326 7 (hardcover)
1. Life on other planets. 2. Life – Origin. 3. Unidentified
QB54.D47 1996
574.999 – dc20 95-36236
CIP

A catalog record for this book is available from the British Library.

ISBN 0 521 34326 7 Hardback

To Terry, Gregory, and Anthony
with love

Glendower: I can call spirits from the vasty deep.

Hotspur: Why, so can I, or so can any man;
But will they come when you do call for them?

Shakespeare
Henry IV, Part I
Act 3, Scene I, 52–58

CONTENTS

List of Illustrations and Tables		*page* x
Acknowledgments		xiii
Abbreviations		xv
Introduction		1
1	From the Physical World to the Biological Universe: Democritus to Lowell	10
	1.1 The Cosmological Connection	11
	1.2 Philosophical Explorations	22
	1.3 Scientific Foundations	29
2	Plurality of Worlds and the Decline of Anthropocentrism	36
	2.1 The Anthropocentric Universe: A. R. Wallace	38
	2.2 The New Universe: Anthropocentrism's Demise?	50
3	Life in the Solar System: The Limits of Observation	59
	3.1 Lowell and Mars: The Search for Intelligence, 1894–1924	62
	3.2 The Search for Martian Vegetation, 1924–1957	105
	3.3 Venus: Last Hope for Intelligence	126
	3.4 The Space Age: Lowell's Legacy Overturned	135
4	Planetary Systems: The Limits of Theory	160
	4.1 Skepticism: Observational Hints and Stellar Encounters	162
	4.2 Turning Point: 1943–1958	180
	4.3 Optimism: Observation to the Rescue?	200
5	Extraterrestrials in Literature and the Arts: The Role of Imagination	222
	5.1 The Invention of the Alien: Verne, Wells, and Lasswitz	223
	5.2 The Development and Uses of the Alien: Burroughs to Bradbury	238
	5.3 The Alien Comes of Age: Clarke, ET, and Beyond	253
6	The UFO Controversy and the Extraterrestrial Hypothesis	267
	6.1 Rise of the Extraterrestrial Hypothesis	268

6.2 The Peak of the Extraterrestrial Hypothesis: 1965–1969 — 288

6.3 Aftermath: The Nature of Evidence and the Decline of the Extraterrestrial Hypothesis in Physical Science — 307

7 The Origin and Evolution of Life in the Extraterrestrial Context — 321

7.1 Arrhenius and Panspermia: An Extraterrestrial Theory of the Origin of Life — 325

7.2 Interlude: The Rise of the Chemical Theory and Its Latent Extraterrestrial Implications, 1924–1957 — 329

7.3 The Integration of Origin of Life and Extraterrestrial Life Studies in the Space Age — 350

7.4 Evolution and Extraterrestrials: Chance and Necessity Revisited — 389

8 SETI: The Search for Extraterrestrial Intelligence — 399

8.1 Prelude: The Era of Interplanetary Communication — 401

8.2 Cornell, Ozma, and Green Bank: The Opening of the Electromagnetic Spectrum for SETI — 414

8.3 A Rationale for SETI: Optimists, Pessimists, and the Drake Equation — 431

8.4 A Strategy for SETI: The Development of Observational Programs — 454

9 The Convergence of Disciplines: Birth of a New Science — 473

9.1 Perceptions of a New Discipline — 475

9.2 Networks: Formation of the Scientific Community — 478

9.3 Institutions: Programs and Funding — 494

10 The Meaning of Life: Implications of Extraterrestrial Intelligence — 502

10.1 Perceptions of Cultural Impact — 503

10.2 Astrotheology — 514

10.3 Life and Purpose in the Universe: The Anthropic Principle — 527

11 Summary and Conclusion: The Biological Universe and the Limits of Science — 537

11.1 The Triumph of Cosmic Evolution — 538

11.2 The Biological Universe as Cosmological Worldview — 541

11.3 The Problematic Nature of Evidence and Inference — 543

11.4 The Limits of Science — 546

11.5 The Cultures of Science — 550

11.6 Exobiology as Protoscience 551
11.7 Cultural Significance of the Debate 552

Select Bibliographical Essay 555
Index 561

ILLUSTRATIONS AND TABLES

Illustrations

1.1	Fontenelle's plurality of solar systems	*page* 19
1.2	R. A. Proctor	31
1.3	Camille Flammarion	33
2.1	Wallace's argument for a single inhabited world	41
2.2	Wallace's anthropocentric image of the universe (1903)	43
2.3	The Andromeda Galaxy, symbol of the new universe	56
3.1	Research publications on Mars (1900–1957)	62
3.2	Lowell at the 24-inch refracting telescope	63
3.3	Most favorable and least favorable oppositions of Mars (1877–1995)	67
3.4	The Mars of Schiaparelli (1877–1888), from Flammarion (1892)	73
3.5	The Mars of Lowell (1895)	80
3.6	E. M. Antoniadi	91
3.7	Antoniadi's map of Mars (1930)	96
3.8	Antoniadi's comparison of his Mars observations with Schiaparelli's	98
3.9	Lowellian canal network compared to Mariner cartography	103
3.10	Kuiper's observation of CO_2 bands on Mars	119
3.11	Sinton's observation of infrared bands, interpreted as evidence of Martian vegetation	124
3.12	Viking biology package	148
3.13	G. Levin's data from the Viking lander	156
4.1	Chamberlin–Moulton hypothesis	169
4.2	Jeans–Jeffreys hypothesis	174
4.3	James Jeans, Edwin Hubble, and Walter S. Adams (1931)	177
4.4	Planet hunter Peter van de Kamp with telescope	184
4.5	Separations of binary stars – Kuiper's argument for planetary systems	192
4.6	Astrometric evidence for planetary systems	204

4.7 Material observed around the star Beta Pictoris 212
4.8 Evidence for planets around a pulsar 216
4.9 Hubble Space Telescope observations of protoplanetary systems in Orion 217
4.10 Classes of objects in the search for planetary systems 218
5.1 Kurd Lasswitz 228
5.2 H. G. Wells 232
5.3 1927 *Amazing Stories* depiction of H. G. Wells's *War of the Worlds* 237
6.1 1929 *Science Wonder Stories* depiction of a flying saucer 273
7.1 The generalized cell in 1922 335
7.2 Building blocks of life 346
7.3 Melvin Calvin 353
7.4 Evolution of chemical and biological complexity 359
7.5 Organized element in meteorites 370
7.6 Scheme for delivery of organics to Earth 376
8.1 A method for communication with Mars (1920) 405
8.2 U.S. Army listening for Martian signals (1924) 409
8.3 Frank Drake and the National Radio Astronomy Observatory's 85-foot telescope 420
8.4 Microwave window 424
8.5 Project Cyclops 461
8.6 The cosmic haystack 465
9.1 Carl Sagan 479
9.2 Joshua Lederberg 480
9.3 Participants in the third U.S.–USSR meeting on SETI 487
9.4 Philip Morrison 489
9.5 Published literature in bioastronomy 493
11.1 NASA depiction of cosmic evolution 540
11.2 1930 *Amazing Stories* cover showing an alien imparting knowledge of the Galaxy 553

Tables

3.1 Observational highlights of Martian canals, 1877–1924 65
3.2 Interpretations of Martian canals, 1877–1924 82
3.3 Milestones in Martian observations related to life, 1924–1957 110
3.4 Important advances in knowledge of Venus 128
3.5 Space Age observations of Mars relevant to life 144

4.1 Planetary companions in the context of angular measurement in astronomy 183
4.2 Estimates of frequency of planetary systems, 1920–1961 199
4.3 Observational milestones in the search for planetary systems 209
6.1 UFO reports received each month by Project Blue Book, 1950–1968 270
6.2 Spectrum of scientific cultures on the UFO question 313
7.1 Selected conferences on the origin of life 365
8.1 Estimates of factors in the Drake Equation for communicative civilizations 441
8.2 Explanations for the apparent absence of extraterrestrials on Earth 451
8.3 Characteristics of selected SETI observing programs 460
9.1 Selected conferences on extraterrestrial life 484
9.2 Selected extraterrestrial life conference papers by discipline 491

ACKNOWLEDGMENTS

I wish to thank Michael J. Crowe (University of Notre Dame) and an anonymous reviewer for Cambridge University Press for reading and commenting on the entire manuscript. For reading parts relevant to their expertise, I am grateful to Ronald Doel and David DeVorkin (National Air and Space Museum), Joshua Lederberg (Rockefeller University), Ronald Schorn, Karl S. Guthke (Harvard University), H. P. Klein (Santa Clara University), Robert Shapiro (New York University), Betty Smocovitis (University of Florida), Philip Klass, David Jacobs (Temple University), Michael Swords (Western Michigan University), and Peter Sturrock (Stanford University).

Among libraries, the unparalleled astronomy collections of the U.S. Naval Observatory Library have been essential for the astronomical portions of this study, as has the assistance of its librarians, Brenda Corbin and Gregory Shelton. In addition, the Library of Congress and the libraries of The American University and Georgetown University have helped fill gaps in nonastronomical literature. I am grateful for access to archives at the British Library, London (A. R. Wallace Papers); the Royal Society, London (James Jeans Papers); the American Philosophical Society Library, Philadelphia (E. U. Condon and D. H. Menzel Papers); the Lowell Observatory, Flagstaff, Arizona; the University of Arizona, Tucson (A. E. Douglass and G. P. Kuiper Papers); Mary Lea Shane Archives of the Lick Observatory (Robert Trumpler Papers); the U.S. Naval Observatory, Washington, D.C. (Clemence Papers); and NASA Ames Research Center and the SETI Institute in Mountain View, California, for access to SETI archives.

I also wish to thank the SETI Institute for support in undertaking oral history interviews and for the cooperation of all those interviewed. These include John Billingham (NASA Ames), Peter Backus (SETI Institute), David Brocker (NASA Ames), Melvin Calvin (University of California, Berkeley), Gary Coulter (NASA headquarters), Frank Drake (University of California, Santa Cruz), Sam Gulkis (JPL), Nikolai Kardashev, Philip J. Klass, H. P. Klein (Santa Clara University), Michael Klein (JPL), Joshua Lederberg (Rockefeller University), Edward Olsen (JPL), Bernard M. Oliver (NASA Ames), Michael Papagiannis (Boston University), Tom Pierson

(SETI Institute), Carl Sagan (Cornell University), Charles Seeger (SETI Institute), Jill Tarter (SETI Institute), and Peter van de Kamp. Oral history interviews of related interest will be found at the Center for the History of Physics of the American Institute of Physics, located at the American Center for Physics in College Park, Maryland. It is a pleasure to acknowledge the usefulness of David Swift's published interviews in *SETI Pioneers* (University of Arizona Press: Tucson, 1990).

I am grateful to Garland Publishing for permission to draw from one of my previous articles for Chapter 1 and to Reidel Publishers for permission to use portions of a previous article for Chapter 8. I wish to thank all those publishers for allowing me to reproduce illustrations, as stated in the credits, and Suzanne Débarbat (Paris Observatory) and the Juvisy Observatory for help in obtaining the Flammarion photograph. Finally, my thanks to Alex Holzman, my editor at Cambridge University Press, for his forbearance with a manuscript that grew beyond original expectations. My thanks, too, to Helen Wheeler for her important role at the beginning and ending stages of this project and to Helen Greenberg for judicious copyediting.

This study was written over a period of 10 years in various locations including New Zealand, California, and Austria, but mostly from my home in Herndon, Virginia, near Washington, D.C. I therefore wish to thank most of all my wife, Terry, and my sons, Gregory and Anthony, to whom this book is dedicated. Through two decades they have wandered with me through what must have seemed to them (as at times it did to me) an infinite number of worlds.

ABBREVIATIONS

A&A	*Astronautics and Aeronautics*
AJ	*Astronomical Journal*
AN	*Astronomische Nachrichten*
ANYAS	*Annals of the New York Academy of Sciences*
ApJ	*Astrophysical Journal*
BMNAS	*Biographical Memoirs of the National Academy of Science*
DSB	*Dictionary of Scientific Biography*
JBAA	*Journal of the British Astronomical Association*
JBIS	*Journal of the British Interplanetary Society*
JGR	*Journal of Geophysical Research*
JHA	*Journal for the History of Astronomy*
JOSA	*Journal of the Optical Society of America*
JRASC	*Journal of the Royal Astronomical Society of Canada*
LOB	*Lowell Observatory Bulletin*
MBAA	*Memoirs of the British Astronomical Association*
MNRAS	*Monthly Notices of the Royal Astronomical Society*
MRAS	*Memoirs of the Royal Astronomical Society*
NYT	*New York Times*
OL	*Origins of Life*
PA	*Popular Astronomy*
PASP	*Publications of the Astronomical Society of the Pacific*
PNAS	*Proceedings of the National Academy of Science*

PTr	*Philosophical Transactions of the Royal Society of London*
QJRAS	*Quarterly Journal of the Royal Astronomical Society*
RMP	*Reviews of Modern Physics*
SAFB	*Bulletin de la Société astronomique de France*
SAJ	*Soviet Astronomical Journal*
SAL	*Soviet Astronomy Letters*
SciAm	*Scientific American*
SciAmS	*Scientific American Supplement*
SFS	*Science Fiction Studies*
SFE	*Science Fiction Encyclopedia*
S&T	*Sky and Telescope*
VA	*Vistas in Astronomy*

INTRODUCTION

The eternal silence of these infinite spaces frightens me.

Pascal, Pensées

Everywhere around us we see the infinite diversity of life on our planet, but nowhere else in the vast universe has any life been found. This remarkable fact – at once a measure of our ignorance and a cry of loneliness echoing through the centuries – is decidedly not a result of indifference on the part of *Homo sapiens*. At least since the time of the ancient Greeks, inquiring minds have wondered whether we are alone in the universe, and since that time the question has been a frequent companion to science and philosophy. We have been to the Moon and Mars; we have explored the solar system to its outer limits; we have begun to listen for intelligent radio signals from our stellar neighborhood. And everywhere we find – in terms of life – desolation and silence. Still, there is no abatement in the quest for cosmic companionship; in many ways, the search has only begun.

The reason for this remarkable interest is no mystery, and in my view can only in part be explained by deep psychological yearnings for companionship, superior wisdom, or "an ineradicable desire to see the skies filled with life."[1] Although all of these elements undoubtedly play their role, the fascination of this question for the historian is that it has for centuries been viewed as amenable to contemporary science, even if at its outermost limits, and that it has had a broad effect on general culture disproportionate to any truly verified claims. As science has removed our home planet successively from the center of the solar system and the center of the Galaxy, and left it bereft of any unique physical position in the universe, the question that has come ever more to the fore in the twentieth century is whether our species is *biologically* unique in the universe, not in the narrow sense of form and function but in the general sense of mind and intelligence. No matter how much we learn about the varied life forms of Earth and the physical nature of the universe of which we are a part, the question of biological uniqueness is central to the quest for who we are and what our role in nature may be, questions as much a part of religion and philosophy as of science. The whole thrust of physical science since the seventeenth-century scientific revolution has been to demonstrate the role of physical law in the

[1] Robert Plank, *The Emotional Significance of Imaginary Beings: A Study of the Interaction between Psychopathology, Literature, and Reality in the Modern World* (Springfield, Ill., 1968), vii.

universe, a mission admirably carried out by Kepler, Galileo, Newton, and their successors. The question at stake in the extraterrestrial life debate is whether an analogous "biological law" reigns throughout the universe; whether there are other histories, religions, and philosophies; whether, in the words of H. G. Wells, "minds that are to our minds as ours are to those of the beasts that perish" ponder the mysteries of the universe. These are questions that have been asked many times and are still unanswered, but the attempts to answer them form one of the most interesting chapters in the history of science and culture. The rich and colorful history of those attempts during the twentieth century is the subject of this book.

Within the last dozen years, the history of ideas about life beyond the Earth has become well known, but only up to the beginning of the twentieth century. My own *Plurality of Worlds: The Origins of the Extraterrestrial Life Debate from Democritus to Kant* (1982) traced the early history of the debate from the perspective of the history of science. Professor Michael J. Crowe's massive study, *The Extraterrestrial Life Debate, 1750–1900: The Idea of a Plurality of Worlds from Kant to Lowell* (1986) brought the story up to the present century. And Professor Karl S. Guthke's valuable contribution, *The Last Frontier: Imagining Other Worlds from the Copernican Revolution to Modern Science Fiction* (1990), examined the problem from the broader perspective of intellectual and literary history. Curiously, the twentieth century – during which scientific approaches to the subject greatly increased in number and complexity, and greater strides were made than in all previous centuries combined – has remained unexamined from the historical perspective. In many ways, the present study may be considered a continuation of those volumes, especially the first two, also published by Cambridge University Press. Taken together, these volumes comprise a detailed history of the debate from the ancient Greeks to the present.[2]

Precisely because the subject is so elusive and yet so compelling for science, and because it is central to any forward-looking concept of religion, philosophy, and numerous other areas of human endeavor, there are many reasons that such a study should be undertaken. Not only is extraterrestrial life a persistent theme of our time, the history of the pursuit of extraterrestrials is also a key to the nature of science, a field that dominates our culture.

[2] Also very useful is the popular treatment by the distinguished *New York Times* reporter Walter Sullivan, *We Are Not Alone* (New York, 1964; revised edition, 1993), as well as a variety of shorter studies cited in these works.

INTRODUCTION

THE BIOLOGICAL UNIVERSE: A PERSISTENT THEME

The idea of extraterrestrial life has been a major theme of twentieth-century science and popular culture, if not yet of religion and philosophy. Indeed, Professor Guthke, in his study, characterizes it as a "religion or quasi-religion," and again as "a 'myth,' or even *the* 'myth' " of the modern age, where the term "myth" is used not in the narrow sense of something untrue but in the broad, positive sense of a symbolic tradition that defines how we understand ourselves. Fully 58 percent of highly educated Americans responding to a Gallup poll in 1973 affirmed their belief in intelligent life on other planets, a number that remained substantially unchanged by 1990.[3] From the furor over Percival Lowell's claim of canals on Mars at the beginning of the century to the Viking spacecraft and NASA's sophisticated Search for Extraterrestrial Intelligence (SETI) at its end, otherworldly life has often titillated and occasionally consumed science and the public. So too have crucially related areas such as the search for planetary systems and the quest for the origin of life. When astronomer Peter van de Kamp announced the existence of a planet around Barnard's star in the early 1960s (now believed spurious), the implication was that this was the vanguard of a raft of inhabited planets. When Stanley Miller produced amino acids in an experiment at the University of Chicago in 1953, it was in an attempt to illuminate the origin of life on Earth under primitive terrestrial conditions; but it was not long before Miller and others extended their discussions to life beyond the Earth. And it was not long before the Space Age reciprocally influenced studies of the origin of life.

Even more titillating from the point of view of popular culture has been the idea that alien intelligence might actually have arrived on Earth; the debate over unidentified flying objects (UFOs) and their possible extraterrestrial nature has not only captured the attention of a broad (and sometimes undiscriminating) public, but has also evoked pointed reactions from the scientific community that shed light on its schizophrenic attitude toward alien life. Major polls show that a majority of Americans in the latter part of the century still believe UFOs are spaceships piloted by extraterrestrials, and during the 1960s the idea received attention from

[3] William Sims Bainbridge, "Attitudes Toward Interstellar Communication: An Empirical Study," *Journal of the British Interplanetary Society*, 36 (1983), 298–304, citing *The Gallup Opinion Index* (January 1974), 20. See also the Gallup Poll of June 1990 in "Mirror of America – The Panorama," press release, August 5, 1990, and the annual volumes of The Gallup Polls from 1978 and 1987. In the polls, 46% of Americans overall affirmed belief in 1973 and 1990. Karl S. Guthke, *The Last Frontier: Imagining Other Worlds from the Copernican Revolution to Modern Science Fiction* (Ithaca, N.Y., 1990), ix–xii; originally published as *Der Mythos der Neuzeit* (Bern, 1983).

the U.S. Congress, the U.S. Air Force, the CIA, and the National Academy of Sciences, to mention only agencies in the United States.[4] This mass popular reaction is mirrored around the world and perhaps indicates that a deep-seated need in the human psyche is *part* of the phenomenon, even if not the primary driving force. Add to this that aliens have increasingly become a prominent theme of that literary genre known as "science fiction," and one begins to have some idea of how the concept of extraterrestrials has pervaded popular culture. The vivid portrayal of such themes in films such as *ET, Alien, Star Wars* and *Close Encounters of the Third Kind* has only heightened this effect. That these movies have been among the most popular in the history of film confirms that the theme repeatedly claims the attention of a broad public, which may view such films not only as entertainment, but also as a reflection of reality.

On the most basic level, then, this book is a study of a persistent theme in twentieth-century Western culture, a theme that is reflected in other cultures to an extent that still needs to be determined. More than that, it is the story of the transformation of human thought from the physical world to the biological universe, a story no less profound than the move from the closed world to the infinite universe that historians well documented decades ago. That the majority of humans now view themselves as surrounded by celestial intelligences, either on Earth or in outer space, surely defines the modern world no less than did the choirs of angels and legions of devils in the medieval world. And the persistence of these supernatural intelligences in some religions of the modern world leads one to believe by analogy that extraterrestrials will persist in human consciousness, even lacking definitive evidence of their existence.

STATUS OF THE HISTORICAL DEBATE

It is no accident that the concept of extraterrestrial life, long the subject of debate among philosophers, scientists, and other inquisitive minds, has only recently come under the scrutiny of historians. The reason for this curious delay is that until recently its history was not considered part of any mainstream intellectual tradition, and certainly not a respectable one in itself. Life on other worlds seemed to be part of the flotsam of curious ideas occasionally spewed forth from mostly erratic minds and, unaccountably, from occasionally brilliant minds. And certainly it seemed to have very little to do with the supposedly rigid rules of reasoning and method that went under the rubric of "science." With a more enlightened

[4] In 1985 54% of Americans believed UFOs were spaceships; the figures were 57% in 1988 and 54% in 1990. Kendrick Frazier, "UFOs as ET Spacecraft? Belief Strong, Steady," *Skeptical Inquirer*, 16 (Summer 1992), 346.

view of the nature of science as including philosophical and social as well as cognitive elements, however, the door was opened to an examination of the place of the idea in a broadened definition of science as the search for knowledge, regardless of preconceived pristine notions of scientific method. But while the door was ajar, the room remained dim; if the historical studies cited previously have surely demonstrated a role for the subject in intellectual history, its status as a science remains controversial even to the present day. Exobiology is "a science looking for a subject," wrote evolutionary biologist George Gaylord Simpson in the 1960s. The subject is a pseudoscience that should not be represented in international scientific societies, physicist Frank Tipler protested during the 1980s. Harvard University evolutionist Ernst Mayr found it a "highly dubious endeavor" and professed himself "astounded" that NASA would fund a SETI program in the 1990s at a time of great federal debt. The U.S. Congress agreed, and terminated the program amid ridiculing remarks about "the great Martian chase" and "little green men."

Along with the basic history of the debate, then, there is clearly a metahistoric issue of the the cognitive status of the debate. As long ago as 1936, the historian Arthur O. Lovejoy concluded that the idea of life beyond the Earth, as well as other conceptions in the new cosmography of the seventeenth century, owed little to any observational facts or "scientific" reasoning, but rather "were chiefly derivative from philosophical and theological premises." In particular, as part of his study of the idea of the great chain of being, Lovejoy held certain cosmographic ideas, including that of extramundane life, to be corollaries of the principle of plenitude, the purely philosophical tenet that the nature of God or the fecundity of Nature demands that anything that can be created must be created, a principle with obvious application to the diffusion of life throughout the universe.[5]

This claim of Lovejoy, while undoubtedly harboring an element of truth, went unchallenged until my 1982 study showed how intimately related the debate was to seminal scientific traditions, including the atomist, Aristotelian, Copernican, Cartesian, and Newtonian worldviews, in which "scientific tradition" encompassed a broadened view of science as an enterprise composed of both philosophical and empirical elements. Professor Crowe, while not denying the role of scientific elements such as observation, came down in his study decidedly on the side of philosophy as having the paramount role, at least in the eighteenth and nineteenth centuries. Professor Guthke's concept of extraterrestrial life as a "myth of modern times" embraces both scientific and philosophical elements. And in yet another

[5] C. O. Lovejoy, *The Great Chain of Being* (Cambridge, Mass., 1971; 1st ed., 1936).

twist to the debate, the Benedictine priest Stanley L. Jaki claimed that the idea has been a detriment to science; in particular, he argued that theories of solar system formation have been influenced by the desire for many solar systems to serve as abodes for extraterrestrials.[6]

Both scientists and historians, and even the popular press, have drawn attention to the important role of metaphysical assumptions in the debate. Few of these studies, however, have taken into account twentieth-century events in any detail because no history of the debate in this century has been written. This book, then, should serve not only to fill a historical lacuna, but also to contribute to historiographic issues that are bound to plague a debate so long as its history remains unknown. Few would deny that if the debate has not always been scientific in the past, it has at least become increasingly so in the twentieth century. But in order to make such a claim, one needs to understand the nature of science. In this study, we find that although controversy remains over the status of exobiology as science, certainly the subject has become more accepted by a wide range of scientists in the course of the twentieth century. If one adopts the operational definition that science is what scientists do (sanctioned in this case even by occasional government funding), then the extraterrestrial life debate may be considered a part of science, if at the outermost limits of its capabilities. Such an inclusive concept leads to a better understanding of the nature of all science.

THE NATURE OF SCIENCE

One of the hallmarks of the extraterrestrial life debate in the present century is the extreme divergence of opinion about whether the question should be pursued and, if so, by what methods. Once certain methods have been adopted, the controversy has been further fueled by widely divergent interpretations of the evidence. Except for the philosophical and scientific import and the pressing interest of the public, scientists might well have refrained from attempting any answer to such an elusive problem. Instead, they have found the temptation irresistible, with results that have generated a literature of considerable interest not only to science but also to science studies.

This situation provides an unusual opportunity to examine scientific activity, inference, and community when the subject of study is at the very limits of science, a status that many scientists themselves claim for the subject as a precautionary prelude to their conclusions. The debate thus

[6] Stanley L. Jaki, *Planets and Planetarians: A History of Theories of the Origin of Planetary Systems* (Edinburgh, 1978).

illuminates how scientists take up controversial research programs, how they differ in their perceptions of scientific method, how they interpret evidence, and how they relate to their peers, both sympathetically and critically. These characteristics vary among scientists to such an extent that it becomes clear in this study that science consists of many "cultures," rather than constituting a monolithic culture to be contrasted with the humanities, as claimed by C. P. Snow more than three decades ago.[7] While the many cultures of science may also be seen in other research problems, nowhere are they so starkly apparent as when science is functioning at its limits. And nowhere are scientists pushed to more extreme limits, by the public and by their own perception of the gravity of the question, than in the extraterrestrial life debate.

Rather than trying to define the limits of science by inventing abstractions, we shall keep constantly in mind how the principals of the debate have dealt with a subject they claim is at the limits of science, and in doing so attempt to illuminate the question "How does science function at its limits?" The Summary and Conclusion in particular summarizes this question in light of the findings of our study.

ORGANIZING PRINCIPLES AND APPROACH

Organization is a perennial problem that every historian, indeed every writer, faces. While a chronological narrative is often preferable, the events in this history are so diverse that a purely chronological treatment of the subject would have lost all coherence. After the first two chapters, in which I bring the reader up to the beginning of the century and stress the importance of worldview to the subject, I have therefore arranged the chapters thematically, but roughly in the order in which the themes first became prominent. Thus, at the beginning of the century, the attention of extraterrestrial enthusiasts and foes alike was focused on the solar system, particularly the planets Venus and Mars. Observational techniques limited to the solar system made this almost inevitable. By the time the first Martian controversy died down, in the 1920s the issue of other solar systems – based on the theories of Chamberlin and Moulton and refined by Jeans and Jeffreys – became a substantial topic of discussion. By midcentury, science fiction and the UFO controversy were becoming increasingly important threads in the story, and in particular affected how popular culture reacted to the possibility of life beyond the Earth. Only in the 1950s, with developments in molecular biology and crucial experiments on chemical evolution, did the study of the origin of life blossom, and the

[7] C. P. Snow, *Two Cultures and the Scientific Revolution* (Cambridge, 1959).

dawn of the Space Age carried that study beyond the Earth. And in the 1960s, the search for extraterrestrial intelligence (SETI) began to receive serious attention as technology finally caught up with concept.

Another reason to proceed thematically is the nature of the scientific interaction among participants in the debate. To a large extent, the participants in the thematic chapters represent different scientific communities, each possessing not only its own techniques and approaches but also its own preconceptions. Optical astronomers dominate the solar system chapter; theoretical astronomers, mathematicians, and, later, observational astronomers with quite different techniques are the principal actors in the planetary systems chapter. Biologists and biochemists populate the chapter on the origin of life, while radio astronomers and engineers are the protagonists of the SETI chapter. Although this approach inevitably leads to retracing the twentieth century several times over, it parallels the significant historical fact that the events here described were for much of the century quite separate in terms of both scientific content and scientific community; when and how they all came together to form a new field of study is the subject of Chapter 9. Finally, it is appropriate to conclude this study with a chapter on theological and philosophical issues, for these issues have never been systematically treated and remain a subject for future exploration.

While this organization reflects my preoccupation with science as knowledge, this study should also be of interest to those who approach science as practice and culture.[8] At the same time, it should be clear that the study is in no way exhaustive in the latter approach, but will perhaps serve as a basis for those who wish to take further an examination along those lines, as indeed it may be elaborated along many other lines historical, philosophical, and sociological.

I am keenly aware that each chapter, and indeed each section within the chapters, could easily form a book of its own, and that I am far from expert on all the varied subjects encompassed by the extraterrestrial life debate. But as the physicist Erwin Schrödinger wrote 50 years ago in tackling biology in his book *What Is Life?*, if we are to hope for unified rather than merely fragmented knowledge, "some of us should venture to embark on a synthesis of facts and theories, albeit with secondhand and incomplete knowledge of some of them – and at the risk of making fools of ourselves."[9] Although for me the risk is considerably greater than for Schrödinger, on the way to the desired goal of a more unified knowledge – to which the extraterrestrial life debate is in a peculiar position to contrib-

[8] See Andrew Pickering, *Science as Practice and Culture* (Chicago, 1992), for an entree to this approach.

[9] Erwin Schrödinger, *What Is Life?* (Cambridge, 1944), Preface.

ute by virtue of its enormous scope – I hope it will be productive to examine how all the events in this particular debate acted in concert, even if only in outline.

I have made no attempt to be encyclopedic, but taking it as a primary task of the historian to separate the wheat from the chaff, trust that I have captured the major trends in the debate while still examining enough detail and context to illuminate the themes mentioned earlier. Even this I may claim to have attempted only for the Western tradition; the extent of participation of other cultures remains a subject for further study. The search for unified and objective knowledge proceeds at many levels, but none more general or striking than may result if the attempts described in this study are one day successful.

1

FROM THE PHYSICAL WORLD TO THE BIOLOGICAL UNIVERSE: DEMOCRITUS TO LOWELL

Astronomy has had three great revolutions in the past four hundred years: The first was the Copernican revolution that removed the earth from the center of the solar system and placed it 150 million kilometers away from it; the second occurred between 1920 and 1930 when, as a result of the work of H. Shapley and R. J. Trumpler, we realized that the solar system is not at the center of the Milky Way but about 30,000 light years away from it, in a relatively dim spiral arm; the third is occurring now, and, whether we want it or not, we must take part in it. This is the revolution embodied in the question: Are we alone in the universe?

Otto Struve (1962)[1]

From the ancient Greek world of Democritus to the eighteenth-century European world of Immanuel Kant, cosmological thinking underwent a revolution that transformed a dead celestial world into a living universe, a transformation no less dramatic than the well-known move from the closed geocentric world to the infinite universe.[2] These two revolutions – from the closed world to the infinite universe and from the physical world to the biological universe – are not unrelated. The shattering of the ancient world of nested spheres not only opened the way for a greatly enlarged and interrelated universe by removing the Earth from its central status, it also opened the question of the uniformity of natural law. And the same logic that argued for the uniformity of physical law throughout the universe might also be applied to the uniformity of biological law, producing not only planets, stars, and stellar systems but also plants, animals, and intelligence. But while the first could be proven by appeal to theory and observation, the latter remains to be proven to the present day. The biological universe therefore is still a revolution of the mind, a transformation that seeks confirmation in the real world, a task that the twentieth century has taken to heart.

How, then, did the transformation from the physical world to the biological universe take place, and how has its central thesis of universal

[1] Otto Struve, *The Universe* (Cambridge, Mass., 1962), 157. Portions of this chapter appeared in Steven J. Dick, "Plurality of Worlds," in Norriss Hetherington, ed., *Cosmology: Historical, Literary, Philosophical, Religious and Scientific Perspectives* (New York and London, 1993), 515–532, and are used by permission of Garland Publishing.

[2] Alexandre Koyré, *From the Closed World to the Infinite Universe* (Baltimore and London, 1970; 1st ed., 1957), documents the latter revolution.

life been propagated into modern consciousness? It is the contention of this chapter that this occurred in three stages over two and a half millennia. The idea of life beyond the Earth was first initiated, and sustained through the seventeenth century, by a variety of cosmological worldviews. Within the framework of the surviving cosmologies, the debate was then refined and propagated through the middle of the nineteenth century by both religious and secular philosophies. And finally, in the second half of the nineteenth century, it received its scientific foundations in two overriding developments – the theory of biological evolution and the rise of astrophysics. At the same time, it became increasingly susceptible to the methods of science, a pattern that leads directly to the twentieth-century debate.

In attempting to characterize such a broad period we must be careful not to draw too rigidly the boundaries of science and philosophy. For many centuries they were inextricably intermixed; indeed, what we today call "science" emerged from "natural philosophy." In claiming that the debate has moved from the great generalizations of physical cosmology, to the exploration of philosophical implications, to more empirical investigations, we must constantly keep in mind that the philosophical is never banished completely and that the cosmological is always present in the background. This is equivalent to stating what is widely accepted: that the subject of extraterrestrial life has become more amenable to the methods of modern science – observation, theory, and experimentation – while still enmeshed in philosophical assumptions that all of science seems unable completely to escape.[3] Indeed, the extraterrestrial life debate itself may be seen as a struggle for a worldview, with all of the problems that this implies. Such status goes a long way toward explaining the emotional nature of the debate, for much more is at stake than another scientific theory. Given its scientific components, the idea of a universe filled with life is in fact a cosmology of its own, incorporating the physical and the biological, and which we may therefore term the "biophysical cosmology."

1.1 THE COSMOLOGICAL CONNECTION

The extraterrestrial life debate began as part of cosmological worldviews, and if we take "cosmological" to have its broadest meaning, this may also include mythical cosmologies. The idea that life might exist beyond

[3] Philosophical influences in science have been especially emphasized since E. A. Burtt, *The Metaphysical Foundations of Modern Physical Science* (New York, 1954; first ed., 1924), and are by now the subject of a large literature in the history and philosophy of science.

the Earth undoubtedly dawned in human consciousness long before it appeared in clearly recognizable form in the Greek cosmology of the fifth century B.C. Perhaps in a rudimentary way it appeared among the gods and goddesses who peopled the heavens of ancient mythology. Sentient beings beyond the Earth were probably part of early human attempts to understand nature through rudimentary theology; in this fact is found another reason for the emotional attachment to the issue in recorded history to the present day.

Such a view might also help to explain why the idea of other worlds was already present in the earliest scientific cosmological worldview, ancient atomism. Constructed in the fourth and fifth centuries B.C. by Leucippus, Democritus, and Epicurus, atomism was, of course, more than a cosmology. But the important point that we shall see borne out again and again in these cosmological worldviews is this: to a large extent, the concept of other worlds was based on the physical principles of the cosmological system. So it was with ancient atomism, in which Epicurus spoke of an infinite number of worlds resulting from an infinite number of atoms.[4] The "world" of Epicurus was the Greek "kosmos," meaning an ordered system, as opposed to chaos. The entire visible universe composed one kosmos; Epicurus here proposed the remarkable idea that *aperoi kosmoi,* an infinite number of such worlds, exist completely beyond the human senses. But not beyond human reasoning. For according to the atomist system, the infinite number of atoms could not have been used in our finite world. As our world was created by the chance collision of atoms in an entirely natural process, so must other worlds be created in like manner.

This atomist doctrine would eventually be spread throughout Europe by the Roman poet Lucretius (ca. 99–55 B.C.), whose *De rerum natura* (On the Nature of Things) supported the belief of Epicurus: Other worlds must exist "Since there is illimitable space in every direction, and since seeds innumerable in number and unfathomable in sum are flying about in many ways driven in everlasting movement."[5] Moreover, Lucretius added that nothing in the world was unique, including the world itself, and "when abundant matter is ready, when space is to hand, and no thing and no cause hinders, things must assuredly be done and completed." This second argument is a manifestation of what historian Arthur O. Lovejoy has called the "principle of plenitude," the idea that what can be

[4] Cyril Bailey, ed. and trans., *Epicurus: The Extant Remains* (Oxford, 1926), 25. Much of what follows in Section 1.1 is discussed in more detail in Steven J. Dick, *The Origins of the Extraterrestrial Life Debate from Democritus to Kant* (Cambridge, 1982).

[5] Lucretius, *De rerum natura,* trans. W. H. D. Rowe, Loeb Classical Library (Cambridge, Mass., 1924), book 2, lines 1052–1066.

done must be done, by the very nature of God or the fecundity of Nature.[6] It is notable, however, that what can be done is defined by the physical principles that one accepts, at least if one banishes the supernatural from the world, as did the atomists. Had the atomists held that a finite number of atoms exist, plenitude could not have been invoked. While plenitude would be used as a separate argument in many areas of thought, including plurality of worlds, it must therefore be secondary to the accepted physical system. It is almost inconceivable that the concept of infinite worlds would have arisen in the atomist system in the absence of the physical principles of the atomist cosmology.

Those physical principles, however, were not destined to win the day or even the millennium; it was almost 2000 years before they would be revived in the sixteenth and seventeenth centuries with the birth of modern science. In the meantime, a far more elaborate cosmology was constructed by Aristotle (383–322 B.C.), student of Plato and founder of the Lyceum in Athens, whose life overlapped that of Epicurus by two decades and who gave new meaning to the word "kosmos." Aristotle's cosmology placed the Earth at the center of a nested hierarchy of celestial spheres, from the spheres of the Moon and planets to the sphere of the fixed stars. The Earth was ever-changing and corruptible, as could be seen by experience, while the region of the celestial crystalline spheres was eternally unchanging. And the Earth was more than the physical center; it was also the center of motion. According to one of the basic tenets of Aristotle's cosmology – the doctrine of natural motion and place – everything in the cosmos moved with respect to that single center: the element earth moved naturally toward the Earth, and the element fire moved naturally away, while air and water assumed intermediate natural places. Aristotle's belief in the impossibility of more than a single kosmos was directly tied to this basic tenet. In his cosmological treatise *De caelo* (On the Heavens) he reasoned that if there were more than one world, the elements of earth and fire would have more than one natural place toward which to move, a physical and logical contradiction.[7]

The issue of a plurality of worlds was thus reduced to a confrontation with the most basic assumptions of Aristotle's system. Either he had to reject his doctrine of natural motion and place, on which he had built his entire physics, along with his belief in four elements, on which his theory of matter rested, or he had to conclude that the world was unique. The choice was not difficult; indeed, he must have taken comfort in reaching a

[6] Arthur O. Lovejoy, *The Great Chain of Being* (Cambridge, Mass., 1971; 1st ed., 1936).

[7] Aristotle, *De caelo* (On the Heavens), trans. W. K. C. Guthrie, Loeb Classical Library (Cambridge, Mass., 1953), book 1, ch. 8, 276b, lines 10–20. For Aristotle's auxiliary arguments, see Dick, *The Origins of the Extraterrestrial Life Debate*, chapter 1.

conclusion so diametrically opposed to that of the atomists, whose system differed from his in so many other ways.

It was Aristotle's system that was transmitted to the Latin West, where it was repeatedly commented on in the context of the Christian system. For Christianity, a plurality of worlds directly confronted its omnipotent God: suppose that God wished to create another world; how could he do so given the principles of Aristotle? Either Aristotle was wrong, and by his own admission wrong in some very basic principles, or God's power was severely limited. For the first century of Aristotle's introduction into the West, his conclusion of a single world was largely accepted; thirteenth-century writers such as the physician Michael Scot in Spain; the bishop of Paris, William of Auvergne; and the Franciscan scholar Roger Bacon in England rejected the plurality of worlds with Aristotle. They noted, for example, that more than one world would necessitate a void between the worlds, and the existence of a void was impossible, according to Aristotle. Writing in about 1273, the great reconciler Thomas Aquinas also accepted Aristotle's conclusion but asserted that a single world did not infringe on the concept of God's omnipotence, since perfection and omnipotence could also be found in the unity of the world.[8]

In contrast, by the end of the thirteenth century the university scholars had begun to weigh in with a more radical idea: Godfrey of Fontaine, Henry of Ghent, and Richard of Middleton at Paris University, and William of Ware, Jean of Bassols, and Thomas of Strasbourg at Oxford claimed that the plurality of worlds was not theologically impossible because God can act beyond the Aristotelian laws of Nature. The famous fourteenth century Paris master Jean Buridan went slightly further: God could create another world and reorder its elements, so that they would move according to Aristotelian law but with respect to their own world. Buridan's Oxford contemporary, William of Ockham – of Ockham's razor fame – went still further by altering Aristotle's doctrine of natural place to state that the elements in each world would return to the natural place within their own world, and this without any intervention from God. By 1377 the Paris master and bishop of Lisieux, Nicole Oresme, had completely reformulated the doctrine of natural place to state that as long

[8] On Michael Scot and William of Auvergne, see Pierre Duhem, *Le système du monde,* 10 vols. (Paris, 1914–1958), vol. IX, 365–369. On Roger Bacon, see *The Opus Majus of Roger Bacon,* trans. Robert Burke, 2 vols. (Philadelphia, 1928), vol. I, 185. Thomas Aquinas, *In Aristotelis libros de caelo et mundo, generatione et corruptione, meteorologicorum expositio* (Rome, 1952), Lectio XX, 96; Dick, *The Origins of the Extraterrestrial Life Debate,* 25–28. For translation of parts of Duhem's study of the medieval doctrine of a plurality of worlds, see Pierre Duhem, *Medieval Cosmology: Theories of Infinity, Place, Time, Void, and the Plurality of Worlds,* ed. and trans. Roger Ariew (Chicago, 1985), Part V, 431–510.

as heavy bodies were situated in the middle of light bodies, no violence would be done to the doctrine of natural place. In one stroke, he thus abandoned the Earth–outer sphere relation so central to Aristotle. Other worlds were possible, and without any supernatural intervention. But while they were possible, almost all the medieval Scholastics stressed, God in reality had not created more than one world. Neither the modern doctrine of the plurality of worlds nor the universe of modern science emerged from the Middle Ages, despite significant advances over Aristotle.[9]

Like the move to an infinite universe, the transition to the biological universe was not made through successive rebuttals to Aristotle's doctrine of a single world. Rather, it stemmed from the complete overthrow of Aristotle's geocentric universe and its replacement with the Copernican system of the world. It is true that Copernicus did not completely discard the old worldview, especially its epicyclic explanations for certain planetary motions. Nor was the Copernican worldview a general cosmology that was meant to explain the entire universe. But by placing the Sun in the center of the system of the planets and making the Earth one of those planets, he laid the foundations for future cosmologies that would be more general and would incorporate the heliocentric system.

It was this heliocentric system that gave birth to the new tradition of the plurality of worlds, where "world" (*mundus*) was redefined to be an Earth-like planet, which now took on the kinematic or motion-related functions of the single Earth in the old geocentric system. Just as the kinematic implications of the decentralization of the Earth led to the birth of a new physics,[10] so the "planetary physics" implications of that move led to the birth of the concept of the biological universe. All discussions of life on other worlds since then, whether consciously or not, recall that fainter echo that Copernicus set in motion: if the Earth is a planet, then the planets may be Earths; if the Earth is not central, then neither is man.

Copernicus himself did not pursue the implications of his system, but the ideas of the Italian philosopher Giordano Bruno showed just how far such implications might go, though he was largely influenced by more metaphysical arguments. An avowed Copernican, Bruno nevertheless

[9] Jean Buridan, *Quaestiones super libris quattuor de caelo et mundo*, ed. E. A. Moody (Cambridge, Mass, 1942), 84–86; William Ockham, *Scriptum super libros sententiarum magistri Petri Lombardi episcopi parisiensis*, 4 vols. (Paris, 1929–1947), vol. I, Distinction 44, question 1, article 2, 1018–1021; Nicole Oresme, *Le livre du ciel et du monde*, eds. Albert Menut and Alexander Denomy, trans. Albert Menut (Madison, Wis., 1968), 167–171.

[10] I. B. Cohen, *The Birth of a New Physics* (New York, 1960), was an early work in what is now a vast literature on this subject. See Richard S. Westfall, *Never at Rest: A Biography of Isaac Newton* (Cambridge, 1980), and further references therein, for the elaboration of this physics by Newton. On the wider aspects of the Copernican system, see Thomas S. Kuhn, *The Copernican Revolution* (Cambridge, Mass., 1957).

gave himself full credit for his universo filled with inhabited worlds. In his *De l'infinito universo e mondi* (On the Infinite Universe and Worlds, 1584), Bruno pointed to the metaphysical concept of unity as the source of his belief. The unity of the universe shattered the old Aristotelian spheres, as well as any celestial–terrestrial dichotomy, and it led to innumerable worlds via his conviction that both the greatness of Divine power and the perfection of Nature lay in the existence of infinite individuals, including infinite worlds. Greatly influenced by the atomists, their principle of plenitude also entered the argument: Nature could not help but produce infinite worlds.[11]

Though not tied directly to his Copernicanism, Bruno's was a view that would undoubtedly have profoundly disturbed Copernicus. And yet, it was the view toward which his system inexorably led, in a march that had not been completed by the twentieth century. Even before the invention of the telescope, which would reveal the roughly Earth-like nature of the moon and at least some of the planets, the young astronomer Johannes Kepler, already a convinced Copernican under the influence of his teacher Michael Maestlin, would ascribe inhabitants to the Moon.[12] The invention of the telescope began the long trend toward attempts at empirical verification of the Copernican implication that the planets were Earth-like. The question was how much like the Earth were they? In the first announcement of his observations in his *Siderius nuncius* (Sidereal Messenger, 1610), Galileo noted that the surface of the Moon was "not unlike the face of the Earth," and that its bright parts might represent land and its darker parts water. But within a matter of weeks, Kepler responded to Galileo with his own publication. Not only did he agree with Galileo's interpretation of the dark and bright lunar spots, he also had a remarkable explanation for a large circular lunar cavity detected by Galileo: it was formed by intelligent inhabitants who "make their homes in numerous caves hewn out of that circular embankment." Kepler also argued for a lunar atmosphere, and in his *Somnium* (Dream, 1634) would later describe the nature of the lunar inhabitants.[13] Beyond the Moon, Kepler could only resort to more philosophical arguments, aside from the general Copernican implication. The newly discovered moons of

[11] Giordano Bruno, *De l'infinito universo e mondi* (1584), translated in Dorothea Singer, *Giordano Bruno: His Life and Thought, with an Annotated Translation of His Work on the Infinite Universe and Worlds* (New York, 1950), especially 229.

[12] For Kepler's ideas see Dick, *The Origins of the Extraterrestrial Life Debate*, 69ff.

[13] Galileo, in *Discoveries and Opinions of Galileo*, ed. and trans. Stillman Drake (New York, 1957), 31, 34; *Kepler's Conversation with Galileo's Sidereal Messenger*, trans. Edward Rosen (New York and London, 1975), 28; and *Kepler's Somnium: The Dream or Posthumous Work on Lunar Astronomy*, trans. Edward Rosen (Madison, Wis., 1967), 151.

Jupiter, for example, could not be meant for the inhabitants of Earth, who never see them, but for the Jovians.

The response of Aristotelians to such ideas was outrageous astonishment, and even among Copernicans predictable caution was the watchword. Galileo himself at first denied such implications, and even in his defense of the Copernican system (*Dialogue on the Two Chief World Systems,* 1632) he sought to downplay the similarities of the Earth and Moon, admitting only that if there was lunar life, it would be "extremely diverse and far beyond all our imaginings."[14] The Copernican tide, however, could not be stemmed. Six years later, in the less repressive atmosphere of Anglican England, Bishop John Wilkins penned his *Discovery of a World in the Moone* (1638), where Galilean caution was thrown to the wind, while avoiding some of the wilder claims of the irrepressible Kepler. Making use of the few observations available, as well as theology and teleology, Wilkins argued strongly for an atmosphere around the Moon and agreed with Galileo that the dark spots were water. And he did not fail to make clear the inspiration for his work; its tenets, he held, could be deduced from Copernicus and his followers, "all who affirmed our Earth to be one of the planets, and the Sun to be the center of all, about which the heavenly bodies did move. And how horrid soever this may seem at first, yet it is likely enough to be true."[15] Copernicanism was not synonymous with inhabited planets, but it did give theoretical underpinning to habitable planets. The proof or disproof of this implication remained a goal of astronomers until the Viking landers touched down on Mars in the late twentieth century.

All important cosmological worldviews of the seventeenth century and later incorporated the Copernican system as a basic truth. Such was the case with the first complete physical system proposed since Aristotle, that of the French philosopher René Descartes. His *Principia Philosophiae* (Philosophical Principles, 1644), also greatly influenced by a revived atomism, offered a mechanical philosophy in which atoms in motion once again formed the basis for a rational cosmology. For the plurality of worlds tradition it did even more, for it was through the Cartesian cosmology that the quest for a biological universe was first carried to other solar systems, and in a fashion so graphic that it remains an ingrained concept to the present day. Unlike the void space of his atomist predecessors (and his Newtonian successors), Descartes proposed that the universe was a plenum filled with atoms in every nook and cranny. A conse-

[14] Galileo, *Dialogue concerning the Two Chief World Systems,* trans. Stillman Drake (Berkeley, Calif., 1962), 100.

[15] Wilkins, *Discovery of a World in the Moone* (London, 1638; facsimile reprint Delmar, N.Y., 1973), proposition 6.

quence of this was that, once set in motion by God, the particles of the plenum formed into vortices, systems analogous to our solar system, centered on every star. Though Descartes himself, again for religious reasons, was careful not to specify that these vortices consisted of inhabited planets, his application of Cartesian laws to the entire universe and the graphic vortex cosmology was plain for all to see.[16]

Descartes's followers were not slow to realize the implications, and some of them elaborated on the nature of the whirling vortices. But none was more bold or more successful than his countryman Bernard le Bovier de Fontenelle. His *Entretriens sur la pluralité des mondes* (Conversations on the Plurality of Worlds, 1686), a treatise that exploits both the Copernican and the Cartesian theories to shed light on the question of life on other worlds, was explicit in its reason for spreading life beyond the solar system:

> If the fix'd Stars are so many Suns, and our Sun the centre of a Vortex that turns round him, why may not every fix'd Star be the centre of a Vortex that turns round the fix'd Star? Our Sun enlightens the Planets; why may not every fix'd Star have Planets to which they give light?[17]

The *Entretriens* spread throughout Europe in an extraordinarily large number of editions, and with it the idea of a plurality of solar systems (Fig. 1.1) became ingrained in the European consciousness. In the same year, the famous Dutch astronomer Christiaan Huygens began to formulate his own ideas on the plurality of worlds, published posthumously in the *Kosmotheoros, sive, de terris coelestibus earumque ornatu conjecturae* (Cosmotheoros, or, Conjectures concerning the Celestial Earths and Their Adornments, 1698). Strongly motivated by his experience as an observational astronomer and explicitly building on the foundation of the Copernican theory, Huygens also proposed a plurality of solar systems, based primarily on the analogy that the fixed stars were suns (another of Descartes's tenets) and only loosely tied to Cartesian vortices.[18] Although

[16] Descartes, *Oeuvres de Descartes*, ed. Charles Adam and Paul Tannery, 11 vols. (Paris, 1897–1903); reprint, 1964), vol. 8. Portions of the *Principles of Philosophy* are translated in *The Philosophical Works of Descartes*, trans. Elizabeth S. Haldane and G. T. R. Ross, 2 vols. (Cambridge, 1972), vol. 1.

[17] Fontenelle, *Entretriens sur la pluralité des mondes* (1686), trans. Joseph Glanville as *A Plurality of Worlds* (London, 1688), reprinted in Leonard M. Marsak, *The Achievement of Bernard le Bovier de Fontenelle* (New York and London, 1970), 125.

[18] Huygens, *The Celestial Worlds Discover'd: or, Conjectures concerning the Inhabitants, Plants and Productions of the Worlds in the Planets* (London, 1698; facsimile reprint, London, 1968).

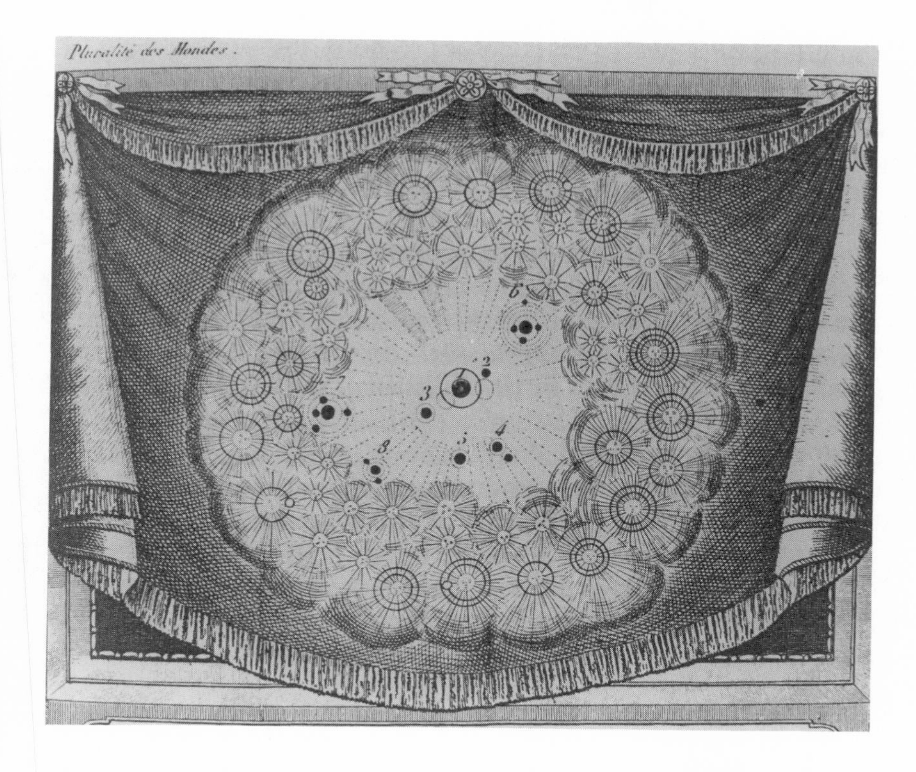

Figure 1.1. Frontispiece to the 1821 French edition of Fontenelle's *Entretriens sur la pluralité des mondes* (1686), depicting the plurality of solar systems. From Steven J. Dick, *Plurality of Worlds* (Cambridge, 1982), by permission of Cambridge University Press.

Cartesian vortices would soon be swept away by the Newtonian system, the general idea of planetary systems would not.

It is ironic that, of all these cosmological worldviews, the scientific principles of the Newtonian worldview – the system that we have inherited in modified form – entailed extraterrestrial life least of all. The ancient atomists held that an infinite number of atoms in an infinite universe must necessarily form an infinite number of worlds, given the example of our finite world; the Copernican principle of the noncentrality of the Earth led directly to the implication of other Earths; and the Cartesian plenum led directly to vortices that common sense dictated were similar to our own solar system. Although a mechanical system like that of Descartes, Newton's atoms and void, with each body subject to universal gravitation according to fixed laws, did not necessarily imply other solar systems. There was no mechanical necessity for the formation of systems, as there had been in Descartes's system; indeed, under Newtonian principles, the whole question has proved to be one of the greatest complexity to the present day.

One could, of course, argue that since our solar system exists and the laws of gravitation are universal, other solar systems should exist. This, however, was only the grossest analogy, almost equivalent to assuming what one wished to prove. Newton himself declined to expound any rational cosmogony that might shed light on the question; in fact, in his letters to the theologian Richard Bentley, he indicated that he did not consider it possible to settle this question based on the principles of his system. Instead, Newton insisted that the formation of all ordered systems was contingent on God's will, contenting himself with the observation in the second edition of his *Principia* (1713) that "if the fixed stars are the centres of other like systems, these, being formed by the like wise counsel, must be all subject to the dominion of the One."[19]

Despite the arguments of Newton and others to retain the Deity in a system subject to natural law, it became increasingly clear that his laws actually lessened the need for a Deity. In this atmosphere it is not surprising that the concept of a plurality of worlds was put to good use – as a proof of God's glory drawn from his works in Nature, the enterprise known as "natural theology." If God could no longer be given a role in maintaining his universe, then the concept of a plurality of inhabited planets could be made to reflect the glory, power, and wisdom of the Creator. In one Newtonian treatise after another, the theological view of an inhabited universe was joined to the physical principles of Newton's system. Again and again, a universe full of inhabited solar systems was applauded as one "far more magnificent, worthy of, and becoming the infinite Creator, than any of the other narrower schemes."[20] Once this decision had been made, overwhelming all Scriptural objections, other arguments could be adduced in its favor. One of the most frequent was the argument of teleology, of purpose in the universe, clearly set forth already by Bentley when he wrote, "All Bodies were formed for the Sake of Intelligent Minds: As the Earth was principally designed for the Being

[19] *Sir Isaac Newton's Mathematical Principles of Natural Philosophy and His System of the World,* Motte's trans., rev. Florian Cajori (Berkeley, Calif., 1960), 544. Dick, *The Origins of the Extraterrestrial Life Debate,* 147.

[20] William Derham, *Astro-Theology: or a Demonstration of the Being and Attributes of God from a Survey of the Heavens* (Edinburgh, 1777); 1st ed. London, 1715.), 61–69.

and Service and Contemplation of Men; why may not all other Planets be created for the like uses, each for their own Inhabitants who have Life and Understanding."[21]

This satisfying vision of the universe, operated by Newtonian laws and reflecting the power of the Deity by spreading intelligence throughout the universe, was transmitted to the modern world. The proof of other solar systems by observation, and the proof of their likely formation by Newtonian principles, remained a desired goal in the centuries to follow. But the basic predisposition toward a universe of inhabited solar systems was set, almost within the lifetime of Newton himself.

In the 1750s and 1760s the Newtonian system was carried to its ultimate cosmological extreme: a hierarchy of ordered systems, stretching from our own solar system to the system of the Milky Way Galaxy itself and even beyond, so that all systems were ordered with respect to one another, all united under the common bond of Newtonian principles. Such a view was expressed by three philosopher-cosmologists: Thomas Wright, Immanuel Kant, and Johann Lambert. Committed Newtonians all, these three eighteenth-century thinkers were also pioneers in stellar astronomy and fervent believers in an inhabited universe. Indeed, both Dick and Crowe have argued that this latter connection was no coincidence, that the sidereal revolution – which focused attention on the distribution of the stars, first explained the gross structure of the Milky Way, and conjectured that other galaxies exist outside of our own – was intimately related to the plurality of worlds debate. More specifically, in Crowe's words, all three authors were "attempting to transform traditional religious notions by interpreting them in physical terms," terms that underlay the sidereal revolution.[22]

The result of a confluence of physical principles, observation, and metaphysics, the universe of Wright, Kant, and Lambert was "animated with worlds without number and without end." Such was the universe passed on to the modern world, a view that brings us back almost full circle to the ancient Greek atomist view of infinite kosmoi, systems of stars separated from one another, each filled with a multitude of inhabited worlds. Thanks to the immense contribution of Newton, the laws of motion of such systems had now been fathomed, somehow extracted from Nature by the very mind that was now projected throughout the

[21] Richard Bentley, *A Confutation of Atheism from the Origin and Frame of the World* (London, 1693), in *Isaac Newton's Papers and Letters on Natural Philosophy*, ed. I. B. Cohen (Cambridge, Mass., 1958), 356–358.

[22] Wright, *An Original Theory or New Hypothesis of the Universe, Founded upon the Laws of Nature . . .* (London, 1750); facsimile reprint, London and New York, 1971); Kant, *Universal Natural History and Theory of the Heavens*, trans. W. Hastie (Ann Arbor, Mich., 1969), 35.

universe. Now even the scattered systems were seen as influenced gravitationally. More important still, the new cosmological view was joined to Christianity and natural theology. That conjunction would take a central place in the plurality of worlds debate in the nineteenth century.

1.2 PHILOSOPHICAL EXPLORATIONS

Following the triumph of the Newtonian system in the mid-eighteenth century, the extraterrestrial life debate was waged not so much on a cosmological scale as on a scale of world views, a level or more below the cosmological. Though sometimes discussed by the elaboration of Newtonian science, such as the Laplacian nebular hypothesis, more often it fell to the domain of philosophical explorations, both secular and religious. If cosmological worldviews gave birth to the idea of extraterrestrial life, then philosophy and literature, in their traditional role of examining the human condition, explored the ramifications of the idea born of that cosmological context.

In particular, much of the late eighteenth- and nineteenth-century plurality of worlds debate – at least in the West – may be understood as a struggle with that widespread philosophical worldview known as "Christianity." If in the Newtonian system the plurality of worlds concept was reconciled with theism via natural theology, this was not equivalent to a reconciliation with Christianity; as Professor Crowe succinctly states, "structures of insects or solar systems may evidence God's existence, but they are mute as to a Messiah."[23] Three choices were logically open to Christians who pondered the question of other worlds: they could reject other worlds, reject Christianity, or attempt to reconcile the two. Historically, all three of these possibilities came to pass in the eighteenth and nineteenth centuries.

Though the Scriptural and doctrinal problems of the issue had been widely discussed throughout the seventeenth century, only to be overwhelmed by natural theology, no one more forcefully expressed the continuing difficulties of the plurality of worlds doctrine for Christianity than Thomas Paine, that agitator for freedom on two continents. In his influential *Age of Reason* (1793), which saw numerous American editions by the end of the century, Paine bluntly stated that "to believe that God created a plurality of worlds at least as numerous as what we call

[23] Michael J. Crowe, *The Extraterrestrial Life Debate 1750–1900: The Idea of a Plurality of Worlds from Kant to Lowell* (Cambridge, 1986), 162. Much of what follows in Sections 1.2 and 1.3 is discussed in more detail in the Crowe volume, as well as in Karl S. Guthke, *The Last Frontier: Imagining Other Worlds from the Copernican Revolution to Modern Science Fiction* (Ithaca, N.Y., 1990).

stars, renders the Christian system of faith at once little and ridiculous and scatters it in the mind like feathers in the air. The two beliefs cannot be held together in the same mind; and he who thinks that he believes in both has thought but little of either."[24] With millions of worlds under his care, Paine argued, could we really believe that the Messiah came to save human beings on this small world? Or did the redeemer hop from one world to the next, when the number of worlds was so great that he would be forced to suffer "an endless succession of death, with scarcely a momentary interval of life"? Despite the force of this argument, few would reject Christianity, as did Paine. But few would reject plurality of worlds either, a testimony to its entrenchment by the end of the eighteenth century. This left but one alternative: the two systems and all they implied would have to coexist.

That other worlds could be incorporated into Christianity, despite Paine, was demonstrated by the Scottish theologian Thomas Chalmers. His *Astronomical Discourses* (1817) incorporated plurality of worlds into evangelical religion, and his countryman Thomas Dick made it a staple of Christianity in a number of works during the first half of the nineteenth century. Even astronomers, who were expected to make pronouncements on the likelihood of other worlds, were strongly influenced by philosophy, as the case of John Herschel shows.[25]

Paine's objections, however, would not disappear. By midcentury the consonance of plurality of worlds with Christianity was once again called into serious question in one of the most interesting intellectual disputes of the nineteenth century. The instigator was William Whewell, philosopher, scientist, and Master of Trinity College, Cambridge. Influenced by Chalmers, Whewell was a pluralist from at least 1827. By 1850, as Professor Crowe has shown from analysis of an unpublished Whewell manuscript, Whewell opposed pluralism.[26] And in 1853 his treatise *Of a Plurality of Worlds: An Essay* appeared anonymously, the most learned, radical, and influential antipluralist treatise of the century.

In his *Essay* Whewell confirmed that ideas about a multiplicity of worlds "are generally diffused in our time and country, are common to all classes of readers, and as we may venture to express it, are the popular views of persons of any degree of intellectual culture, who have, directly

[24] *Basic Writings of Thomas Paine* (New York, 1942), 67–78; Crowe *The Extraterrestrial Life Debate,* 163.

[25] On Chalmers, see Crowe, *The Extraterrestrial Life Debate,* pp. 182–190; on Thomas Dick, see ibid., pp. 195–202; on Herschel, see ibid., pp. 216–221.

[26] Crowe, *The Extraterrestrial Life Debate,* pp. 277–282. Professor Crowe discusses Whewell and the reaction to his work on the plurality of worlds at length in chapters 6 and 7.

or derivatively, accepted the doctrines of modern science."[27] Such ideas – that the planets are earths, the stars suns, these suns centers of systems of planets, these planets the seats of intelligence – were widely seen as the doctrines of science, according to Whewell. But they raised difficulties for religion. In particular, Whewell noted, they overwhelmingly reinforced the cry of the Psalmist, "What is man, that thou art mindful of him?" a question valid enough before the vast extent of the universe and the possible existence of widespread inhabitants was conceived. Impressed by Chalmers's attempt to quiet the infidel who used this view of nature against Christianity, Whewell nevertheless addressed his treatise to the friend of religion, who still found the difficulties of the Psalmist compelling, despite the efforts of Chalmers. It is clear that Whewell numbers himself among those so disturbed, and in fact that his own Christian concerns are the source of the treatise.[28] Before we are forced to make radical changes to Christian doctrines such as Redemption and Incarnation, Whewell argued, should we not be certain that the plurality of worlds doctrine that gives rise to the difficulties in the first place is firmly grounded in modern science? This is the task that Whewell set himself in *Of a Plurality of Worlds*. The result is that, unlike Paine (as well as Shelley, Emerson, and Walpole, among others), Whewell would argue that it was pluralism, not Christianity, that should be rejected.

Driven by his concerns for Christianity, Whewell carried out the task of disproving the now entrenched doctrine of a plurality of worlds with both scientific and philosophical arguments. Significantly, he made every attempt to analyze the scientific arguments on their own merits, following this with an attempt to show that a single inhabited world would not prove disastrous to philosophy or natural theology. In both cases, he tried to keep his Christian concerns a separate issue.

As if to stress the utility of science in his cause, Whewell used a scientific argument to dispose quickly of one of the chief philosophical arguments of the pluralists: the argument, from teleology, that all the vast space must have some purpose. He argued that geology reveals human existence on Earth to be but a short "atom of time" compared to the age of the Earth; therefore, why could not intelligence be confined to the "atom of space" that was the Earth? If the Earth has been devoid of inhabitants for most of the known immensity of time, then this cannot be against God's nature; neither, then, can we argue that a vast uninhabited universe is against his nature.[29]

But what about the well-grounded scientific arguments for other

[27] Whewell, *Of a Plurality of Worlds: An Essay* (London, 1853), 16.
[28] Crowe, *The Extraterrestrial Life Debate,* 281.
[29] Whewell, *Of a Plurality of Worlds,* pp. 52–112, especially 103.

worlds? Whewell agreed that the universe was indeed vast (about 3000 light years by his estimate), but he argued that the possible locales for inhabitants had been vastly overrated. One by one he dealt with the supposed abodes of celestial inhabitants, from the far-flung nebulae that represented for him the outermost regions of the known universe, to the fixed stars, and finally to the planets in our own solar system. It is true, he admitted, that some nebulae have been resolved into "lumps" of luminous matter, but that these are stars at all like the brighter stars or the Sun, or that they have planets, are "very bold structures of assumption." Many other nebulae were not resolvable, and Whewell argued that they probably consist of tenuous gas and thus can never be the abodes of intelligence.[30] Moreover, Whewell argued that while clusters of stars, especially globular clusters, are generally regarded as being composed of so many suns, there is again no proof that they are in any way like our Sun. Nor was there even any evidence to support the analogy that the naked-eye stars were like the Sun; they might be more tenuous or less stable than our Sun.[31] Gathering momentum, Whewell dismissed Goodricke's 1782 conjecture that the variability of Algol was due to an opaque body around it by pointing out that such a body would not be the size of a planet.[32] Nor, he believed, would planetary orbits around binary stars be conducive to conditions for life. Finally, Whewell argued that even in our own solar system no planets other than the Earth are known to be habitable. We are certain that the Moon is not habitable despite its proximity to the Earth, and this represents an important negative case; we are thus not allowed to reason which case truly represents the rest of the bodies in the universe. Of the other planets, Mars comes closest to Earth in conditions, but it is possible that Mars is in a condition of preintelligence, just as the Earth was for so many millennia.[33]

Altogether, Whewell presented a serious challenge to a doctrine that had come to be cherished by both science and religion. Much of his treatise can be seen as an argument against the unbridled use of analogy in science, which he saw as greatly exaggerated in the case of other worlds: "The argument that the Fixed Stars are like the Sun, and therefore the centers of inhabited systems as the Sun is, is sometimes called an argument from ANALOGY; and this word Analogy is urged, as giving great force to the reasoning. But it must be recollected, that precisely the point

[30] Ibid., 121.
[31] Ibid., 147.
[32] Ibid., 152–154. On Goodricke, see also *Journal for the History of Astronomy,* 10, pt. 1, 27ff.
[33] Whewell, *Of a Plurality of Worlds,* 170–171.

in question is, whether there IS an analogy." Professor Guthke found that Whewell's "great, historic achievement was that he drew attention to the need to put the analogical argument for plurality more precisely and in greater detail than had been usual up to that time." No longer was the Copernican fact that the planets were Earths a sufficiently precise argument; greater attention had to be given to the details of their physical conditions.[34]

Having disposed of the scientific arguments for other worlds, Whewell argued that a universe with a single inhabited world is not opposed to natural theology. In particular, God might work according to general laws that make it seem that much of the universe has no purpose, but in reality the worth of a single human soul is commensurate with all the vastness of the universe. Occasioned by Christian concerns, Whewell's treatise was nevertheless an important step forward in the plurality of worlds debate. It recognized that the truth of this doctrine is "a point which rather belongs to science than to religion." It stressed that the question must be "examined by all the light which modern science can throw upon it." And it insisted that "nothing can be more opposite to the real spirit of modern science, and astronomy in particular," than to accept weak arguments "professing to be drawn from science and from astronomy" when in fact they have very little ground in science.[35]

Whewell's treatise generated a tremendous amount of debate, but in the end it did little to weaken support for a plurality of worlds among scientists or the religious. Crowe documents 20 books and some 54 articles and reviews in response to Whewell; of these, about two-thirds still favored pluralism despite Whewell's arguments. Moreover, when broken down into scientific and religious authors, scientists were more inclined to accept plurality of worlds by a margin of 83 to 64 percent.[36] And not always for the best of reasons: one of the most outspoken critics of Whewell was the Scottish physicist Sir David Brewster, who in 1854 published *More Worlds Than One: The Creed of the Philosopher and the Hope of the Christian*. Brewster shows that he had learned little from Whewell, either by way of methodology or content, for though he employed scientific arguments, they were clearly driven and overwhelmed by his strong attachment to teleology and natural theology.[37]

Innumerable other discussions of the relation of plurality of worlds to

[34] Ibid., 49, 161. Guthke, *The Last Frontier*, 331–332. Trudy E. Bell has also drawn attention to the importance of analogy in "The Grand Analogy: History of the Idea of Extraterrestrial Life," *Griffith Observer* (August 1978), 2–16.

[35] Whewell, *Of a Plurality of Worlds*, iii–iv, 46.

[36] Crowe, *The Extraterrestrial Life Debate*, 351–352.

[37] David Brewster, *More Worlds Than One: The Creed of the Philosopher and the Hope of the Christian* (London, 1854); Crowe, *The Extraterrestrial Life Debate*, 300–305.

Christianity were penned throughout the nineteenth century. Almost without fail, the point of all of them is some kind of reconciliation, with an occasional conclusion that one or the other would have to go. Reconciliation with the doctrines of Incarnation and Redemption was never achieved, however, with most insisting that Christ's incarnation on Earth was of great enough force to save any extraterrestrials. This prevented a planet-hopping Christ, but it strained credulity. In addition, plurality of worlds became a central doctrine for at least two nineteenth-century religious groups: the Mormons and Seventh-day Adventists. Yet another group, the Swedenborgians, had held it as one of their beliefs since the mid-eighteenth century.[38]

Thus Christianity, along with its well-known influence on many other areas of human thought, holds the distinction of being the philosophical worldview that most influenced the plurality of worlds doctrine in the nineteenth century, at least in the Western world. Despite all the discussion, the long-sought resolution between Christianity and pluralism was elusive. Efforts to achieve that resolution would continue in the twentieth century, but in a far less dominant role as science and religion went their separate ways.

Secular philosophies also grappled with the concept of other worlds, though none so strongly or persistently as Christianity. Already in the seventeenth century the British empiricist John Locke had pointed out, in his *Essay Concerning Human Understanding* (1689), that human ideas are limited by the human senses and that extraterrestrials might have no such limitations, or at least different ones. The German philosopher Gottfried Wilhelm Leibniz, well known for his belief that ours is the best of all possible worlds, may have been influenced in that belief by the possibility of actual worlds. This view was satirized by the most famous of the *philosophes,* Voltaire, who nevertheless also made much use of extraterrestrials in his writings. Such views, while not expounded at length by any of these philosophers, could clearly have had an important effect on their philosophy.[39]

By contrast, many German philosophers of the nineteenth century were opposed to a plurality of worlds, not because of science or religion but because of their anthropocentrism. G. W. F. Hegel held that the Earth is the most excellent of all planets, and his students Carl Michelet and Ludwig Feuerbach argued strongly against the pluralist position. Against the teleological argument that all the stars must have some purpose,

[38] On the Mormons see Crowe, *The Extraterrestrial Life Debate,* pp. 241–246, and Erich R. Paul, *Science, Religion and Mormon Cosmology* (Urbana, Ill., 1992); on the Swedenborgians, see Crowe, ibid., pp. 97–101.

[39] Crowe, *The Extraterrestrial Life Debate,* pp. 27–28.

Michelet compared them to the multitude of barren islands on Earth, concluding that the spiritual richness of Earth with its life more than compensates for any barrenness in the rest of the universe. Similarly, Feurbach held that the Earth "is the soul and purpose of the great cosmos," and as for the rest of the cosmos, and even on Earth, purposeless existence is evident everywhere. Friedrich Schelling and his disciple Heinrich Steffens, though not Hegelians, also opposed other worlds with anthropocentric arguments, and the ever-pessimistic Arthur Schopenhauer, though he accepted extraterrestrials, also believed that humanity was at the pinnacle of creation.[40]

Finally, in its role of exploring the meaning of life, literature played an important role in responding to the challenge of other worlds. It did so in diverse ways extending back at least to the seventeenth century. Some authors, such as John Milton in England, cautioned early on

> Dream not of other worlds; what creatures there
> Live in what state, condition, or degree . . .

reflecting his Christian concerns. But Alexander Pope's *Essay on Man* (1734), reflecting the Enlightenment attitude, used extraterrestrials to give proper perspective to humanity:

> He, who through vast immensity can pierce,
> See worlds on worlds compose one universe,
> Observe how system into system runs,
> What other planets circle other suns,
> What varied Being peoples every star,
> May tell why Heaven has made us as we are

A century later Tennyson expressed similar sentiments, and the Romantic poets Byron, Shelley, and Coleridge used other worlds in a religious context. In prose no less than in poetry the extraterrestrial perspective became entrenched. From the cosmic voyage genre of the seventeenth century to its birth as a classic theme of science fiction in the work of H. G. Wells, the implications of extraterrestrials and their worlds began to be explored at an increasing pace in ever more subtle and expressive form.[41]

By the last quarter of the nineteenth century, plurality of worlds had won the day, despite the difficulties with Christianity. It was so firmly

[40] Ibid., pp. 258–260.
[41] See ibid. on Milton, Pope and others, as well as Guthke, *The Last Frontier.*

entrenched as part of the Newtonian worldview that Whewell's Christian objections were unable to destroy the concept, despite his best efforts. Not only did plurality of worlds and Christianity live on in fragile coexistence, other worlds were also incorporated into many secular philosophies. Thus, Professor Crowe's characterization of the dominance of philosophical considerations in the extraterrestrial life debate seems appropriate for the nineteenth century.

At the same time, however, science kept chipping away at the problem with its limited empirical tools, until two unexpected developments laid the scientific foundations for all subsequent discussion of life beyond the Earth.

1.3 SCIENTIFIC FOUNDATIONS

Two fundamental developments shook science just as the Whewell debate reached its height: in 1859 Charles Darwin (prodded by A. R. Wallace's independent discovery) published his theory of the origin of species and evolution by natural selection, and in the early 1860s the new technique of spectroscopy was applied to astronomy. Though these and less sweeping developments in other fields did not cause an immediate and radical change in the character of the plurality-of-worlds debate, they did signal the beginning of a long-term change that would bring the subject of other worlds increasingly under the purview of science. Harvard Professor Karl Guthke wrote:

> What is new about this period is that in those discussions of the subject that are historically relevant, the theological and metaphysical approaches [of Christian dogma and teleology] are absent; at most they are very marginally present, as premises, usually not expressed explicitly, that are no longer felt to be legitimate and conclusive. The method of argument is, so far as is possible in philosophical discussion, of a purely scientific nature: To begin with, teleology is no longer accepted as a principle by which natural phenomena in astrophysics and biology may be interpreted. More significantly, however, it is science (unconcerned with any metaphysical purposes underlying the Creation) that supplies a radically new way of thinking that invests the idea of the plurality of worlds with a previously unimagined credibility that allows it to penetrate deep into the consciousness of the age.[42]

[42] Guthke, *The Last Frontier,* 324. Although Guthke viewed this new era as beginning with Whewell, the latter's theological motivations and polemical stance, mixed with his un-

Much of the progress in the twentieth-century debate may be seen as beginning with evolution and astrophysics. Natural selection not only provided the basis for a discussion of the evolution of life under differing conditions beyond the Earth, it also gave impetus to the idea of the physical evolution of the universe.[43] And by examining the chemical fingerprints present in the analysis of starlight, spectroscopy produced a tool to study the nature of the planets and stars in ever-increasing detail. This, in turn, provided for the first time a means for determining the possibility of life on the basis of physical conditions and for demonstrating the evolutionary universe.

Evolution and spectroscopy represented, respectively, breakthroughs in theory and technique. Of the two, spectroscopy would have the more immediate and profound effect on the fate of the biological universe. Though the arguments of analogy and uniformity of nature had long given credence to the belief that the building blocks for matter – and life – were alike throughout the universe, now for the first time this great truth could be observationally proven. Many of the spectroscopic pioneers themselves did not fail to see the connection of their subject to life in the universe. When Sir William Huggins reported the first major results of his researches with the chemist William Allen Miller in 1864, they believed that their discoveries provided "an experimental basis on which a conclusion, hitherto but a pure speculation, may rest – viz. that at least the brighter stars are, like our sun, upholding and energising centres of systems of worlds adapted to be the abode of living beings." While Huggins later admitted that theology played a role in that inference, spectroscopy certainly brought the problem of other worlds a step closer to the realm of science. So too did Huggins's discovery that not all nebulae were resolvable into stars and in fact consisted of tenuous gas. And his early attempts to probe planetary atmospheres spectroscopically marked the first step toward yet another research program that would be increasingly central to the extraterrestrial life debate in the future. Other pioneers in the new science did not fail to make similar connections, and their early

mistakable desire to examine the problem from the perspective of "modern science," makes Whewell more of a transition figure, around whom pivots the whole nineteenth-century change in view toward the plurality of worlds. Crowe (*The Extraterrestrial Life Debate,* pp. 359ff.) found the "new approaches" beginning at about 1860 with the "new astronomy."

[43] In *The Study of Stellar Evolution* (Chicago, 1907), p. 2, astronomer George Elery Hale wrote that while evolution was not a new idea to astronomers in the nineteenth century, "it has occupied a more important position since Darwin published his great work." Similarly, Guthke (*The Last Frontier,* p. 339) finds that while evolution of worlds and life was discussed before Darwin, after Darwin "this was to become a *leitmotif* in the plurality debate, often with explicit reference to Darwin."

Figure 1.2 British astronomer R. A. Proctor.

work and that of their successors put the astrophysical component of the extraterrestrial life debate on a firm footing.[44]

The Darwinian theory of evolution had a more gradual effect, but one eventually no less significant. Its earliest effect was in the support it lent to the idea of the physical evolution of the universe. This evolutionary universe – along with the application of the first results of spectroscopy – is apparent already in one of the most prominent treatises on the plurality of worlds, Richard A. Proctor's *Other Worlds Than Ours* (1870). Proctor (Figure 1.2) conceived his book as a way of bringing the latest astronomical discoveries to the public; it may also be seen as reflecting a trend toward a more scientific approach for the debate. Subtitling his work *The Plurality of Worlds Studied under the Light of Recent Scientific Researches* did not ensure the banishment of all metaphysical arguments, which in any case has not been accomplished even to the present day. But it did indicate the

[44] *The Scientific Papers of Sir William Huggins*, eds. Sir William Huggins and Lady Huggins (London, 1909), 60. For comments on plurality of worlds by other pioneers in astrophysics, see Crowe, *The Extraterrestrial Life Debate*, 359–367.

closer relationship to empirical science that the late nineteenth century sought to bring to a subject heretofore seriously deficient in this approach, Whewell notwithstanding.

Influenced by teleology, Proctor had originally hoped to show how all the planets were inhabited, but "found the ground crumbling under my feet. The new evidence . . . was found to oppose fatally . . . the theory I had hoped to establish."[45] This willingness to suspend belief by the weight of physical evidence was the absolute prerequisite for a scientific approach to the question. But it was also joined by a nascent evolutionary view: While Jupiter, Saturn, and the outer planets were probably not now inhabited, Proctor argued, in the course of their evolution they would one day become as habitable as Venus, Earth, and Mars now were. In *Our Place among Infinities* and *Science Byways,* both published in 1875, the evolutionary view, in which all planets would attain life in due time, assumed a central role. This evolutionary universe still left Proctor with "millions of millions of suns which people space," of which "millions have orbs circling round them which are at this present time the abode of living creatures."[46] Unlike Brewster, Proctor had learned the lessons of Whewell; he took seriously that 1853 treatise and its forceful arguments. In Proctor the role of teleology was replaced by the eons of time required for evolution. With seven editions by 1893, 29 printings by 1890, and a number of other pluralist treatises, Proctor was undoubtedly the most widely read astronomy writer in the English language.

In France that role was assumed by the astronomer Camille Flammarion (Figure 1.3), who, like Proctor, professed to take a scientific approach to the problem of other worlds. It is clear that he did not achieve his goal entirely, any more than did Proctor. Analogy and plenitude, for example, were mixed with data from the new astronomy in Flammarion's arguments. And if anything, in subsequent publications Flammarion's pluralism became even more radical, taking on the quality of philosophy. Yet, "for all his tendency to enthusiasm and verbosity, Flammarion adheres consistently to this principle of experimental, inductive science," Guthke found. Moreover, at least by the time of the 1872 edition, Flammarion had been deeply influenced by Darwin. Life began by spontaneous generation, evolved via natural selection by adaptation to its environment, and was ruled by survival of the fittest. In this scheme of cosmic evolution, anthropocentrism was banished; the Earth was not unique, and humans were in no sense the highest form of life. Though Flammarion was led to

[45] Proctor, *Other Worlds Than Ours* (New York, 1871); Crowe, *The Extraterrestrial Life Debate,* 371.

[46] Crowe, *The Extraterrestrial Life Debate,* p. 375. In these volumes, Proctor also criticized teleology and the method of analogy that populated planets in our solar system.

Figure 1.3 French astronomer Camille Flammarion, at age 20 in 1862, the year of publication of his *Pluralité des mondes*. Flammarion and Proctor (Figure 1.2) were both crucial in turning the debate toward a more modern approach. Photo courtesy of the Observatoire de Juvisy. Société Astronomique de France.

almost religious heights by his pluralism, it was in no sense a Christian view, and was in any case a conclusion reached after the fact rather than an assumption on which his pluralism was based. *La Pluralité* reached 33 editions by 1880 and was reprinted until 1921. Through this and numerous other pluralist writings, both fiction and nonfiction, Flammarion would exercise considerable influence on the twentieth-century debate, not only through his nineteenth-century works but more directly through personal enthusiasm until his death in 1925.[47]

By the end of the century it was the scientific approach of Proctor and Flammarion, stripped of the latter's radical pluralism (not to say evangelical enthusiasm), that ruled the day. The writings of the Irish Astronomer

47 Flammarion, *La pluralité des mondes habités* (Paris, 1862); Guthke, *The Last Frontier*, 344–349; Crowe, *The Extraterrestrial Life Debate*, pp. 378–386.

Royal Robert S. Ball, the Irish science popularizer J. E. Gore, the American astronomer Simon Newcomb, and many others have much in common with Flammarion and Proctor on the subject. In them Laplace's nebular hypothesis – which not only provided the cosmogonical framework for our own solar system but also pointed to an abundance of planetary systems – became an integral part of the evolutionary approach to nature, with Gore even estimating the probability of planetary systems, a task that would prove a favorite pastime for the twentieth century. Further developments, such as G. Johnstone Stoney's application of the kinetic theory of gases to planetary atmospheres, not only added further scientific content to the debate but also illuminated a later stage of planetary evolution whereby (based on their masses) it was shown that the Moon could have no atmosphere and Mars no aqueous atmosphere.[48]

The evolutionary view could, however, be carried too far. In the United States, Percival Lowell concluded that because Mars was an older and more highly evolved planet, so were its superhuman inhabitants. The dangers of the scientific approach, when applied to such extremely difficult problems as the observation of planetary surface features, the search for planetary systems, and the origin and evolution of life, constitute one of the leitmotifs of our volume. And as we shall see in the next chapter, the further dangers inherent even in a professed "scientific" methodology are evident in the turn-of-the-century volume of the famous evolutionist A. R. Wallace, *Man's Place in the Universe* (1903).

Transcending all problems of observation and methodology is the historical fact that from the ancient Greeks to the nineteenth century (and even by the time of Kant a century before), the physical world had been transformed into the biological universe, one of the great revolutions of Western thought. By the middle of the eighteenth century, cosmological worldviews had brought the idea of life beyond the Earth into the mainstream of European consciousness, where it was for a time reconciled with Christianity because its vision of universal life was compatible with an omnipotent God. A century later it was accepted, reconciled, or rejected by a plethora of authors who explored more fully its compatibility not with God, but with Christian dogma. And at the threshold of the twentieth century the theory of an evolutionary universe, the development of astrophysics, and refined observational techniques combined to lay the scientific foundations on which the twentieth-century extraterrestrial life debate would rest.

By the beginning of the twentieth century, the biological universe had

[48] Venus, on the other hand, could retain an atmosphere potentially like the Earth's, and the giant outer planets' thick atmospheres could retain all known gases.

achieved the status of a worldview, a "biophysical cosmology" that asserted the importance of both the physical and biological components of the universe. Like all cosmologies, it made a claim about the large-scale nature of the universe: that life is not only a possible implication but also a basic property of the universe. Like all cosmologies, it redefined our place in the universe. And most important, like other cosmologies, in the twentieth century the biophysical cosmology became increasingly testable – even if it still embodied philosophical assumptions along with scientific theory and observation. The remainder of this volume documents the attempts of our century to undertake the exceedingly difficult task of testing the biophysical cosmology and coming to grips with the biological universe.

2

PLURALITY OF WORLDS AND THE DECLINE OF ANTHROPOCENTRISM

Our position in the material universe is special and probably unique, and . . . it is such as to lend support to the view, held by many great thinkers and writers today, that the supreme end and purpose of this vast universe was the production and development of the living soul in the perishable body of man.

A. R. Wallace (1903)[1]

We see the Earth as a small planet, one member of a family of planets revolving round the Sun; the Sun, in turn, is an average star situated somewhat far out from the centre of a vast system, in which the stars are numbered by many thousands of millions; there are many millions of such systems, more or less similar to each other, peopling space to the farthest limits to which modern exploration has reached. Can it be that throughout the vast deeps of space nowhere but on our own little Earth is life to be found?

Harold Spencer Jones (1940)[2]

"The question of questions for mankind – the problem which underlies all others, and is more deeply interesting than any other," T. H. Huxley wrote with regard to Darwin's theory of evolution, "is the ascertainment of the place which Man occupies in nature and of his relations to the universe of things."[3] Whatever the scientific merits of the extraterrestrial life debate, the same emotional issue of human status is inextricably linked to all discussions of inhabited worlds. Long before Huxley took up the cause of Darwin, pluralism and anthropocentrism had been locked in a battle that had not been completely decided by the dawn of the twentieth century. Committed anthropocentrists were likely to be the staunchest foes of pluralism, no matter what the evidence, and pluralists – whether they liked it or not – contributed significantly to the demise of anthropocentrism. For all the appeal to scientific argument, the continuing battle between

[1] A. R. Wallace, "Man's Place in the Universe, as Indicated by the New Astronomy," *The Fortnightly Review,* March 1, 1903, 396.

[2] Sir Harold Spencer Jones, *Life on Other Worlds* (New York, 1940), 22–23.

[3] T. H. Huxley, *Man's Place in Nature* (1863; Ann Arbor, Mich., 1971 edition), chapter 2, "On the Relations of Man to the Lower Animals," 71. Although cosmology is related to the issue of the nobility of humanity, it is not equivalent. The place of humanity in the medieval geocentric cosmos was open to interpretation, so that geocentrism was not equivalent to anthropocentrism: while the Earth was central, it was also far removed from the perfection of the celestial region. On the tensions of this dichotomy, see C. S. Lewis, *The Discarded Image* (Cambridge, 1964), 120, and Hans Blumenberg, *The Genesis of the Copernican World* (Cambridge, Mass., 1987).

anthropocentrism and the belief in life on other worlds pervades twentieth-century discussions of extraterrestrial life and carries the Darwinian debate on the status of humanity into the universe at large.

The roots of the connection were deep and irreconcilable. In the words of the Italian historian Paolo Rossi, the idea of inhabited worlds in the late Renaissance represented "a profound threat to a 'terrestrial' and anthropocentric conception of the universe, removing all meaning from the traditional discourse of the Humanists concerning the nobility and dignity of man."[4] One could always take the position of Kepler that the Earth still harbored the best creatures in the universe. Or one could claim with some degree of plausibility, as William Herschel did in the late eighteenth century, that our solar system was situated in the center of the system of stars. But just as Darwin's theory of biological evolution endangered the image of man because of his descent from lowly terrestrial creatures, the even more encompassing concept of a plurality of inhabited worlds belied an anthropocentric view of the world and threatened to make of man only one link in a chain of being that stretched far beyond the Earth.[5]

During the twentieth century this tug of war between anthropocentrism and pluralism was profoundly affected by radical changes in the astronomical worldview. While it was still possible, as the century began, for scientists to argue for an anthropocentric universe based on the Earth's privileged physical position in the cosmos, by 1930 advances in astronomy had destroyed this argument. The resultant worldview – an expanding universe of enormous dimensions in which the solar system was at the periphery of one galaxy among millions – tipped the scales strongly toward the presumption of other worlds for the rest of the century. "The assumption of mediocrity" became an underlying current of thought favoring other inhabited worlds, superseding the assumption of uniqueness that had opposed it.[6] The hopes for anthro-

4 Paolo Rossi, "Nobility of Man and Plurality of Worlds," in Allen G. Debus, *Science, Medicine and Society in the Renaissance*, ed. Allen G. Debus (New York, 1972), 157. For a succinct overview of the humanist "dignity of man" theme, see Paul O. Kristeller, *Renaissance Concepts of Man and Other Essays* (New York, 1972), 1–21, and Charles Trinkaus, "The Renaissance Idea of the Dignity of Man," *Dictionary of the History of Ideas*, vol. 4 (New York, 1973), 136–147, and references therein.

5 Kepler held "that we men find ourselves on the globe which is most suited to the most important and most noble rational creature among all physical bodies" (Rossi, "Nobility of Man," 142). For Herschel see "On the Construction of the Heavens," (1785), reprinted in M. A. Hoskin, *William Herschel and the Construction of the Heavens* (London, 1963), 82–106, especially 96. On the chain of being see Arthur O. Lovejoy, *The Great Chain of Being* (Cambridge, 1936).

6 I. S. Shklovskii and Carl Sagan, *Intelligent Life in the Universe* (San Francisco, 1966), chapter 25, "The Assumption of Mediocrity," pp. 356–361. This assumption by other names begins to become prominent in the 1940s.

pocentrism at the beginning of the century, and its rapid demise thereafter, constitute one of the profound shifts in twentieth-century thought. Before proceeding to the details of the extraterrestrial life debate, we must emphasize how important that change is to a complete understanding of our subject, and therefore how completely the proponents and opponents of extraterrestrial life champion not only a scientific theory, but an entire philosophy.

2.1 THE ANTHROPOCENTRIC UNIVERSE: A. R. WALLACE

Nowhere is the effect of the anthropocentric worldview clearer than in the work of Alfred Russel Wallace (1823–1913), cofounder with Darwin of the theory of evolution by natural selection. His influential work *Man's Place in the Universe: A Study of the Results of Scientific Research in Relation to the Unity or Plurality of Worlds* (1903) is in many ways a heretical volume that repudiated much of the past plurality-of-worlds tradition. Yet it embodied many of the problems that would be elaborated in ever more subtle form throughout the century. In many ways Wallace's treatise is much more typical of the plurality of worlds tradition than Percival Lowell's infamous theory of canals on Mars, reflecting both the time-worn arguments of the past and one of the chief trends of the future, an approach to the subject from the biological point of view.

The man who now raised the greatest threat to the case for life beyond the Earth since William Whewell exactly 50 years before was an octogenarian who had lived an eventful life.[7] Born in Wales in 1823, veteran of adventurous and productive expeditions to the Amazon (1848–1852) and the Malay Archipelago (1854–1862), Wallace is known primarily for his work on evolution. The joint presentation of the evolutionary theory of Darwin and Wallace before the Linnaean Society of London in 1858 is a legendary landmark in the history of science, as is Wallace's divergence from Darwin (beginning in 1869) on the subject of the descent of humanity. Though a strong proponent of natural selection in the animal world, Wallace believed natural selection alone was not enough to produce the human brain, that humanity was set apart from the animal

[7] For a comparison of Whewell and Wallace on this subject see William C. Heffernan, "The Singularity of Our Inhabited World: William Whewell and A. R. Wallace in Dissent," *Journal of the History of Ideas,* 39 (1978), 81–100. See also James J. Kevin, Jr., "*Man's Place in the Universe*: Alfred Russel Wallace, Teleological Evolution, and the Question of Extraterrestrial Life," M.A. thesis, University of Notre Dame, 1985.

world, and "that some higher intelligence may have directed the process by which the human race was developed."[8]

As an evolutionist it was only natural that Wallace, like Huxley, should be interested in the issue of humanity's place in nature; in effect he was now expanding the scope of Huxley's *Man's Place in Nature* (1863) from the Earth to the heavens. In order to do so, he added minuscule knowledge of astronomy to a lifetime's experience in biology, with a result that was predictably controversial. Wallace had become aware of recent astronomical advances while writing four new chapters for *The Wonderful Century* (1898), and it is here that we find the source of his anthropocentric view of the universe. Here Wallace related his amazement at discovering the view of John Herschel, Simon Newcomb, and Sir Norman Lockyer that our Sun was situated near the center of the Milky Way system. Other research had shown that this system was finite, implying that our Sun and its accompanying planets were situated in the center of the entire universe. The startling fact of this privileged position, together with the recent indication that the Earth was the only inhabited planet in our solar system, led Wallace to wonder if the Earth was the only inhabited planet in the whole universe. In addition, Wallace added:

> For many years I had paid special attention to the problem of the measurement of geological time, and also that of the mild climates and generally uniform conditions that had prevailed throughout all geological epochs; and on considering the number of concurrent causes and the delicate balance of conditions required to maintain such uniformity, I became still more convinced that the evidence was exceedingly strong against the probability or possibility of any other planet being inhabited.[9]

[8] For a succinct view of Wallace see H. Lewis McKinney, *Dictionary of Scientific Biography* (New York, 1976), XIV, 133–140. A complete biography is Wilma George, *Biologist Philosopher: A Study of the Life and Writings of Alfred Russel Wallace* (New York, 1964), while H. Lewis McKinney, *Wallace and Natural Selection* (New Haven, Conn., 1972), concentrates on his early life and works. On Wallace's spiritualism and its effect on his views on the origin of humanity see Malcolm Jay Kottler, "Alfred Russel Wallace, the Origin of Man, and Spiritualism," *Isis*, 65 (1974), 145–192, and Michele D. Malinchak, "Spiritualism and the Philosophy of A. R. Wallace," Ph.D. dissertation, Drew University, 1987. On his broader philosophy see Roger Smith, "Alfred Russell Wallace: Philosophy of Nature and Man," *British Journal for the History of Science, 6* (1972), 177–199.

[9] A. R. Wallace, *Man's Place in the Universe: A Study of the Results of Scientific Research in Relation to the Unity or Plurality of Worlds* (New York, 1903), Preface, v–vi. On Wallace's early interest in astronomy and the immediate background to his book on other worlds see James Marchant, *Alfred Russel Wallace: Letters and Reminiscences* (New York and London, 1916), 401–406. In developing his astronomical views, Wal-

Wallace presented his views in more systematic form in an article published simultaneously in Britain and the United States in early 1903.[10] Beginning with the Copernican system and continuing with the vast Newtonian universe and the revelations of larger and larger telescopes, Wallace argued that the importance of the Earth and its inhabitants has diminished. But during the last quarter century, observations tended "to show that our position in the material universe is special and probably unique, and that it is such as to lend support to the view, held by many great thinkers and writers today, that the supreme end and purpose of this vast universe was the production and development of the living soul in the perishable body of man."[11] In support of his striking claim, Wallace argued that our Sun is located in the center of a cluster of suns, itself at the center of a finite stellar universe, and that only this central position in the stellar universe is suitable for life.

After he strengthened his arguments, by the end of August 1903 Wallace's book was ready for the press.[12] The "connected argument" for the Earth as the unique home of life in the universe began only after five chapters on the background of contemporary astronomy, chapters that the historian of astronomy Agnes Clerke – no mean critic – later praised as a brilliant summary of the latest results in astronomy, especially considering that their author was outside the field.[13] From that point on, Wallace's main argument for the uniqueness of the Earth is based on three indispensable premises: (1) life can exist only around our Sun or the cluster of suns surrounding it; (2) no life exists on planets around other suns in the solar cluster; and (3) no life exists in our solar system beyond the Earth. The bulk of the rest of the book may be seen as aimed at supporting these premises, namely, the arguments from Position, Planetary Systems and Planetary Conditions (Figure 2.1).

It is no exaggeration to say that the Position Argument assumed major, if

lace was particularly influenced by Agnes Clerke, *The System of the Stars* (1890) and *History of Astronomy* (1st ed., 1885); Simon Newcomb, *The Stars* (1902); Sir John Herschel, *Outlines of Astronomy* (1869); and Lord Kelvin. He was also familiar with the works of R. A. Proctor and J. E. Gore but did not always accept their views.

10 Wallace, "Man's Place in the Universe"; *The Independent*, February 26, 1903.

11 Wallace, "Man's Place in the Universe," 396.

12 The manuscript of the book, along with other Wallace papers, is in the British Library in London, ADD MSS 46420, 263 folios. The original title, "Universe for Man" has been deleted and replaced with "Man's Place in the Universe." Parts of chapters 3 and 6 are typescript inserted verbatim from *The Wonderful Century*. On the crucial role of Wallace's literary agent, see Add MS 55221, Wallace to MacMillan & Co., January 11, 1905, folios 52–53.

13 Agnes Clerke, "Life in the Universe," *Edinburgh Review*, 200 (1904), 60. The Wallace papers in the British Library (Add MSS 46437, folios 89–92) show that Clerke and Wallace corresponded about the book beginning March 15, 1901. On Clerke see M. T. Bruck, "Agnes Mary Clerke, Chronicler of Astronomy," *QJRAS*, 35 (1994), 59–79.

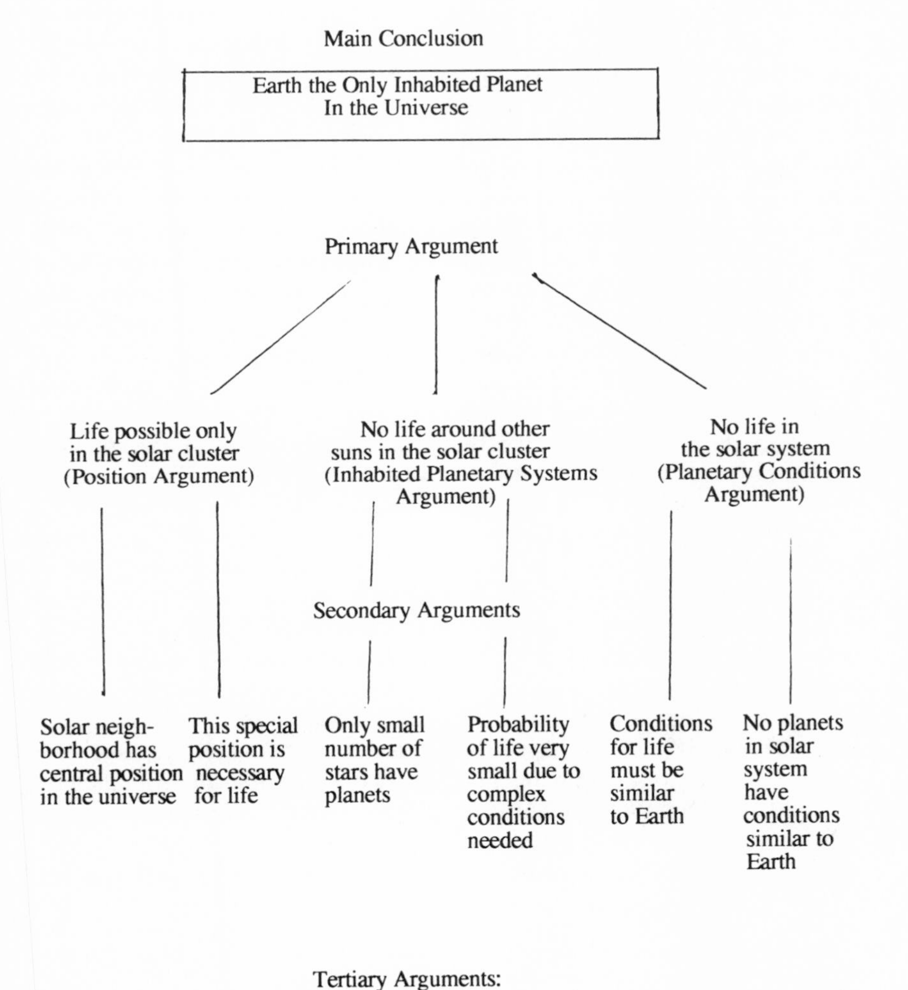

Figure 2.1. Main components of the "connected argument" in A. R. Wallace's *Man's Place in the Universe* (1903).

not predominant, importance in Wallace's book; indeed, it was the catalyzing argument for the entire volume. In supporting it, Wallace claimed that he was simply espousing the view of the most eminent astronomers of his day, a claim exaggerated but not completely wide of the mark. When

Wallace wrote in 1903, all stars, and indeed all observable phenomena in the universe, were widely believed to be part of a single system, physically associated through the gravitational force, perhaps several thousand light years in diameter (compared to the 100,000 light years now estimated), with the Sun in a nearly central position. The "island universe" theory, which postulated many such systems, had been in gradual decline since 1864 when William Huggins, in one of the pioneering observations in the rising new science of astrophysics, turned his spectroscope on a nebula in the constellation Draco and determined that it was gaseous, and not a system of stars analogous to our own system. By the late 1880s the island universe theory had completely fallen from favor. And though in 1898 the American astronomer James E. Keeler had begun a research program on spiral nebulae that would result in the eventual revival of the island universe theory, in 1903 this conclusion was far from certain.[14]

It is therefore not surprising that Wallace viewed the universe as a single system of stars with our solar system at the approximate center. The unity of the stellar universe, he believed, had been established by such a vast body of facts, especially in the last 30 years, that it is "now hardly questioned by any competent authority."[15] From the equality of star counts on both sides of the plane formed by the Milky Way – "the fundamental phenomenon upon which the argument set forth in this volume primarily rests" – Wallace argued that we reside in the central part of this plane and that the stellar system is spherical. Embedded in this general spherical structure of stars is a "solar cluster" consisting of a group of several hundred to several thousand stars surrounding our sun, and which seem to form a condensed group of stars separate from the rest of the Milky Way. Wallace set the dimensions of the entire system at 3600 light years, placed the Sun at the outer margins of the solar cluster, and put it in orbit around the center of gravity of the cluster, resulting in the image of the universe shown in Figure 2.2.[16]

[14] The definitive treatments of the history of the island universe theory are Robert Smith, *The Expanding Universe: Astronomy's Great Debate, 1900–31* (1982), and Richard Berenzden, Richard Hart, and Daniel Seeley, *Man Discovers the Galaxies* (1976). More popular treatments are C. A. Whitney, *The Discovery of Our Galaxy* (1971) and Timothy Ferris, *The Red Limit* (1977; rev. ed., 1983).

[15] Wallace supported the view of a single finite system of stars by various arguments that the stars themselves are not infinite in number, including Olber's paradox that if the stars were infinite in number, their total radiation would make the night sky at least as bright as the day sky; that larger telescopes using photography did not show the increase in the number of stars expected if the system extended indefinitely; and that star counts show us the number of stars to increase 3.5 times for each magnitude up to the 10th, beyond which this increase began to diminish very rapidly.

[16] Wallace, *Man's Place in the Universe,* 156–165. He cites Newcomb, Clerke, and Lockyer in favor of the view of a central position for Earth. On the spherical form, see pages 165–169 and 291–301. Wallace cites William Herschel, Gould, Clerke, Kapteyn,

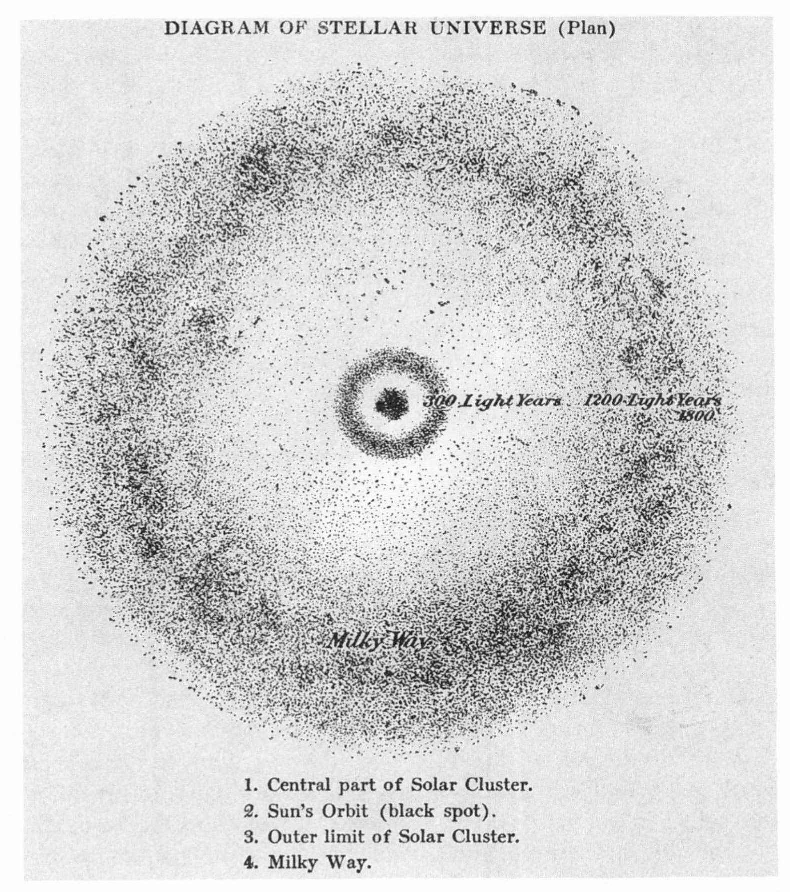

Fig. 2.2. The anthropocentric image of the universe according to A. R. Wallace (1903), showing the Milky Way stellar system, 3600 light years in diameter, with the Sun (a tiny black dot almost invisible here) near the center.

The centrality of the Sun was "the very heart of the subject" of Wallace's inquiry for more than philosophical reasons, for on that position rested much of his argument against other worlds. The importance

and Newcomb in favor of the solar cluster, despite the fact that Clerke had informed him that Kapteyn had abandoned the idea (Clerke to Wallace, April 17, 1903, British Library ADD MSS 46437, folio 125). For the dimensions of his system, Wallace adopted Lord Kelvin's approach of trying to explain stellar velocities beginning with a universe 6400 light years in diameter with 1000 million suns like ours, which had now contracted to about half that size and produced the stellar velocities observed.

of centrality for Wallace was that only in such a position could a star such as the Sun generate a uniform heat supply over the long period needed for the development of life on Earth, which he put at several million years. The origin of solar power was a long-standing problem whose solution lay two decades in the future with an understanding of atomic processes in the Sun. Wallace's theory must be judged to be as plausible as any other at the time. Drawing on Lord Kelvin's meteoric hypothesis, he argued that the solar cluster would, through gravitation, draw inflowing matter toward it from the outer region of the star system. This matter first "furnishes the energy requisite for bringing these slowly aggregating stars to the required intensity of heat for forming luminous suns" and then maintains this energy "by a constant and steady inrush of matter" possessing such high velocities as to aid in maintaining the temperature of the sun over such long periods are as required for life on Earth.[17] The same mechanism of energy production would apply to other stars in about the same position in the solar cluster, while the stars of other regions of the system would presumably have much shorter lives, not requisite for the development of life. This view of the universe thus greatly reduced, to perhaps several hundred, the number of candidate stars that might harbor life-bearing planets. Thus did Wallace arrive at the first of the three major premises needed for his final conclusion that the Earth was the only inhabited planet in the universe.

This argument for a nearly central position for our solar system left open the question of life on other planets in the sun's system and around the few hundred long-lived stars surrounding the Sun. If Wallace could prove that life had not arisen on those few hundred stars in the solar cluster, and had not arisen within our solar system beyond the Earth, he would have his final argument clinched. As an essential part of his discussion of whether life can exist on any planet in our solar system (the Planetary Conditions Argument), Wallace launched into a detailed discussion of the complexity of life on Earth, for he realized that if there is any prerequisite to understanding life beyond the Earth, it is to understand first life on Earth. Here Wallace brought to bear some of the problems on which he had spent a lifetime of contemplation, thus offering a unique view, since (as Clerke commented) "No great biologist has ever before seriously considered the possibilities of cosmic life, and they can only be fully discussed in the light of expert biological science." This was a difficult task; in many ways biology did not even become a science until the nineteenth century, in part through events surrounding Darwin and Wal-

[17] Wallace, *Man's Place in the Universe,* 304–305.

lace themselves, and the study of physiology was still at the organic rather than the microscopic level.[18]

Nevertheless, Wallace plunged ahead with one of the greatest questions of all, one that will haunt the extraterrestrial life debate throughout the century: what is life? Rejecting philosophical definitions given by Aristotle, Spencer, and others as vague and abstract, and dissatisfied as well with definitions based on general properties, Wallace quickly focuses on protoplasm, which Huxley had called the "physical basis of life." He marveled that it is composed basically of only the four elements hydrogen, carbon, nitrogen, and oxygen, and that the variety of life is produced by the amazing ability of carbon to form compounds. The chemical reactions in the protoplasm, Wallace argued, determine the physical conditions necessary for the development and maintenance of life on Earth: a regular heat supply, resulting in a limited range of temperatures; a sufficient amount of solar light and heat; water in great abundance and universally distributed; an atmosphere of sufficient density; and alterations of day and night to keep the temperature in balance. Life, he believed, must have a temperature range from 32 to 104 degrees Fahrenheit (°F), while solar light is necessary for its essential role in decomposing CO_2 ("carbonic acid") into carbon and oxygen in plants. Water not only constitutes a large proportion of the living body of both plants and animals, it is also essential in producing a limited range of temperatures. And the atmosphere must be of sufficient density to store heat and to supply oxygen, carbon dioxide, and water vapor for life; a reduction of density by one-fourth would probably render the Earth uninhabitable. Thus "Considering the great variety of physical conditions which are seen to be necessary for the development and preservation of life in its endless varieties, any prejudicial influences, however slight, might turn the scale, and prevent that harmonious and continuous evolution which we know *must* have occurred." In order to achieve these conditions on a planetary scale, the distance from the Sun, the obliquity of the ecliptic (causing the seasons), the persistence of a mild climate through geological time, and the distribution of water were essential factors.[19]

The purpose of this detailed description of the complexity of life on Earth, the delicate balance of conditions needed to maintain it, and the planetwide features needed to generate the conditions is clear: it is offered as a justification for the claim that a complex combination of conditions is necessary to develop and maintain life on Earth. The question, then, is

18 For the state of biology in the early twentieth century see Garland Allen, *Life Science in the 20th Century* (New York, 1975). Clerke's statement is in a letter to Wallace of October 22, 1903, British Library ADD MSS 46437, folio 141.

19 Wallace, *Man's Place in the Universe,* 191–194, 205–239.

whether this complex combination of conditions is found beyond the Earth. In contrast to his elaborate discussion of complex conditions, Wallace's *a priori* opinion on their existence beyond the Earth is given in a single sentence:

> The combination of causes which lead to this result [the development and maintenance of life over long periods] are so varied, and in several cases dependent upon such exceptional peculiarities of physical constitution, that it seems in the highest degree improbable that they can *all* be found again combined either in the solar system or even in the stellar universe.[20]

The complexity of life on Earth could have stood alone as a commonsense argument against life having developed elsewhere in the solar system or being widespread in the universe. But combined with the arguments from Position, Planetary Conditions, and Planetary Systems, it became much more compelling, for these arguments greatly narrowed the number of possible life sites. Nor could it be plausibly argued that life of an entirely different nature might exist under very different conditions. Spectroscopy, meteorites, and other considerations showed the composition and laws of the universe to be everywhere the same: "We may therefore feel it to be an almost certain conclusion that – the elements being the same, the laws which act upon and combine, and modify those elements being the same – organised living beings wherever they may exist in this universe must be, fundamentally, and in essential nature, the same also." And as a corollary, "Within the universe we know, there is not the slightest reason to suppose organic life to be possible, except under the same general conditions and laws which prevail here.[21] The question of whether biological beings must be constructed of the same chemical elements as beings on Earth Wallace did not yet tackle.

Turning to the planets of our solar system, Wallace now argued that planetary habitability primarily depends on the mass of the planet, for this determines whether or not it can retain the molecules that compose an atmosphere. Thus Mercury and Mars are too small to retain water vapor. Because of their low density, the larger planets "can have very little solid matter" on which life might develop. This left only Venus, but Venus was believed to always keep the same face toward the Sun, so one side is too cold, the other too hot. Moreover, the lifetime of the Sun

[20] Ibid., 310.
[21] Ibid., 182–189.

renders it impossible that these planets now unsuitable for life might have been suitable in the past or could be in the future:

> We are, therefore, again brought to the conclusion that there has been, and is, no time to spare; that the whole of the available past life-period of the sun has been utilised for life-development on the earth, and that the future will be not much more than may be needed for the completion of the grand drama of human history, and the development of the full possibilities of the mental and moral nature of man.[22]

In arguing against planetary systems, Wallace played a game of successive elimination. Stars too small would not give enough heat. Stars in the Milky Way were ruled out because of the "excessive forces there in action." We are therefore back to the solar cluster "variously estimated to consist of a few hundred or many thousand stars." Among these, some would be too small or too large for the appropriate long-term heat required for the long development of life. Others were just in the process of forming. And recent research, especially that of W. W. Campbell at Lick Observatory, showed that many stars were actually binary stars, which must be inimical to the development of planets. The solar cluster, Wallace argued, might harbor only two or three single stars:

> But we do not really *know* that any such suns exist. If they exist we do not *know* that they possess planets. If any do possess planets these may not be at the proper distance, or be of the proper mass, to render life possible. If these primary conditions should be fulfilled, and if there should possibly be not only one or two, but a dozen or more that so far fulfil the first few conditions which are essential, what probability is there that all the other conditions, all the other nice adaptations, all the delicate balance of opposing forces that we have found to prevail upon the earth, and whose combination here is due to exceptional conditions which exist in the case of no other known planet – should *all* be again combined in some of the possible planets of these possibly existing suns? I submit that the probability is *now* all the other way.[23]

[22] Ibid., 258–277; the quote is on p. 275. On planetary atmospheres Wallace cites G. Johnstone Stoney, "Of Atmospheres upon Planets and Satellites," *Transactions of the Royal Dublin Society,* vol. 6, series ii, part xiii.

[23] Wallace, *Man's Place in the Universe,* 278–285.

Taken together, Wallace had composed a formidable array of arguments based on the best science available at the beginning of the twentieth century. The Position Argument narrowed the possible habitats for life from millions to a few hundred stars. The Planetary Systems Argument achieved the same purpose from a different angle, perhaps with less observational evidence. The Complexity of Life Argument, which might on its own have eliminated the possibility of life elsewhere in the universe, did so more forcefully in conjunction with the first two arguments. And the Planetary Conditions argument showed empirically that no other life existed in the solar system.

Although this ended Wallace's "connected argument," he also put forth very briefly a philosophical argument from purpose. If the Earth is the only inhabited planet in the universe, Wallace asserted, this may be seen either as a coincidence or as a very important conclusion indicating that the universe was brought into existence for the ultimate purpose of the development of man on Earth. While Wallace believed the majority of scientific men would call it coincidence, he left no doubt that he was of the opposite opinion, and that this opinion was contrary neither to science nor to religion. Life on every planet

> would introduce monotony into a universe whose grand character and teaching is endless diversity. It would imply that to produce the living soul in the marvellous and glorious body of man . . . was an easy matter which could be brought about anywhere, in any world. It would imply that man is an animal and nothing more, is of no importance in the universe, needed no great preparations for his advent, only, perhaps, a second-rate demon, and a third or fourth-rate earth.

The immensity of space and time, Wallace believed, "seem only the appropriate and harmonious surroundings, the necessary supply of material, the sufficiently spacious workshop for the production of that planet which was to produce, first, the organic world, and then, Man."[24] Although isolated and briefly stated, this argument from purpose perhaps played a greater role as a driving force than Wallace would admit.

A final argument, "An Additional Argument Dependent on the Theory of Evolution," was added to the 1904 edition of Wallace's book. Especially interesting because Wallace was so closely involved with the evolu-

[24] Ibid., 317–318. This argument is different from Wallace's other arguments in that the hypothesis to be proven is assumed true, and it is then seen whether any false conclusions may be drawn as a consequence.

tion arguments of his day, it is independent of the three connected scientific arguments and may be seen as another aspect leading to the same conclusion. Wallace argued that since humanity is the result of a long chain of modifications in organic life, since these modifications occur only under certain circumstances, and since the chances of the same conditions and modifications occurring elsewhere in the universe were very small, the chances of beings in human form existing on other planets was very small. Moreover, since no other animal on Earth, despite the great variety and diversity of forms, approaches the intelligent or moral nature of humanity, Wallace concluded that intelligence in any other form was also highly improbable. He was even willing to say how improbable it was:

> If the physical or cosmical improbabilities as set forth in the body of this volume are somewhere about a million to one, then evolutionary improbabilities now urged cannot be considered to be less than perhaps a hundred millions to one; and the total chances against the evolution of man, or an equivalent moral and intellectual being, in any other planet, through the known laws of evolution, will be represented by a hundred million of millions to one.[25]

By its deletion of the idea of purpose from his central argument and by its inclusion of biological aspects, Wallace's book marked a signal advance in the debate about other worlds. Its failure, however, is marked by the dominance of the anthropocentric worldview over all other arguments. Despite his best intentions and protestations to the contrary, in the end Wallace's theory was simply too ad hoc, less driven by probabilities than by deep-seated anthropocentrism. Convinced by 1898 of the nearly central position of the Sun, Wallace first sought – and found – the significance of this fact in the uniqueness of life, and then adduced arguments in favor of the view that life was found beyond the Earth neither in our solar system nor in others. Although his theory was based on what he believed to be the fact of the Sun's centrality, the pioneer of evolution nevertheless fell victim to the the effect of a worldview with insufficient proof. In his concept of the plurality of worlds, no less than in his ideas about the evolution of humanity, he found it necessary to set humanity apart. Although professing a scientific approach, Wallace's book serves as a lesson on the limits of science when worldviews dominate empirical evidence. It is a lesson the twentieth century should take to heart.

25 Wallace, *Man's Place in the Universe* (London, 1904), Appendix, 326–336: 334–335.

2.2 THE NEW UNIVERSE: ANTHROPOCENTRISM'S DEMISE?

The skepticism with which Wallace's thesis was received was barely disguised. Denying that the universe is finite and that the Sun is central, British astronomer E. Walter Maunder concluded that "Every one . . . of Dr. Wallace's demonstrations falls to the ground." H. H. Turner, Savilian Professor of Astronomy at Oxford, charged that Wallace "seems to me to have unconsciously got his facts distorted, and to indicate practically nothing wherewith to link them to his conclusions." In the United States, Harvard astronomer W. H. Pickering rejected Wallace's argument, characterizing it as "not science, and . . . not very satisfactory." And in France, pluralist par excellence Camille Flammarion surprised no one with his conclusion that "An examination of Mr. Wallace's plea in favor of his geocentric and anthropocentric theory has not convinced me; on the contrary, it seems to me to give a more solid basis than ever to the opposite opinion."[26]

Simon Newcomb, at the apex of his career as America's most famous astronomer and the source of so many of Wallace's "facts," offered both praise and damnation for "the Nestor of scientific authors": "As a piece of analytical reasoning, in which the greatest variety of facts are, with artistic logic, arrayed in support of the conclusion, the book is well worthy of its distinguished author, and yet we doubt if many readers will accept his conclusions." Characterizing the book as an extreme reaction against the ideas of men like Brewster and Thomas Dick, he nevertheless agreed that "it must be confessed that the general trend of the discoveries of modern science has been in the direction of the reaction which Mr. Wallace carries to an extreme point." While it was true that conditions on Earth are not approached anywhere in our solar system except perhaps on Mars, and that no other solar systems like ours are known (though a few hundred stars are known to have "dark planets" around them), Newcomb emphasized there are "probably more than a hundred million stars in the heavens," so that "looking at

[26] E. Walter Maunder, "The Earth's Place in the Universe," *The Journal of the British Astronomical Association,* 13 (1903), 227–234: 234; H. H. Turner, "Man's Place in the Universe: A Reply to Dr. Wallace," *Fortnightly Review,* 73 (April 1903), 598–605; William H. Pickering, "Man's Place in the Universe," *The Independent,* 55 (1903), 597–600; Camille Flammarion, "The Earth and Man in the Universe," *The Independent,* 55 (1903), 958–968. These reviews were all based on Wallace's initial article; Maunder expanded his criticism after publication of the book in "Man's Place in the Universe," *Knowledge* (December 1903), 268–270. Professor Michael J. Crowe has found more than 40 reviews of Wallace's volume, and I am indebted to him for bringing many of them to my attention. I discuss only a few of the major ones here.

the question as one of probabilities, the chances are very much against Mr. Wallace's theory." More significantly, Newcomb rejected the Position Argument: While "our position does not deviate from the center to an extent that can be determined by any method yet known," the term "center" is ill-defined, and the 10 miles per second motion of the Sun would change this in any case. Newcomb also did not accept the proposition that terrestrial conditions were necessary for the development of life; only experiment could show this. In the end the position of the dean of American astronomers was cautious: "Wide though our knowledge of the universe has become, it is infinitesimal when compared with the range it will have to include before anything positive can be said on the subject of life in other worlds."[27]

Of all the reviews, none was more well considered and sympathetic to Wallace than that of Agnes M. Clerke, the historian of astronomy, who had corresponded with Wallace about his book before publication. Unlike her colleagues, she strongly supported Wallace's premise that the Sun is near the center of a finite stellar universe: "We gather, then, from all the evidence at our disposal that the sidereal universe – the only universe within our ken – has boundaries; and that it has a center of symmetry, or of gravity, inevitably follows. Further, our position cannot be very far removed from that center."[28] Though she accepted Wallace's reasoning that stability is necessary for the long development of life, she did not accept Wallace's claim for a solar cluster. While stability may not be found in the great ring of the Milky Way itself, she felt that anywhere within this ring (rather than only at its center) "a family of planets might be reared in perfect security."

Clerke also moved beyond the anthropocentric argument to consider life in the solar system. "If we are to think on the subject at all, our thoughts must be guided by analogy; they can transcend experience only at the risk of falling into the quagmire of extravagance. Life is inconceivable to us save under conditions to some extent similar to those with which we are familiar." Given this premise there could be no life beyond Earth in our solar system, with the doubtful exception of Mars, a conclusion supported by Stoney's work on planetary atmospheres. Nor, in her opinion, were other habitable planetary systems likely, since variable and double stars did not provide suitable conditions. Clerke thus accepted Wallace's conclusion that most stars do not have planetary systems. The probability that they did "seemed considerable only through ignorance. With the growth of knowledge regarding the real nature of the stars, it

[27] [Simon Newcomb], "Wallace on Life in the Universe," *The Nation* (January 14, 1904), 34–35. The author is identified as Newcomb from internal evidence.

[28] Agnes Clerke, "Life in the Universe," *Edinburgh Review*, 200 (July 1904), 59–74:71.

shrinks towards evanescence." In the end, Clerke's conclusion was cautious – and similar to Newcomb's:

> Unquestionably the trend of modern research is to encourage the opinion that the solar system is set apart among the stars, and the earth among the planets, as if for the express purpose of harbouring in safety the frail craft bearing the burthen of life. But demonstrative evidence on the point is not at hand, and cannot be looked for. Arguments *a priori* are futile. They rest on arbitrary assumptions.[29]

On one point all of the participants in the Wallace controversy agreed: reaching any definitive conclusion was extremely difficult. Wallace himself admitted that his conclusion was "a point as to which absolute demonstration, one way or the other, is impossible. But in the absence of any direct proofs, it is clearly rational to inquire into probabilities; and these probabilities must be determined not by our prepossessions for any particular view, but by an absolutely impartial and unprejudiced examination of the tendency of the evidence." Maunder nevertheless charged that Wallace "has attempted to prove what is either in itself beyond proof, or is beyond our present ability to prove." Pickering viewed the whole subject as "at the outermost bounds of our knowledge" and felt that many of Wallace's facts were open to interpretation. And Newcomb characterized the book as "an attempt to prove a negative in a case where no such proof is possible."[30] The reaction to Wallace shows that while anthropocentrism could still be intelligently argued early in the twentieth century, in the cosmological context, the opinion of astronomers was moving away from anthropocentrism.

The flight from anthropocentrism was soon accelerated by events in astronomy. How quickly the fate of humanity could change is indicated by the fact that within 15 years of Wallace's death in 1913, most of his central assumptions were rendered obsolete by an emerging new cosmology.[31] Einstein's theory of relativity, while a radical departure from previous ideas of space and time, had no immediate impact on the view of our place in the universe, except insofar as it fostered the idea of an enormous universe. But other discoveries did. In 1918 the American astronomer Harlow Shapley reported, based on his study of the distribution of globu-

29 Ibid., 61, 74.

30 Wallace, *Man's Place in the Universe,* Preface, vi; Maunder, "Earth's Place," 233; Pickering, "Man's Place," 597; Newcomb, "Wallace on Life," 34.

31 Wallace's book went through seven editions by 1908 and another in 1914. It was translated into German in 1903 and into French in 1907. Beyond the second decade of the twentieth century, however, it was not reprinted and its earlier editions had little influence.

lar clusters of stars, that our solar system was located in a very eccentric position in the Galaxy, at its periphery rather than its center. By 1924 Edwin P. Hubble had demonstrated to the satisfaction of most astronomers that many other galaxies exist outside our own, galaxies that he showed a few years later are fleeing from one another in an "expanding universe." And beginning in the 1920s, Eddington and others devised energy-producing mechanisms for stars that swept away any need for an infalling matter theory such as Wallace proposed.[32]

Once again cosmology was radically transformed. And once again this change in worldview affected the status of humanity and the debate over a plurality of worlds. The questions now posed were reminiscent of those generated by the Earth's displacement in the Copernican theory but applied to a much wider field of stars congregated in innumerable galaxies spread throughout an infinite universe. If life could arise on a planet at one extreme edge of the Galaxy, why should it not arise anywhere in the Galaxy? And if the Galaxy was but one of millions, should not life be spread throughout the universe, perhaps even one of its fundamental properties? Clearly, these new developments turned the previous logic on its head, with the presumption in favor of inhabited worlds. This reversal of fortune put the burden of proof on the anthropocentrists. The "assumption of mediocrity" replaced the assumption of centrality, this time as a very powerful, commonsense argument for other worlds in the absence of other evidence.

This logic, however, did not have immediate impact but only gradually worked its way into public and scientific consciousness. Without a doubt, this is largely due to a more specific theory indicating that life might *not* be widespread throughout the universe. In particular, the British astronomer James Jeans as early as 1916 had argued that solar systems were formed by stellar collisions, and that because such collisions were extremely rare, planets on which life might develop would also be rare. During the 1920s and 1930s the belief in life beyond the solar system was at a low point for just this reason. In his many popular works Jeans propagated the view that "astronomy . . . begins to whisper that life must necessarily be somewhat rare" in the universe and that it was just possible that only the Earth could support life.[33] Influenced by Jeans, Henry Norris Russell, the dean of American astronomers following the death of

[32] These developments are treated in Robert Smith, *The Expanding Universe* (Cambridge, 1982).

[33] James Jeans, *The Nebular Hypothesis and Modern Cosmogony* (Oxford, 1923). Among Jeans's other popular works propagating this theory are *The Universe Around Us* (Cambridge, 1929), *The Mysterious Universe* (Cambridge, 1930), and *Eos, or the Wider Aspects of Cosmogony* (London, 1930). We shall treat this story in detail in Chapter 4.

Newcomb, argued in his widely used textbook that close stellar encounters were "extremely rare," planetary systems "infrequent," and the existence of habitable or actually inhabited planets "matters of pure speculation."[34] And in 1923, 6 years after his famous work showing the eccentric position of the solar system in the Galaxy, Shapley himself agreed that planetary systems were unlikely and habitable planets "very uncommon," even as he held out the possibility that the Earth was not entirely unique.[35] It is notable that, given a consensus theory that planetary systems were rare, astronomers jumped to no philosophical conclusion about life in the universe based on the discovery of the eccentric position of the Earth in the Galaxy.

This situation began to change as early as 1940, as is evident in the book of Astronomer Royal Sir Harold Spencer Jones, *Life on Other Worlds*. The most influential work on the subject since Wallace's volume, it remained the standard for a quarter century, read by scientists, lay persons, and students alike until superseded by Shklovskii and Sagan's *Intelligent Life in the Universe* in 1966. The "picture of the universe" with which Spencer Jones began his book differed radically from that of Wallace. Unlike Wallace, the Astronomer Royal was under no illusions about a unique physical location for the abode of humanity. Fully familiar with the work of Shapley and Hubble, by 1940 he also knew that the extent of the observable universe, as seen by the great 100-inch telescope at Mount Wilson, was 500 million light years, compared to the 3600 light years that Wallace had postulated. Thus "we see the Earth as a small planet, one member of a family of planets revolving round the Sun; the Sun, in turn, is an average star situated somewhat far out from the centre of a vast system, in which the stars are numbered by many thousands of millions; there are many millions of such systems, more or less similar to each other, peopling space to the farthest limits to which modern exploration has reached."[36] Given this view, Spencer Jones asked the leading question that would become the rhetorical clarion call of extraterrestrial life proponents: "Can it be that throughout the vast deeps of space nowhere but on our own little Earth is life to be found?" Answering his own question, he wrote that "with the Universe constructed on so vast a scale, it would seem inherently improba-

[34] Henry Norris Russell, Raymond S. Dugan, and John Q. Stewart, *Astronomy: A Revision of Young's Manual of Astronomy*, vol. 1, *The Solar System* (Boston, 1926), 468.

[35] Harlow Shapley, "The Universe and Life," *Harper's Monthly Magazine*, 146 (1923), 716–722. Similar views are given in Shapley, "The Origin of Life," a symposium by Professors Harlow Shapley, Edward C. Jeffrey and Kirsopp Lake, held at Harvard on November 6, 1923, *Forum*, 72, 100–109, and "Life in Other Worlds?" *The Universe of Stars*, based on Radio Talks from the Harvard Observatory, Harlow Shapley and Cecilia H. Payne, eds. (Cambridge, Mass., 1929; 1st ed., 1926).

[36] Sir Harold Spencer Jones, *Life on Other Worlds*, 22–23.

ble that our small Earth can be the only home of life."[37] Still constrained by Jeans's theory, Spencer Jones was unable to argue for widespread life in this first edition, but in subsequent editions he certainly did.

By the late 1930s Russell and others had already found Jeans's encounter theory untenable, and when in the 1940s the nebular hypothesis was revived, the formation of planetary systems was held to be common and the floodgates were open to other worlds. This was observationally very difficult to prove, but as early as 1943 observational evidence was believed to be at hand. In an article entitled "Anthropocentrism's Demise," Russell seized on this evidence as a disproval of Jeans's theory and an indication that a very large number of planetary systems might exist, a "radical change – indeed practically a reversal – of the view which was generally held a decade or two ago."[38] By 1942 Jeans himself was forced to admit that even if collisions are rare, the extent of the universe, with its many galaxies and stars, requires the number of planets to be far from insignificant, and on a great number life will have arisen.[39] And after ignoring the subject for decades, by the 1950s Shapley reversed his opinion stated in the 1920s, a reversal that surpassed even that of Jeans. As early as 1953, but especially in his 1958 book *Of Stars and Men* (significantly subtitled *Human Response to an Expanding Universe*), Shapley acknowledged the adjustment required by the fact that the Earth and its life are "on the outer fringe of one galaxy in a universe of millions of galaxies. Man becomes peripheral among the billions of stars in his own Milky Way; and according to the revelations of paleontology and geochemistry he is also exposed as a recent, and perhaps an ephemeral manifestation in the unrolling of cosmic time." The new cosmology, Shapley now asserted, requires us to believe that "millions of planetary systems exist, and billions is the better word." And just as crucially, Shapley held that biochemistry now indicated that "whenever the physics, chemistry and climates are right on a planet's surface, life will emerge, persist and evolve."[40]

[37] Ibid., vii.

[38] Henry Norris Russell, "Anthropocentrism's Demise," *Scientific American* (July, 1943), 18–19.

[39] Jeans, "Is There Life on Other Worlds?" *Science,* 95 (June 12, 1942), 589, reprinted in Donald Goldsmith, *The Quest for Extraterrestrial Life: A Book of Readings* (Mill Valley, Calif., 1980), 81–83; "Origin of the Solar System," *Nature* (June 20, 1942), 695; and "Non-Solar Planetary Systems," *Nature* (December 18, 1943), 721.

[40] Harlow Shapley, *Of Stars and Men* (Boston, 1958), 108–114, in a chapter adapted from an article in *The American Scholar* (Autumn 1956). Shapley had returned to the subject of extraterrestrial life in "On Climate and Life," in *Climatic Change: Evidence, Causes and Effects* (Cambridge, Mass., 1953), the result of a conference sponsored by the American Academy of Arts and Sciences in May 1952. The article, in slightly different form, was originally printed in *The Atlantic Monthly* earlier in 1953.

Figure 2.3 The Andromeda Galaxy, symbol of the new universe in which the Milky Way Galaxy is only one of billions of galaxies. Courtesy Mount Wilson Observatory.

By the time Shklovskii and Sagan's *Intelligent Life in the Universe* was published in 1966, the view of a vast universe was entrenched, and the authors' invocation of the principle of mediocrity as an argument in favor of many inhabited worlds was in no way surprising. Not all astronomers had come to the same conclusion. To the end of his life, Eddington, for example, seems to have held to his earlier view that while "a few rival earths" may exist, "at the present time our race is supreme." Nor did all astronomers, including the pioneering Edwin Hubble, express their views on extraterrestrial life in detail.[41] But like the anthropocentric image represented by Wallace, the Einstein–Shapley–Hubble image of an enormous universe with the Earth stuck in a corner of one galaxy among billions formed an important background to all arguments on the plurality of worlds after the 1930s. Shapley was no Dante, but the worldview that he expressed in *Of Stars and Men* was an influential image for the twentieth century no less than Dante's *Divine Comedy* was for the Middle Ages. Impressive photographs of galaxies (Figure 2.3) belied Wallace's worldview and came to symbolize the new universe, without a center or even the possibility of a center.[42]

And yet, the twentieth century held a final surprise in the intimate dance of anthropocentrism and plurality of worlds. During the last quarter of the century, advocates of the "anthropic principle" pointed out that if the structure and foundations of the universe were the slightest bit different, life could not have formed at all, for planets and stars could not have formed. Did this mean that, after all, life was the end purpose of the entire universe? And since the only life we knew to exist with certainty was terrestrial, did this mean that the entire purpose of the universe was life on Earth? In an argument reminiscent of Wallace's statement that "the supreme end and purpose of this vast universe was the production and development of the living soul in the perishable body of man," at the end of the century anthropocentrism once again became part of the debate over life in the universe, a development we shall examine in Section 10.3.[43]

[41] Arthur S. Eddington, *The Nature of the Physical World* (New York, 1928), 169–178. The statement remained unchanged in subsequent editions. Hubble was quoted in 1939 as follows: "It seems reasonable to assume – doesn't it? – that among the myriads of stars that we now know to be in the grand universe there are innumerable ones that have planets associated with them. Many of these planets must be suitable for supporting life. On the whole, one is inclined to think that there may be countless other worlds with life, even life such as we do not know or cannot conceive on the basis of our earthly experience." *The Sky* (May 1939), 3, reprinted in Thornton Page and L. W. Page, *The Origin of the Solar System* (New York, 1966), 176.

[42] For the story of this change in worldview as seen by astronomers, see Peter van de Kamp, "The Galactocentric Revolution, A Reminiscent Narrative," *Publications of the Astronomical Society of the Pacific,* 77 (October 1965), 325–335; Bart J. Bok, "Harlow Shapley and the Discovery of the Center of Our Galaxy," in J. Neyman, ed., *The Heritage of Copernicus* (Cambridge, Mass., 1974), 26–62; and R. Berenzden, "Geocentric to Heliocentric to Galactocentric to Acentric: The Continuing Assault to the Egocentric," *Vistas in Astronomy,* 17 (1975), 65–83, also found in A. Beer and K. Strand, eds., *Copernicus Yesterday and Today* (Oxford, 1975), 65–83.

[43] For a comparison of the anthropic principle to Wallace's view, see Stephen Jay Gould, "Mind and Supermind," *Natural History* (May 1983), 34–38, reprinted in Gould, *The Flamingo's Smile: Reflections on Natural History* (New York, 1985), 392–402.

While anthropic arguments were problematic by any measure, scientists and the public had to contend with the geography of the universe established with more certainty by 1930. In Shapley's universe humanity could no longer be considered physically in a special position; in Hubble's universe it lived in an expanding universe of billions of galaxies; in Einstein's universe it was afloat in an enormous sea of space-time. None of these advances proved anything about extraterrestrial life. But the anthropic principle aside, it was clear for much of the century that the only avenue left for anthropocentrism was through biological uniqueness, an assertion that the laws of biology were not universal in the same way as were physical laws. In its broadest aspect, this is precisely the core of the twentieth-century extraterrestrial life debate: to demonstrate through science the uniqueness or mediocrity of humanity. The extraterrestrial life debate represents many things to many people, but among its most infamous roles, it is the last battle of anthropocentrism. We need to keep this constantly in mind in evaluating the more scientific aspects of the debate, and the passionate nature of the arguments, in the chapters that follow.

3

LIFE IN THE SOLAR SYSTEM: THE LIMITS OF OBSERVATION

Not everybody can see these delicate features at first sight, even when pointed out to them; and to perceive their more minute details takes a trained as well as an acute eye, observing under the best conditions. . . . These are the Martian canals.

Percival Lowell (1906)[1]

The hypothesis of plant life . . . appears still the most satisfactory explanation of the various shades of dark markings and their complex seasonal and secular changes.

Gerard P. Kuiper (1955)[2]

There are doubtless some who, unwilling to accept the notion of a lifeless Mars, will maintain [that] the interpretation I have given is unproved. They are right. It is impossible to prove that any of the reactions detected by the Viking instruments were not biological in origin. It is equally impossible to prove from any result of the Viking experiments that the rocks seen at the landing sites are not living organisms that happen to look like rocks. Once one abandons Occam's razor the field is open to every fantasy. Centuries of human experience warn us, however, that such an approach is not the way to discover the truth.

Norman H. Horowitz (1977)[3]

The Moon having long been considered dead by all but a few eccentrics, it is not surprising that the beginning of the twentieth century found the search for life focused on the planets, especially the Earth's nearest neighbors. The search, in its earliest form in our century, is thus intimately connected with the study of the physical characteristics of the planets, a field that received its theoretical underpinnings from the Copernican theory (which conferred a similar status on Earth and the planets) and was observationally launched with the invention of the telescope. Logically, extraterrestrial life and "planetary science," as the study of the planets is known today, are separable concepts; one can, after all, conceive a study of the planets in which the search for extraterrestrial life might have played no part.[4] Historically, they are in fact closely inter-

[1] Percival Lowell, *Mars and Its Canals* (New York, 1906; reprinted 1911), 175.

[2] Gerard P. Kuiper, "On the Martian Surface Features," *PASP* (October 1955), 281.

[3] Norman H. Horowitz, "The Search for Life on Mars," *SciAm*, 237 (November 1977), 52–61:61.

[4] The term "planetary science" is a product of the Space Age. In 1900 the physical study of the planets was discussed under headings like "planets and satellites" or "solar system astronomy" and was considered by some a part of astrophysics, as opposed to celestial

twined. Why this should be so may have to do with the illogic of humans, but it will be evident in this chapter that the question of life imparted continual impetus to planetary science. The importance of that question inspired scientists from Lowell to those of the Viking era to undertake research that otherwise might have been delayed or diverted to other interesting problems.

If hopes for planetary biology provided a spur to planetary science, by the same token, for most of the century, the techniques of planetary science provided the best hope for observationally resolving the question of life beyond the Earth. The key to progress on the question of life on other planets was the development and refinement of observational techniques that reveal planetary conditions relevant to life, recognized since the seventeenth century in the related factors of temperature, water, and atmosphere. The three sections in this chapter dealing with Mars reveal the progression of those techniques from visual to physical to spacecraft methods as the search was pressed at ever more refined levels in turn for intelligence, vegetation, and, finally, microorganisms or just organic molecules. Visual observation was aided in the twentieth century by the ever-increasing size of telescopes; the physical methods of photography, photometry, and spectroscopy – rooted in the nineteenth century – were greatly refined and brought to bear with a vengeance on planets as well as stars; and the Space Age had hardly dawned before direct investigation of the planets was proposed, with the question of life a top priority. The extension of astronomy to new regions of the spectrum, notably the radio and infrared regions, would also prove of great importance.

While planetary science and the search for life in the solar system are thus closely intertwined, it should be immediately clear that they are not coextensive; the search for life is a subset that applies only to the planets on which the conditions for life might exist.[5] Although these conditions themselves were subject to dispute, in practice astronomers and biologists alike imposed limits on likely parameters for life. When Venus was found in the 1930s to have no oxygen and was postulated in 1940 to have a temperature in excess of 100° C (the boiling point of water), their interest in it as an abode of life largely ceased, though it continued as a lively

mechanics or geodetic and nautical astronomy, which dealt with motions. See, for example, the categories used in the standard index of astronomical research papers at the time, *Astronomische Jahresbericht,* 1 (Berlin, 1900).

[5] Although standard histories of astronomy contain information on solar system studies, a critical and analytical history of planetary science remains to be written. A beginning has been made with Ronald E. Doel, *Solar System Astronomy in America: Communities, Patronage, and Interdisciplinary Research, 1920–1960* (Cambridge, 1995), and J. N. Tatarewitz, *Space Technology and Planetary Astronomy* (Bloomington, Ind., 1990). Further NASA-sponsored historical studies are underway by Tatarewicz and Ronald Schorn.

subject in planetary science. Similarly, until the 1960s Jupiter was bereft of interest for those searching for life, until it was realized that organic molecules might exist even in its ammonia/methane atmosphere. And only with the golden age of planetary exploration could planetary satellites such as Titan and Europa, previously mere points of light, become objects of interest to those searching for life.

This rigid screening of planets for vital conditions narrows considerably the scope of this chapter, largely to Mars and Venus. A final distinction serves to sharpen our focus even more, and indeed is necessary to a proper evaluation of the mass of writings on this subject. We must distinguish between the primary researchers in the field, those who with new instruments or new detectors or new techniques add new data to the argument, and secondary researchers, who react to or interpret the new data and are themselves a step removed from the original observational research. Even more must we distinguish between this second tier of actors and a third tier, which reports, interprets, and speculates, often with little scientific acumen, in an enormous popular literature. For our analysis in this chapter the primary and secondary researchers are our primary interest. The voluminous product of their research on the planet Mars prior to the Space Age is shown graphically in Figure 3.1, which not only gives an idea of the increasing volume of Mars literature during the century, but also of its ebb and flow.[6] This ebb and flow is due in part to the celestial dance of Earth and Mars in their orbits, which results in particularly favorable approaches, or "oppositions" as we shall seen in our first section.

Finally, it will be clear throughout this chapter that the extremely contentious arguments hinging on planetary observations provide a prime study of the problematic nature of observation in science. This problematic nature characterizes not only the observations undertaken at the immense distances of the planets, but even those garnered when Space Age technology allowed spacecraft to land on planetary surfaces. Yet – and this is a crucial point – despite the many problems of observation, the debate progressed throughout the century from the question of intelligence in the solar system, to vegetation, and finally to microorganisms and organic molecules. False leads, emotional debate, and wishful thinking notwithstanding, the historical record shows that twentieth-century science did resolve the question of life in the solar system to the satisfaction of most. Precisely how it reached its momentous conclusion, through a thicket of subtle problems, is the subject of the present chapter.

[6] Figure 3.1 is based on data from the annual volumes of *Astronomische Jahresbericht* (1900–1957). Each item in the "Mars and Its Moons" section was counted as one article, except for the "Brief Reports of Mars Observations," which I have consistently counted collectively as one item.

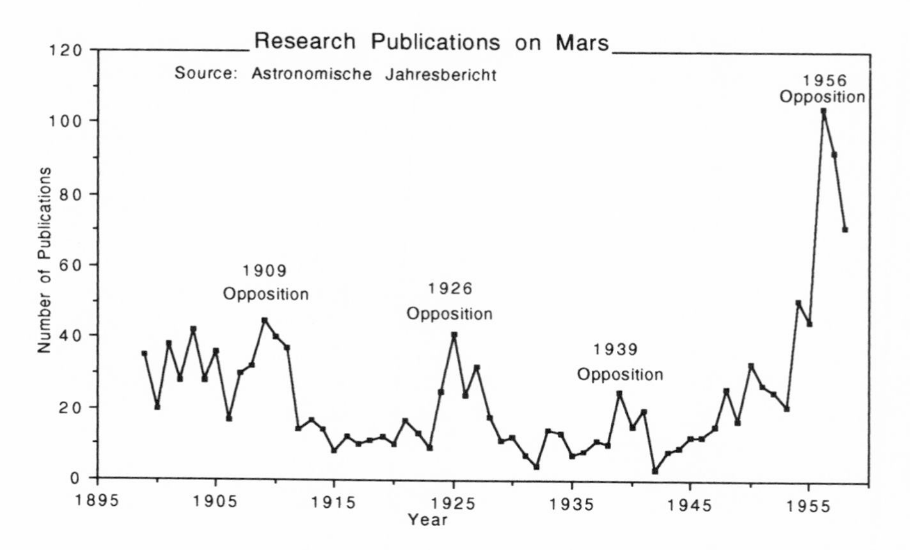

Figure 3.1. Research publications on Mars, 1900–1957. Publications increased with favorable approaches of Mars and increased in general as the Space Age approached. For source of data, see footnote 6.

3.1 LOWELL AND MARS: THE SEARCH FOR INTELLIGENCE, 1894–1924

The central character in the search for intelligence on Mars, surely one of the strangest and most contentious episodes in the history of science, was the unlikely figure of Percival Lowell (Figure 3.2). Thanks especially to the researches of William Hoyt (1976) and Michael Crowe (1986), we now know a great deal about Lowell's work on Mars, though no full biography of him exists.[7] A wealthy Bostonian businessman with a science and mathematics background from Harvard, Lowell neither initi-

[7] Percival's brother A. Lawrence Lowell, president of Harvard University, wrote a *Biography of Percival Lowell* (New York, 1935), which is useful but incomplete. For a concise biography and an entree to other literature see Brian G. Marsden, "Lowell," *Dictionary of Scientific Biography*, 8 (New York, 1973), 520–523. Historian David Strauss is preparing a detailed study of Lowell, parts of which have been published (see footnotes 24, 27, and 60). William G. Hoyt, *Lowell and Mars* (Tucson, Ariz., 1976), places Lowell's ideas on Mars in the context of his life, and chapter 10 of Michael J. Crowe's *The Extraterrestrial Life Debate, 1750–1900: The Idea of a Plurality of Worlds from Kant to Lowell* (Cambridge, 1986) places them in the context of previous ideas on extraterrestrial life. My debt to Hoyt and Crowe in this section will be evident.

Figure 3.2. Lowell at the 24-inch refracting telescope, purchased from Alvan Clark in 1896, with which he made many observations of Mars, often with decreased aperture. In this view, Lowell is observing Venus by daylight. Lowell Observatory photograph.

ated nor concluded the debate about canals on which his theory of Martian intelligence depended. But for 23 years, from 1894 to his death in 1916, he was at once its leader and most eloquent spokesman, and thus the eye of the storm that raged around intelligence on Mars. At the height of the controversy, which both Hoyt and Crowe place at around 1910,[8] his name was known to scientists and lay persons alike, in America and throughout Europe, where his writings were avidly read and his lectures invariably popular. Drawing on both primary and secondary sources, we examine in this section two primary questions: How did Lowell arrive at his startling theory that rectilinear features of Mars were proof that the planet was not only habitable but actually inhabited by intelligence? And how was it that science was unable to put this theory to rest for almost two decades?

The outlines of the story, its key participants, and several crucial characteristics of the debate are evident in Table 3.1. We see here that the canal controversy began with the Italian astronomer Giovanni Schiaparelli in 1877, was given impetus by the confirmation of Henri Perrotin and Louis Thollon in France in 1886 and 1888, took on its sensational aspect with Lowell's theory in the United States in 1894, and saw the beginnings of final resolution with Eugene M. Antoniadi's observations in France in 1909. We note too that the initial observations in 1877 were made with a very modest telescope, that not everyone saw canals even with larger telescopes, and that those who did see them did not always see them in exactly the same way. Moreover Table 3.1 reveals that most of the controversy centered on visual observations made by astronomers at the eyepiece of the telescope, though by 1905 photographic confirmation was believed to be at hand. Finally, the table shows that the debate was carried out not by a scientific subculture, but by leading astronomers representing reputable institutions around the world. What Table 3.1 does not show are the hundreds of lesser participants in the debate and the press hype that made Mars a public and scientific cause célèbre. Nor does it show the ancillary arguments over planetary atmospheres and water that were central to that debate; we shall see their place in due course.

The approximately 15-year intervals between Schiaparelli's detection of a Martian canal network and the elaboration of Lowell's theory, and again between Lowell's theory and the resolution of the canals question, are no coincidence. Indeed, these intervals reflect a basic astronomical fact that is also the cause of the ebb and flow in Martian research graphically apparent in Figure 3.1: approximately every 2 years the distance

[8] Hoyt, *Lowell and Mars*, 230–231; Crowe, *The Extraterrestrial Life Debate*, 539.

Table 3.1. *Observational highlights of Martian canals, 1877-1924*

Observer	Date	Observatory	Instrument	Method	Result/remarks
Schiaparelli	1877	Brera (Milan)	8 in.	Visual	First observation of system of *canali*
Green	1877	Madeira	13-in. reflector	Visual	Shaded area boundaries; no canals
Hall/ Harkness	1877	U.S. Naval	26 in.	Visual	Moons of Mars found but no canals
Schiaparelli	1879/ 1880	Brera (Italy)	8 in.	Visual	First report of double canals
Maunder	1882	Greenwich	28 in.	Visual	Some canals
Perrotin/ Thollon	1886 1888	Nice (France)	15 in. 30 in.	Visual	Many canals; first confirmation of double canals
Holden/ Keeler et. al	1888	Lick	36 in.	Visual	Some canals but no doubles
Pickering/ Douglass	1892	Harvard Arequippa (Peru)	13 in.	Visual	Canals and "lakes" at canal junctions
Barnard	1892	Lick	36 in.	Visual	Some canals but not fine lines
Lowell/ Pickering/ Douglass	1894 1895	Lowell	12 in. 18 in.	Visual	Canals artificial
Antoniadi	1894/ 1896	Juvisy (France)	9.6 in.	Visual	42 canals; 1 double doubling is illusion
Cerulli	1896	Teramo (Italy)	15.5 in.	Visual	illusion - optical origin
Lampland	1905	Lowell	24 in.	Photographic	Canals photographed
Todd/ E. Slipher	1907	Chile	18 in.	Photographic	Canals photographed
Antoniadi	1909	Meudon	33 in.	Visual	Canals resolved
Hale	1909	Mount Wilson	60-in. reflector	Visual	Much detail, no canals
Trumpler	1924	Lick	36 in.	Visual/ photographic	Strips of vegetation

Sources of data: Crowe, *The Extraterrestrial Life Debate*; Hoyt, *Lowell and Mars*, primary sources as given in footnotes. Instruments are refractors unless specified otherwise.

between Earth and Mars is minimized because they are "in opposition," that is, in a straight line as seen from the Sun, and approximately every 15 years they are not only in opposition but also especially close. As a result, the apparent size of the disk of Mars viewed from Earth varies from about 14 arcseconds at its most unfavorable oppositions to 25 arcseconds at its closest oppositions (and only 4 arcseconds when it is most distant from the Earth); when we consider that even during oppositions, on average Mars is only one one-hundredth the diameter of the full Moon (one-half degree, or 1800 arcseconds), we begin to see the difficulties that Martian observations faced from the beginning. The year 1877 marked one of the 15-year close approaches of Mars, and that year saw not only Schiaparelli's first observation of an extensive system of canals, but also Asaph Hall's discovery at the U.S. Naval Observatory of the two moons of Mars. Aside from distance, another very important factor was the altitude of Mars in the sky, which could vary enormously from one opposition to the next and affect the quality of observations at observatories at different latitudes.[9] It was this factor that led to several expeditions to observe Mars south of the equator, where its altitude was greater at near oppositions. The elaborate dance between altitude and distance, graphically evident in Figure 3.3, determined the best observing conditions, with high altitude and close distance optimal. Even under the best conditions, however, one still had to contend with viewing through the Earth's turbulent atmosphere, as with all astronomical observations.

Genesis of the Theory

How then did Lowell arrive at his theory? It is essential to note at the outset that, even if Lowell is the central character in this story, he was neither the first to suggest intelligence on Mars nor the first to observe the canal phenomenon on which he based his theory. The years following Schiaparelli's observations generated considerable debate and established one primary point: even under the best observing conditions, not all observers could see canals. Schiaparelli saw an extensive network of canals with his 8-inch refractor and made it known to the world in his *Osservazione* (1878). But in the very year that Hall found the moons of Mars, he and a fellow astronomer at the Naval Observatory, William Harkness, did not see any canals with their 26-inch telescope, the largest refractor in the world.[10] Even before the question of life entered the scene,

[9] The maximum "declination" (celestial latitude) of Mars can vary by more than 50 degrees over a series of oppositions, giving rise to the effect seen in Figure 3.3.

[10] Schiaparelli's 1877 observations have been discussed in Crowe, *The Extraterrestrial Life Debate,* 480–486. The observations of Hall and Harkness are discussed briefly in

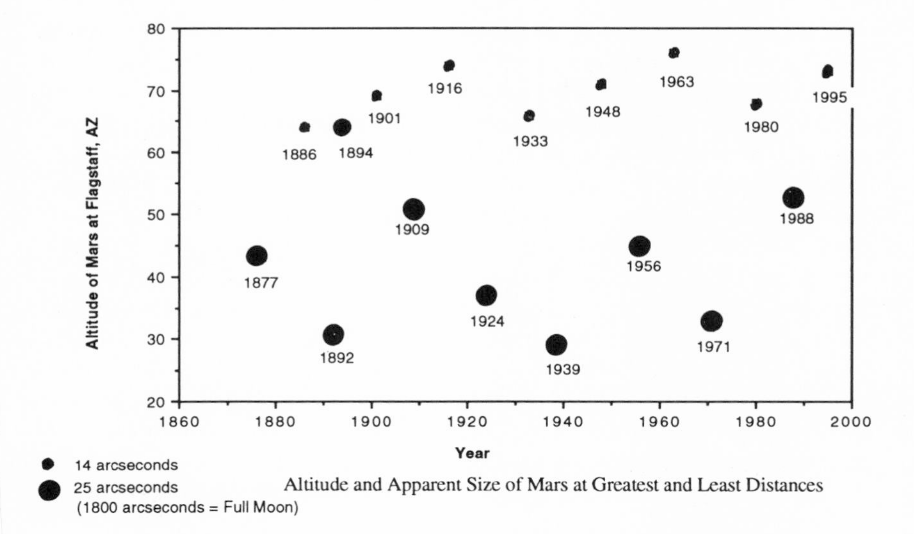

Figure 3.3. As Mars approaches and recedes, its apparent size changes as viewed from Earth, from about 25 arcseconds to 14 arcseconds during oppositions and as small as 4 arcseconds at its most distant. The full Moon is 1800 arcseconds by comparison. With the exception of 1894, only maximum and minimum opposition data are graphed here. The relatively large disks in 1877, 1892, and 1909 made those important years for observing Mars. The low altitude of Mars in 1892 spurred the Harvard southern observations of W. H. Pickering in Peru. The opposition following the closest approach finds the planet much higher in the sky but the size of the disk already reduced by 20%. Lowell began his observations at Flagstaff in 1894 during such a time. Altitudes are plotted here for Flagstaff; Schiaparelli's observatory in Milan was 10 degrees farther north, so all the altitudes shown here would have been lower by 10 degrees in the sky. Antoniadi would have viewed Mars almost 14 degrees lower than would observers at Flagstaff.

the very existence of the canals, and the advantages of large versus small telescopes in discerning planetary detail, were subjects of controversy.

Schiaparelli himself continued to see the canals, and indeed at the next opposition in 1879 announced the phenomenon of "gemination" or doubling of the canals, an observation he confirmed in 1882.[11] He must have

Flammarion, *La planète Mars* (Paris, 1892), 282. Since Milan is 6 degrees farther north than Washington, Mars would have been 6 degrees lower than for Hall and Harkness, not enough to make much difference.

11 Crowe, *The Extraterrestrial Life Debate;* Schiaparelli, "Osservazioni sulla topografia del pianeta Marte," *Opere,* vol. 1, 379–388, English translation in *Popular Science*

been relieved, if somewhat surprised during the unfavorable opposition of 1886, when Perrotin and Thollon at the Nice Observatory saw many canals and confirmed some doubles with their 15-inch telescope.[12] But any hope that observation might quickly resolve the issue were dashed in 1888, when the largest telescope in the world, the Lick 36-inch, came into operation and gave an ambiguous verdict: some canals and no doubling.[13]

The year 1888 did see the entry of another prime player on the scene, one destined to have a considerable, and perhaps crucial, effect on Lowell: William H. Pickering, brother of Harvard College Observatory Director E. C. Pickering. Based on the observations of others, Pickering weighed in with his theory that the canals were made visible by adjacent strips of vegetation, an idea that Lowell would later adopt.[14] Pickering's real impact began to be felt in 1892, which saw a close approach of Mars similar to that of 1877. Since the planet was too low in the sky for optimal observing by northern observatories, which contained the vast majority of all telescopes, some of the most interesting work came from one of the few southern stations, that at Arequippa, Peru. Set up by Harvard in 1889, it was now headed by William Pickering, whose Martian observations (assisted by A. E. Douglass) gave the world a taste of the controversy to come. Pickering reported, among other things, the existence of 40 large lakes, and clouds 20 miles above the Martian surface. Observers at Lick, who had the largest telescope but could see Mars only 30 degrees off the horizon, saw none of this. Thus the battle lines were drawn years before Lowell came on the scene.[15]

By 1892 the French astronomer Flammarion had produced a massive volume of historical observations of Mars, more than half of which covered the years since 1877. The title of the volume, *La planète Mars et ses conditions d'habitabilité,* as well as Flammarion's own conclusion that Mars was indeed habitable, indicate the burning question and the stan-

Monthly, 24 (December 1883), 249–253; and "Descouvertes nouvelles sur la planète Mars," *Opere,* vol. 1, 389–394.

[12] Henri Perrotin, "Observations des canaux de Mars," *Bulletin astronomique,* 3 (July 1886), 324–329. These observations reached Britain and America in Anonymous, "The 'Canals' of Mars," *Nature,* 34 (June 3, 1886), 104, and *SciAm Supplement,* 22 (July 10, 1886), 8774.

[13] E. S. Holden, "Physical Observations of Mars during the Opposition of 1888, at the Lick Observatory," *AJ,* 8 (September 14, 1888), 97–98.

[14] W. H. Pickering, "The Physical Aspect of the Planet Mars," *Science,* 12 (August 17, 1888), 82–84.

[15] On Pickering's observations see "Mars," *Astronomy and Astro-Physics,* 11 (October–December 1892), 668–675, 849–852; also "Photographs of the Surface of Mars," *Sidereal Messenger,* 9 (June 1890), 254–255. For the Harvard context see B. Z. Jones and L. G. Boyd, *The Harvard College Observatory* (Cambridge, Mass., 1971), 297–308, especially 306–307. For the comments of Lick Director E. S. Holden see "The Lowell Observatory in Arizona," *PASP,* 6 (June 1894), 160–169.

dard answer of the time.[16] But matters would not be left there, even before the advent of Lowell. The name most closely linked with the planet Mars at this time was still Schiaparelli, who had continued to make observations and offer interpretations. In an important lengthy article on the eve of the 1894 opposition, Schiaparelli described his view of Mars, and in it we can see the starting point of the theory of Lowell, who had Pickering translate it in 1894.[17] The Schiaparellian Mars was a planet with two polar caps composed of snow and ice, seas and continents arranged very differently from those on Earth, and an atmosphere rich in water vapor. The vaporous atmosphere he believed was supported not only by changes in the polar caps requiring a transportation mechanism for water vapor, but also by the spectroscopic observations of the German astronomer Hermann Vogel. It was also a planet of change, for the melting polar caps seemed to produce a temporary sea around the northern cap. Schiaparelli believed this water was distributed over great distances by "a network of canals, perhaps constituting the principal mechanism (if not the only one) by which water (and with it organic life) may be diffused over the arid surface of the planet."[18]

Moreover, Schiaparelli described these canals in detail. The continents, he wrote, are "furrowed upon every side by a network of numerous lines or fine stripes of a more or less pronounced dark color whose aspect is very variable." They range in length from 500 to thousands of kilometers and in width from 30 to 300 kilometers. Some, such as Nilosyrtis, are easy to see; others "resemble the finest thread of spider's web drawn across the disc." The canals usually appear as "a nearly uniform stripe, black, or at least of a dark color, similar to those of the seas." From their similar appearance to seas and from the phenomena of the melting snows, Schiaparelli concluded that they constituted "a true hydrographic system."[19]

But the nagging question was how such a system of canals could originate, a question that the Italian astronomer did not fail to consider. Schiaparelli was well aware that their singular geometric aspect, "as if

[16] C. Flammarion, *La planète Mars et ses conditions d'habitabilité* (Paris, 1892). Volume 2 appeared in 1909, but a projected third volume was never published. These illustrated volumes represent the most comprehensive compilation of Martian observations through 1901.

[17] Schiaparelli's 1893 article, "Il pianeta Marte," first appeared in *Natura ed Arte* for February 15, 1893, and can be found in Schiaparelli's *Opere*, vol. 2, 47–74. Pickering's English translation first appeared in *Astronomy and Astro-Physics*, 13 (October 1894), 632–640 and 714–723. Crowe, *The Extraterrestrial Life Debate*, 633, note 140, gives further English references. It also appeared in W. H. Pickering's collection entitled *Mars* (Boston, 1921), 63–96, from which the quotations here are taken.

[18] Schiaparelli, in W. H. Pickering, *Mars* (Boston, 1921), 69–76.

[19] Ibid., 83–87.

they were the work of rule and compass," had led some to see them as the work of intelligent inhabitants. He himself held that the canal network was not necessarily artificial, "and notwithstanding the almost geometrical appearance of all of their system, we are now inclined to believe them to be produced by the evolution of the planet, just as on the Earth we have the English Channel and the Channel of Mozambique." But he left the door open to the artificial hypothesis: "I am very careful not to combat this supposition, which includes nothing impossible."[20]

Moreover, while Schiaparelli held that intelligence was not necessary to explain the gemination or doubling of canals that he had observed, he also noted that "extensive agricultural labor and irrigation upon a large scale" might produce such a phenomenon. Though in the end he concluded that nature itself can produce geometric regularity such as crystalline forms, his essay was very suggestive. He ended with an exhortation for further study of the physical state of Mars in order to determine the cause of these observations. "Let us therefore hope and study" was his rallying cry.[21]

The first piece of the puzzle in understanding the genesis of Lowell's theory, then, is that such speculations were in the air, and particularly in Schiaparelli's essay, which appeared in Italian in February 1893 and not only discussed in detail the canals of Mars but also held open the possibility of their artificial construction. As Lowell entered the debate, the general opinion was that the canals were cracks in the Martian crust made during the planet's solidification, but some argued for their artificiality.[22]

But what precipitated Lowell's entry into the debate? Was he responding directly to Schiaparelli's challenge or to some more deeply rooted interest? Why did Lowell, a businessman and traveler with a penchant for the exotic, enter astronomy at all? Born into a wealthy family in 1855, Lowell was attracted to astronomy from boyhood, when he scanned the heavens with a 2-inch telescope. He attended Harvard from 1872 to 1876 and studied with the famous mathematician Benjamin Peirce, who described him as one of his most brilliant students.[23] Another Harvard influence may have been John Fiske (1842–1901), who presented his evolutionary cosmic philosophy in *Outlines of a Cosmic Philosophy Based on the Doctrine of Evolution* in 1874, 2 years before Lowell graduated. That Lowell embraced such a philosophy is evident from his com-

[20] Ibid., 88–92. Pickering emphasized this passage in his translation by including the original Italian.

[21] Ibid., 92–96.

[22] J. R. Holt, "The Canals of Mars," *Astronomy and Astro-Physics,* 13 (May 1894), 347–354. This article was published in the same month that Lowell began his observations.

[23] A. Lawrence Lowell, *Biography of Percival Lowell;* Hoyt, *Lowell and Mars,* 16.

mencement address on "The Nebular Hypothesis." But until 1893 he spent his time managing family business affairs and traveling, the latter in which he achieved fame as an author of books on the Far East. When Lowell's first book on Mars appeared in 1895, it was therefore not books on astronomy that were advertised on the flyleaf, but *The Soul of the Far East* (1888), *Choson: The Land of the Morning Calm. A Sketch of Korea* (1885), *Noto: An Unexplored Corner of Japan* (1891), and *Occult Japan: the Way of the Gods* (1895).[24]

The year 1893, and the appearance of Schiaparelli's provocative article, found Lowell still in the Far East. His continued, or perhaps rekindled, interest in astronomy at this time is evident from several facts. While in Japan in July 1891, Lowell had written his brother-in-law William Lowell Putnam that he planned to write a book on cosmic philosophy, with illustrations from celestial mechanics. It is also well known that when Lowell left in 1892 for his final trip to the Far East, he took with him a 6-inch telescope, ample testimony to a more concrete interest in astronomy.[25] We know from several of his friends that he had kept up with Schiaparelli's work, and some stated that he turned toward Mars because of reports of Schiaparelli's failing eyesight. In autumn 1893 Lowell returned to Boston from Japan; within a few months, the *Boston Herald* reported that he was financing an expedition to Arizona in conjunction with Harvard University, to be headed by Harvard astronomer William H. Pickering.[26]

It is reasonable to suppose that Lowell, a Bostonian and graduate of Harvard, should approach Harvard College Observatory after his interest in Mars was piqued, and historian David Strauss has shown that Lowell was in correspondence with William Pickering about Mars already in 1890, and with William's brother, Observatory Director E. C. Pickering, by late 1892, asking the latter for Schiaparelli's drawings of Mars. When Edward recalled William from Arequippa in 1893, the latter made plans

24 On the influence of Fiske and Lowell's early commitment to an evolutionary philosophy, see Crowe, *The Extraterrestrial Life Debate,* 507. On Fiske, who ironically was one of the nineteenth century's best-known skeptics of what he called "eccentric literature" on science, see Milton A. Rothman, "Two 19th-Century Skeptics: Augustus De Morgan and John Fiske," *Skeptical Inquirer,* 16 (Spring 1992), 292–297. On Lowell's place in Far Eastern studies see David Strauss, "The 'Far East' in the American Mind, 1883–1894: Percival Lowell's Decisive Impact," *Journal of American–East Asian Relations,* 2 (Fall 1993), 217–241.

25 This was the "6-inch Clark refractor with a portable tripod mounting" later used by A. E. Douglass for site testing in Arizona and mentioned by Douglass in his "Lowell Observatory Reminiscences" (see footnote 28), 3. See also Hoyt, *Lowell and Mars,* 32.

26 *Boston Herald,* February 12, 1894. Most Lowell obituaries mention the Schiaparelli connection and the link to his failing eyesight. The link has been brought into question in Crowe, *The Extraterrestrial Life Debate,* 507.

for a Martian expedition to Arizona, which had desert conditions similar to those of Peru. Lack of financing, however, left him searching for funding. By December 1893 Lowell had not yet decided on his course of action, but perhaps influenced by a Christmas gift of Flammarion's *La planète Mars* in which Schiaparelli's canals were clearly shown (Figure 3.4), in January 1894 he spoke to both Pickerings about his interest in Mars.[27] Though the exact sequence of events has always been unclear, A. E. Douglass, the young assistant of Pickering in Arequippa, who had returned from Peru with Pickering at about the same time that Lowell returned from Japan, later recalled the origin of the Pickering–Lowell collaboration. Douglass wrote many years later:

> W. H. Pickering desired to continue the Mars work we had been doing in South America On his return to Cambridge, he was invited to dine with Percival Lowell at the house of a friend, Mr. Foncelli. At this dinner, Pickering described our successful South American work on Mars and Lowell became very much interested indeed. He proposed to finance an expedition that Pickering thought should go to Arizona for the purpose of carrying on observations of Mars, in the clear atmosphere of that country during the next opposition in 1894. Such an expedition would need a preliminary survey of different parts of the state of Arizona to be made with a 6″ telescope. And when a site was selected, we would have to build a dome and secure a big lens and telescope.[28]

The picture that emerges from the scattered evidence, then, is that Lowell certainly knew of Schiaparelli's work and perhaps his failing eyesight; that in a more immediate sense William Pickering played a crucial role in bringing Mars and its problems to the attention of Lowell; and that it was Pickering who first suggested the Arizona site, a long-cherished idea of his. It was in mid-January 1894 that Lowell made his offer to finance the expedition; Pickering immediately began to assemble the necessary equipment with the help of the telescope maker Alvan Clark, and Douglass was assigned the task of site testing. Moreover, when we recall Pickering's previous claims of vegetation, lakes, and

[27] David Strauss, "Percival Lowell, W. H. Pickering and the Founding of the Lowell Observatory," *Annals of Science,* 51 (1994), 37–58.

[28] A. E. Douglass, "Lowell Observatory Reminiscences," 1–2, Douglass Archives, University of Arizona. While these reminiscences may date from the 1940s, the fact that Douglas cited Foncelli by name indicates that he either had a very good memory or access to notes. For more details see Strauss, "Percival Lowell."

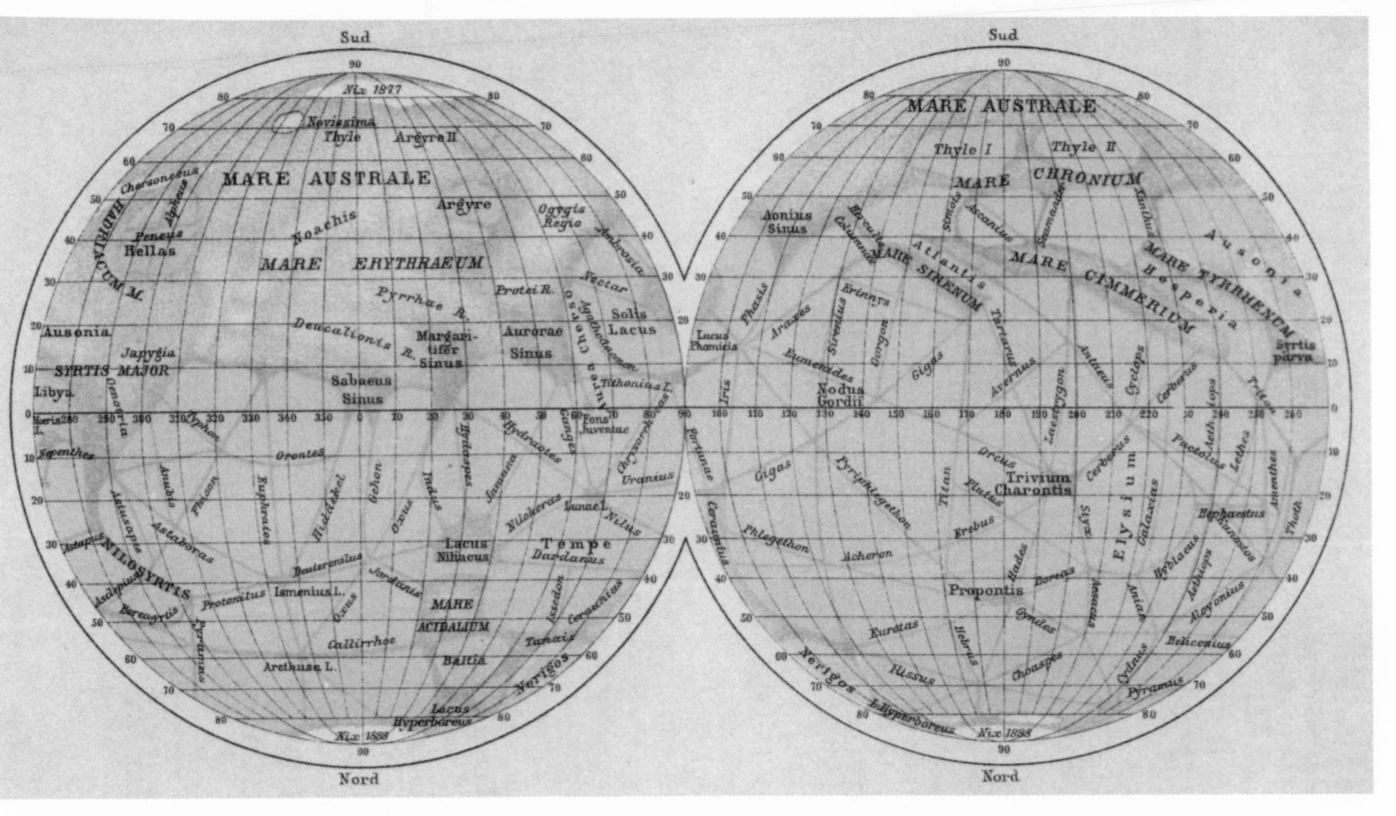

Figure 3.4. The Mars of Schiaparelli based on his observations during six oppositions between 1877 and 1888, inherited by Lowell. The map, as well as others showing Martian canals, was reproduced in Flammarion, *La planète Mars et ses conditions d'habitabilité* (1892), of which Lowell received a copy just before his decision to build Lowell Observatory.

clouds on Mars, it becomes clear that Pickering's influence on Lowell's thinking likely surpassed his role in introducing Lowell to the practical aspects of astronomy.[29]

We cannot say with certainty whether Schiaparelli's 1893 article played a part in his decision to focus on Mars, but within a few months Pickering had published an English translation of it at Lowell's request. Douglass's reminiscences are further suggestive with regard to the influence of Schiaparelli: "The purpose that Mr. Lowell had in carrying on this expedition was to extend the studies of Mars under the most favorable conditions with the possibility of demonstrating the existence of life on that planet. The name Schiaparelli had given to the long dark markings on the planet, was canals, which immediately suggested living people on the planet."[30] This account is given further credence by Douglass's statements, which agree with other sources, of how the Harvard connection fell through: "This plan of an expedition from Harvard Observatory was described in the local newspapers with scarcely a reference to Mr. Lowell's generous proposal to support it. Accordingly, he realized that it would be better for him to take complete charge of the expedition and that Pickering and the writer should get a leave of absence from Harvard University. This was done and I started out on the preliminary survey the 28th of February 1894."

What began as a Harvard expedition financed by Lowell ended as a new observatory, one where, according to both Douglass and Lowell himself, the search for life on Mars was clearly the goal. In a paper read before the Boston Scientific Society in May 1894, Lowell noted that the observatory's work was to be "an investigation into the condition of life in other worlds, including last, but not least, their habitability by beings like (or) unlike man. This is not the chimerical search some may suppose. On the contrary, there is strong reason to believe that we are on the eve of pretty definite discovery in the matter." Lowell went on to make clear that the source of this belief was a nonanthropocentric worldview and the nebular hypothesis: "If the nebular hypothesis be correct, and there is good reason at present for believing in its general truth, then to develop life more or less distantly resembling our own must be the destiny of every member of the solar family which is not prevented by purely physi-

[29] Though the Harvard connection soon fell through, Lowell's observatory was long afterward mistakenly associated with Harvard in the minds of some. The *San Francisco Chronicle,* in an article dated February 12, 1894, from Cambridge, Mass., reported that Harvard was sending an expedition to Arizona to establish a new station, that this had been the plan of W. H. Pickering for years, and that Lowell was providing the financial backing. The article is reprinted in *PASP,* 6 (March 31, 1894), 122.

[30] Douglass, "Reminiscences," 16.

cal considerations, size and so forth, from doing so."[31] With an early interest in astronomy, a taste for the exotic, and knowledge of the work of Schiaparelli on Mars; intrigued by the notion of that planet as part of an evolutionary scheme that might include life; and with a direct push from W. H. Pickering, Lowell's subsequent actions begin to make more sense.

Few men were in a position to found their own observatory to explore an idea that fascinated them – though James Lick may have been fascinated by a similar idea in founding his observatory a few years earlier. This is exactly what Lowell did, and as a result, to a remarkable extent his works are different from other treatises on the plurality of worlds. No Proctor or young Flammarion here, enthusiastically assessing, mostly secondhand, the chances of life in the universe. Not even a more sober and scholarly Flammarion, concentrating on the planet Mars to compile observations and assess "its conditions of habitability." Rather, Lowell's books and articles are reports of observations that he himself largely made or sponsored, culminating already in his first work with the claim that Mars is not only habitable but actually inhabited. Not that Lowell should be accused of being provincially Martian; difficult as the observations were, Mars was simply the closest planet which opened the door to more general claims. While others sought a general treatment heavily laced with philosophical arguments, Lowell's three major works – *Mars* (1895), *Mars and Its Canals* (1906), and *Mars as the Abode of Life* (1908) – as well as three volumes of annals issued from the Lowell Observatory during his lifetime and a considerable number of its 73 *Bulletins,* concentrated all of his considerable energies on Mars as a case study of the more general belief. In this may be found his unique contribution, his drive, and the source of his reputation as a man of controversy.

Though with his financial means Lowell could easily construct an observatory, one would not think it would be so easy to construct and sustain a theory unless there was a considerable observational basis for doing so. Yet Lowell's first paper on Mars appeared in August 1894, and his theory was already well advanced. After describing his observations of canals, but no clouds, Lowell wrote, "Here we have a *raison d'etre* for the canals. In the absence of spring rains a system of irrigation seems an

[31] Excerpts from Lowell's paper, read May 26, 1894, are reprinted in E. S. Holden, "The Lowell Observatory in Arizona," *PASP,* 6 (June 9, 1894), 160–169:161. Holden (the director of Lick Observatory) objected that astronomers were very unlikely to be on the brink of any such discovery and stated that "the very essence of the scientific habit of mind is conscientious caution." These comments on Lowell's paper were the beginning of lengthy contentiousness between the Lick and Lowell observatories.

absolute necessity for Mars if the planet is to support any life upon its great continental areas."[32] In his first book on the subject the following year, that theory was set forth in full. Perhaps the speed of the theory's formation is itself an indictment. But we now arrive at the final and most crucial point in our quest for the genesis of Lowell's theory: with what data, and by what chain of arguments, did he support the theory that the canals of Mars represented artificial constructions?

Fortunately, we have Lowell's own words, in the form of the conclusion of his 1895 volume, by which he summarized his deductions in retrospect:

> To review, now, the chain of reasoning by which we have been led to regard it probable that upon the surface of Mars we see the effects of local intelligence. We find, in the first place, that the broad physical conditions of the planet are not antagonistic to some form of life; secondly, that there is an apparent dearth of water upon the planet's surface, and therefore, if beings of sufficient intelligence inhabited it, they would have to resort to irrigation to support life; thirdly, that there turns out to be a network of markings covering the disk precisely counterparting what a system of irrigation would look like; and, lastly, that there is a set of spots placed where we should expect to find the lands thus artificially fertilized, and behaving as such constructed oases should. All this, of course, may be a set of coincidences, signifying nothing; but the probability points the other way. As to details of explanation, any we may adopt will undoubtedly be found, on closer acquaintance, to vary from the actual Martian state of things; for any Martian life must differ markedly from our own.[33]

This passage is perhaps the most important in the book for understanding how Lowell came to develop his theory and how his mind worked, for it shows his "chain of reasoning." It may not have been the actual route he took to the theory, not the actual "context of discovery," but it was at least the way he justified it to the public shortly after he had formulated the theory. We note, first, that Lowell *began* with the idea of life on Mars.

[32] Lowell, "Mars," *Astronomy and Astro-Physics,* 13 (August 1894), 538–553:550. This was followed quickly by articles of the same title in the same journal for (October 1894), 645–652; (November 1894), 740; and (December 1894), 814–821.

[33] Percival Lowell, *Mars* (Boston and New York, 1895), 201. The Preface is dated November 1895. The first chapter clearly shows Lowell's acceptance of a nonanthropocentric philosophy; the general framework of the book concerns Mars as a case study of nonanthropocentrism, with all the power of that philosophical argument that we have emphasized in Chapter 2.

This is not surprising, since we have seen that it was "in the air" in the years prior to his first observations, from no less an authority than Schiaparelli, "the Columbus of a new planetary world," as Lowell would later call the Italian astronomer.[34] Lowell then turned to the planetary conditions – atmosphere and water – to see if the idea was possible. Finally, he attempted to explain the nature of the surface features known as canals and oases. Around these concepts – atmosphere, water, canals, and oases – Lowell structured his 1895 book, and in his discussion of them we must judge his success or failure as a fledgling astronomer.

It is clear that the conditions for life were crucial, for if life was impossible, Lowell's theory of artificial canals was impossible. The arguments for a Martian atmosphere were straightforward and entirely in line with the ideas of his time. The long-known fact of changes in the polar caps and Schiaparelli's recent detection of changing tints on the Martian surface, Lowell argued, was proof of an atmosphere, for a planet's atmosphere is the agent of change. This was backed up by Douglass's measurement of variations in the diameter of the planet, which Lowell argued could only be due to an atmosphere. As for the nature of this atmosphere, Lowell pointed to G. Johnstone Stoney's kinetic theory of gases to infer that its molecules must be similar to those of the Earth. Lick astronomer W. W. Campbell's observations showing no Martian water vapor, made that very summer, Lowell did not yet take into account.[35]

As to surface water, Lowell argues that the polar cap is composed of aqueous snow, as opposed to frozen carbon dioxide ("carbonic acid gas"), and that a polar sea is produced after the cap melts in the summer. In his discussion of "areography" Lowell argues that the blue green areas, considered to be seas by some, are actually areas of vegetation that come to life with the melting of the polar caps, so that far from open seas in abundance on the planet, water is actually very scarce. Lowell argues that this is what one would expect for a planet more evolved than Earth, as Mars must be since it is smaller than Earth. He sees Martian seas as midway in their evolution between the aqueous seas of Earth and the arid seas of the Moon. Lowell concludes the chapter by saying that any inhabitants on the planet would need a system of irrigation – and this is just what is seen.[36] Not only did he find Martian conditions conducive to life, he also found in those conditions an explanation for the canals.

Those canals, Lowell was quick to point out, were not invented by him. "The first hint the world had of their existence," by Lowell's account, was 18 years earlier from Schiaparelli, who labored alone for 9 years before the

[34] The reference to Schiaparelli is in Lowell's dedication in *Mars and its Canals* (1906).
[35] *Mars*, chapter 2, 31–75.
[36] Ibid., chapter 3, 76–128.

confirmation by Perrotin and Thollon in 1886.[37] Since then, more astronomers had seen them, but by no means all, a situation that Lowell attributes primarily to one factor: a steady atmosphere. This, he explains, is how Schiaparelli saw them with an 8.5-inch telescope, while observers using the Washington 26-inch telescope could not see them at all. Even with steady air, "attentive perception" on the part of the observer is essential to catch those moments that reveal the canals "with the clear-cut character of a steel engraving." Though some who have not seen them are still skeptical, Lowell notes that skeptics had also objected to the discovery of Jupiter's satellites and Huygens's explanation of Saturn's ring.[38]

Lowell and his staff at Flagstaff had seen about four times as many canals as had Schiaparelli, and Lowell now proceeded to map and catalog the 183 canals detected there (Figure 3.5). But it was from his explanation of them – as artificial constructions of Martian inhabitants – that Lowell drew his fame. This theory Lowell built on their "supernaturally regular appearance," based on their straightness, uniform width, and systematic radiation from special points. Lowell believed no natural explanation could account for these features; thus the origin of his theory. Though Schiaparelli believed the canals natural, Lowell noted, he had not completely ruled out intelligence.[39]

Lowell had no illusions that he was seeing the canals themselves; instead, he thought he was seeing the strips of fertilized land bordering them, an explanation he credited to William Pickering. And therein lay for him further proof of their nature. For Lowell noted that the canals change appearance, growing progressively from the pole toward the equator as the polar caps melt. This does not happen suddenly, but with a delay that might be seen if vegetation needed time to renew and make itself visible. And the oases are seen as further evidence that the canals are constructed so as to fertilize them.[40]

By 1895, then, Lowell's theory of artificial canals on Mars was fully set forth, not in the cautious and brief manner of Schiaparelli 2 years earlier but in an ebullient, highly readable book that threw caution to the wind. Thrust into the open, an issue full of public interest, a golden opportunity now presented itself for science either to confirm an earthshaking theory or to crush it quickly under the glare of objective argument. To the dismay of the public and the embarrassment of astronomers, science was unable to do either for almost two (some would say seven) decades.

[37] Ibid., chapter 4, 136.
[38] Ibid., chapter 4, 138–140.
[39] Ibid., chapter 4, 135–175, 202.
[40] Ibid., chapters 4 and 5, 162–165; 186.

Resolution of the Conflict

Having seen the genesis of Lowell's theory, we now come to the second major question of the canal controversy: why, despite their best attempts, did astronomers take so long to resolve this issue? We shall see that the answer lies not only in the disputed appearance of the canals, but also in disputed ideas of the proper route from observation to theory and in disputed observations of planetary conditions that, had they been more definitive, might have settled the controversy themselves by precluding intelligence. Far from the objectivity that the public expected from science, over the next two decades or more, the opinions of scientists on the Martian canals would cover the entire spectrum of possibility from those who denied even the reality of the canals, on the one hand, to those who accepted Lowell's theory of artificial Martian constructions, on the other. Whereas Table 3.1 listed chronologically some of the highlights in the observation of Martian canals, Table 3.2 displays highlights in the *interpretation* of canals, arranged along the spectrum of opinion from illusory to artificial. Among its important points are that the interpreters (Wallace, Eddington, etc.) were certainly more numerous than the serious observers (a fact that led Lowell to label some of them "armchair critics"), that occasionally differences of opinion arose even at the same institution (Lick Observatory is a prime example), that individuals changed their minds over the course of the controversy (Antoniadi), and that there is no strong pattern of American versus European attitudes toward canals.

Among the most illuminating immediate reactions to Lowell's theory was W. W. Campbell's review of *Mars,* which he penned for the respected journal *Science.* This review places Campbell in the middle of the spectrum of opinion of Table 3.2, for in it he accepted the reality of the canals – no small point coming from an observational astronomer, and a Lick astronomer at that – but he did not accept their artificial nature. Lowell could not escape comparison with Schiaparelli, and Campbell compared with a vengeance: Whereas Schiaparelli observed the planet from 1877 to 1892 and gave the world his results in a few technical papers and one short popular article, Lowell's observations covered only one-quarter of a Martian year and had already produced lectures, articles, and the entire book under review, all little more than a year after his observations had begun; Schiaparelli was concerned with establishing facts, Lowell with establishing a theory (and not, Campbell points out, a completely new one at that); Schiaparelli holds in Martian studies the place Darwin does in evolution,

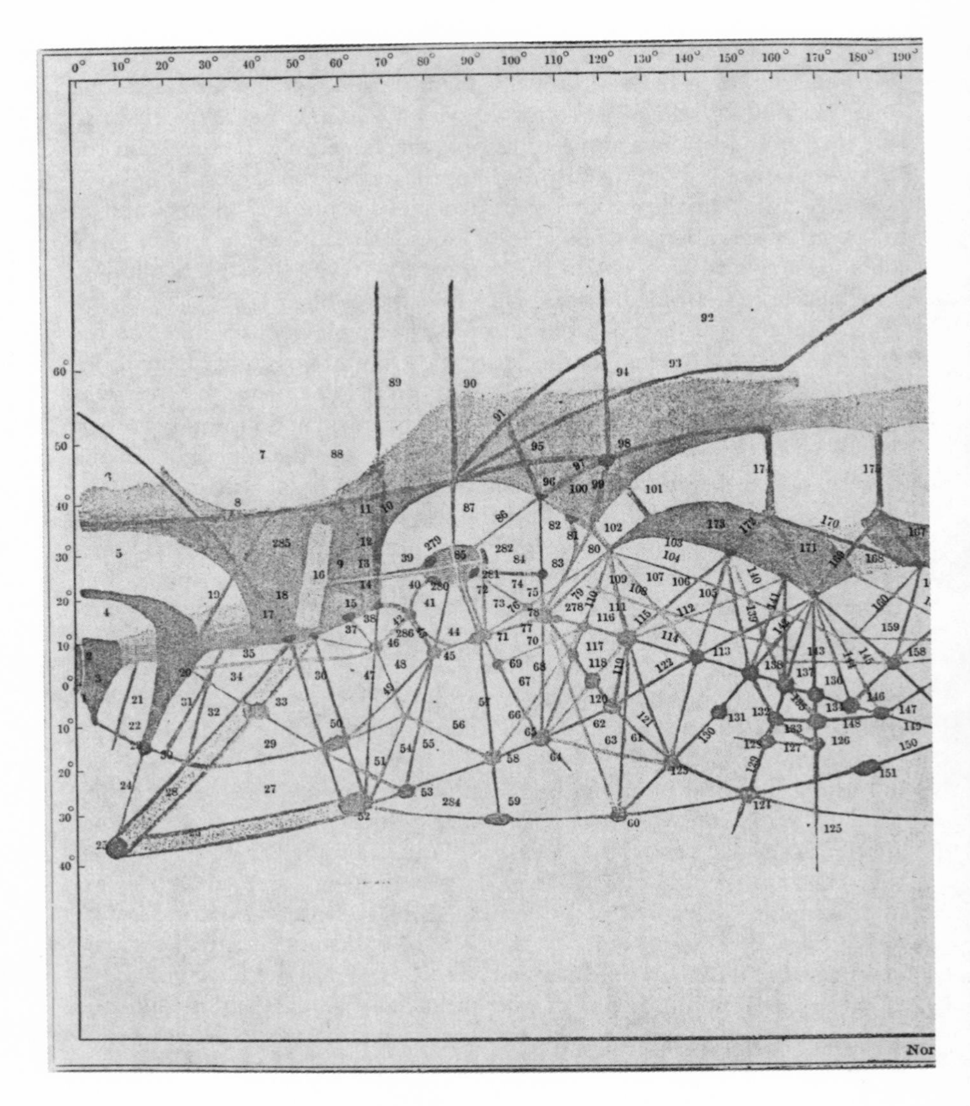

Figure 3.5. Lowell's first map of Mars (1895).

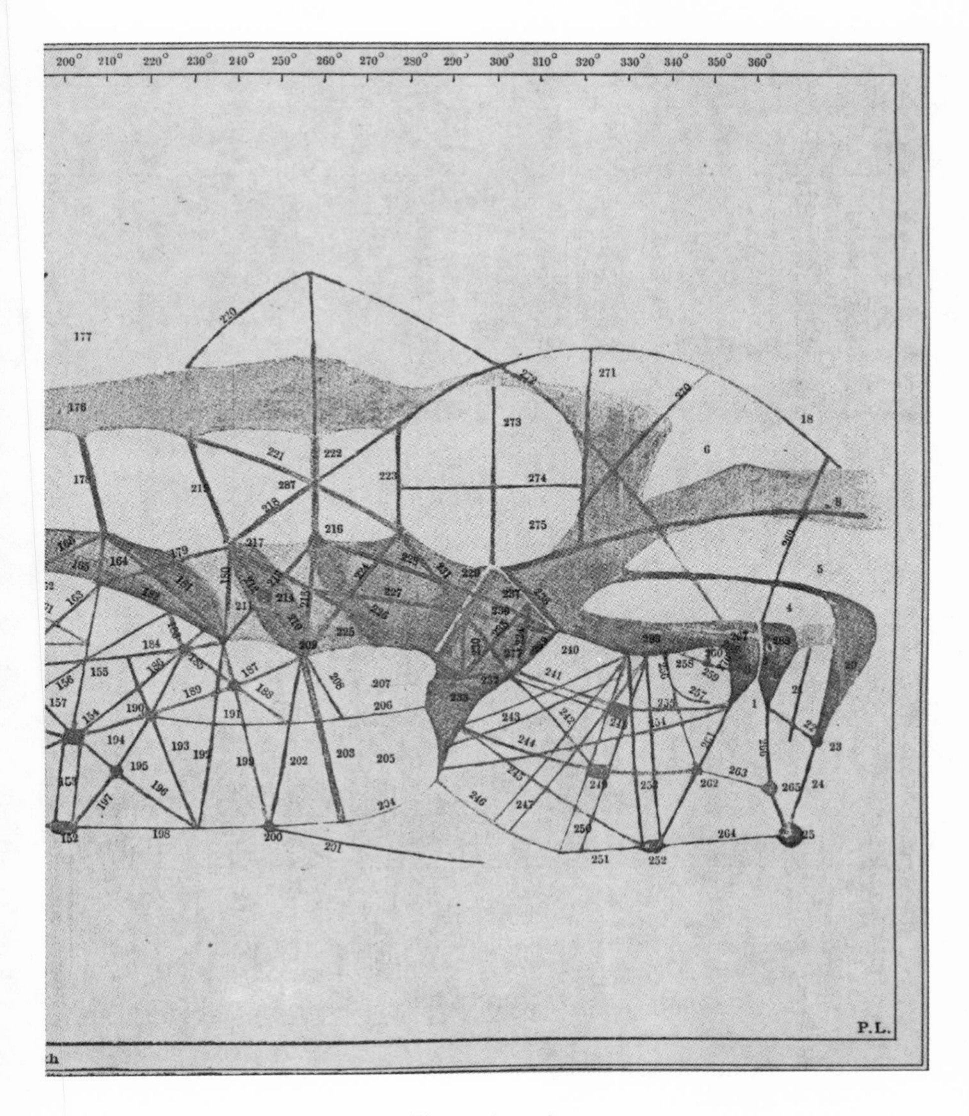

Fig 3.5 (*cont.*)

Table 3.2 *Interpretations of Martian canals, 1877-1924*

Interpretation	Interpreter	Institution/Location
	Green (1878)	Madeira
Illusory	Maunder (1894, 1903)	Greenwich
Rows of spots or	Holden (1895)	Lick
boundaries of	Cerulli (1898)	Italy
shadings; no	Newcomb (1907)	Washington, D.C.
continuous lines	Antoniadi (1909-1930)	Meudon, France
	Hale (1910)	Mount Wilson
Real and continuous	W. H. Pickering (1892)	Harvard (Peru station)
	Barnard (1894)	Lick
But natural cracks	W. W. Campbell (1895, 1896)	Lick
and/or strips of	Antoniadi (1898)	Juvisy (France)
vegetation	A. R. Wallace (1907)	England
	Eddington (1907)	England
	Trumpler (1924)	Lick
Fine lines, natural	Schiaparelli (1877 ff.)	Milan
Nature unknown	E. Slipher (1930)	Lowell
Fine lines, artificial	Lowell (1894 ff)	Lowell
	Flammarion (1892)	Juvisy
	N. Lockyer	England
	Russell (1901)	Princeton, N. J.
	most of public	worldwide

while Lowell shows "an apparent lack of familiarity with the literature of the subject."[41]

Lowell's observation of the wave of darkening from the south pole to the equator Campbell sees as a confirmation and extension of Schiaparelli's results for the northern Martian hemisphere. Regarding the canals, Campbell draws attention to differences among the Flagstaff observers even on the same night with the same telescope, but he notes that Douglass's observation of canals across dark areas confirms Schaeberle's Lick observations. If true, this would mean that the dark areas are not seas. Campbell accepts Schiaparelli's theory of the organic origin of these dark areas.

[41] W. W. Campbell, "Mars, by Percival Lowell," *PASP*, 8 (August 1896), 207–220: 220. The original appeared in *Science*, 4 (August 21, 1896), 231–238.

But as for Lowell's more breathtaking journey from observation to theory, Campbell is insistent: "Mr. Lowell went direct from the lecture-hall to his observatory in Arizona; and how well his observations established his pre-observational views is told in his book."[42] While Campbell accepts the canals as real surface features, he ridicules the artificial theory as impossible from a hydraulic point of view, and no more of intelligent origin than the long, straight markings on the Moon that radiate from the crater Tycho. Though he thinks that Lowell's clouds are more likely mountains, it is notable that Campbell does not argue that conditions on Mars are unsuitable for intelligent life; rather, he states that Lowell's view of Mars's atmosphere based on Stoney's theory, is in accord with his spectroscopic results of 1894, which failed to detect water vapor at a certain level. Campbell therefore concludes not by emphasizing harsh conditions, but with an observational caveat bearing directly on the canals themselves: "Drawings of Mars by different observers, even on the same night and with the same telescope, are proverbially different. So far as the drawings by the three Flagstaff observers have been published, the proverb still seems to be in force."[43]

In short, Campbell objects to some of Lowell's observations and certainly to Lowell's pathway from observation to the theory of artificiality, but he is unable to undermine completely Lowell's position by appeal to impossible conditions on the planet. For all of these reasons the case against Lowell was inconclusive, as Lowell soon demonstrated by his response (via his assistant, Douglass). Campbell replied in return in the pages of *Science,* and the battle was joined.[44] It was only the beginning of a long struggle.

As the controversy picked up steam, the combatants became so numerous, and their opinions so varied, that it is easy in retrospect to lose sight of the trajectory of argument. In order to portray the variety and subtlety of opinion underlying this trajectory – the heart of two decades of controversy – it is best to keep in mind the spectrum of opinion in Table 3.2. We may begin at the conservative end of this spectrum with what came to be called the "illusionist arguments." These we can trace back all the way to 1877 and those who doubted that Schiaparelli really saw what he claimed to have seen. One important example is the British artist Nathaniel Green, who suggested in 1878 that either the hard, sharp lines of Schiaparelli did not exist and were really a problem of drawing technique or that "the atmospheric vibrations have a tendency to cause a

42 Ibid., 208.
43 Ibid., 219.
44 The Douglass–Lowell reply is in *Science,* 4 (September 11, 1896), 358–359, and Campbell's brief rejoinder is in *Science,* 4 (September 25, 1896), 455–456.

series of points in a line to appear connected together."[45] That Lowell and his staff were themselves worried about the objectivity of their canal observations is clear from experiments performed by Douglass early in the controversy.[46] Lowell and his staff recognized this problem and tried to grapple with it almost from the beginning.

The champion of the illusion theory would be E. W. Maunder (1851–1928), astronomer at the Royal Greenwich Observatory, where he specialized in sunspot work and discovered the famous "Maunder minimum" of sunspot activity.[47] It is of particular interest that Maunder's illusion theory grew out of this sunspot work. Having had his attention drawn to an 1888 paper of Flammarion regarding naked-eye observations of sunspots, Maunder was surprised to note how often small groups of spots were detected while a larger single spot was invisible. Making such observations himself using only a dark glass a few years later, Maunder found that this really was true, and undertook experiments to determine the smallest black marks visible on white paper illuminated by dull, diffused daylight. The results, which he applied to the canals of Mars issue in 1894, indicated to him that they were "the summation of a complexity of detail far too minute to be ever separately discerned."[48]

Spurred on to further investigation of this same issue by experiments in 1902 by B. W. Lane indicating that the canals were purely subjective, Maunder pursued further experiments with J. E. Evans, headmaster of the Royal Hospital School.[49] Here the experiments were conducted with schoolboys averaging 13 years of age who knew nothing of the Mars controversy. Circular disks ranging in diameter from 3.1 to 6.3 inches, bearing images based on actual telescopic drawings of the surface of Mars, were shown to the students, who sat in rows at various distances and thus saw the images with differing angular diameters. After 13 experi-

45 N. Green, "On Some Changes in the Markings of Mars, since the Opposition of 1877," *MNRAS*, 40 (March 1880), 331–332; Crowe, *The Extraterrestrial Life Debate*, 487–88.

46 The experiments at the Lowell Observatory are described in Lowell, *PA*, 3 (February 1896), 299. Even before the Lowell controversy, Douglass and W. H. Pickering had carried out experiments of this kind at Arequippa in 1892. Pickering describes them in detail in "Visual Observations of the Moon and Planets," *Annals of the Astronomical Observatory of Harvard College*, 32, part 1 (Cambridge, Mass., 1900), 117–317.

47 On Maunder see Crowe, *The Extraterrestrial Life Debate*, 490 and references cited therein.

48 Maunder, "Some Experiments on the Limits of Vision for Lines and Spots as Applicable to the Question of the Actuality of the Canals of Mars," *JBAA*, 13 (1903), 344–351. Maunder's earlier articles on the illusion theory of Martian canals appeared in *Knowledge* (November 1894), 249–252, and *Knowledge* (March 1895), 54–59. The Flammarion article on sunspots appeared in *L'Astronomie*, 7 (1888), 121–133.

49 J. E. Evans and E. Walter Maunder, "Experiments as to the Actuality of the 'Canals' on Mars," *MNRAS*, 63 (June 1903), 488–499. B. W. Lane's article appeared in *Knowledge* (November 1902), 250–251.

ments, using different combinations of disks, and drawings with and without lines representing canals, the boys often drew "canals" in places where they had actually been claimed to exist. Maunder and Evans concluded that the canals of Mars "may in some cases be, as Mr. Green suggested, the boundaries of tones or shadings, but that in the majority of cases they are simply the integration by the eye of minute details too small to be separately and distinctly defined."

Here, at last, was a seemingly logical and scientific explanation for the observations first claimed by Schiaparelli and proclaimed by Lowell:

> It would not therefore be in the least correct to say that the numerous observers who have drawn canals on Mars during the last twenty-five years have drawn what they did not see. On the contrary they have drawn, and drawn truthfully, that which they saw; yet, for all that, the canals which they have drawn have no more objective existence than those which our Greenwich boys imagined they saw on the drawings submitted to them.[50]

In Italy the astronomer Vincenzo Cerulli made similar arguments, but in "somewhat different terms," according to Maunder, who cited Cerruli's result.[51]

Was this, then, to be the end of the controversy? Far from it. Those at the other end of the spectrum, including more than Lowell, were not impressed. As early as 1896 Lowell had argued against the Maunder school, and he did so again now.[52] The young Henry Norris Russell, destined to become the dean of American astronomers, wrote as late as 1901, "Perhaps the best of the existing theories, and certainly the most stimulating to the imagination, is that proposed by Mr. Lowell and his fellow workers at his observatory in Arizona."[53] Moreover, Lowell set his staff to work on photographing the canals, and when in 1905 C. O. Lampland succeeded, Lowell believed he finally had his objective evi-

50 Evans and Maunder, "Experiments," 499. Maunder and his wife undertook other experiments to try to explain why small markings gave the impression of straight lines, as reported in *JBAA*, 13 (1903), 344–351. See also *Scientia*, 7 (1910), 253–269.

51 Cerulli, who had observed canals with his 15.5-inch refractor beginning in 1896, formulated his theory in 1898 that they were optical illusions. On Cerulli, see Crowe, *The Extraterrestrial Life Debate*, 521–522. See also G. J. Stoney, *Philosophical Magazine*, 6th series, 16 (1908), 318–339, 796–811, and 950–979.

52 Lowell's further arguments against the illusionist school may be found in "Experiment on the Visibility of Fine Lines in Its Bearing on the Breadth of the 'Canals' of Mars," *Lowell Observatory Bulletin*, no. 2 (1903), 1–4.

53 Henry Norris Russell, "The Heavens in March," *SciAm*, (March 2, 1901), 131. Russell saw serious difficulties with Lowell's theory, but did not rule out the hypothesis of artificiality.

dence. He was quick to exploit this new argument in *Mars and Its Canals* (1906), where he proclaimed that "the canals at last speak for their own reality themselves."[54] A 1907 expedition to the Andes, headed by David Todd, accompanied by E. C. Slipher from Lowell, claimed to photographed even more canals.[55]

Surely photography was as objective an arbiter as one could hope for. But the fact was that photography had only succeeded in showing a few canals, not an entire network, and even if one accepted the few canals photographed, the illusionists could still question the reality of the vast majority. Not only was Maunder himself skeptical, in 1907 America's most famous astronomer, the elderly Simon Newcomb, only 2 years away from death but still intellectually active, joined the fray on the side of the illusionists. Well aware of Maunder's experiments, Newcomb was nevertheless dissatisfied with them for two reasons: experienced observers were not equivalent to Maunder's Greenwich schoolboys, and the pencil sketches of the latter needed to be supplemented by descriptions.[56] This Newcomb attempted to do with his own experiments and an analytical approach appropriate for one whose life had been devoted to celestial mechanics. Newcomb argued that the observations of a single practiced observer under favorable conditions should completely outweigh those of less practiced observers, and he divided the problem into two parts: optical and psychological. The first included all causes that affect the formation of an image on the retina on the eye (atmosphere, instrument, eye), the second those that affect the observer's perception of this image, including the stimulus and perception of the stimulus (visual inference). Under optical principles Newcomb concentrated mainly on the instrumental contribution, noting the role of aberration in widening any lines seen on Mars. Under psychological principles, he pointed out that "seeing" involved not only sensation (the stimulus of the optical nerve by light), but also visual inference, a rational but unconscious interpretation of the stimulus based on previous experience. Having set the stage, Newcomb described his own experiments, similar to those of Lowell and Maunder but using in one case broken lines drawn with ink on paper, then placed on a window and observed with the transmitted light, and in another case

[54] Lampland's photographic results were announced in Lowell, "First Photographs of the Canals of Mars," *Proceedings of the Royal Society of London,* 77A (1906), 132–135. See also Lowell, *Mars and Its Canals* (New York, 1911), 271–277: 276. On photography and the canals see Hoyt, *Lowell and Mars,* chapter 11.

[55] The Todd expedition to Peru is described in D. P. Todd, "The Lowell Expedition to the Andes," PA, 15 (1907), 551–553, and "Professor Todd's Own Story of the Mars Expedition," *Cosmopolitan Magazine,* 44 (March 1908), 343–351.

[56] Newcomb, "The Optical and Psychological Principles Involved in the Interpretation of the So-Called Canals of Mars," *Astrophysical Journal,* 26 (July 1907), 1–17: 11.

a circular disk with markings observed by the astronomers W. H. Pickering and S. I. Bailey at Harvard, as well as E. E. Barnard and Philip Fox at Yerkes Observatory. In all cases the astronomers drew canal-like features.

Applying these results to the planet Mars, Newcomb noted that taking into account aberration in the best achromatic telescope, each canal appeared to the eye to have a width of about 55 miles, even if its actual width on Mars is only 10 miles. Taking Lowell's system of 400 canals, each with an average length of 1500 miles, gave a surface area (as seen with the telescope) of 33 million miles for the canals, or more than half of the total surface area of Mars. Under such circumstances, Newcomb concluded, it was a wonder that the canals could be differentiated from each other. And it was not at all surprising that on a disk only 20 arcseconds in diameter, such a complex system is interpreted according to the experience and habits of the observer. This conclusion, similar to Maunder's but much more cautiously drawn, was influential by virtue of Newcomb's reputation but hardly clinched Maunder's argument. While adding to an understanding of image formation and perception, it was hardly the desired devastating argument that demolished Lowell's contentions with deductive rigor. Newcomb himself stated that although the result "may weaken the probability of the reality of the entire canal system, it does not disprove its possibility. In fact, it is quite consistent with Lowell's fundamental explanation of the phenomena."[57] Newcomb did not question what he termed the "subjective reality" of the canal system – that Lowell and other Mars observers really saw what they claimed – but he believed that the "objective reality" awaited investigation into the process of visual inference, a study that might be aided by further experiments of the kind he had discussed.

Lowell, while unconvinced, suggested that such experiments should be undertaken through a telescope so as to keep the conditions of the experiment as close as possible to the conditions of observations of Mars; that the area of Newcomb's disks in the experiments was much smaller than the area of Mars's disk; and that Newcomb had omitted several technical points about aberration that mitigate his results, as actual observation of other planetary features proves. Lowell described his own experiments made of lines on paper but viewed with a telescope and an image more closely approximating the real Mars image, with conclusions opposing those of Newcomb and Maunder. In short, one more round in the Martian ring did nothing to strengthen the illusionist position.[58]

[57] Ibid., 17.
[58] Lowell, "The Canals of Mars, Optically and Psychologically Considered: A Reply to Professor Newcomb," *ApJ*, 26 (October 1907), 131–140. Brief rejoinders appeared in

In a hundred variations, these arguments were echoed around the scientific world and reechoed and amplified in newspapers and the popular press. That neither side had won a full victory by 1907 is evidenced not only by Newcomb's inconclusive results, but also by the fact that not all scientists shared either the extreme illusionist view of Maunder nor the artificial construction view of Lowell and his followers. A middle ground of interpreters, who argued that the canals exist but are not artificial, had existed almost from the beginning and picked up supporters during the 1907 opposition. Among these were the distinguished naturalist A. R. Wallace, still fresh from his book arguing that the Earth was the only inhabited planet in the *universe* (see Chapter 2), and Lowell's former assistant William Pickering. Both of them saw canals as real but natural formations composed of cracks in the Martian surface.[59]

In his final two books on the subject, Lowell rejected this middle ground no less than the illusionist position. *Mars and Its Canals* is Lowell's most detailed case for the reality of the canals and his interpretation, while *Mars as the Abode of Life* (1908) was his last hurrah. In the latter he argued that the canals of Mars are unlike the natural formations seen on other planetary bodies. But perhaps tiring of the fight and wishing to expand his vision, Lowell spent little time on these arguments and went beyond them. Based on lectures at the Lowell Institute, this book – and *The Evolution of Worlds,* published the following year – took a broader perspective on the genesis of worlds. Showing the pervasive worldview of an evolutionist, they linked the nebular hypothesis to the Darwinian theory in a setting of comparative planetology. Ironically, in these books Lowell adopted the Chamberlin–Moulton theory of the formation of the solar system by the close encounter of stars, a theory that in the 1920s would lead to the widespread belief that life was rare in the universe because of the rarity of such meetings and thus of their resulting planetary systems.[60]

If the interpretations of canal observations were indecisive, the study of the Martian atmosphere and surface conditions held the promise of excluding the possibility of life by virtue of a harsh environment. Though the full power of spectroscopic techniques for planetary study still lay decades in the future, the crucial oppositions of 1907 and 1909 produced several

Newcomb, "Note on the Preceding Paper," *ApJ,* ibid., 141, and in Lowell, "Reply to Professor Newcomb's Note," *ApJ,* ibid., 142.

[59] A. R. Wallace, *Is Mars Habitable?: A Critical Examination of Professor Percival Lowell's Book "Mars and Its Canals," with an Alternative Explanation* (London, 1907). See the review comments on Wallace's book on page 111.

[60] Lowell, *Mars as the Abode of Life* (New York, 1908). On Lowell as the founder of comparative planetology see David Strauss, " 'Fireflies Flashing in Unison': Percival Lowell, Edward Morse and the Birth of Planetology," *JHA,* 24 (iAugust 1993), 157–169.

important results in this regard. In 1908 V. M. Slipher reported from Lowell Observatory the observation of water vapor in the spectrum of Mars, igniting that controversy once again.[61] Slipher had attempted this feat in 1905, with "indecisive" results, but the recent improvement of photographic plates more sensitive in the orange and red parts of the spectrum promised more definitive results. In this part of the spectrum was found the "a band," the strongest accessible band caused by the absorption of water vapor; previous studies had been limited to the "D band" region. Using a single prism spectrograph on the 24-inch Lowell refractor, Slipher reported in the *Astrophysical Journal* that he had found water in the atmosphere of Mars. Although he cautioned that further observations were needed before the amount could be determined, he was quick to point out that this favored a mild climate on Mars and polar caps composed of snow rather than hoarfrost. No mention was made of Martian canals, but the implications of these results for Martian hydrography were not lost on anyone who followed the controversy.

As Slipher delighted in pointing out, this result, while in agreement with early observations, opposed that of W. W. Campbell at Lick Observatory. Not only had the photographic technique improved, this time there was another difference: astronomer Frank W. Very claimed to have proved Slipher's result quantitatively. In *Science* for January 1909, Very stated that while previously the presence of water in the Martian atmosphere had been a matter of opinion, now, "with material aid from Mr. Lowell," he had built an instrument capable of measuring the strength of lines, called a "spectral band comparator." With it, he compared the intensity of the a band of water in a Martian spectrum with that of the Moon, where there is no atmosphere and that band is entirely due to absorption in the Earth's atmosphere. The result was that the Martian band was more intense by a factor of 1.2. (A comparison of two other control lines showed no intensity change.) Very even went on to estimate the amount of water present in the Martian atmosphere.[62]

This result was of great interest to W. W. Campbell at Lick Observatory, for he was preparing his own expedition to observe the same phenomenon, also using the new red-sensitive plates. Campbell was quick to point out that Slipher's spectra confirmed his own past results in that region of the spectrum previously observed while tacitly admitting the

[61] V. M. Slipher, "The Spectrum of Mars," *ApJ*, 28 (December 1908), 397–404. In this article, Slipher also gives a brief review of previous attempts to find water vapor on Mars.

[62] Very's results were reported in full in "Measurements of the Intensification of Aqueous Bands in the Spectrum of Mars," *Lowell Observatory Bulletin* no. 36 (1909), 207–212, and summarized in "The Presence of Water Vapor in the Atmosphere of Mars Demonstrated by Quantitative Measurements," *Science* (January 29, 1909), 191–193.

possibility that the newly observable "a band" could now be indicating otherwise.[63] Campbell went to great lengths to make the most of the opportunity afforded by the favorable 1909 opposition and mounted an expedition to Mount Whitney, funded by Crocker.[64] The observations, made September 1–2, showed no intensification of the "a band" for Mars.[65] Campbell thus showed that water vapor was very scarce on Mars even compared to the amount above Mount Whitney. While these results were directly opposed to those of Slipher at Lowell and to Very's supposedly quantitative measurements, "these measurements," Campbell concluded, "do not prove that life does not or can not exist on Mars. The question of life under these conditions is the biologist's problem rather than the astronomer's."[66]

How can we reconcile Campbell's observations with those of Slipher? In a review of spectroscopic observations of Mars, Campbell tried to do just this. He pointed out that Slipher had observed the Moon later in the night than Mars, when the water vapor content might have decreased. Very's results then would measure not necessarily the difference in the "a band" of the Moon and Mars due to a Martian atmosphere but rather the difference due to decreasing water vapor in the Earth's own atmosphere in the 2 to 8 hours that separated their exposures.[67] Very not only answered these criticisms, but also in September 1909 reported that he had used his method to show that the "B band" of oxygen indicated that this gas was present on Mars. And in 1914 he measured new plates of Mars taken by Slipher, with similar results.[68]

[63] Campbell, "Note on the Spectrum of Mars," *Science*, 29 (March 26, 1909), 500.

[64] Campbell's overall work on the Martian atmosphere is examined in David H. DeVorkin, "W.W. Campbell's Spectroscopic Study of the Martian Atmosphere," *QJRAS* (1977), 18, 37–53. The 1909 expedition is discussed in more detail in Donald E. Osterbrock, "To Climb the Highest Mountain: W. W. Campbell's 1909 Mars Expedition to Mount Whitney," *JHA*, 20 (June 1989), 77–97.

[65] Campbell's results from the 1909 Crocker expedition are reported in detail in "The Spectrum of Mars as Observed by the Crocker Expedition to Mt. Whitney," *Lick Observatory Bulletin*, 5, no. 169 (1909), 149–156.

[66] Campbell, "Water Vapor in the Atmosphere of the Planet Mars," *Science*, 30 (October 8, 1909), 474–475.

[67] Campbell, "A Review of the Spectroscopic Observations of Mars," *Lick Observatory Bulletin*, 5 no. 169 (1909), 156–164.

[68] Very, "Water Vapor on Mars: Reply to Campbell's Criticism," *Lowell Observatory Bulletin*, no. 43 (1909), 239–240 and *Science* (August 5, 1910). Very reported intensification in the oxygen band in "Quantitative Measurements of the Intensification of Great B in the Spectrum of Mars," *Lowell Observatory Bulletin*, no. 41 (September 16, 1909), 221–229, summarized in "Oxygen as Well as Water Proved to Exist in the Atmosphere of Mars," *Science*, n.s. 30 (November 12, 1909), 678–679. The 1914 confirming results are given in Very, "Intensification of Oxygen and Water Vapor Bands in the Martian Spectrum: Abstract of Measures for 1914," *Lowell Observatory Bulletin*, no. 65 (April

Figure 3.6. E. M. Antoniadi, courtesy Royal Astronomical Society Library, photo by Phébus, Constantinople.

The disparate conclusions of Slipher, Very, and Campbell left the conditions for life on Mars still uncertain, though with Campbell's increased reputation his results were now taken more seriously than in 1894. And so, 15 years after Lowell had first published his hypothesis, astronomers worldwide prepared once again to observe the surface features of Mars under the most favorable conditions since 1894. Contemporary astronomers and modern historians agree that it was Eugene M. Antoniadi (Figure 3.6) whose observations contributed most to resolving the riddle of the canals of Mars. Antoniadi (1870–1944), a Greek-born astronomer known later in life for his studies on ancient Greek and Egyptian astronomy, serves as vivid testimony to one of the unappreciated truths of this controversy: that many of its participants could, and did, change their minds on the nature and existence of the canals as observational evidence improved and its interpretation became more sophisticated.

16, 1914), 71–72, reported in more detail in "Intensification of Oxygen and Water-Vapor Bands in the Martian Spectrum," *AN,* 199 (October 1914), 153–170.

It is difficult to escape a number of ironies in Antoniadi's clinching role: that it was in Flammarion's France, hotbed of pluralism, that Lowell's thesis would meet its Waterloo; that Antoniadi himself had been inspired by the great pluralist Flammarion and was actually working at Flammarion's observatory at Juvisy in 1893–1894 as Lowell entered astronomy; and that many of Antoniadi's results would be published in the journal of the French astronomical society that Flammarion founded. Above all, there is the apparently anomalous fact that Antoniadi resolved the controversy using a 33-inch refractor – smaller than the 36-inch refractor available at Lick Observatory in 1888. Both observatories were at mountain sites, and Lick astronomers were much more skeptical of canals than Antoniadi. Why, we may well ask, did not the Lick astronomers resolve the canals in 1888 (or Yerkes astronomers with the 40-inch telescope after 1897) and spare the world the entire Lowell controversy? For that matter why did the Meudon 33-inch telescope not resolve the canals shortly after its erection in 1891?

A closer look at Antoniadi does not entirely explain the anomaly, despite the fact that in 1909 Mars was higher in the sky and better situated for observation in Europe than during previous favorable oppositions, rising to an altitude of 37 degrees at Meudon compared to 13 degrees in 1907. Antoniadi had first observed Mars in 1888, at the age of 18, on the Greek island of Prinkipo with a small 3-inch telescope. In his association with Flammarion's observatory at Juvisy, the young Antoniadi had used a 9.6-inch equatorial telescope to observe Mars and its canals in 1894.[69] He did so with such impressive results that in 1896 he was appointed head of the Mars section of the British Astronomical Association, where his results and those of others were published in that society's *Journal* and *Memoirs*. By 1898 Antoniadi (himself still observing at Juvisy) had made the "Reports of the Mars Section" a major forum of discussion for observations and interpretations of the canals. In the last Martian oppositions of the nineteenth century he argued that the canals were not fine lines, advocated a geologic explanation for their origin, and believed that some were the boundaries of shaded areas. But as late as

[69] On Antoniadi see the entry by Giorgio Abetti in *Dictionary of Scientific Biography*, 1, 172; Richard McKim, "The Life and Times of E. M. Antoniadi, 1870–1944," *JBAA*, 103 (1993), 164–170, 219–227. See also Fernand Baldet, *L'astronomie*, 58 (1944), 58, and P.M.R., "E. M. Antoniadi," *JBAA*, 55 (September 1945), 163–165. Antoniadi summarizes his history of observing Mars in *La planète Mars* (Paris, 1930), 61, note 1. In 1909 Antoniadi wrote that the present apparition of Mars was the most favorable for many years because, unlike 1892 and 1907, "the very large disc now subtended by the planet is raised for European observers above the tremors and chromatic dispersion of the southern horizon," *JBAA*, 19 (1909), 427. Similarly, he had written that the altitude of Mars during the 1896 opposition "more than compensated" for the reduced disk compared to 1892 or 1894. *MBAA*, 6 (1898), 75.

1903 he still believed in "the incontestable reality" of many of them, even as he held that their apparent doubling was illusory.[70]

Antoniadi continued to observe Mars at every opposition, either at Juvisy or with his own 8.5-inch telescope, and so great was his reputation that by the 1909 opposition he was given use of the 33-inch refractor at Meudon, the largest telescope in Europe. Antoniadi approached these observations with "an open mind," believing in Schiaparelli's observations but doubting their reality because of the difficulty of reconciling the canal phenomenon with logic. For 2 months, from September to November, he observed the planet, but by October he was ready with the result. Of 36 canals examined "steadily under good seeing," all corresponded to something "real," such as diffuse streaks or borders of shaded areas. "They are not canals at all, but complex shadings – the integration of irregular details, too minute to be accessible to our means, as Mr. Maunder argued 15 years ago," Antoniadi wrote. Other canals, seen only in "flashes" lasting about one-third of a second, corresponded to nothing on the surface. "The geometrical canal network is an optical illusion; and in its place the great refractor shows myriads of marbled and chequered objective fields, which no artist could ever think of drawing." The planet, Antoniadi concluded, looked very much as Green had depicted it in 1877 and like his own map of Mars published in 1903. By the time of Antoniadi's Fifth Interim Report in December 1909, Maunder, savoring victory and his role in it, rejoiced that scientists and the public "need not occupy their minds with the idea that there were miraculous engineers at work on Mars, and they might sleep quietly in their beds without fear of invasion by the Martians after the fashion that Mr. H. G. Wells had so vividly described." [71]

Antoniadi's results received quick support from the United States, where the era of reflectors had dawned and the new 60-inch telescope of Mount Wilson had come into use in 1908. George Elery Hale, founder

[70] Antoniadi's "Interim Reports" are found in the *JBAA*, while the much longer final reports are published in the Society's *Memoirs*. See especially *MBAA*, 6 (1898), 55–102; *MBAA*, 11 (1903), 85–142; *MBAA*, 17 (1910), 65–112; and especially the crucial "Report of the Mars Section, 1909," *MBAA*, 20 (1915), 25–92. Crowe, *The Extraterrestrial Life Debate,* discusses some of Antoniadi's Reports on 519–527, 536–537, and passim, and Hoyt, *Lowell and Mars,* on 168–172. The Mars map of Flammarion and Antoniadi found in volume 2 of Flammarion's *La planète Mars* (1909), dating from 1901, is still crisscrossed by many canals.

[71] Antoniadi's first observations of Mars with the 33-inch telescope are reported and discussed in "Third Interim Report for 1909, Dealing with the Nature of the so-called 'canals,' " *JBAA,* 20 (October 1909), 25–28. Of the 36 canals observed, he reported that 31% were broad and diffuse, 28% knotted, 16% the edges of faint shadings, 11% dark bands, 6% short, irregular black lines, and 8% detached, irregular "duskiness." Maunder's remark is on page 123 of volume 20. Antoniadi continually refers to Maunder's "profound" and prophetic paper of 1894 in *Knowledge*.

and Director of the Mount Wilson Observatory, wrote to Antoniadi in January 1910 to report the accord of Antoniadi's visual drawings with observations of the 60-inch telescope. "I was able to see a vast amount of intricate detail – much more than has been shown on any drawings with which I am acquainted," Hale wrote. "In spite of the very fine seeing on certain occasions, which required a power of 800 to show the smallest details with an aperture of 44 inches, no trace of narrow straight lines, or geometrical structure, was observed. A few of the larger 'canals' of Schiaparelli were seen, but these were neither narrow nor straight." Reporting further that he had resolved two "canals" into minute twisted filaments, he concluded, "I am thus inclined to agree with you in your opinion (which coincides with that of Newcomb) that the so-called 'canals' of Schiaparelli are made up of small irregular dark regions."[72] Adding that the 60-inch telescope showed smaller details than Lowell's telescope; that under good atmospheric conditions he had found a (stopped-down) aperture of 44 inches or even a bit more "highly advantageous for the study of planetary details"; and that Abbot, W. S. Adams, Babcock, and A. E. Douglass, among others, all fully agreed with his 60-inch observations, based on their own experience, Hale, using his new instrument, did much to consign canals to oblivion. Citing other observations from large telescopes, Antoniadi concluded that "the frail testimony of small refractors has vanished before the decisive evidence of giant instruments; and the telescopes of Princeton, Lick, Yerkes, Mount Wilson, and Meudon have settled the question for ever."[73]

After 1909 Antoniadi continued his observations and solidified his claim of resolved canals.[74] Not surprisingly, Lowell argued valiantly

[72] Hale's letter of January 3, 1910, is reprinted in Antoniadi, "Sixth Interim Report for 1909, dealing with some further Notes on the so-called 'Canals,' " *JBAA*, 20 (January 1910), 189–192: 191–192. Hale also cited Barnard, "Micrometrical Measures," *MNRAS*, 56 (January 1896), 166, where Barnard reported irregular "short diffused hazy lines" rather than "straight hard sharp lines," as in agreement with this interpretation. On Barnard's early interest in Mars see William Sheehan, "E. E. Barnard and Mars: The Early Years," *JBAA*, 103 (1993), 34–36. For the full context see William Sheehan, *The Immortal Fire Within: The Life and Work of Edward Emerson Barnard* (Cambridge, 1995).

[73] Antoniadi, "Sixth Interim Report," 192. E. E. Barnard, now at Yerkes, also supported Antoniadi's observations. See Barnard to Antoniadi in "Fifth Interim Report," 137. Both Barnard and Hale also photographed Mars in 1909, with results showing no canals. Barnard reported briefly on his photographs with the 40-inch Yerkes refractor in *MNRAS*, 71 (March 1911), 471, and Hale's photographs with the 60-inch telescope were discussed at the BAA meeting of December 29, 1909, *JBAA*, 20 (December 1909), 119–125. Antoniadi reported on the Mars photographs of Hale and Barnard in "On Some Drawings from Photographs of Mars taken in 1909 by Professor Barnard and Professor Hale," *MNRAS*, 71 (June 1911), 714–716.

[74] Antoniadi continued to observe with the Meudon 33-inch until at least 1930, publishing his results in the *JBAA* and French journals.

against Antoniadi right up to his death in 1916, stating that Antoniadi was seeing the blurring of continuous lines.[75] But the handwriting was on the wall for canals, and everywhere astronomers hailed the observations of Antoniadi and Hale. This makes all the more surprising the conclusions a few years later of Robert J. Trumpler, a Lick astronomer on the verge of fame for his work on interstellar "reddening." Based on a lengthy visual and photographic study of Mars during both the 1924 and 1926 oppositions, Trumpler concluded that while the canals were not the artificial fine lines of Lowell, they were real, were not less than 25 miles wide, and were made visible by vegetation growing along them. Ironically, the Swiss-born Trumpler had been brought in 1919 to Lick Observatory by none other than Campbell himself, the arch enemy of Lowell's canals. By 1924 Campbell was president of the University of California, which ran the observatory, and was still nominal director of Lick. But the observatory was really run now by R. G. Aitken, and what he and his Lick colleagues thought of Trumpler's work on Mars we can only guess.[76] Still, it is a telling fact that even an astronomer of Trumpler's reputation was unable to reignite the canal controversy by the mid-1920s.

Despite Trumpler, Antoniadi held firm to his illusionist conclusions. A lifetime of work on the subject was summed up in his magnum opus, *La planète Mars* (1930), the capstone to the visual era in the canals of Mars controversy, initiated with Schiaparelli's 1878 *Osservazione* and punctuated with Lowell's 1895 *Mars*.[77] Three quarters of the book dealt with the topography of Mars, which for purposes of discussion Antoniadi divided into three 120 degree zones of longitude and two polar zones. This portion of the book constitutes the state of the art in Martian topography and its result, seen in Figure 3.7, marks the end of an era. But it is the first quarter of the book that helps us understand how Antoniadi arrived at that result. In the chapter on "Instruments and

75 Antoniadi had written to Lowell on October 9, 1909 regarding his resolution of the canals, and Lowell responded by letter on November 2. The Lowell–Antoniadi correspondence of 1909 is discussed in Hoyt, *Lowell and Mars*, 168–172. E. C. Slipher was still arguing for canals shortly before his death in 1964: E. C. Slipher, *The Photographic Story of Mars* (Cambridge, Mass., 1962), 163. But the demise of canals can be followed by the decreasing number of articles devoted to them, as indexed in *Astronomische Jahresbericht*.

76 The details of Trumpler's work may be found in Trumpler, "Observations of Mars at the Opposition of 1924," *Lick Observatory Bulletin*, no. 387 (1927), 19–45, summarized in "Visual and Photographic Observations of Mars," *PASP*, 36 (October 1924), 263–268. Follow-up results for 1926 are found in "Visual and Photographic Observations of Mars Made at the Opposition of 1926," *PASP*, 39 (April 1927), 103–111, and "Observations of Mars at the Opposition of 1924," *Lick Observatory Bulletin* no. 387 (March 1927), 19–45.

77 Antoniadi, *La planète Mars* (Paris, 1930). English translation, *The Planet Mars*, Patrick Moore, trans. (Shaldon Devon, 1975).

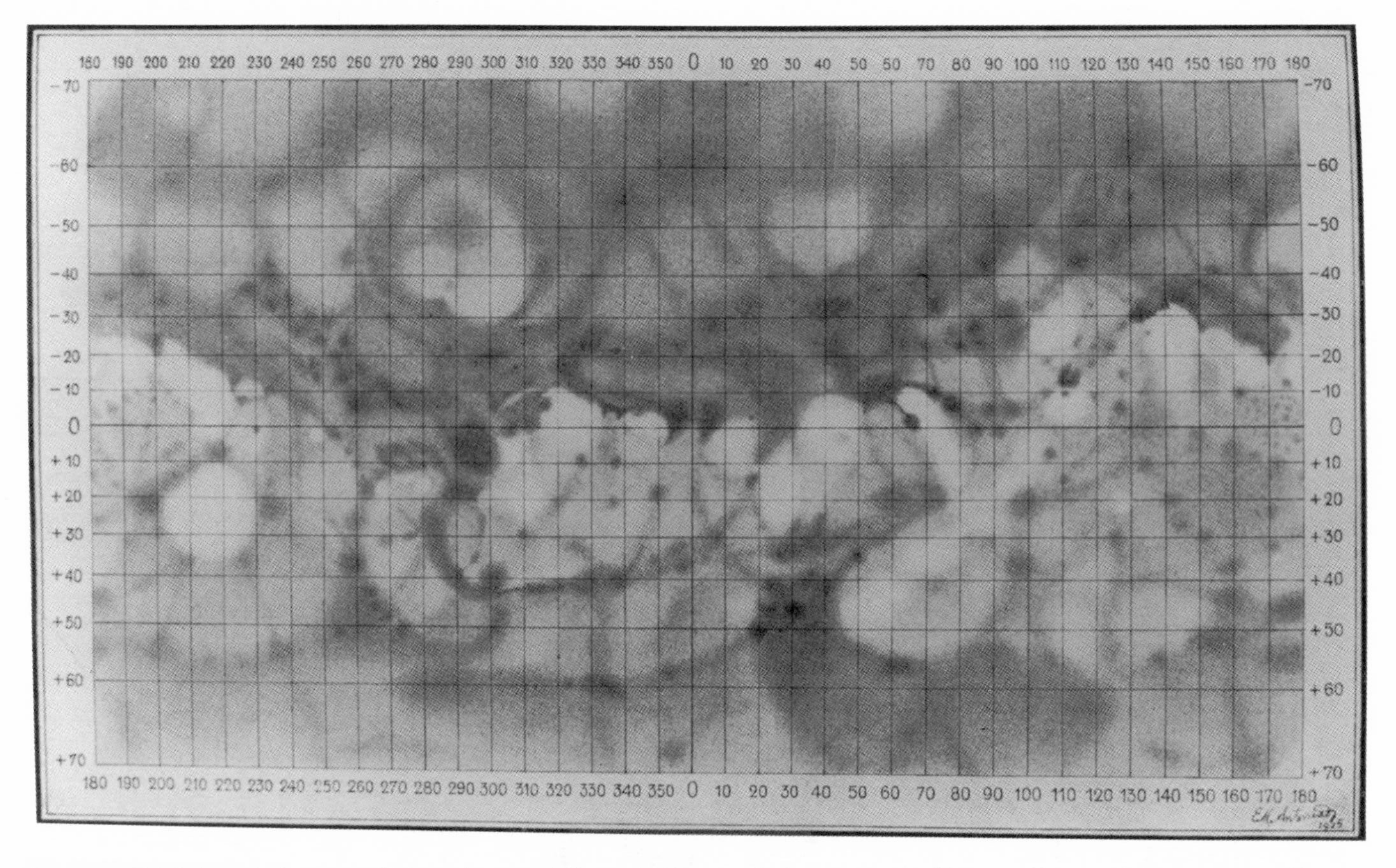

Figure 3.7. E. M. Antoniadi's map of Mars, showing dark patches but not the lines interpreted by Lowell as canals. From Antoniadi (1930).

Observatories," Antoniadi states, based on his own experience, that telescopes of large aperture are "vastly superior to smaller ones" for studying planetary surfaces. Among the 10 reasons he gives for this conclusion are their greater resolving and light-gathering power; the gain in contrast, intensity, and dimensions of the surface features; and the visibility of faint half tones, colors, and tints. While agreeing that atmospheric turbulence may at times reduce the superiority of large instruments, under the best conditions Antoniadi found that they reveal details completely inaccessible to smaller telescopes. And while high-altitude observatories such as Lowell's at Flagstaff are optimal, lower-altitude sites such as Meudon at Paris will, during their less frequent times of good seeing, produce images just as good with the same aperture. Antoniadi thus criticizes Lowell's "pernicious habit" of stopping down the aperture of the 24-inch refractor to 15 inches. And he again applauds the accord of the Mount Wilson results – now confirmed with its 100-inch reflector – with his own visual drawings.[78]

It was through such instruments that we could finally realize "the illusion of canals" as he titles one of his chapters. After giving a history of Martian canal observations, drawing largely on Flammarion's massive tomes, Antoniadi developed his thesis that in addition to the canals having been resolved into separate spots, the laws of diffraction prevented the canals from being real, since telescopes could not have seen them. His conclusion:

> Nobody had ever seen a genuine canal on Mars, and the more or less rectilinear, single or double 'canals' of Schiaparelli do not exist as canals or as geometric pattern; but they have a basis of reality, because on the sites of each of them the surface of the planet shows an irregular streak, more or less continuous and spotted, or else a broken, greyish border or an isolated, complex lake. Thus the details on Mars present everywhere a structure which is infinitely irregular and natural, so characteristic of the patches on all the bodies in the solar system.

[78] Ibid., chapter 2, 6–10; pages 14–19 in the English translation; see page 35 on the Mount Wilson results. Antoniadi examined the question of telescope size and Martian canals, including the opinions of Schiaparelli, Campbell, Barnard, Hale, Deslandres, and others, in "Les Grande Instruments," *SAFB*, 41 (1927), 145–155. He had earlier expressed his opinion in "On the Advantage of Large over Small Telescopes in Revealing Delicate Planetary Detail," *JBAA*, 21 (1910), 104–106. Antoniadi's discussion was part of a broader controversy over the relative merits of telescope size extending back to Victorian astronomy; see John Lankford, "Amateurs versus Professionals: The Controversy over Telescope Size in Late Victorian Science," *Isis*, 72 (March 1981), 11–28.

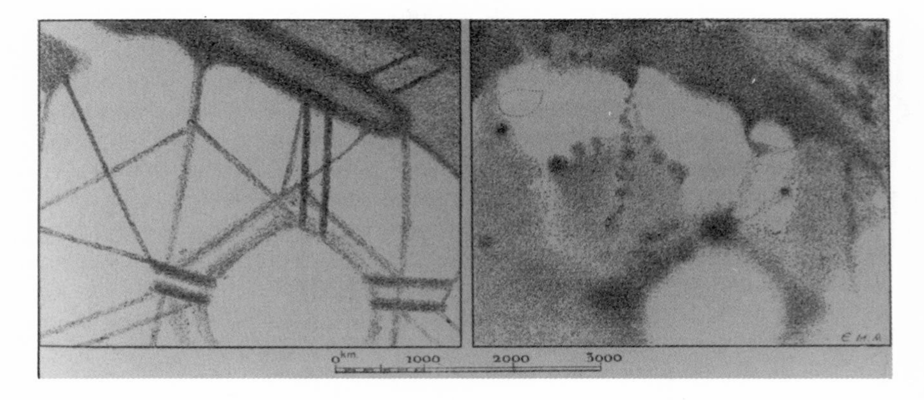

Figure 3.8. Antoniadi's comparison of his Mars observations (right) with those of Schiaparelli, showing how dark splotches were interpreted by Lowell as fine lines. Schiaparelli's observations (left) were made with telescopes much smaller than the 33-inch instrument Antoniadi used. From Antoniadi (1930).

This claim he supported with one of his drawings compared to Lowell's (Figure 3.8).[79] Antoniadi made a sharp distinction between Schiaparelli's canals and those of Lowell. The latter, he asserted, were largely illusory, as were those of E. C. Slipher and Trumpler:

> What can one think of the totally non-existent reseau of Trumpler, contradicted in the most categorical way by his own photographs, and allegedly seen – as we have noted – with the great Lick refractor: that refractor which Burnham considered superior to that of Yerkes, and with which the keen-eyed Barnard never saw any straight features on the planet? The 'canals' of Slipher and Trumpler defy perspective; this is absolute and permanent evidence that they have absolutely nothing to do with real features of the planet.[80]

One could hardly call Trumpler a bad scientist, considering his subsequent achievements; his claims of canals made with the Lick telescope go a long way toward exonerating Lowell of the the charges of fraud and delusion that have sometimes been leveled.

Despite his conclusion on canals, Antoniadi believed that Mars possessed an atmosphere which, though rarer than ours, was proven to exist by the growing and shrinking polar caps, which required an atmospheric

[79] Antoniadi, *La planète Mars,* chapter 5, 22–34: 28–29; English translation, pages 33–46: 40.
[80] Ibid., 30; English translation, 41.

blanket.[81] He also realized that Campbell's spectroscopic results indicated a desertlike Mars, a world "in a state of advanced decrepitude," and in a chapter on physical conditions and habitability Antoniadi joined in accepting what was already becoming the new view of Mars – that life is probable on that planet but it is in the form of flora, not fauna.

Thus, Antoniadi provides us with a final lesson in the canals of Mars controversy: that despite his demolition of the canal myth, his Flammarionesque faith that life was an essential part of the universe remained unshaken:

> If, then, we consider also life's marvellous power of adaptation, one of the aims of the Creator of the Universe, we can see that the presence of animals or even human beings on Mars is far from improbable. This was the view held by Flammarion, Schiaparelli, Moulton and others. However, it seems that advanced life must have been confined to the past, when there was more water on Mars than there is now; today we can expect nothing more than vegetation around the vast red wildernesses of the planet.[82]

Aftermath of the Canal Controversy

Armed now with the historical facts, we return a final time to our initial question: why did the canals of Mars controversy, like the canals themselves, take so long to resolve? The simplest answer is that the observations were extremely difficult and the stakes extremely high, the latter ensuring all possible attempts at resolution. The immense distance of Mars was itself enough to discourage any observation of its fine details, but the favorable oppositions, those close approaches about every 15 years during which the distance of Mars was almost halved, tantalized observers and in fact gave them the opportunity for their greatest discoveries: the moons of Mars and the Schiaparellian canal network in 1877, the Lowellian elaboration in 1894, and the Antoniadian resolution in 1909 (despite the Trumplerian surprise of 1924). The planetary dance between Earth and Mars, courtesy of the laws of celestial mechanics, almost precluded from the start any quick resolution as the object of study was perversely withdrawn from optimal

[81] Ibid., chapter 7, 39–49; English translation, 52–64.

[82] Ibid. chapter 8, 50–53: 52; English translation, 65–69: 67. Antoniadi discusses the vegetation hypothesis in his chapter 4 on surface features, 18; English translation, 29. For Antoniadi's convictions on life in the universe, see "La vie dans l'univers," *SAFB*, 52 (1938), 1–14.

view. Mars was, as Henry Norris Russell would say, a "heartbreaking" object for the observer.[83]

But that was just the beginning. Even at the closest approaches there was the necessity of observing through the Earth's atmosphere, the bane of all ground-based astronomy. Even when Mars reached high altitudes in the sky, the ceaseless turmoil of the terrestrial atmosphere allowed only fleeting glimpses of the finest details of the Martian surface, forcing observers to suspend these glimpses in their minds rather than study them at leisure, leaving the mind to its own imperfect methods of visual inference based on scarce data. What is more, this same atmosphere also complicated attempts to determine spectroscopically the presence of oxygen and water vapor on Mars, fooling V. M. Slipher and others before him into believing that the atmosphere of Mars harbored both, when in fact they were seeing the insidious effects of the air of their home planet. Because of these known atmospheric effects, Lowell had located his observatory on the high plateau of Flagstaff, giving him confidence in his observations, confidence that the Lick observers at their mountain observatory tried to dash. Because of the atmosphere, too, Campbell climbed Mount Whitney in 1909 and disproved Slipher's spectroscopic claims.

Then there were the telescopes themselves – the controversy over whether a large or small aperture was better. How was it that the sharp-eyed Schiaparelli could discover an entire network of finely etched canals with an 8.5-inch refractor, while the equally sharp-eyed Barnard could see only a few canals, and those as broad streaks that would belie Lowell's claim of artificiality? If large was better (as Antoniadi concluded), why did not the Lick 36-inch refractor resolve the canals in 1888 or the Yerkes 40-inch refractor after 1897? Why was it left to Antoniadi in 1909 with only a 33-inch refractor? That his atmospheric conditions were consistently better than those of the American observatories is difficult to believe, though it is possible that a combination of persistence and large telescope aperture made the difference for Antoniadi. But that his diligence was greater than that of Trumpler, whose detailed 1924 study of Mars with the Lick 36-inch telescope yielded broad but unresolved canals, again strains credulity. Even taking into account such factors as the altitude of Mars during favorable oppositions, a satisfactory explanation seems elusive.

The experiments of Maunder, Newcomb, and others indicating that the canals were illusory form an instructive chapter in Martian studies, an important attempt to understand the problematic nature of observation

[83] Henry Norris Russell, "Percival Lowell and His Work," *The Outlook,* 114 (1916), 781–783:781.

and the process of visual inference. They showed how the mind could connect spots into a straight line; what they did not show was that this was at all relevant to the case of Mars. If so, how to explain the differences among observers apparent in Table 3.2? Why did not all observers connect the same spots and interpret them as sharp straight lines? The incredible distance, the churning atmosphere, the telescope itself, the individual: all these variables combined to prolong the debate.

And yet, by 1910 a verdict had been reached by the scientific community, a view of "resolved canals" that Antoniadi led and others consolidated, and that became increasingly accepted with every confirmatory observation. In assessing the controversy, the reflections of astronomers after the heat of battle are particularly revealing. Following Lowell's death in 1916, Henry Norris Russell, after "a somewhat careful investigation," concluded that "only one explanation appears reasonable, namely, the influence of what is called the 'personal equation.' "[84] As Russell noted, personal bias was present in some of the most routine and precise astronomical observations, including the determination of stellar magnitudes and positions. For this reason, many observations were taken by many people at many observatories and the results averaged, with a final value and an accompanying "standard deviation." If this was true of observations made on objects clearly visible, Russell argued, the effect of personal equation on delicate and evanescent planetary details must be even greater. What is worse, physical descriptions of planetary surfaces do not lend themselves to quantitative averages and standard deviations.

After Lowell's death, W. W. Campbell too took one last stab at the Martian question in 1918, in a stinging comparison of the discordant observations of "Schiaparelli's successors," William Pickering and Percival Lowell. Though Pickering had observed Mars for 27 years and Lowell for 23, their observations of the canals were about as different as two observers could make them, Campbell argued. Campbell's verdict, in the form of a leading question, was an indictment of 40 years of Martian canal observations: "Is it possible that anyone has been trying to see surface features on Mars which exceed the powers of existing telescopes and human eyes?"[85] In Campbell's view, too many astronomers had tried to resolve a problem beyond the limits of the science of the time.

Antoniadi, who resolved the canals of Mars, and Trumpler, who did

[84] Ibid., 782. See also Russell, "The Lowell Observatory," Appendix II to A. Lawrence Lowell, *Biography of Percival Lowell,* for his positive opinion of Lowell and the work of his staff.

[85] Campbell, "The Problem of Mars," PASP, 30 (March 1918), 133–146. Lowell was dead, but Pickering replied in ibid., 299–300.

not, would scarcely agree with Campbell that the canals were beyond the powers of their telescopes. But both Russell and Campbell were right: with the hindsight of history there is no doubt that the instruments of the time were inadequate to the task before them, and that more than the normal latitude was left for individual error in the form of personal equation. They were right even more than they knew, for history has now provided us with the "truth" about the canals. With few exceptions, we know that Lowell's canals correspond to no actual surface features on Mars (Figure 3.9).[86] Not only was Lowell wrong, so was Antoniadi in the sense that he believed he had resolved the canals.

But the problem of the canals was not entirely one of observational fact; beyond inadequate instruments and personal equation, there was the problem of turning this observational fact into evidence for a theory. That Lowell and many others sincerely believed they saw canals on Mars, and with no taint of fraud, is without question. Campbell and others believed that Lowell had a preconceived theory, and there is no doubt, considering that the artificial hypothesis for canals had already been proposed, that he did. It is equally indubitable that every scientist has some preconceived idea of what will be seen when he or she makes an observation, for no mind is a blank slate that awaits pure data, a fact clearly understood by Maunder, Newcomb, and others. Every mind is affected by past experience and harbors some expectation of what will be seen. This is why confirmation and consensus are crucial steps in the dynamics of science.

The real question in the problem of Mars – beyond the difficulty of observation – is not about preconceived notions but about the path that scientists take from observation to theory. Lowell's theory was clearly not a deduction from hypothesis; given Martians, even given Martians on an arid planet, it did not follow with deductive rigor that they would build canals for irrigation. It was only one of the possibilities, a "probability" Lowell claimed, given all the observations. Even if the appearance of the canals could have been agreed on by taking the majority opinion of an increasing number of observers, there would still have been the issue of the proper route from observation to theory, of the elusive "unadorned fact" versus its use as evidence for a theory. And important peripheral issues that might have settled the controversy – the existence of water

[86] C. Sagan and P. Fox, "The Canals of Mars: An Assessment after Mariner 9," *Icarus*, 25 (August 1975), 602–612. This analysis superseded an assessment based on earlier photographs taken by Mariners 6 and 7 in 1969, which suggested that a significant number of canals corresponded to surface features: R. B. Leighton, "Mars Pictures from Mariner 6 and 7," *Sky and Telescope*, 38 (October 1969), 212. An interesting comparison of the Martian canal network with networks produced by intelligence on Earth is W. A. Webb, "Analysis of the Martian Canal Network," *PASP*, 67 (1955), 283–292.

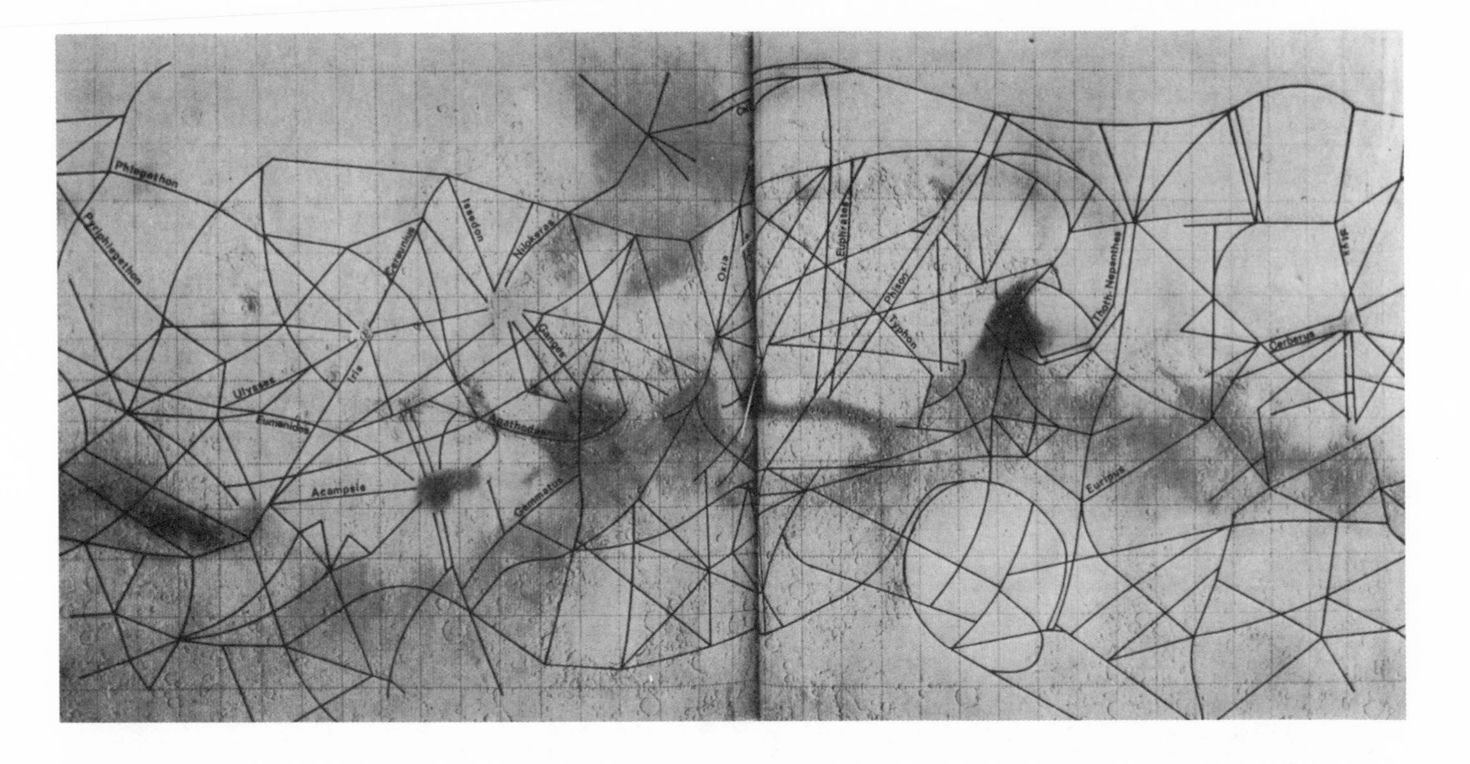

Figure 3.9. Lowellian canal network compared to Mariner cartography, according to the study of Sagan and Fox (1975). They found virtually no matches between Lowell's canals and real surface features. From *Icarus*, 254 (1975), 610, by permission of Academic Press and the authors.

and oxygen in the Martian atmosphere – were filled with their own observational problems.

In assessing Lowell, we need not be as harsh as one critic who claimed that Lowell was an individual "requiring the subordination of perception to imagination, [who] seized on the Martian canals and, as it were, bent them to his own inner needs." Nor need we question whether Lowell was really a professional scientist in the sense of following some ideal scientific method, an elusive method that in reality few scientists have ever followed. Rather, whatever his psychological makeup, what stands out is that Lowell was a scientist with ample imagination. As A. Lawrence Lowell, the president of Harvard, who knew his younger brother well, later wrote, "He had a highly vivid imagination, compared with many men of science who proceeded more cautiously; and hence he sought, not only to ascertain new facts, but to draw conclusions from them more freely than is customary with experts of that type."[87]

Nor was Lowell alone in drawing imaginative conclusions in the canals debate or in clinging to those conclusions. In this respect, the canals of Mars controversy points toward at least two extreme "cultures" of scientists, the one concentrating on gathering the hard facts and unwilling to go beyond them, the other seeing the hard facts as a basis for a larger theory, with all of the imaginative leaps that this implies. In this sense, Table 3.2 may be seen as segregating men of similar scientific styles, the conservative and the liberal, the fact gathering and the imaginative. Some, like Schiaparelli himself, teetered between the two, as his 1893 article shows. Others eschewed the controversy altogether. But of those who engaged in battle, two markedly different types appear, undoubtedly the extreme ends of a spectrum of scientific types. And if this is so, surely Lowell and his allies and opponents were neither the first representatives of two scientific cultures nor the last.

Once the canal controversy was prolonged, the reluctance of Lowell and his champions to give up a long-cherished theory – the work of a lifetime – also became a factor. Nor is it unique among scientists that he

[87] A. Lawrence Lowell, *Biography of Percival Lowell,* Preface. William Sheehan, *Planets and Perception: Telescopic Views and Interpretations, 1609–1909* (Tucson, Ariz., 1988), a book useful for the question of the problematic nature of observation, gives his psychological estimate of Lowell on page 156. The author is a physician specializing in psychiatry. On the issue of Lowell as an amateur versus a professional scientist, see Norris Hetherington, "Amateur versus Professional: The British Astronomical Association and the Controversy over Canals on Mars," *JBAA,* 86 (1976), 303–308, and "Percival Lowell: Professional Scientist or Interloper?" *Journal of the History of Ideas,* 42 (January–March 1981), 159–161, and the reply in William C. Heffernan, "Percival Lowell and the Debate over Extraterrestrial Life," *Journal of the History of Ideas,* 42 (July–September 1981), 527–530.

fought for his theory even in the face of considerable contrary evidence.[88] It is not our task to prescribe in retrospect what Lowell should have done; rather, our interest is in what these events tell us about the nature of the scientific enterprise. We may well speculate that had Lowell never entered the debate – had he remained more enamored of the Far East than of the Far Out, as some have put it – Campbell's work and the controversy over the Martian atmosphere would have run its course much as it did. Undoubtedly, the canal issue would have received attention, though probably not at the popular and at times vitriolic scientific level that it did with Lowell as its champion. As it is, Lowell's role shows the power of the purse and the power of the pen, and reveals much about the problems of perception, inference, and how science functions at its limits. Lowell's persistence forced many scientists, whether cautious or imaginative, to state their opinions on the controversial subject of canals, demonstrating how science functions with uncertain data and high stakes.

As the conclusions of Antoniadi, Campbell, and Trumpler show, abandonment of Lowell's theory did not mean that Mars was devoid of all life. To the contrary, if belief in intelligence on Mars was in steep decline, the possibility of vegetation was still very much alive. Campbell believed in life on Mars and in the solar system; Lowell had simply not proven it to his satisfaction. And while Antoniadi had broken away from Flammarion's view on Mars, he still believed in Martian vegetation and in fact championed the idea of life throughout the universe.

With Schiaparelli dead since 1910, Lowell since 1916, and Flammarion in the last year of his life, science was set for a new generation of Martian investigations during the favorable oppositions of 1924 and 1926. Though Antoniadi's 1930 treatise gave no indication of it, in the search for vegetative life on Mars new astrophysical methods were supplementing the visual topographic tradition, dominant in Martian studies for 50 years, of which Antoniadi represents the apex.

3.2 THE SEARCH FOR MARTIAN VEGETATION, 1924–1957

The extent to which hopes for life on Mars were alive by the late 1920s was well summarized for the public in a "symposium" published in the *New York Times Magazine* for December 9, 1928. Subtitled "Eminent Astrono-

[88] For one example of the extended dying of an astronomical research program in the face of entrenched ideas, see E. Robert Paul, "The Death of a Research Programme: Kapteyn and the Dutch Astronomical Community," *Journal for the History of Astronomy*, 12 (June 1981), 77–94.

mers Give Their Reasons for Belief That Life Exists on the Great Red Planet," the article spoke of the "traditional theory that Mars is dead" and held that scientists skeptical of Martian life a few years before had now reached some consensus that life was part of the riddle of Mars after all.[89] But the contrast to the Lowellian period was sharp: of the scientists polled, only W. H. Pickering believed it to be "almost certain" that intelligence existed on Mars. C. G. Abbot, director of the Smithsonian Astrophysical Observatory, held the opposite extreme view: that physical conditions limited Martian life to very low forms of vegetation. Most of the other astronomers, including Henry Norris Russell at Princeton, Harlow Shapley at Harvard, William Coblentz at the National Bureau of Standards, Walter S. Adams at Mount Wilson, E. B. Frost at Yerkes, and E. C. Slipher at Lowell, affirmed a belief in vegetation and admitted the possibility of low types of animal life. Their opinions can be traced back to the favorable oppositions of 1924 and 1926, during which Mars approached the Earth within 35 million miles. In this section, we examine to what extent results from those years justified the belief that the red planet harbored at the very least vegetation, and how this hope was kept alive even at the dawn of the Space Age. We also analyze the sometimes conflicting roles of visual and astrophysical observations, and how astronomers balanced this evidence in attempting to answer the question of life on Mars.

It is first necessary to emphasize that while Lowell's theory of artificial canals on Mars was largely dead, the question of the natural explanation for these canals was not. The persistent ambiguity about canals is evident in Russell's viewpoint in 1928:

> At the present time it is generally recognized that there exists an objective basis for the canals in the form of fine detail on the surface of Mars, and it is widely believed that these details have, in a general way, the streaky character of the canals. But the existence of a geometrical network is doubted or denied by a large majority of astronomers. It is therefore necessary to render a verdict [of] not proved with regard to this theory."[90]

[89] H. Gordon Garbedian, "Mars Poses Its Riddle of Life," *New York Times Magazine*, December 9, 1928, 1–2, 22. In addition to his continued belief in artificially constructed canals, Pickering pointed to possible radio signals of intelligent origin from Mars. On this aspect of the question, see Section 8.1 in Chapter 8 of this volume.

[90] Ibid., 2. Russell himself seems not to have completely accepted the illusion theory of Maunder and Antoniadi. He stated further in 1928 (in Garbedian, "Mars," 22) that Lowell's canal observations "rest essentially upon visual observations so delicate and difficult that the results obtained by sincere and impartial observers are in disagreement. No method of solving the difficulty can be found with present telescopes and other apparatus, wonderful as they are." He thus denies that Antoniadi has resolved the canals – or the issue.

While the Lowellian theory of a geometric, artificially constructed network was thus rejected (or at least suspended), the canals themselves, viewed as natural features, were still very much a part of the new period of Martian studies. Even if one accepted the theory of Maunder and Antoniadi that continuous canals were an illusion caused by disconnected patches, the nature of the patches and their apparently rectilinear alignment still required some explanation. And if one believed, as some still did, that the canals were continuous natural surface features, one needed to explain what natural physical forces would produce lines some 20 miles wide and a few hundred to several thousand miles long. In the latter regard, the old theory that the canals represented strips of vegetation was not only alive, it would receive surprising support at the favorable opposition of 1924 with the work of Robert Trumpler.

But the remarkable and novel feature of the period beginning with the 1924 opposition is that the story of life on Mars also moved beyond the canals to the physics of the planet, from areography to areophysics, as the astronomer Gérard de Vaucouleurs later put it. "Physical methods were first applied on a large scale to the study of Mars during the perihelic oppositions of 1924 and 1926," he wrote in 1954, and in so doing he perceptively delineated a new era in Martian studies.[91] This move from visual observations to astrophysical techniques also ushered in a new era in the study of Martian biology, an era when the search for intelligence gave way to the search for vegetation, bringing new possibilities for resolution of an age-old question but, alas, also a whole new set of observational problems.

To be sure, the pioneering methods of astrophysics had been applied earlier to Mars; we need only recall Campbell's work on Martian water vapor in 1894 and 1909, among others. But by the early 1920s the field was maturing, instrumentation was being refined, and promising new techniques were being brought to bear on a wide variety of questions related to the physical conditions on Mars: its temperature, its atmospheric pressure and composition, and its changing surface features. Those features included possibly aqueous polar caps, tantalizing indications of vegetation in the form of seasonal changes and the "wave of darkening" as the polar caps melted, and continued interest in the nature of the infamous canals.

Two types of Martian observations were undertaken during this period: those to determine the general physical conditions of Mars and

[91] Gérard DeVaucouleurs, *Physics of the Planet Mars: An Introduction to Areophysics* (London, 1954). This volume, which holds a similar place in the physical tradition of Mars similar to that of Flammarion's *La planète Mars* in the visual tradition, carried on the tradition of French fascination with Mars.

those designed specifically to determine the actual existence of life. It is not our purpose in this section to examine the entire field of areophysics, fascinating as such a task would be. Rather, while describing briefly the new methods employed, we must ask to what extent the observers studying the physical conditions on Mars seem to have been motivated by the question of life, what inferences about indigenous Martian life they drew from their own data, and how in general their conclusions about physical conditions affected the Martian life debate. Second, we must examine in detail some of the more important observations of the second type: those designed specifically to search for life on Mars. As we shall see, the path from observation to conclusion in both types was problematic, and fraught with assumption and interpretation. For this reason, we will pay particular attention to technique, inference, and the interplay between visual and astrophysical observations.

We recall that as the favorable oppositions of 1924 and 1926 approached, Mars was believed to have an average temperature well below the freezing point of water, a thin atmosphere of unknown composition, and an enigmatic surface subject to change. As in the past, those enigmatic surface features were scrutinized again. Especially notable were the conclusions from Robert Trumpler's detailed study of Mars undertaken at Lick Observatory, which indicated to him that, while the canals were the result of natural topography, vegetation caused the dark areas and made the canals visible. The observations of E. C. Slipher at Lowell and of W. H. Wright at Lick also strengthened the hypothesis of Martian vegetation. At least part of Lowell's legacy was thus still alive at the close approach of Mars in 1924. The visual and photographic case for vegetation was summarized by Slipher, who argued that the observations

> strongly evidence that the dark markings of the planet are all due to the same cause and obey the same law of change. The seasonal date that these dark markings begin to darken or make their appearance; the rate and manner of their development, the seasonal date at which they mount to the highest intensity – which is the summer solstice and thereafter – their color and appearance, and in turn the time of their fading out again – all obey the law of change that we should expect of vegetation.[92]

[92] E. C. Slipher, "Atmospheric and Surface Phenomena on Mars," *PASP*, 39 (August 1927), 209–216. While Slipher used standard methods, Wright at the Lick Observatory used color filters, beginning in 1924, to examine the planet in the spectrum extending from ultraviolet to infrared. See W. H. Wright, "Photographs of Mars Made with Light of Different Colors," *PASP*, 36 (October 1924), 239–254. See also C. D. Shane, "W.H. Wright," *BMNAS*, 50 (Washington, DC, 1979), 377–396. On Trumpler, see the previous section in this chapter.

But as we have said, what was new with the 1924 opposition of Mars was the possibility, made real by technical advances in physical methods, of more accurately determining Martian conditions. The new experiments, the highlights of which are given in Table 3.3, now sought to confirm, quantify, or extend the visual and photographic results. Central to this new era, and important in understanding the transfer of the new physical methods from the laboratory to astrophysics, was the work of two independent teams attempting to measure the temperature of Mars. Both William W. Coblentz and C. O. Lampland at Lowell Observatory, and Edison Pettit and Seth Nicholson at Mount Wilson, used the newly invented vacuum thermocouple for their experimental determinations. Both teams came to similar conclusions about the temperatures on Mars, but only one of the four astronomers extensively discussed the implications of his results for the question of life on Mars. In order to illuminate the difference in scientific style among these astronomers, as well as the new physical methods they employed, it is useful to examine their work in some detail.

A 1903 graduate of Cornell, where his interests were first turned toward spectroscopy by the head of the Physics Department, E. L. Nichols, Coblentz became a pioneer in the fields of infrared spectroscopy and radiometry. His early work was on the infrared spectroscopy of organic molecules, first at Cornell and then at the National Bureau of Standards, where he served as head of the radiometry section from 1905 to 1945. In his autobiography Coblentz documents his interest in astronomy from an early age, but it was only in 1914 that his scientific work turned toward astrophysics.[93] Stimulated by the work of Johns Hopkins Professor August Pfund a year earlier in measuring a star's heat using a thermocouple on the 26-inch refractor of the U.S. Naval Observatory, Coblentz in that year used the 36-inch Crossley Reflector of the Lick Observatory to measure the radiation from 110 stars.[94]

In the summer of 1921, 7 years after the work at Lick, Coblentz's interest turned toward planetary temperatures, an interest he pursued in collaboration with Lampland using the 40-inch reflector at Lowell Obser-

[93] W. W. Coblentz, *From the Life of a Researcher* (New York, 1951). On Coblentz see also *BMNAS,* 39 (1967), 54–101 (with a full bibliography); D. J. Lovell, *DSB,* 3 (New York 1971), 327–328; and the commemorative issue of *Applied Optics,* 2, no. 11 (November 1963), especially John Strong, "On the Astrophysical Work of W. W. Coblentz," 1101–1102.

[94] Reported in Coblentz, "A Comparison of Stellar Radiometers and Radiometric Measurements on 110 Stars," *Bulletin of the Bureau of Standards,* 11 (1915); 613–56, and "Radiometer Measurements of 110 Stars with the Crossley Reflector," *Lick Observatory Bulletin,* 8 (1915), 104–123. On early testing of his thermocouples in the laboratory see Coblentz, *PA,* 31 (February 1923), 105–121, and the accompaying bibliography.

Table 3.3. *Milestones in Martian observations related to life, 1924-1957*

Observer	Date	Location	Instrument	Method	Result
Temperature					
Coblentz/ Lampland	1924	Lowell	40-in. reflector	Thermocouple	-28C for whole disk -10C to 5C for bright 10C to 20 C for dark
Pettit/ Nicholson	1924	Mount Wilson	100-in. reflector	Thermocouple	-13C disk, by first method -33C disk, by second method
Atmospheric Pressure and Composition (Earth = 1000 mb)					
Adams and St. John	1925	Mount Wilson	60-in. reflector	Prism spectrograph	H_2O 6% of Mount Wilson O_2 16% of Mount Wilson
Menzel	1926			Visual and photographic albedo	Pressure <5 cm Hg
Lyot	1929	Meudon	33 in.	Visual polarimetry	Pressure <1.8 cm Hg
Adams and Dunham	1933	Mount Wilson	100-in. reflector	Grating spectrograph	No oxygen No water vapor
Barabashoff	1934	Kharkov	8 in.	Photographic photometry	50 mb
DeVaucoleurs	1945	Perdier		Visual photometry	93 mb
Kuiper	1947	McDonald	82-in. reflector	IR spectroscopy	CO_2
Hess	1948	Theoretical		Meteorology	80 mb
Dollfus	1948-1951	Pic du Midi	24 in.	Visual polarimetry	83 mb
Surface Features/DirectIndications of Life					
Slipher	1924	Lowell	24 in.	Spectroscopy	No chlorophyll
Trumpler	1924	Lick	36 in.	Visual/photographic	Canals, but not artificial
Pettit	1939			Visual	Canals
E. Slipher	1927-62		24 in.	Photographic	Canals, vegetation
Dollfus	1946-1948		24 in.	Polarimetry	Clouds
Tikhov	1947-1949			Laboratory studies of reflection spectra	Vegetation
Sinton	1957	Harvard	61-in. reflector	IR spectroscopy	Organic molecules vegetation

Instruments are refractors unless otherwise specified

vatory. Though tests were made in 1921 and some measurements in 1922,[95] the results of the 1924 opposition were most definitive.[96] The results of these experiments contain much of interest, not the least of which is the pathway from the raw data of radiation measurement (the radiometric data) to an actual temperature. Coblentz and Lampland analyzed their data using at least three methods, which in general showed good agreement, despite the many assumptions they embodied. At Henry Norris Russell's suggestion, they were aided in using one of the most important methods by Russell's student, Donald Menzel, just launched on the road that would lead to the post of director of the Harvard Observatory.[97] The resulting temperatures differed depending on what part of the planet was observed, but in 1925 Coblentz summarized his results as yielding temperatures of −10° to 5°C for the bright regions and 10° to 20°C "or even higher in the dark regions," a result that Coblentz noted was at variance with the prevailing opinion that temperatures on Mars could not rise above freezing.[98]

In a popular article, "Climatic Conditions on Mars," published in 1925 (in contrast to his more technical publications), Coblentz did not hesitate to draw inferences about life on Mars from his results, in particular from the surface temperatures of the dark areas: "the observed high surface temperatures on the dark areas of Mars may be explained on the basis of the presence of living vegetation superposed upon a dry vegetable mold which is a non-conductor of heat." With practically no water on Mars, Coblentz reasoned, "the decay and disintegration of vegetable

95 Coblentz, "Further Tests of Stellar Radiometers and Some Measurements of Planetary Radiation," *Scientific Papers of the National Bureau Standards,* 18 (1922), 535–558; "Recent Measurements of Stellar and Planetary Radiation," *JOSA,* 6 (1922), 1016–1029; "Further Measurements of Stellar Temperatures and Planetary Radiation, *NAS,* 8 (1922), 330–333; "Thermocouple Measurements of Stellar and Planetary Radiation," *PA,* 31 (February 1923), 105–128; Coblentz and Lampland, "Measurements of Planetary Radiation," *LOB,* no. 85 (1923), 91–134. The Lowell Observatory archives contain considerable correspondence between Lampland and Coblentz.

96 W. W. Coblentz, "Climatic Conditions on Mars," *PA,* 33 (May 1925), 310–316, 363–382: 367. Also Coblentz and Lampland, "New Measurements of Planetary Radiation," *Science,* 60 (September 26, 1924), 295. Also reported in "Radiometric Measurements on Mars," *PA,* 32 (November 1924), 570–572 and *PASP,* 36 (October 1924), 272–274; "Measurement of Spectral Components of Planetary Radiation," *PASP,* 36 (October 1924), 220–221, and "The Temperature of Mars," *Science,* 60 (November 7, 1924), 429. "Temperature Estimates of the Planet Mars," *Scientific Papers of the Bureau of Standards,* 20 (1925), 371–397, contains the fullest report. Coblentz, "Climatic Conditions," cites Pickering as having recently arrived at similar results based on visual phenomena, *PA,* 33 (May 1925), 365.

97 Menzel first reduced the 1914 and 1922 observations, and then the 1924 and 1926 observations, using the "water cell transmission theory," proposed by Russell and developed by Menzel in "Water Cell Transmissions and Planetary Temperatures," *APJ,* 58 (September 1923), 65–74. See Doel, *Solar System Astronomy,* chapter 1.

98 Coblentz, "Climatic Conditions on Mars," 366–367.

matter would be slow and there would be a slow accumulation of the dry matter of the preceding season which would protect the living plants from the extreme cold of winter."[99]

With conditions similar to those of north temperate and frigid zones of the Earth, except for atmosphere, Coblentz compared Mars with Siberian conditions, where moss and lichen tundra existed, as well as color changes with season. In addition to temperature, Coblentz noted, visual observations of coloring and darkening in the southern hemisphere of Mars by Lowell, Pickering, Slipher, and others were "strikingly similar" to the observations of areas like Siberia near the terrestrial polar regions. Because certain types of moss have the capacity to raise the temperature above a barren surface as a result of their high absorption of solar radiation, the presence of lichen and moss would explain the high temperatures. But, Coblentz cautioned, entirely different types of plants might have evolved on Mars due to environmental conditions. While vegetable and perhaps animal life is possible on Mars, any animal life that cannot migrate "must be trogdolytic [*sic*], able to burrow deep and hibernate, or able to withstand the long sieges of intense cold in a benumbed state, as do, for example, the torpid insects which one finds on warm days in winter."[100]

Clearly, Coblentz was very interested in the question of life on Mars. He noted that previous opinions on Martian temperatures were segregated into two groups, one concluding from mathematical calculation that the temperature cannot rise above 0°C, the other concluding from visual observations that it must. Among the latter were William Herschel, Schiaparelli, Lowell, and Pickering, whom Coblentz admired "for their courage in maintaining their views in the face of mathematical calculations which seemed to demonstrate the opposite view." Coblentz did not fail to see the surprisingly close parallelism of the conclusions of this group, based on visual observations, with his own, based on measurements by quantitative physical methods. Lest he be accused of confirming Lowell's ideas, Coblentz commented regarding his own work that "Perhaps it is just as well that these measurements were undertaken and completed without preconceived notions and without a thorough knowledge of what had already been observed."[101]

The combination of physical radiometric measurements with visual observations shows the continuing importance of visual observations of Mars in reaching conclusions about Martian vegetation. Physical methods alone sufficed to prove the *possibility* of vegetation, but visual observations went a long way toward confirming that this possibility might

[99] Ibid., 376.
[100] Ibid., 381.
[101] Ibid. See also Coblentz's 1925 papers on life on other planets, cited in *BMNAS*.

actually be realized. Conversely, physical methods now confirmed that conditions gave credence to the vegetation hypothesis. No such conclusion was reached about Venus, which Coblentz and Lampland, with similar experiments, had determined to have a temperature of 50°C. Unlike Mars, no wave of darkening gave credence to Venusian plant life.

Coblentz's observations formed a turning point in Martian studies. For the first time, temperatures were determined by experiment to rise well above freezing on Mars. What is more, for the first time, the dark areas were found to have a higher temperature than the light areas. These observed facts led Coblentz to an "assumption of the existence of plant life, in the form of tussocks, whether grass or moss," a form of vegetation having properties that could account for the higher temperature of the dark areas. But "beyond this assumption the writer withholds speculation," Coblentz declared. He had gone far enough in his deductions, further than either of his collaborators, Lampland and Menzel, went in their published writings, and he would go no further until more was known about the composition and extent of the Martian atmosphere.

In the meantime, in 1921 Edison Pettit had also begun work with the vacuum thermocouple in collaboration with Seth Nicholson. Nicholson, who later became a member of the National Academy of Sciences, edited the *Astronomical Journal,* discovered four moons of Jupiter, and did pioneering solar work, had assisted Coblentz in his early experiments at Lick in 1914 and so formed a link between the work of the two teams.[102] Pettit, who had received his degree in astronomy from the University of Chicago in 1920, was interested in the Sun, particularly solar prominences and the temperature of sunspots. Later described by a Mount Wilson colleague as having "a remarkable facility for working with his hands," he built his own thermocouples so successfully that he could also measure the much feebler radiation from stars and planets.[103] The results of Pettit and Nicholson began to appear in 1922, first on stars, by 1923 on the planet Mercury, and by 1924 on Mars.[104]

Unlike Coblentz, Pettit and Nicholson came to their work from astronomy rather than physics. They used a similar technique, but as astronomers were able to command the 100-inch reflector at Mount Wilson. Their results were similar to those of the Coblentz team, with Martian temperatures found "a little above freezing" for a point in the tropics at

[102] On Nicholson see Paul Herget's obituary in *BMNAS,* 42 (1971), 201–227, including a complete bibliography.

[103] On Pettit see the obituaries by his colleagues S. B. Nicholson, *PASP,* 74 (December 1962), 495–498, and Ralph S. Richardson, *Griffith Observer* (July 1962), 90–97.

[104] Petit and Nicholson, "Radiation Measures on the Planet Mars," *PASP,* 36 (October 1924), 269–272. For photographs of the device, see *PA,* 32 (December 1924), 605.

noon.[105] A few months before Coblentz, they too wrote an article for *Popular Astronomy,* but unlike Coblentz they refused to draw any conclusions regarding the possible existence of vegetation on Mars. We might take this as a model of scientific restraint were it not for the fact that Pettit later wrote papers claiming to have photographed the canals of Mars!

Pathbreaking as these observations were, the possibility of life on Mars depended on much more than the surface temperature of the planet; as Coblentz had noted, atmospheric pressure and composition would enable the possibilities of plant life to be further pinned down. Here a measure of the amount of planetary radiation, whether from a part of the planet or from its integrated whole, was not enough. This was a problem for spectroscopy, and 14 years after Campbell's expedition to Mount Whitney, the 60-inch reflector at Mount Wilson was pressed into service for just this purpose, using the same Doppler technique with even higher dispersion. Using a six-prism spectrograph on February 2, 1925, Walter S. Adams and Charles E. St. John found that the water vapor in the atmosphere of Mars was 6 percent of that over the observing site at Mount Wilson (3 percent of that over Pasadena at sea level), indicating "extreme desert conditions over the greater portion of the Martian hemisphere toward us at the time." Similar measurements gave oxygen on Mars at 16 percent that above Mount Wilson or two-thirds that above Mount Everest, the beginning of a series of downward revisions in the values of both Martian water vapor and oxygen. Like Petit and Nicholson, Adams and St. John drew no conclusions about Martian life from their work in their technical publications. But Adams was one of those in the *New York Times* symposium of 1928 who still lent support to the possibility of vegetation and even low animal life.[106]

By the end of the 1920s, then, both visual observation and physical experiments were at one in strengthening the vegetation hypothesis. Visual methods still provided the best evidence, now supported by higher temperature measurements and less so by indications of very small amounts of oxygen and water vapor. Some of the participants, notably Trumpler, Coblentz, and Russell, were quick to discuss the implications

[105] Ibid., 271.

[106] "An attempt to Detect Water-Vapor and Oxygen Lines in the Spectrum of Mars with the Registering Microphotometer," *ApJ,* 43 (January–June 1926), 133–137. Donald Menzel also published calculations, based on albedo, that gave a value of about 6 cm of mercury, or 80 mb for the maximum atmospheric pressure at the surface. "The Atmosphere of Mars," *ApJ,* 43 (January–June, 1926), 48–59. The French astronomer Lyot, using visual polarimetry, found a value of 1.8 cm of mercury. These attempts, the highlights of which are given in Table 3.3, are reviewed in DeVaucouleurs, *Physics of the Planet Mars,* 99–127.

of these results for life; others, like Petit and Nicholson, Lampland and Menzel, were not. Still others rendered an opinion only if asked, and for this information newspaper accounts provide a unique source. In whatever form it was expressed, whether research report or popular press article, we can now better understand the fragile nature of the evidence that prompted many prominent astronomers to favor not only vegetation, but also low forms of animal life on Mars.

This fragile optimism of the 1920s was, however, soon jeopardized, most notably in 1933 by the renewed search of Mount Wilson Director Walter S. Adams, now in conjunction with Theodore Dunham, for oxygen on Mars. While Adams and St. John had used a microphotometer to study the alpha band of oxygen in 1926, advances in photographic emulsions now enabled Adams and Dunham to photograph the much stronger B band of oxygen. Instead of a six-prism spectrograph, they now used a grating spectrograph, and instead of the Mount Wilson 60-inch telescope, they used the 100-inch one. Undoubtedly to their dismay, in contrast to the small but detectable amounts of the 1926 observations, this technique detected no oxygen and yielded an upper limit for Martian oxygen abundance of less than 1/10th of 1 percent that of Earth.[107]

The reaction to this result is revealing in light of contrary evidence from visual observation. Mount Wilson astronomer Robert S. Richardson noted that after the observations of Adams and Dunham, interest in Mars and the planets went into decline, even if (as Doel has shown) the larger field of solar system astronomy remained active at a modest level. The reason for any decline of interest in the planets, however, cannot be laid to the negative results of a single set of observations. The oppositions of 1937 and 1939, neither particularly favorable, "passed almost unnoticed," Richardson recalled. The most notable Martian event of the immediate pre–World War II years was not any novel observation, but rather the public reaction of panic to the October 1938 Orson Welles radio broadcast of "War of the Worlds," a panic that shows how far popular culture lagged behind scientific theories of Mars. And as Figure 3.1 indicates, research publications reached an all-time low during World War II as many astronomers were diverted to other tasks. By 1944 Richardson was calling attention to the deplorably low state into which planetary observing had fallen; his call for a cooperative effort to observe Mars had to be published in a science fiction magazine.[108]

[107] Walter S. Adams and Theodore Dunham, Jr., "The B Band of Oxygen in the Spectrum of Mars," *ApJ*, 79 (January–June 1934), 308–316.

[108] Robert S. Richardson, *Exploring Mars* (New York, 1954), 19; Richardson, "A Postwar Plan for Mars," *Astounding Science Fiction* (January 1944). On the 1938 Orson Welles broadcast see Hadley Cantril, *The Invasion from Mars: A Study in the Psychology of*

Yet, despite the lack of new research (or perhaps because of it) and despite the negative 1933 result, the Martian vegetation hypothesis survived the 1930s and the war.[109] Though the result of the 1933 Adams and Dunham observations was discouraging for the hypothesis of Martian life, it did little to change the belief in Martian vegetation during this period. The reason, demonstrated by one of the experiment's own participants, bears on the relative weight astronomers gave to the visual/photographic versus the new physical methods. Reviewing knowledge of the planets in 1938, Dunham himself wrote that, despite the lack of spectroscopic evidence for water or oxygen on Mars, the visual observation of a thin sheet of ice covering a small surface area was a more sensitive test than the spectroscopic attempts to find water spread over the entire atmosphere. And, as Russell had suggested, oxygen may have existed in the past and been locked in rocks to form ferric compounds, giving Mars its rusty color. Dunham's conclusion: "It would therefore be unwise to say that there may not be enough of both water vapor and oxygen to support life in some form which may have become gradually adapted to the rigorous conditions existing on Mars."[110] This gradual adaptation by evolution was precisely the opinion of Russell himself in the aftermath of the observations of Adams and Dunham.[111]

That the hypothesis of Martian vegetation survived, damaged but intact, between the late 1920s and 1945 is evident in the 1926 and 1945 editions of the standard textbook of that time, Russell, Dugan, and Stewart's *Astronomy*. Whereas Russell had written in 1926 that recent observations make it "very probable" that conditions on Mars support vegetable life, in 1945 he replaced those words with "not impossible," concluding in both editions that "it seems more likely than not that it does." But spectroscopic

Panic (Princeton, 1940), reprinted 1982. Declining interest in the planets does not imply a lack of ongoing work in broader solar system astronomy. Doel, *Solar System Astronomy*, argues that the study of the Moon, comets, meteors, celestial mechanics, the origin and evolution of the solar system, and even some work on planetary atmospheres, "while not a core program of the discipline [of astronomy], was nevertheless far from abandoned."

[109] It should be noted that Fournier (1939), Pettit (1939, 1947), and E. C. Slipher (1921, 1931, 1940) continued to detect canals, though no longer claiming they were artificial.

[110] "Knowledge of the Planets in 1938," *PASP*, 51 (October 1939), 253–273: 270. After three decades of development of remarkable techniques, a value of 80 mb at the surface was generally accepted at the end of the 1950s.

[111] Henry Norris Russell, "Fading Belief in Life on Other Planets," *SciAm* (June 1934), 296–297: 297. During this period, visual and photographic observations continued to play a role in detecting surface and atmospheric features. The vegetation hypothesis was supported by seasonal variations observed on the surface (E. C. Slipher, "Martian Phenomena Observed from Our Southern Hemisphere," *The Telescope*, 7, [September–October 1940], 102–109), and by colors indicating green plants on Mars. See DeVaucouleurs, *Physics of the Planet Mars*, p. 18, regarding further controversies around 1940.

observations had taken a greater toll on the belief in animal life. Whereas in 1926 animal life had been viewed as "not impossible, or indeed, even improbable," by 1945 Russell wrote that "the great scarcity of oxygen makes the existence of animal life appear improbable." That the situation was in flux in both years Russell indicated by stating in both editions, "Recent evidence has greatly changed our estimate of the situation, and further evidence may change it again."[112]

The wisdom of Russell's statement is seen in events following World War II, due in part to advances in technology made during the war for wartime purposes. Undoubtedly the premier event of the decade in Martian studies was the discovery in 1947 of carbon dioxide in the atmosphere of Mars, a discovery of biological significance because it is a principal gas in the process of photosynthesis. The discoverer was Gerard P. Kuiper (1905–1973), one of several American astronomers working in the field of planetary atmospheres in the immediate postwar era.[113] A graduate of the University of Leiden in 1933 under Ejnar Hertzsprung, Kuiper for the first part of his career worked on the problems of stellar astronomy, especially double and multiple star systems. As we shall see in the next chapter, this led him to a deep interest in the possibility of planetary systems. It was during World War II that Kuiper became interested in solar system studies and discovered the presence of methane on Titan. After the war he learned of the new lead sulfide infrared detectors developed on both sides, and after September 1946 he collaborated on such a device useful for astronomical work in the 1- to 3- micron region with R. J. Cashman and W. Wilson.[114] It was this work that led in 1947 to the first actual detection of a component of the Martian atmosphere – carbon dioxide. This remarkable feat, carried out during one of the less

[112] Henry Norris Russell, Raymond S. Dugan, and John Q. Stewart, *Astronomy: A Revision of Young's Manual of Astronomy,* vol. 1, *The Solar System* (Boston, 1926), 344 and (Boston, 1945), 344. See also Russell's "Physical and Chemical Properties of the System" in *The Solar System and Its Origin* (New York, 1935), 44–92.

[113] On Kuiper see Dale P. Cruikshank, "Gerard Peter Kuiper," *BMNAS,* 62 (Washington, D.C., 1993), 259–295, an earlier version in *Sky and Telescope,* 47 (March 1974), 159–164, and Tobias Owen and Carl Sagan, "Planetary Astronomer, Gerard Peter Kuiper (1905–1973)," *Mercury,* 3, nos. 2–3 (March–June 1974), 16–18, 37:16. Although Owen and Sagan call Kuiper "the only full-time professional planetary astronomer engaged in North America, and one of the few in the world" from the mid-1940s to the mid-1960s, Doel (*Solar System Astronomy,* chapter 2) has shown that work on planetary atmospheres was ongoing at Lowell Observatory and that upper atmosphere research at Harvard was an important part of solar system astronomy broadly conceived.

[114] G. P. Kuiper, W. Wilson, and R. J. Cashman, "An Infrared Stellar Spectrometer," *ApJ,* 106 (September 1947), 243. Doel (*Solar System Astronomy,* chapter 2) gives the full background and postwar milieu of Kuiper's work, as well as Doel, "Evaluating Soviet Lunar Science in Cold-War America," *Science After '40, Osiris,* 7 (1992), 238–264.

favorable oppositions, is a testimony to the ingenuity of Kuiper and the power of the new physical techniques.

This and related research were summarized during a Symposium on Planetary Atmospheres held in September 1947 on the occasion of the 50th anniversary of Yerkes Observatory. Much of this symposium dealt with research on the best-known planetary atmosphere – Earth's. But Arthur Adel discussed selected aspects of infrared spectroscopy, Dunham summarized the results of planetary spectroscopy at Mount Wilson, and Kuiper himself surveyed research in planetary atmospheres. In his section on planetary spectra, Kuiper summarized the state of research and described how he was able to detect the CO_2 bands using the new PbS cell as the receiver on the infrared spectrometer of McDonald Observatory. As Figure 3.10 shows, to the untrained eye the evidence is not overwhelming; but to Kuiper it was, and his discovery has withstood the test of history.[115]

In his study Kuiper could not resist addressing broader aspects of the Martian problem, of which the atmosphere was only one element. The bright desert regions of Mars, he believed, were composed of igneous rock and the polar caps of water frost. As for the nature of the green areas, "often held to be vegetation because of the observed seasonal changes," Kuiper's strategy from the start was to compare the Martian spectra with laboratory spectra of various plants on Earth – a technique pioneered by in the Soviet Union by Tikhov.[116] Applying this method, he found no evidence of chlorophyll characteristic of higher plants on Earth and no evidence of plants containing water. But the spectrum was consistent with the reflection spectrum of lichens. As Kuiper realized, this was by no means proof, and he summarized by saying, "It is too early to conclude that the green areas on Mars are covered with lichens or mosses; different forms of plant life may well have developed under the very different conditions existing on Mars."[117] He advocated further study of seasonal color variations on Mars, as well as experiments on lichens under simulated Martian conditions – that is, low oxygen, prolonged drought, temperature extremes, and ultraviolet (UV) light exposure.

By the second edition of 1952, Kuiper had obviously given considerably more thought to integrating all facets of research on the subject. With his discovery of CO_2 in the atmosphere and the identification of the polar caps as frozen water, he wrote, "the minimum conditions for photo-

[115] Gerard P. Kuiper, ed., *The Atmospheres of the Earth and Planets: Papers Presented at the Fiftieth Anniversary Symposium of the Yerkes Observatory, September, 1947* (Chicago, 1949), 321–342, especially 334 and 336. The diagram is on page 328.

[116] Ibid., 337–338.

[117] Ibid., 339. Kuiper does not mention Tikhov.

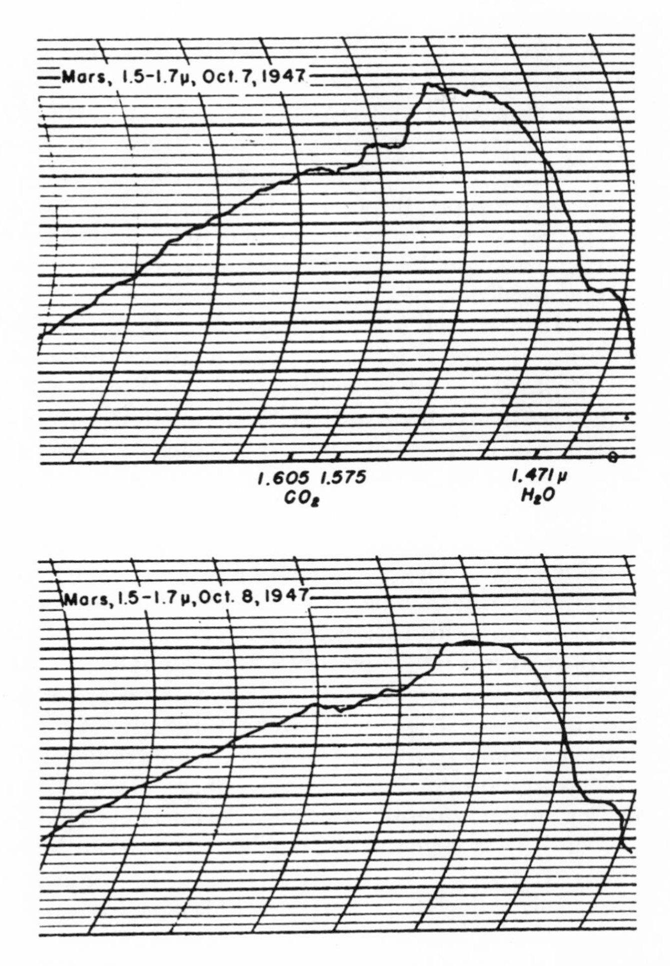

Figure 3.10. G. P. Kuiper's evidence of CO_2 bands on Mars in the 1.6-micron region of the infrared, an observation confirmed by later evidence. CO_2 is indicated by the two small dips on each of the plots, which were taken on two successive days. The observations were made with the infrared spectrometer on the 82-inch telescope of McDonald Observatory in Texas. From Kuiper's *The Atmospheres of the Earth and Planets* (Chicago, 1952), 360, by permission of The University of Chicago Press.

synthesis appear to be fulfilled." Though the spectrum did not indicate the presence of chlorophyll, one would not expect it to do so for lichens. Kuiper proceeded to build a case for the possibility of lichenlike plants, which would not show chlorophyll, on Mars. Given the small amount of atmospheric water vapor, he argued, one would expect only a very thin layer of vegetation, consistent with lichens in dry climates on Earth. The same dearth of water vapor disposed of Arrhenius's nonbiological hypothesis as due to salts with varying amounts of moisture. The idea of Öpik that dust storms would have covered the green areas if they had no regenerative powers also helped.

The continued important role of visual observations is evident in Kuiper's reference to Antoniadi as the best authority on visual color variations, which Kuiper used to compare with the known properties of terrestrial lichens.[118] After discussing the literature on lichens in relation to Martian conditions, he concluded with a much stronger statement than he had 3 years earlier: "The evidence that these [Martian green] areas are covered by a living substance appears very good" for four reasons: the colors were not those expected from inorganic substances under the observed conditions; the colors showed regional patterns like those of vegetation on Earth; the colors changed with the seasons and resembled those found in lichens; and Öpik's argument for regenerative powers from dust storms.[119] Kuiper relegated the physical observations to the status of "circumstantial evidence," including the presence of CO_2 and H_2O and possibly N_2; a temperature range large but not extreme; and a UV exposure that "may be serious but not definitely ruinous." Furthermore, a milder and more humid climate may have prevailed earlier, he argued, enabling slow adaptation; volcanic vents may have played a role as a heat source; and Mars "has had as much time for the evolution of life as the earth has."

Still, Kuiper realized the tentative nature of his conclusions:

> On the basis of the accumulated evidence the hypothesis of plant life might be accepted unreservedly if it were not for the fact that it implied two parallel developments, one on Mars and one on the earth, of a complexity that still defies description and understanding. For this reason, final judgment should probably still be withheld. Particularly, the comparison with lichens must be regarded to have only heuristic

[118] Ibid., 2nd edition (Chicago, 1952), 362, 401–405. V. M. Slipher found no evidence for chlorophyll in the reflection spectra of the dark areas, *PASP*, 36 (October 1924), 261–262.

[119] Ibid., 2d edition (1952), 404.

> value; it would be most surprising if similar species had developed on Mars as on the earth.[120]

This caution notwithstanding, by the early 1950s the hypothesis of Martian vegetation had once again gained considerable ground.

That these conclusions were not merely wishful thinking – or at least not an unusual method of scientific inference – is evident in the similar conclusions reached outside the Western world, based largely on independent research. Thus, taking the standard visual approach, one of Kuiper's counterparts in the Soviet Union, N. P. Barabashev, concluded after 30 years of observations of Mars that its color changes were due to vegetation.[121] At the same time, taking a more original approach, the lifetime work of Soviet astronomer Gavriil Adrianovich Tikhov also culminated with similar conclusions. Tikhov (1875–1960), a Russian Lowell with a passion for Martian vegetation rather than Martian canals, employed color filters (as would Wright and others later in the United States) to study Martian features. In 1909 he made 1000 photographs using filters on the 30-inch Pulkovo refractor, finding that the south polar cap resembled terrestrial ice in color and the Martian canals resembled the Martian seas, with obvious implications.[122] He also investigated whether Martian vegetation might contain chlorophyll but (like Kuiper) found none.[123]

By the mid-1940s Tikhov's studies led to a more organized effort. "By 1945 it had become clear to the author . . . that in order to understand the difference between the optical properties of terrestrial and Martian

120 Ibid., 404.

121 N. P. Barabashev, *A Study of the Physical Conditions of the Moon and Planets* (Kharkov, 1952). Soviet astronomers undoubtedly had some knowledge of ideas in the Western world; for example, Harold Spencer Jones's volume, *Life on Other Worlds* (1940), was translated into Russian in 1946. Barabashev was not alone in solar system astronomy in the Soviet Union at this time, but many astronomers working in the United States at the beginning of the Space Age found the enormous mass of work churned out in the Soviet Union and elsewhere essentially worthless due to inadequate techniques, instrumentation, and theory. Ron Schorn to S. Dick, private communication, October 1994. For the context in the Soviet Union see Loren R. Graham, *Science in Russia and the Soviet Union* (Cambridge, 1993), especially Appendix A, and Graham, *Science and Philosophy in the Soviet Union* (New York, 1972), chapter 5.

122 G. A. Tikhov, "Is Life Possible on Other Planets?," *JBAA*, 65 (1955), 193–204. Tikhov had graduated from the mathematical section of Moscow University in 1897, 3 years after Lowell began his work on Mars in the United States. From 1898 to 1900 he studied at the Sorbonne and worked with Jules Janssen at his Meudon astrophysical observatory. He became an adjunct member of the Pulkovo Observatory staff in 1906 and received the degree of master of astronomy and geodesy in 1913, 4 years before the Bolshevik Revolution. On Tikhov see P. G. Kulikovsky, "Tikhov," *DSB*, 13 (New York, 1976), 408–409.

123 Reported in *Bulletin de l' Academie Imperiale des Sciences de St. Petersbourgh*, 4 (1910), 881–890.

vegetation, it was necessary to study these properties in terrestrial planets which live under the conditions of a rigorous alpine and sub-arctic climate," Tikhov wrote. "Thus the problem itself indicated the necessity of combining astronomy and botany. And so the new science of astrobotany was born."[124] In 1947 an Astrobotanical Section of the Academy of Sciences of the Kazakh Republic opened. Its purpose was to study the optical properties of terrestrial vegetation in harsh climates and application of the results to Mars. Original as it may have been, Tikhov's work, like Lowell's, provoked great criticism in his own country as well as abroad.[125]

As the favorable oppositions of 1954 and 1956 approached, interest in Mars, driven by interest in the Martian vegetation hypothesis, was increasing. In 1953 an informal International Mars Committee was formed, and seven observatories around the world, including Lowell, began a "Mars Patrol" for 1954. Veteran Mars observer E. C. Slipher, sponsored by the National Geographic Society, obtained 20,000 photographs with the 27-inch refractor at the Lamont-Hussey Observatory in South Africa. Pettit and Richardson at Mount Wilson claimed to have photographed canals and identified changes in them. Harvard astronomer William M. Sinton and Johns Hopkins physicist John Strong made the first extensive measurements of planetary temperatures in 30 years. Now using not only the 200-inch reflector at Palomar, but also a "Golay detector" (developed in World War II) more sensitive than the thermocouples used earlier, they found a maximum temperature of 25°C.[126] And the primary planetary astronomer in the United States, Gerard P. Kuiper, continued to maintain that "the hypothesis of plant life . . . appears still the most satisfactory explanation of the various shades of dark markings and their complex seasonal and secular changes," though he emphasized that vegetation on Mars might be quite different from terrestrial lichens.[127]

These conclusions were a mere prelude to the results of 1956. As the 1956 opposition approached, Sinton planned a direct search for vegetation by spectroscopic methods. Keenly aware that previous tests for infra-

[124] Tikhov, "Is Life Possible on Other Planets?" 195. That Tikhov's work aroused some interest in the United States is evident in the translation of some of his writing into English. See G. A. Tikhov, *Principal Works: Astrobotany and Astrophysics, 1912–1957* (Washington, D.C., 1960).

[125] See the Troitskaya summary in *Astronomicheski zhurnal* (Astronomical Journal), 29 (1952), and the violent objection in the same journal, 738–741. For a description of the controversy see Tikhov, *Principal Works*, 60–76.

[126] The activities of the International Mars Committee are described in Sinton, "New Findings about Mars," *S&T*, (July 1955), 360–363. See also *S&T* (June 1954), 251.

[127] Kuiper, "On the Martian Surface Features," *PASP*, 67 (October 1955), 271–282: 281.

red reflectivity characteristic of plants – including the work of Slipher (1924), Millman (1939), Tikhov (1947), and Kuiper (1949) – had been negative, Sinton's own search had a new element: it depended on the fact that organic molecules have absorption bands at about 3.4 microns in the infrared part of the spectrum – beyond Kuiper's work that had been done in the 1- to 2.5-micron region. Sinton still used a lead sulfide photoconductive cell, as had Kuiper, but now cooled to 96 Kelvin (K) with liquid nitrogen to increase its sensitivity to 3.6 microns. The difficulties of the observations can be appreciated from the fact that the sensitive area of this cell was only .16 mm square and the diameter of Mars was less than 1 mm. Nevertheless, after 4 nights of observations at the Newtonian focus of the 61-inch Wyeth reflector, Sinton had enough evidence for his conclusion: "The probability is therefore very high that an organic spectrum is required to account for the data."[128]

Sinton was very much aware of previous visual evidence for vegetation in the form of seasonal changes in the size of the Martian dark areas. In fact, he saw the dip at 3.4 microns as "additional evidence for vegetation" and concluded that "this evidence, together with the strong evidence given by the seasonal changes, makes it seem extremely likely that plant life exists on Mars." Thus, his graph of infrared absorption did not constitute direct visual confirmation, but depended on the interpretation of spectrograms, an interpretation undoubtedly affected by preconceived ideas.

Sinton's result caused considerable excitement, especially when it was confirmed by Sinton with equipment 10 times more sensitive on the 200-inch Palomar telescope during the 1958 opposition. Although the image of Mars was only 2 mm, this time Sinton separated the dark areas from the bright areas on Mars and confirmed his previous conclusion of absorption bands near 3.5 microns (Figure 3.11). Again he concluded that "the observed spectrum fits very closely . . . that of organic compounds and particularly that of plants."[129] In addition, a 3.67-micron absorption band was confirmed, which Sinton attributed to carbohydrate molecules in plants, analogous to tests on plants on Earth.

Sinton's results were widely hailed and cited in the literature. But his results were open to interpretation: not only were other biological inter-

128 William M. Sinton, "Spectroscopic Evidence for Vegetation on Mars," *ApJ*, 126 (September 1957), 231–239: 237. Various statistical tests had to be performed before the conclusion was reached.

129 William M. Sinton, "Further Evidence of Vegetation on Mars," *Science* (November 6, 1959), 1234, 1237. Sinton still used a nitrogen-cooled lead sulfide cell, but now custom made by Infrared Industries.

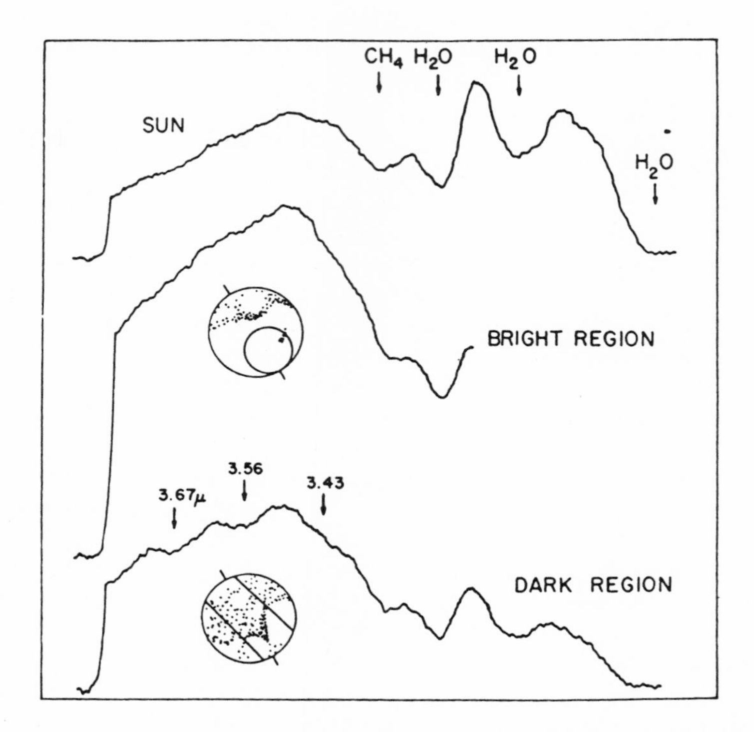

Figure 3.11. Sinton's evidence for Martian vegetation. The top curve shows the solar spectrum with absorptions by methane and water in the Earth's atmosphere. The middle curve shows a bright desert area of Mars where no vegetation was expected. The bottom curve shows absorption from one of the dark areas of Mars, interpreted as due to vegetation. Unlike the data shown in Figure 3.10, this evidence turned out to be spurious in terms of revealing real characteristics of the planet. With permission from Sinton, "Further Evidence of Vegetation on Mars," *Science* (November 6, 1959), 1234. Copyright 1959 American Association for the Advancement of Science.

pretations possible, by 1963 D. G. Rea, T. Belsky, and M. Calvin had done extensive work on infrared reflection spectra of terrestrial compounds and were critical of Sinton's interpretation.[130] And by 1965 Rea, O'Leary, and Sinton himself suggested that two of the Sinton bands were

[130] N. B. Colthup, "Identification of Aldehyde in Mars Vegetation Regions," *Science* (August 25, 1961), 529, suggests another biological interpretation. For criticism of Sinton's interpretation see D. G. Rea, T. Belsky, and M. Calvin, "Interpretation of the 3- to 4-Micron Infrared Spectrum of Mars," *Science*, 141 (September 6, 1963), 923–927.

due to heavy "deuterated" water (HDO) in the Earth's atmosphere, with the remaining band still possibly organic.[131]

In the end, Sinton's refined methods had been mitigated by refined problems. Whereas V. M. Slipher a half century before believed he had found oxygen and water vapor on Mars, and Very believed he had quantitatively measured Slipher's spectrograms, only to find that their results were contaminated by the Earth's own atmosphere, Sinton's results too were contaminated, this time by heavy water.

As with the canals of Mars, the search for Martian vegetation again demonstrates differences in approach and worldview among scientists, with one extreme group much more likely to go out on a limb and to extrapolate than the other. Some astronomers probing the physical conditions on Mars presented their data and left it at that. Others used their data – indeed, were probably first inspired to gather their data – in the service of the question of extraterrestrial life. Still others rendered no opinion at all. Given the inherent difficulties of Martian observations and their implications for life, it is little wonder that many Mars experts tried to avoid the controversy. In his book *Physics of the Planet Mars* (1954), Gérard de Vaucouleurs promised to "allude only seldom to the problem of life on Mars." His reason was clearly stated but not universally shared:

> It is our belief that such a problem is still, to a large extent, beyond the limits of our positive knowledge and can only be the subject – either way – of vague speculations in which general "principles" of a metaphysical nature have always to be taken as a guide. Such speculation, we believe, can amount to nothing more than mere conjecture and under the circumstances it seems useless to pursue the discussion at the present time.[132]

De Vaucouleurs thus declared his intention to return to the unbiased facts as far as possible. He saw two groups coexisting in the 1950s, one emphasizing the difficulties for the existence of life and the dangers of extrapolation from our planet, the other emphasizing the variety of life forms on Earth and the adaptability and universality of life. De Vaucouleurs left little doubt as to his own position: the "overwhelming majority" of life forms on Earth could not survive on Mars. It is uncertain whether low forms of plant life could adapt themselves to Martian conditions, and to say anything more about generalized life "remains to

131 D. G. Rea, B. T. O'Leary, and William M. Sinton, "The Origin of the 3.58 and 3.69-Micron Minima in the Infrared Spectrum," *Science*, 147 (1965), 1286–1288.

132 DeVaucouleurs, *Physics of the Planet Mars*, 19.

a large extent beyond the realm of positive science."[133] Beyond this de Vaucouleurs left the most pressing Martian question to the reader to decide, citing only the biologist Hubertus Strughold for further reading.

To many, such a cautious attitude was not satisfying. They undoubtedly realized that the stakes in the debate extended far beyond Mars: as Kuiper wrote, "If life truly exists on the only two planets of the solar system that are at all suitable to sustain it, it is tempting to conclude that, after enough time has elapsed, it will develop spontaneously wherever conditions permit. Since planetary systems are presumed to be very numerous, life would then be no exception in the universe."[134]

At the dawn of the Space Age, then, the canal controversy had receded and much was known about the physical conditions of the planet Mars. Vegetation of some sort was still a very real possibility, dependent to some extent on what one saw as the limits to the adaptability of life. Vegetation did not have the popular appeal of intelligence, but to the scientist it was still a holy grail that held the promise of revealing the secrets of life. That promise was to play no small role in making Mars an important target for interplanetary probes of the Space Age.

3.3 VENUS: LAST HOPE FOR INTELLIGENCE

Before moving on to the final act in the drama of life on Mars and the extraordinary revelations of the Space Age, we must pause and consider the fate of the brilliant and mysterious planet Venus. Closer to the Earth than Mars, and so similar in many of its gross characteristics to the Earth that it was for much of the century known as the Earth's "sister planet," its fate by the dawn of the Space Age was very different from that of Mars. No serious expectations for life accompanied Venus spacecraft, although the planet held plenty of other mysteries. For the historian of science, Venus provides a good control experiment of scientific attitudes toward extraterrestrial life when (unlike Mars) evidence accumulated that the physical conditions on a planet were so extreme that life would be impossible by anyone's definition. Would scientists push on, with ever more inventive scenarios of how life might survive despite harsh conditions? Or would they drop Venus like the proverbial hot potato from the list of potential exobiological sites? And if the latter, how quickly? Unlike Mars, no Lowell pushed the idea of intelligence on Venus or claimed direct evidence of intelligence. And yet, in the decades after hope had been given up for intelligence on Mars, several prominent scientists held

[133] Ibid., 17, 41.

[134] Kuiper, *The Atmospheres of the Earth and Planets*, 2d ed. (Chicago, 1952), 404–405.

out hope for intelligence on Venus. Why this was so, and how this view changed with new evidence, is the subject of this section.

The heritage of the nineteenth century was an inhabited Venus. R. A. Proctor had written in 1870 that "in size, in situation, and in density, in the length of her seasons and of her rotation, in the figure of her orbit and in the amount of light and heat she receives from the sun, Venus bears a more striking resemblance to the earth than any other orb within the solar system." Its year, he believed with other astronomers, was 224 days and 17 hours, its day about 35 minutes shorter than the Earth's. With the exception of a possibly highly tilted axis, Proctor found that "on the whole, the evidence we have points very strongly to Venus as the abode of living creatures not unlike the inhabitants of earth."[135]

Although these statements remained unchanged in the fourth "carefully revised" edition of 1894, a new claim made in 1890, if proven true, promised drastic consequences for the climate of Venus. The first in a long series of observations of rotation rate, temperature, and atmospheric pressure that would affect the twentieth-century view of life on Venus (Table 3.4), this claim was made by none other than Giovanni Schiaparelli, supported by Perrotin, of a Venusian day as equal in length to approximately 225 Earth days. Since this was about equal to its period of revolution around the Sun, Venus (like the Moon) would have remained locked with one face perpetually toward the Sun, a circumstance that would have made it uninhabitable.[136] Supported by further observations in 1895, and by Slipher and Lowell in 1903 and 1910, most observers of Venus nevertheless confirmed the previous 1-day rotation period; the problem was not resolved until the early 1960s at the longer period – and retrograde compared to the other planets.[137]

[135] Proctor, *Other Worlds Than Ours: The Plurality of Worlds Studied under the Light of Recent Scientific Researches* (New York, 1871; 1st ed., 1870), 84, 87, 94. On the history of Venus studies prior to the Space Age see Dale P. Cruikshank, "The Development of Venus Studies," in D. M. Hunten, L. Colin, T. M. Donahue, and V. I. Moroz, eds., *Venus* (Tucson, Ariz., 1983), 1–9, and Patrick Moore, *The Planet Venus* (New York, 1961; 1st ed., 1956).

[136] G. V. Schiaparelli, "Considerazioni sul moto rotatorio del pianeta Venere," *Rendiconti del Reale Istituto Lombardo di Scienze e Lettere,* XXIII (Milan, 1890), reprinted (with his other work on Venus) in Schiaparelli, *Le Opere,* vol. 5 and reviewed in J. L. E. Dreyer, "Schiaparelli's Researches on the Rotation of Venus and Mercury," *MNRAS,* 51 (February 1891), 246–249. The observations of Schiaparelli were first made in 1877–1878; his 1890 work also reviews other determinations of Venus's rotation period.

[137] See Moore, *The Planet Venus,* 80–89. Anyone who doubts the problematic nature of observation in astronomy – and possibly the power of preconceived notions – should peruse Moore's table of observed Venusian rotation rates, 133–35. Beginning with Cassini in 1666–1667 through 1958, some 65 independent determinations gave values of about 24 hours or less; some 20 others gave a rate of about 224 Earth days. The actual value accepted today is 243 days – in a retrograde direction! – as first reported by R. L. Carpenter, "Study of Venus by CW radar," *AJ,* 69 (February 1964), 2–11, and

Table 3.4. *Important advances in knowledge of Venus*

Year	Author	Instrument	Result
1890	Schiaparelli	Refractor	224-day rotation rate
1897	Lowell	Refractor	Surface markings
1922	St. John and Nicholson	Mount Wilson 100-inch	No oxygen or water in atmosphere
1932	Adams and Dunham	Mount Wilson 100-inch	CO_2 in atmosphere
1940	Wildt	Theory	Greenhouse effect; surface temp > 408 K
1958	Mayer et al.	Radio telescope	Surface temp. 600 K
1963	Barath et al.	Mariner II	Surface temp. 700 K
1964	Carpenter	Radar astronomy	- 250-day ± 50-day rotation rate
1964	Goldstein	Radar astronomy	- 250-day rotation rate
1973	Young	Polarimetry	Sulfuric acid clouds

The overriding concern among astronomers interested in Venus at the turn of the century was thus this uncertainty about its rotation period. "Upon it turns our whole knowledge of the planet's physical condition," Lowell wrote in 1909. In light of his claims about Mars, it is notable that Lowell's observations of the "fingerlike streaks" on the surface of Venus, as well as spectroscopic observations made at his observatory, convinced Lowell of the truth of Schiaparelli's conclusions, even though this circumstance left one side of Venus baked and one side frozen and completely uninhabitable.[138] In 1910 T. J. J. See argued that the study of Venus was

R. M. Goldstein, "Venus Characteristics by Earth-based Radar," *AJ*, 69 (February 1964), 12–18. For a review of modern rotation rate determinations see I. I. Shapiro, "Spin and Orbital Motions of the Planets," in *Radar Astronomy*, J. V. Evans and T. Hagfors, eds. (New York, 1968), 143–185.

[138] Lowell, *The Evolution of Worlds* (New York, 1909), 73–90. See also Lowell, "Detection of Venus' Rotation Period and of the Fundamental Physical Features of the Planet," *PA*, 4 (December 1896), 281–285; "Determination of the Rotation and Surface Character of the Planet Venus," *MNRAS*, 57 (March 4, 1897), January 148; *Lowell Observatory Bulletin*, 6 (January 1904), 31; *Nature*, 55, 421 and many other sources. Lowell

much more important than the study of Mars, not only for the question of habitability but also because it held out the hope of shedding light on the role of tidal friction in the solar system. Despite Lowell's recent argument, See argued for a rapid rotation based on the observations of Schroeter, a rate that would not cause the temperature to be excessive and that "would go a long way towards establishing the habitability of the Earth's twin sister in space."[139] In 1911 R. G. Aitken of Lick Observatory also pointed out in a discussion of planetary habitability that the rotation rate of Venus was the "critical point in our present investigation. If it rotates on its axis once in about 24 hours, we have reason to believe that it is habitable, for the conditions we named as essential to life – air, water in its liquid form and a moderate temperature – are undoubtedly realized. But if its day equals its year, then it must be utterly desolate." Citing the observations of Schiaparelli, Lowell, and Slipher, Aitken concluded that the evidence favors the longer period, a conclusion he believed to be the opinion of most astronomers; but he agreed that the question of the habitability of Venus had to remain open until the rotation rate was definitely settled.[140]

Given this situation, the popular and scientific picture that emerged of Venus in the early twentieth century is somewhat surprising. Significantly, it was not the same as the picture of Mars, despite the attempt of C. E. Housden, an active member of the British Astronomical Association, to argue that Lowell's claimed markings on Venus represented channels through which intelligent Venusians pumped water. Despite Housden's claim that Lowell's markings had been "generally confirmed by the Mercury and Venus Section of the British Astronomical Association," Lowell himself, now in the year before his death, must have cringed at Housden's theory of intelligent Venusians.[141] And most astronomers, far from accepting surface markings on Venus, continued to believe it was surrounded by clouds. It was therefore not Housden's theory, but one only slightly less surprising, that caught the popular and scientific imagination. Developed by the Nobel Prize–winning chemist Svante Arrhenius, who presented his

was careful to point out that the Venus markings were completely different from the canals of Mars. Antoniadi was one of those who accused Lowell of seeing canals everywhere in *JBAA,* 8 (1897), 45. For A. E. Douglass's defense of Lowell's Venus observations see "The Markings on Venus," *MNRAS,* 58 (May 1898), 382–387. Hoyt, *Lowell and Mars,* discusses Lowell's Venus markings, 108–121. The spectroscopic observations were reported in *Popular Science Monthly* (December 1909).

139 T. J. J. See, "The Rotation of Venus and Life on Planets Other Than the Earth," *PA,* 18 (January 1910), 1–3.

140 R. G. Aitken, "Life on Other Worlds," *JRASC,* 5 (September–October 1911), 291–308: 300–303 on Venus. This article was an address delivered to the Royal Astronomical Society of Canada at Toronto, January 23, 1911.

141 C. E. Housden, *Is Venus Inhabited?* (London, 1915), 8–9.

view of Venus in his popular lectures on planetary evolution *The Destinies of the Stars* (1918), the new version argued for a lush, steamy Venus with only low forms of life. In support of this view, Arrhenius cited the work on planetary albedos of Henry Norris Russell as evidence of an extensive atmosphere on Venus.[142] Admitting the uncertainty about the rotation rate, he sided with those who argued for a rapid rotation rather than Lowell and Slipher's slow one, noting that otherwise all gases but hydrogen and helium would have condensed on the cold side and those would have boiled away on the hot side. Arrhenius asserted that the relatively higher temperature compared to Earth would vastly increase the proportion of moisture in the air.[143]

> We must therefore conclude that everything on Venus is dripping wet. . . . A very great part of the surface of Venus is no doubt covered with swamps, corresponding to those on the Earth in which the coal deposits were formed, except that they are about 30 degrees C (54 F) warmer. . . . The temperature on Venus is not so high as to prevent a luxuriant vegetation. The constantly uniform climatic conditions which exist everywhere result in an entire absence of adaption to changing exterior conditions. Only low forms of life are therefore represented, mostly no doubt belonging to the vegetable kingdom; and the organisms are nearly of the same kind all over the planet. The vegetative processes are greatly accelerated by the high temperatures.[144]

At the poles, Arrhenius asserted, the temperature might be only 10°C, and "the organisms there should have developed into higher forms than elsewhere, and progress and culture . . . will gradually spread from the poles toward the equator." Perhaps sometime in the distant future, he surmised, the highest beings in the solar system may dwell on Venus. Arrhenius thus accepted an evolutionary view of worlds in the solar system, but his views nevertheless diverged from those of Lowell in the case of Venus.

Arrhenius's view was taken a step further by Charles Greeley Abbot (1872–1973), the distinguished director of the Smithsonian Astrophysical Observatory. Rejecting Lowell's view of Mars and impressed by W.

[142] Svante Arrhenius, *The Destinies of the Stars,* J. E. Fries, trans. (New York and London, 1918), 130–131. The work cited is Russell, *PNAS,* (1916); the same article appeared as "On the Albedo of the Planets and Their Satellites," *ApJ,* 43 (April 1916), 173–196.

[143] Ibid., 138, 152–153. Arrhenius accepts a value of 47°C (117°F) for the temperature of Venus, 250. He rejects Lowell's view that Venus is not surrounded by clouds, and indeed uses the existence of clouds as an argument against a slow rotation rate.

[144] Ibid., 250–253.

W. Campbell's failure to detect spectroscopically water vapor on Mars, he claimed as early as 1920 that the most likely abode of extraterrestrial intelligence in our solar system was not Mars but Venus. "A twin planet to the earth in size and mass, its high reflecting power seems to show that Venus is largely covered by clouds indicative of abundant moisture; probably at almost identical temperatures to ours, our sister planet appears lacking in no essential to habitability."[145] Abbot's reasoning was similar to that of Arrhenius 2 years earlier: the reflecting power of Venus has been carefully measured to be 60%; this apparently demanded the existence of clouds; these clouds "can hardly be of any other substance than water"; these clouds could not exist on the assumption of a long rotation period, which locks the planet to the Sun and would cause all of the water to be distilled from the hot side to the cold side. This alone, Abbot concluded, "seems to be sufficient to overcome the observational evidence which indicates the equality of periods of rotation and revolution."[146] Moreover, Abbot asserted, the spectroscopic evidence was only accurate enough to show that the rotation period is more than 10 earth days, and the visual evidence based on markings "can only be regarded with the greatest doubtfulness."

This view was not a passing fancy of Abbot but was reiterated in his popular astronomy text *The Earth and the Stars* in 1926. By this time Abbot had new evidence to draw on, for the same physical techniques being applied to Mars were also being brought to bear on Venus. Even as his 1920 paper was being published, Abbot heard from Charles E. St. John that preliminary spectroscopic studies of Venus threw doubt on the existence of water vapor there. By 1922 Slipher at Lowell Observatory and St. John and Nicholson at Mount Wilson had confirmed this result, but failing to see how a cloud-covered planet could lack water, Abbot reasoned in 1926 that perhaps only high cirruslike clouds of ice crystals were being viewed, while the water vapor was in clouds closer to the surface.[147] He had no rebuttal to the failure of St. John and Nicholson to

[145] C. G. Abbot, "The Habitability of Venus, Mars, and Other Worlds," *Annual Report of the Board of Regents of the Smithsonian Institution . . . for 1920* (Washington, D.C., 1922), 165–171: 170. See also the *English Mechanic* (March 5, 1920), summarized in "Dr. Abbot on Life on Venus," *JBAA*, 30 (1920), 198. On Abbot see Deborah J. Warner, "Charles Greeley Abbot (1872–1973)," *Yearbook of the American Philosophical Society* (1975), 111–116, and Abbot's autobiography *Adventures in the World of Science* (Washington, D.C., 1958). For his controversial claims about variations in the solar constant and their effect on terrestrial weather see David H. DeVorkin, "Defending a Dream: Charles Greeley Abbot's Years at the Smithsonian," *JHA*, 21 (February 1990), 121–136.

[146] Abbot, "Habitability," 170. Abbot cites not Arrhenius but rather Alexander Graham Bell as the source of this argument!

[147] C. G. Abbot, *The Earth and the Stars* (New York, 1926), 74.

find any oxygen on Venus, nor did he follow their suggestion that the lack of oxygen was due to lack of plant life, itself made impossible by a deficiency of water. Nor did he comment on their prescient remark that on Venus "it may be that the exacting conditions for the origin of life have not been satisfied so that the existing atmosphere may consist of other permanent or semipermanent gases such as nitrogen or carbon dioxide."[148] He did, however, latch on to Pettit and Nicholson's observations of 1924 that the temperature of Venus was uniform and similar to the Earth's, indicating a rapid rotation.[149]

Despite the uncertainties, Abbot concluded that "the conditions seem likely to be nearly as favorable for life there as on our earth." He cautioned that favorable conditions were no guarantee for the origin of life and that the veil of clouds left little chance for confirmation. But in a passage that shows the influence of Lowell and Housden, he left little doubt about the level of life he had in mind: "Whether there are intelligent works of engineering there, we may never observe. Unless there should be wireless communication, we probably never shall know whether we have intelligent neighbors on Venus or not."[150] In both his 1920 paper and his 1926 book, it is clear that the idea of intelligence had caught Abbot's imagination:

> It makes the blood stir to imagine what might follow if ever our people could come into fluent communication by wireless with a race brought up completely separate, having their own systems of government, social usages, religions, and surrounded by vegetation and animals entirely unrelated to any here on earth. It would be a revelation far beyond the opening of Japan, or the discoveries of Egyptologists, or the adventures of travellers in the dark continent.[151]

During the 1930s and 1940s, the impact of results from spectroscopic studies would gradually take their toll on the Venusian views of Arrhenius and Abbot. Already in 1926, the authors of the famous Russell, Dugan, and Stewart textbook took the lack of oxygen on Venus much more seriously, and were in basic agreement with Nicholson and St. John's 1922

[148] Charles E. St. John and Seth B. Nicholson, "The Absence of Oxygen and Water-Vapor Lines from the Spectrum of Venus," *ApJ*, 56 (December, 1922), 380–399: 399.

[149] Edison Pettit and Seth B. Nicholson, "Radiation from the Dark Hemisphere of Venus," *PASP*, 36 (August 1924), 227; *PA, 32 (December 1924), 14*. See also their later paper, "Temperatures on the Bright and Dark Sides of Venus," *PASP*, 67 (October 1955), 293.

[150] Abbot, *The Earth and the Stars*, 75. See also Moulton, *Introduction*, 273, for the idea that life was more likely on Venus than on Mars.

[151] Abbot, *The Earth and the Stars*, 75. Abbot elaborates more on the history and current discussion of wireless communication in his 1922 paper, "Habitability," 171.

conclusion that life appears never to have originated on that planet. James Jeans in his popular books spread the same view.[152] The discovery in 1932 by Adams and Dunham of CO_2 in the atmosphere of Venus, confirmed by experiments of Dunham, Adel, and Slipher, gave support to a Venus in a primitive prebiotic state.[153] Though some argued that the CO_2 might have been given off by plants and animals, others continued to point to the apparent lack of oxygen.[154] By 1940 Rupert Wildt was postulating a "greenhouse effect" on Venus due to the CO_2, an effect that he concluded would make the surface temperature of Venus higher than 100°C, the terrestrial boiling point of water.[155] Sir Harold Spencer Jones, in his influential *Life on Other Worlds* (1940), accepted this view of the powerful ability of CO_2 to produce a greenhouse effect, and considering also the lack of detectable oxygen, he concluded that Venus was perhaps a young Earth on which life might develop in the future.[156] Jeans also followed this view of a boiling hot Venus in 1942. Though in 1946 Abbot qualified his statements on Venus only slightly, saying that "the conditions may possibly be as favorable for life there as on our earth," by the 1940s the momentum was clearly developing in the opposite direction.[157]

By 1952 Harold Urey, taking the view of a geochemist, argued that no water could be present on the surface of Venus because CO_2 readily reacts with water to form carbonates. Any water present on Venus would therefore have been locked up in rocks, and Venus would now be a desert.[158] This was counterbalanced in 1955 by Menzel and Whipple's suggestion

[152] Henry Norris Russell, Raymond S. Dugan, and John Quincy Stewart, *Astronomy: A Revision of Young's Manual of Astronomy* (Boston, 1926), 319–320. For Jeans's opinion see *The Universe around Us* (New York, 1929), 323, and *The Stars in Their Courses* (New York, 1931), 49–50.

[153] Walter S. Adams and Theodore Dunham, Jr., "Absorption Bands in the Infra-Red Spectrum of Venus," *PASP,* 44 (August 1932), 243–245. On the experiments see Dunham, *Carnegie Institution of Washington Year-Book,* no. 31 (1932) 154, and Arthur Adel and V. M. Slipher, "Concerning the Carbon Dioxide Content of the Atmosphere of the Planet Venus," *Physical Review,* Ser. 2, 46 (1934), 240.

[154] Edwin P. Martz, Jr., "Venus and Life," *PA,* 42 (March 1934), 165–167 argues that the presence of CO_2 indicates life. Martz was a former colleague of William H. Pickering. In the same year, Russell continued to argue for the primitive prebiotic model, and this view was accepted by others, e.g., Harold Spencer Jones, *Worlds without End* (London, 1935), 88–89.

[155] Rupert Wildt, "Note on the Surface Temperature of Venus," *ApJ,* 91 (January–June 1940), 266–268. Earlier, others had suggested that the "blanketing effect" of CO_2 could raise the surface temperature to the boiling point of water. See Spencer Jones, *Worlds without End,* 88.

[156] H. Spencer Jones, *Life on Other Worlds* (New York, 1940), 177–199, especially 190–199.

[157] Abbot, *The Earth and the Stars* (1946), 109. James Jeans, "Is There Life on the Other Worlds?," *Science,* 95 (June 12, 1942), reprinted in Donald Goldsmith, *The Quest for Extraterrestrial Life: A Book of Readings* (Mill Valley, Calif., 1980), 81–83.

[158] Harold P. Urey, *The Planets* (New Haven, Conn., 1952), 222.

that Venus might be completely covered with water on the surface, so that this fixation of CO_2 into carbonates could not proceed if all the land was covered.[159] This latter idea of Venus was received with great skepticism; Otto Struve pronounced Menzel and Whipple's theory too revolutionary to appraise. And it was soon discounted by new observations at the dawn of the Space Age. Mayer and his colleagues, using radio techniques beginning in 1956, found a temperature of 600 K, and though it was still uncertain whether this originated at the surface or in the Venus atmosphere, there was little surprise when a microwave radiometer aboard the Mariner 2 spacecraft confirmed the hot surface model.[160]

The idea of life on the cloud-shrouded Venus thus began the twentieth century as a good possibility, was considered unlikely by 1930 with the failure to detect oxygen, was held to be even less likely by 1940 with the discovery of abundant CO_2 and the realization of a possible greenhouse effect, and was widely seen to be impossible by 1960 with the determination of surface temperatures.[161] Carl Sagan's review article on Venus, published a month after the first Venus spacecraft was launched in 1961, held out no hope for life on the surface. Pointing to the microwave temperature of 600 Kelvin deduced since 1956, he concluded, "At such high temperatures, and in the absence of liquid water, it appears very unlikely that there are indigenous surface organisms at the present time." Moreover, though he mentioned the possibility of life below the surface or in the atmosphere, the fact that no period seemed to have existed when Venus had both extensive bodies of water and surface temperatures below the boiling point of water led Sagan to conclude, "it is unlikely that life ever arose on Venus."[162]

Finally, the Panel on Planetary Atmospheres of the Space Science Board

[159] Donald H. Menzel and Fred L. Whipple, "The Case for H_2O Clouds on Venus," *PASP,* 67 (June 1955), 161–168; *S&T,* 14 (November 1954), 20. Menzel and Whipple rejected Wildt's hypothesis of a greenhouse effect, pointing to recent temperatures by William Sinton as given in his Ph.D. dissertation "Distribution of Temperature and Spectra of Venus and Other Planets," Johns Hopkins University, 1953. For the context of Menzel and Whipple at Harvard, see Doel, *Solar System Astronomy,* chapter 6.

[160] C. H. Mayer, T. P. McCullogh, and R. M. Sloanaker, "Observations of Venus at 3.15 cm Wavelength," *ApJ,* 127 (January 1958), 1–10; F. T. Barath et al., "Microwave Radiometers," part of "Mariner II: Preliminary Reports on Measurements of Venus," *Science,* 139 (March 8, 1963), 908–909, and "Mariner 2 Microwave Radiometer Experiment Results," *AJ,* 69 (February 1964), 49–58. For Struve's comment on Menzel and Whipple see Otto Struve, "Life on Other Worlds," *S&T,* 14 (February 1955), 137–140: 140.

[161] As late as 1962, the British writers Francis Jackson and Patrick Moore, however, still held out hope for aquatic life if the Menzel–Whipple model was true "unless the surface temperature is prohibitively high." Francis Jackson and Patrick Moore, *Life in the Universe* (London, 1962), 95.

[162] Carl Sagan, "The Planet Venus," *Science,* 133 (March 1961), 849–858: 857.

of the National Academy of Sciences found no hope for life in any of the three models of the Venusian atmosphere believed to be most likely in 1961 as the first spacecraft sped toward that planet. Their view was borne out not only by the confirmation of extremely high temperatures and pressures, but also by the later discovery that the clouds were composed of sulfuric acid. By contrast, "The Question of Life on Mars" was very much on the panel's agenda.[163]

3.4 THE SPACE AGE: LOWELL'S LEGACY OVERTURNED

The question of life on Mars might have forever remained shrouded in mystery and debate were it not for one of the truly epochal events of the twentieth century – the dawn of the Space Age. In this section we examine how the Space Age altered traditional approaches to the search for life in the solar system and finally brought consensus (though not unanimity) among scientists intent on providing a final solution to the riddle of Mars. We shall see that the search for life via spacecraft required fundamental questions to be asked about the nature of life, for the first time bringing substantial numbers of biologists into an enterprise that had long been the province of astronomers. We shall see that although controversy marked the risky decision to search for life on Mars via spacecraft, once adopted that decision was a driving factor in the American space program. And though consensus on the startling results was eventually reached, we shall see that observation via spacecraft – even when undertaken from the surface of the planet – still had problems.

Despite its novel methods, the beginning of the Space Age did not imply a rejection of all that went before or a diminution of the importance of ground-based studies. To the contrary, planning for expensive spacecraft created a voracious appetite for whatever Martian data could be obtained, by whatever means. Canals aside, Lowell's legacy continued to play a significant role both as a driving force for exploration of the red planet and as a source of information, a fact that unifies the twentieth-century search for life on Mars. Lowell left at least two legacies in his studies of Mars, one thoroughly discredited, the other still very much alive. The discredited legacy was a planet crisscrossed by canals built by inhabitants intent on making the best of the remaining water on their arid planet. The other legacy has been described by several of the biologists who developed experiments to search for life on Mars:

[163] Space Science Board, *The Atmospheres of Mars and Venus*, National Academy of Sciences Publication 944 (Washington, D.C., 1961), 33, 37–50. For the Space Age results of investigations on Venus see Hunten et al., *Venus*.

> When the Space Age dawned in 1957, the generally accepted picture of Mars was that of an arid but fundamentally Earth-like planet. At that time Mars was considered to have an atmospheric pressure of 85 millibars and ice caps of water ice that waxed and waned with the seasons. Water was transferred seasonally from one pole to the other, and its movement across the planet in the springtime was accompanied by a darkening (often described as a "greening") of large areas of the surface. Spectroscopic evidence suggested the existence of organic matter on the Martian surface.[164]

The persistence into the Space Age of Lowell's second legacy – a harsher but still Earth-like Mars with possible vegetation – and its final demise are the closing dramatic elements in a story never short on drama.

Even before planetary spacecraft became an imminent possibility with the launch of Sputnik in 1957, the effect of Lowell's second legacy is clear in some of the earliest space biology discussions, ideas that emerged from the discipline of space medicine. Born out of the experience of four decades of aviation medicine, this field had as its goal to establish the effect on humans of the peculiar conditions of high-altitude flight.[165] But by 1950 – in a development itself significant for our history – space medicine was expanding its horizons to consider the effects of harsh conditions on organisms other than humans. One of the leaders in this expansion was Hubertus Strughold, a pioneer in the field of aviation medicine.[166] His book *The Red and Green Planet: A Physiological Study*

[164] H. P. Klein, N. H. Horowitz, and Klaus Biemann, "The Search for Extant Life on Mars," in *Mars,* H. H. Kieffer et al., eds. (Tucson, Ariz., 1992), 1222–1223. For more on Lowell's legacy see N. H. Horowitz, "The Legacy of Percival Lowell," in *To Utopia and Back: The Search for Life in the Solar System* (New York, 1986) 78–82; Bruce Murray, "The Ghost of Percival Lowell," in *Journey into Space: The First Thirty Years of Space Exploration* (New York, 1989), 31–45; and J. N. Tatarewicz, "The Legacy of Percival Lowell," in *Space Technology and Planetary Astronomy* (Bloomington, Indiana, 1990), 2–6. Tatarewicz (60ff.) also emphasizes the importance of continued ground-based study. One could also consider other elements of Lowell's legacy, including firing the imagination of science fiction writers such as H. G. Wells (see chapter 5 in this volume), as well as pioneering in the planetary astronomy techniques of spectroscopy, photography, and polarimetry.

[165] For a history of space medicine as related to manned spaceflight see John A. Pitts, *The Human Factor: Biomedicine in the Manned Space Program to 1980* (Washington, D.C., 1985).

[166] Strughold (1898–) received his Ph.D. from the University of Munster in 1922 and his M.D. the following year, specializing early in aviation medicine. Prior to coming to the School of Aviation Medicine of the U.S. Air Force in 1947, he had served successively as professor of physiology at the universities of Wuerzburg, Berlin, and Heidelberg. Shortly after his appointment in 1949 as the first head of the Department of Space Medicine of that school, Strughold's views on Martian life began to appear in the literature of aviation medicine. On Strughold see Shirley Thomas, "Hubertus Strughold," in *Men of Space,* vol.

of the Possibility of Life on Mars (1953) was strongly influenced by the classic work of Lowell, Pickering, Antoniadi, and Kuiper, as well as by Harold Spencer Jones's *Life on Other Worlds* (1940).[167] Though as a biologist his approach was markedly different, the Lowellian "green planet" was the backdrop to Strughold's work, and indeed its inspiration. But instead of astronomy, Strughold drew on a tradition of biology, medicine, and physiology in attempting to determine if life might exist on Mars. He spoke of "planetary ecology" and "ecospheres" and discussed the structure of plants to find "a reasonable basis for the assumption that a form of vegetation can survive and maintain itself" on Mars.[168] Strughold agreed with Kuiper that Mars would most likely harbor only primitive plant life similar to lichens, but he also offered "an outside chance that we might find on Mars a lowly type of animal existence: bacteria of some kind, for example." Astronomers, physicists, physiologists, and students of aviation medicine, Strughold noted, all had raised this possibility, but none could confirm it. A research laboratory in which planetary conditions might be reproduced might solve this problem, he suggested, and a prototype could be found in the low-pressure chambers and climatic chambers used in aviation medicine over the last 20 years.[169]

This idea of simulated planetary conditions would prove fertile and would be given increasing impetus as the space program gained momentum. Of particular interest here is that when Strughold and his colleagues presented the results of their early experiments to the astronomical community less than 4 months before the dawn of the Space Age, it was in a symposium, organized and chaired by Strughold, held in Lowell territory – Flagstaff, Arizona. Billed by Lowell Observatory Director Albert G. Wilson as "the first American symposium in Astrobiology," it included the reports of Sinton and Dollfus on their latest astrophysical research on Martian vegetation, along with the experiments of Strughold and his colleagues.[170] By 1959 the integration of this work in space medicine into the

4 (Philadelphia, 1962), 233–272. For an example of Strughold's work while in Germany, see *German Aviation Medicine in World War II*, vol. 2 (Washington, D.C., 1950).

167 Strughold, *The Red and Green Planet* (Albuquerque, N.M., 1953).

168 Ibid., 66.

169 Ibid., 92–95. The French physiologist Paul Becquerel pioneered in determining the limits of life under harsh conditions, but Strughold was apparently not familiar with this work.

170 "Problems Common to the Fields of Astronomy and Biology," *PASP*, 70 (February 1958), 41–78, including William M. Sinton, "Spectroscopic Evidence of Vegetation on Mars," 50–56; Audouin Dollfus, "The Nature of the Surface of Mars," 56–64; John A. Kooistra, Jr., Roland B. Mitchell, and Hubertus Strughold, "The Behavior of Microorganisms under Simulated Martian Environmental Conditions," 64–69; and Strughold, "General Review of Problems Common to the Fields of Astronomy and Biology," 43–50.

more general field of space science is evident in Strughold's participation in the Lunar and Planetary Exploration Colloquium, a seminal series of discussions in which Kuiper and numerous influential scientists also participated. Here Strughold outlined the scope of astrobiology, including new experiments in simulated planetary conditions.[171] Clearly, the experiments on simulated planetary conditions begun in the space medicine tradition constituted one of the streams flowing into the torrent of excitement about life on Mars. Such experiments became a major activity carried out at many laboratories.[172]

Meanwhile, the actuality of spaceflight and the imminent possibility of sending probes to other planets had set in motion another chain of events rising out of the concern of the larger biological science community, some of whose members realized the exciting possibilities now within their grasp but also foresaw that the very spacecraft designed to search for life might unwittingly destroy those possibilities. The Lowellian preoccupation with life – though now applied to microorganisms and much harsher conditions than Lowell had envisioned – was still at the heart of an idea that became a major concern of the space program: biological contamination. Contamination of the planets by some hardy terrestrial microorganism clinging to a spacecraft, or conversely and even more disastrously, "back contamination" of the Earth by any returning samples or astronauts, was a prospect that had to be seriously considered, even in the context of such a seemingly hostile environment as the Moon. No longer was extraterrestrial life merely an academic question; it was not quite "The War of the Worlds," but perhaps some science fiction-minded space scientists had not forgotten that H. G. Wells's Martians were in the end destroyed by terrestrial microorganisms.

As early as 1957 Joshua Lederberg, soon to receive the Nobel Prize for his work in genetics, brought the issue of contamination before the scientific community.[173] At Lederberg's urging, by February 1958 the U.S.

[171] "Advances in Astrobiology," *Proceedings of Lunar and Planetary Exploration Colloquium*, vol. 1, no. 6 (April 25, 1959), 1–7. For more on these colloquia see Tatarewitz, *Space Technology*, 34–38. See also Strughold, "Space Medicine and Astrobiology," *XIth International Astronautical Congress* (Stockholm, 1960). See W. Vishniac, *Aerospace Medicine*, 31 (1960), 678, for early interaction of extraterrestrial life with the space medicine tradition.

[172] That the tradition of space medicine continued to play a role in exobiology is evident, for example, in *Foundations of Space Biology and Medicine*, Melvin Calvin and O. G. Gazenko, eds. (Washington, D.C., 1975).

[173] Edward C. Ezell and Linda Neuman Ezell, *On Mars: Exploration of the Red Planet, 1958–1978* (Washington, D.C., 1984), 55–56. See also Charles R. Phillips, *The Planetary Quarantine Program: Origins and Achievements, 1956–1973* (Washington, D.C., 1974). For more details on how the contamination issue led Lederberg into the field of exobiology, see chapter 9.

National Academy of Sciences was urging the International Council of Scientific Unions (ICSU) to evaluate the problem and, if necessary, develop recommendations to prevent contamination. This they did through a Committee on Contamination by Extraterrestrial Exploration, which developed standards adopted by the ICSU in October 1958.[174] But this was only the beginning. The contamination problem, as well as the problems of planetary life detection via spacecraft, was an important part of the discussion of two National Academy of Sciences Panels on Extraterrestrial Life, one on the East Coast (EASTEX) and one on the West Coast (WESTEX), which met from 1958 to 1960. Their efforts emphasized the seriousness of the problem. The ICSU's Council on Space Research (COSPAR) also addressed the issue.[175]

The concern for contamination seemed to some hard-nosed engineers an obsession of wild biologists, but in the end it was taken very seriously. The contamination question was a contentious one that translated into real issues of economics, for spacecraft sterilization and astronaut quarantine were expensive programs. By the mid-1960s an extensive literature had developed on the subject,[176] and one need only recall the quarantine of the returning astronauts and their samples to appreciate the attention given to back contamination of Earth. In order to shed light both on issues of contamination and on the conditions under which life might survive on another planet, scientists spared no efforts in studying the question of the adaptability of life in hostile environments on Earth.[177]

Such programs to avoid biological contamination of the planets show how seriously the prospect of extraterrestrial life was taken. But no-

[174] Ezell and Ezell, *On Mars,* 56; "Development of International Efforts to Avoid Contamination of Extraterrestrial Bodies," *Science,* 128 (October 17, 1958), 887–889. See also G. A. Derbyshire, "Resumé of Some Earlier Extraterrestrial Contamination Activities," in National Academy of Science, *A Review of Space Research* (Washington, D.C., 1962), chapter 10, 11–13.

[175] "Report of the Committee on the Exploration of Extraterrestrial Space (CETEX) 1959," *ICSU Review,* 1, (1959), 100; second report, "Contamination by Extraterrestrial Exploration," *Nature,* 183 (April 4, 1959), 925–928.

[176] P. J. Geiger, L. D. Jaffe, and G. Mamikunian, "Biological Contamination of the Planets," in Mamikunian and Briggs, *Current Aspects of Exobiology* (Oxford, 1965), 283–322. This paper was part of a symposium at JPL in February 1963 and contains 86 references to support its well-developed arguments. See also Carl Sagan and Sidney Coleman, "Decontamination Standards for Martian Exploration Programs," *Biology and the Exploration of Mars* (Washington, D.C., 1966), 470–481. On later contamination issues see Norman H. Horowitz et al., "Planetary Contamination I: The Problem and the Agreements," *Science,* 155 (March 24, 1967), 1501–1505; Bruce C. Murray et al., "Planetary Contamination II: Soviet and U.S. Practices and Policies," *Science,* 155 (March 24, 1967), 1505–1511, cited in Horowitz, *To Utopia and Back,* 156–157; and Sagan et al., "Contamination of Mars," *Science,* 159 (March 15, 1968), 1191–1196.

[177] See Horowitz et al., "Microbiology of the Dry Valleys of Antarctica," *Science,* 176 (April 21, 1972), 242–245, reporting results following Vishniac's work.

where was the Lowellian view of Mars more evident than in the exciting prospect that spacecraft provided the means for in situ examination of the planets for life. The biological interest in Mars was by no means confined to a small group of scientists expanding the boundaries of space medicine or to those worried about planetary contamination. Practitioners of the wider biological sciences were quick to realize the opportunities brought by the space program. As Lederberg wrote in 1960, while space biology and medicine had been traditionally associated with the problems of humanity in space, "it is becoming apparent that space flight may furnish a unique instrument for studying the most fundamental problems of biology: the origin of life and its progress in independent evolutionary systems. To distinguish this aspect of space biology – the evolution of life beyond our own planet – the term 'exobiology' has been introduced."[178]

NASA itself was not slow to embrace exobiology as an important goal. In July 1959 NASA's first administrator, T. Keith Glennan, appointed a Bioscience Advisory Committee, which reported in January 1960 that NASA should not only be involved in a traditional and obviously necessary space medicine role in support of manned spaceflight, but should also undertake "investigations of the effects of extraterrestrial environments on living organisms including the search for extraterrestrial life."[179] The committee, whose chairman and members represented various fields of the biological sciences, did not fail to see the same opportunities as Lederberg had earlier. "For the first time in history," they wrote with reference to the search for life and its origins, "partial answers to these questions are within reach." In the spring of 1960, NASA set up an Office of Life Sciences; by August, with the possibility of planetary missions on the horizon, it had authorized the Jet Propulsion Laboratory (JPL) to study the type of spacecraft needed to land on Mars and search for life.[180] In order to study chemical evolution, the conditions under which life might survive, and a variety

[178] Joshua Lederberg and H. Keffer Hartline, "The Biological Sciences and Space Research," in L. V. Berkner and Hugh Odishaw, eds., *Science in Space* (New York, 1961), 403–406: 403. Lederberg, "Exobiology: Experimental Approaches to Life Beyond the Earth," ibid., 407–425, also appearing in *Science* and reprinted in Goldsmith, *Extraterrestrial Life*. In chapter 9, we shall examine the detailed aspects of the birth of exobiology, while concentrating here on the scientific results.

[179] "Report of National Aeronautics and Space Administration Bioscience Committee" (January 25, 1960); Ezell and Ezell, *On Mars*, 57.

[180] Ezell and Ezell, *On Mars*, 57. The chairman of the Biosciences Advisory Committee was Seymour S. Kety of the Public Health Service. NASA set up the Life Science Office under Orr Reynolds and an exobiology division under Freeman Quimby.

of related issues, NASA's first life sciences lab was also set up at its Ames Research Center in 1960.

In the Soviet Union, by comparison, the Russians had shown an even more focused interest in Mars as part of the larger "space race." The Soviets in fact had attempted unsuccessfully to send their first probe to Mars in 1960 and would persist in their attempts in the coming decades. Because the Cold War was in full swing, U.S. scientists had little knowledge of any details of the Russian Mars probes. They believed at the time that Soviet goals were similar to those of the United States, "with particular attention to the possibility of life." For reasons still unclear, however, the Russians seem never to have attempted any life detection experiments of the complexity contemplated by the Americans. Years later, it was reported that considerable effort was put into the idea of detecting the Sinton bands during the earliest flybys, but the Russian Mars landers carried only cameras and instruments to detect the chemical composition of the atmosphere and surface soil.[181]

By contrast, interest in the United States was not only real, but was followed up by feverish action. Support for NASA's involvement received strong endorsement from the prestigious National Academy Space Science Board Summer Study at Iowa State University in 1962, which reviewed the full scope of space biology. In setting forth the general philosophy for the three areas of space biology – exobiology, environmental biology, and humanity in space – the authors were unequivocal in "setting the search for extraterrestrial life as the prime goal of space biology." In doing so, they realized not only the scientific but also the philosophical import of exobiology:

> it is not since Darwin – and before him Copernicus – that science has had the opportunity for so great an impact on man's understanding of man. The scientific question at stake in exobiology is, in the opinion of many, the most exciting, challenging, and profound issue, not only of this century but of the whole naturalistic movement that has characterized the his-

[181] For a contemporary American view of the goals of Soviet exploration of Mars see B. G. Murray and Merton Davies, "A Comparison of U.S. and Soviet Efforts to Explore Mars," *Science,* 151 (February 25, 1966), 945–954. For tables of Mars spacecraft and their instruments as revealed by 1992, and mention of interest in the Sinton bands, see G. W. Snyder and V. I. Moroz, "Spacecraft Exploration of Mars," in Hugh H. Kieffer et al., *Mars* (Tucson, Ariz., 1992), 71–119. The work of G. A. Tikhov may have spurred interest in the Sinton bands; but the anomaly is that the country of Oparin – the acknowledged pioneer in origin of life studies – apparently did not develop experiments to test for life on the Martian surface.

> tory of western thought for three hundred years. What is at stake is the chance to gain a new perspective on man's place in nature, a new level of discussion on the meaning and nature of life.[182]

With this statement, and the detailed recommendations that followed, exobiology was firmly entrenched as a major NASA effort and – not coincidentally – one with high public interest.

All of this activity led to a study initiated in June 1964, again by the Space Science Board of the National Academy of Sciences, a landmark study in many ways a culmination of previous discussions. Convened at the request of NASA, which was seeking guidance for its space programs, it included on its steering committee Lederberg and Calvin, veterans of previous studies, who knew that their recommendations were likely to be taken very seriously by NASA. The purpose of this study was "to recommend to the government, through the Academy's Space Science Board, whether or not a biological exploration of Mars should be included in the nation's space program over the next few decades; and further, to outline what that program, if any, should be."[183] As in the 1962 study, the conclusion was that "the exploration of Mars – motivated by biological questions – does indeed merit the highest scientific priority in the nation's space program over the next decade."

With a strong consensus on carrying out a search for life on Mars via spacecraft, the next question was exactly how to implement this consensus. Already in March 1959 University of Rochester biologist Wolf Vishniac, a veteran of the EASTEX meetings of 1958, had received a NASA grant to develop "a prototype instrument for the remote detection of microorganisms on other planets."[184] Proposals had proliferated so much by 1963 that NASA's Ames Research Center set up a team to evaluate the concepts. Reviews of life detection instruments were given at a JPL symposium on "Current Aspects of Exobiology" in 1963; the subject received a good deal of attention during the Space Science Board study in 1964, and

[182] National Academy of Sciences, *A Review of Space Research: The Report of the Summer Study Conducted under the Auspices of the Space Science Board of the National Academy of Sciences at the State University of Iowa, Iowa City, Iowa, June 17–August 10, 1962* (Washington, D.C., 1962), 9–2, 9–3.

[183] Colin S. Pittendrigh, Wolf Vishniac, and J. P. T. Pearman, eds., *Biology and the Exploration of Mars: Report of a Study Held under the Auspices of the Space Science Board, National Academy of Sciences, National Research Council, 1964–1965* (Washington, D.C., 1966), viii.

[184] Ezell and Ezell, *On Mars*, 63. Vishniac's first thoughts on a life detection system were stimulated by a challenge of astronomer Thomas Gold at the EASTEX meeting in December 1958. On Vishniac see Shirley Thomas, *Men of Space*, 6, 276.

increasingly, it became a subject of greater urgency throughout the 1960s.[185] And with good reason: photographic reconnaissance spacecraft to Mars were in the design stage in the early 1960s, and spacecraft with real experimental capability in the realm of life detection were the ultimate goal.[186]

In short, for the first time, substantial government resources were being invested in the search for extraterrestrial life. This was not Percival Lowell and his small staff chasing an eccentric theory, nor even a scattering of astronomers around the country at their observatories trying to prove or disprove a theory. This was nothing less than a national government proposing to spend substantial sums of taxpayer money to launch a concerted effort to search for life beyond the Earth. For the first time, therefore, the question of extraterrestrial life entered the arena of science policy. It is not surprising in a democratic society that this NASA goal – like many others – was challenged and criticized. Among the critics were two prominent scientists: chemist Philip H. Abelson and microbiologist Barry Commoner. Both questioned the goal not only for scientific reasons, but also as a matter of priorities in a world filled with poverty. Abelson used his position as editor of the influential journal *Science* to argue in 1965 that "In looking for life on Mars we could establish for ourselves the reputation of being the greatest Simple Simons of all time."[187]

Nevertheless, spacecraft exploration of Mars enjoyed widespread public support in the United States and had developed tremendous momentum that could not be halted even when new ground-based studies and the earliest spacecraft yielded surprisingly discouraging results (Table 3.5). We recall that by 1963 Sinton's claim for Martian vegetation had been brought into question, and that by 1965 Sinton himself believed the infrared bands were not due to Martian vegetation at all, but rather to

185 Ezell and Ezell, *On Mars,* 72–73; NASA Ames Research Center Life Detection Experiments Team, "A Survey of Life Detection Experiments for Mars" (August 1963), 70–71; *Concepts for Detection of Extraterrestrial Life,* Freeman H. Quimby, ed., NASA SP-56 (Washington, D.C., 1964); G. L. Hobby, "Life Detection Experiments," in G. Mamikunian and M. H. Briggs, *Current Aspects of Exobiology* (Oxford, 1965); four articles in Pittendrigh et al., *Biology and the Exploration of Mars,* and volume 22 of *Advances in the Astronautical Sciences,* ed. James Hanrahan (Washington, D.C., 1967).

186 On the history of the Voyager project to Mars and Venus see Ezell and Ezell, *On Mars,* 83–119.

187 Philip H. Abelson, *Proceedings of the National Academy of Sciences,* 47 (April 1961), 575–580; "Scientists' Testimony on Space Goals," *Hearings before Senate Committee on Aeronautical and Space Sciences,* 88th Congress, 1st session, June 10–11, 1963, 3ff.; "The Martian Environment," *Science,* 147 (February 12, 1965), 683; Ezell and Ezell, *On Mars,* 80. For background on Abelson see Thomas, *Men of Space,* 6 (1963), 1–27.

Table 3.5. *Space Age observations of Mars relevant to life*

Observer	Date	Location	Method	Result
Sinton[a]	1959	Palomar 200 in.	IR spectroscopy	Vegetation
Spinrad, Munch, Kaplan[b]	1963	Mount Wilson 100 in.	IR spectroscopy	Water vapor 25 mb pressure
Mariner 4[c]	Nov. 28, 1964	Mars flyby	Photography	Impact craters
Mariner 6	Feb. 24, 1969	Mars flyby	Photography	20 % of surface photographed
Mariner 7[d]	Feb. 18, 1969	Mars flyby	Photography	
Mariner 9[e]	May 30, 1971	Mars orbit	Photography	Entire surface photographed; channels
Viking 1 lander	July 20, 1976 (landed)	Mars surface (Chryse Planitia)		
Viking 2 lander	Sept. 3, 1976 (landed)	Mars surface (Utopia Planitia)		
Viking Biology Experiments:				
Oyama[f] and Berdahl			Gas exchange	O_2 and CO_2 liberated no life
Levin[g] and Straat			Labelled release	Positive result but ambiguous
Horowitz[h] Hobby Hubbard			Pyrolitic release (carbon assimilation)	Positive result but not biological
Biemann[i] Oro et al.			Gas chromatograph mass spectrometer (GCMS)	No organic compounds
Hess et al.[j]			Meteorological	Pressure 7 mb average at surface; temperature180 K-240 K over 1 day

deuterated water in the Earth's own atmosphere. In 1963 too Spinrad, Münch, and Kaplan reported from spectroscopic observations at Mount Wilson small amounts of water vapor and an extremely low Martian atmospheric pressure of 25 millibars. And when in July 1965 Mariner IV passed within 6118 miles of Mars and relayed 22 photos showing a cratered and apparently dead planet, its instruments measured an even lower atmospheric pressure of 10 millibars at the surface of the planet. This was only 1% of the Earth's 1000-millibar surface atmosphere, the equivalent of Earth's atmosphere at 90,000 feet. Though all of these developments supported Abelson and other critics, the National Academy of Sciences Board, a summary of its conclusions just published, did not alter the recommendations of its final report, and NASA pressed

Notes to Table 3.5 (*cont.*)

[a] William M. Sinton, "Further Evidence of Vegetation on Mars," *Science* (November 6, 1959), 1234-1237.

[b] H. Spinrad, G. Munch and L. D. Kaplan, "The Detection of Water Vapor on Mars," *ApJ*, 137 (1963), 1319-1321; Kaplan, Munch and Spinrad, "An Analysis of the Spectrum of Mars, *ApJ*, 139 (1964), 1-15.

[c] Mariner 4 photographic results are given in R. B. Leighton et al., "Mariner IV Photography of Mars: Initial Results," *Science*, 149 (August 6, 1965), 627-630, and Richard K. Sloan, *Proceedings of the Conference on the Exploration of Mars and Venus, August 23-27, 1965* (Blacksburg, Va., 1965), 9-1 to 9-37.

[d] Mariner 6 and 7 results have been discussed together in *Journal of Geophysical Research* (*JGR*), 76 (January, 1971), 293-472.

[e] Mariner 9 results are given in *Icarus* (October 1972, January 1973 and July 1973), and in *JGR* (July 1973 and September 1974).

[f] V. I. Oyama and B. J. Berdahl, "The Viking Gas Exchange Experiment Results from Chryse and Utopia Surface Samples," *JGR*, 82 (1977), 4669-4676.

[g] G. V. Levin and P. A. Straat, "Recent Results from the Viking Labeled Release Experiment on Mars," *JGR*, 82 (1977), 4663-4667.

[h] N. H. Horowitz, G. L. Hobby, and J. S. Hubbard, "Viking on Mars: The Carbon Assimilation Experiments," *JGR*, 82 (1977), 4659-4662.

[i] K. Biemann et al., "The Search for Organic Substances and Inorganic Volatile Compounds in the Surface of Mars," *JGR*, 82 (1977) 4641-4658.

[j] S. L. Hess, R. M. Henry, C. B. Leovy, J. A. Ryan, and J. E. Tillman, "Meteorological Results from the Surface of Mars: Viking 1 and 2," *JGR*, 82 (1977), 4559-4574.

ahead.[188] By early 1969 Mariners 6 and 7 had taken high-resolution photographs of about 20 percent of the Martian surface, with results that did nothing to improve the bleak outlook for life on Mars.[189]

But as if to underscore the need for caution in making preliminary conclusions, the mission of Mariner 9 in 1971 showed a very different Mars – one with many kinds of terrain, including channels that resembled nothing so much as dry river beds. If there had once been water on Mars, might not life have existed – and perhaps survived?[190] Mariner 9 – a major milestone as the first spacecraft to orbit another planet, returning numerous images of high resolution – revived such questions, but at the same time it permanently laid to rest the Mars of Percival Lowell. As one project scientist wrote:

> Mariner 9 discovered all the major geologic features on Mars and permanently deleted the word "canal" from the Martian vocabulary. One of the classical canals, Agathodaemon, was found to coincide with Valles Marineris, and traces of a few others were identifiable, but the vast bulk of them corresponded neither to topographic nor to albedo features and appear to have no relation to the real Martian surface.[191]

The culmination of the search for life in the solar system was the landing of two Viking spacecraft on the surface of Mars in 1976, surely one of the great adventures in the history of science and technology. The Viking project, initiated in 1968 after the demise of the Mars Voyager project and now managed by NASA's Langley Research Center, was an example of "big science" at its best in terms of budget, staff, goals, and results. The cost of the Viking spacecraft, including the orbiters, landers and support (but not launch) vehicles, was $930 million. It was headed by Project Manager James S. Martin. Thousands of scientists and engineers at NASA and its contractors participated in the construction of the Viking spacecraft and its subsystems. Although the usual funding hurdles had to be overcome and many critics answered, in the end two Viking

[188] Pittendrigh et al., *Biology and the Exploration of Mars,* Postscript: October 1965, 19.

[189] See Table 3.4. The detection of Martian water vapor by Spinrad and colleagues had an enormous impact in the media. Many astronomers remained skeptical, according to Ron Schorn, chief of Ground-Based Planetary Astronomy at NASA at the time. Schorn to S. Dick, private communication, October 1994, and Tatarewitz, *Space Technology,* 60ff.

[190] Mariner 9 results appeared in *Icarus* (October 1972, January 1973, July 1974), and *Journal of Geophysical Research* (July 10, 1973, and September 10, 1974).

[191] Conway Snyder, "The Planet Mars as Seen at the End of the Viking Mission," *JGR,* 84 (December 30, 1979), 8487. The most detailed study comparing Mariner data with Lowell's canals is C. Sagan and P. Fox, "The Canals of Mars: An Assessment after Mariner 9," *Icarus,* 25 (August 1975), 602–612.

orbiters arrived at the planet on June 19 and August 7, 1976. After suitable reconnaissance, as the United States celebrated its bicentennial back on Earth, two Viking landers set down on Mars in July and September. Under the guidance of project scientist Gerald A. Soffen, 13 teams with a total of 78 scientists undertook 13 separate investigations, including 3 mapping experiments from the orbiter, 1 atmospheric experiment, 1 radio and radar experiment, and 8 surface experiments. The total cost for development and execution of these experiments was another $227 million.[192] The results increased knowledge of Mars far beyond all previous investigations combined, finally providing definitive answers to age-old questions, including the issues of temperature, atmospheric composition, and pressure so crucial to life.

From beginning to end, though the various science teams grappled with the myriad problems of meteorology, seismology, chemistry, imaging, and physical properties of the planet Mars, the Viking biology experiments were the driving force behind the project, as evidenced by both budget and public, congressional, and even scientific interest. A total of $59 million was spent on the Viking biology package and another $41 million on the molecular analysis experiment that was relevant to the question of life because of its ability to detect organic molecules. Harold P. Klein of NASA's Ames Research Center headed the Viking biology science team; Klaus Biemann of MIT headed the separate molecular analysis team. While appreciating the relevance of many of the teams to the question of Martian biology, we must here focus on the work of the 2 of the 13 teams whose experiments were most directly relevant to the question of Martian biology.

The Viking biology package (Figure 3.12) embodied in one piece of technology the most sophisticated thinking of the twentieth century on the subject of extraterrestrial life in the solar system. The assumptions behind its experiments, the results obtained, and the ensuing controversies over the interpretation of these results are therefore of considerable importance. The diverse ideas about the nature of Martian life led to three different biology experiments aboard Viking, each representing a different approach to the problem of life. Indeed, biology team leader Klein later stated that had it not been for the constraints of 15 kilogram weight and about 1 cubic foot volume for the biology package, even more of the approaches conceived during the previous two decades would have been included on the spacecraft. The idea was that the three experiments,

[192] Gerald A. Soffen, "The Viking Project," *JGR*, 82 (September 30, 1977), 3959–3970. The official NASA history of the Viking project is Ezell and Ezell, *On Mars*. See also R. S. Young, "The Origin and Evolution of the Viking Mission to Mars," *Origins of Life*, 7 (1976), 271–272.

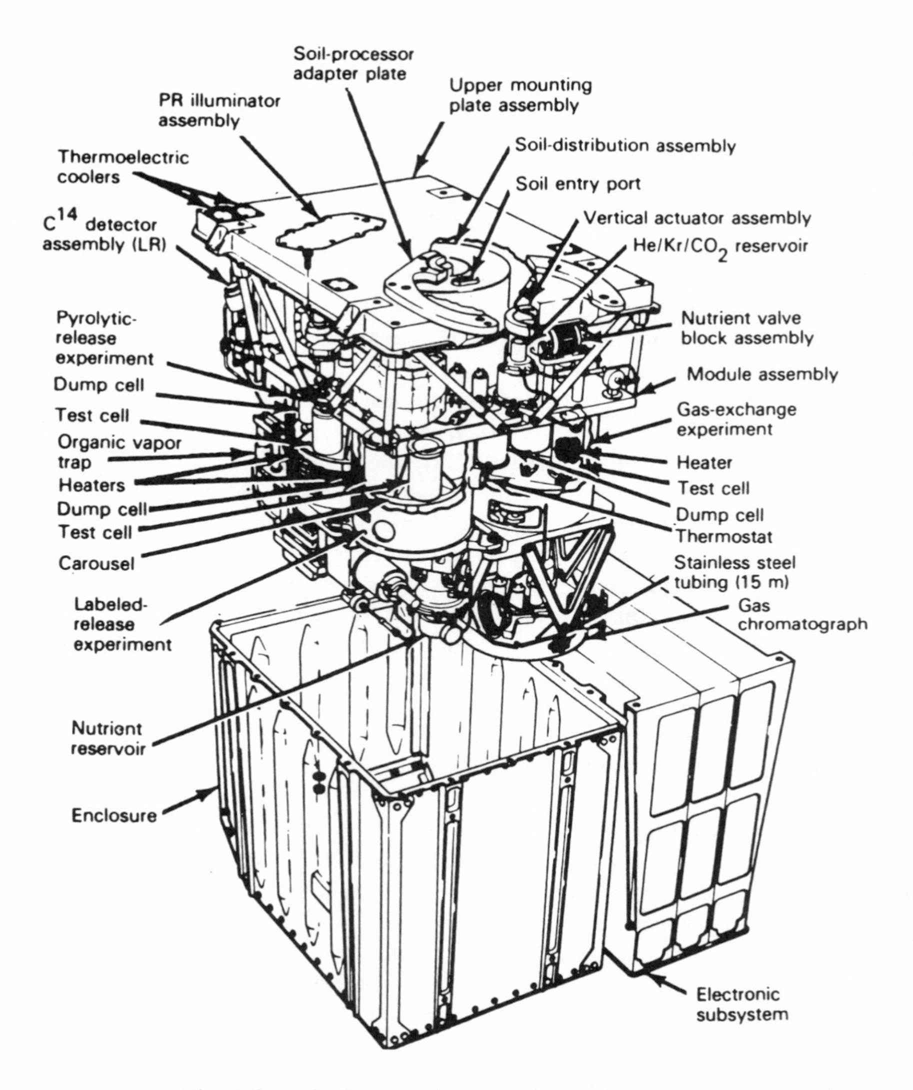

Figure 3.12. The Viking biology package, with a volume of approximately 1 cubic foot, showing the three biology experiments and associated equipment. The pyrolitic release experiment is at the upper left, the labeled release experiment is at the lower left, and the gas exchange experiment is at the center right. Courtesy NASA.

singled out and recommended by the Space Science Board of the National Academy of Sciences in 1968, would test for life using different philosophies, environmental conditions, and detectors.[193]

One approach, which came to be known as the "labeled release" experiment, was developed by Gilbert Levin, who had spent much of the 1950s trying to improve methods for the detection of bacterial contaminants in city water supplies and believed his method could be applied to the search for life on Mars. He was awarded a NASA contract for his "Gulliver" concept in 1961 and was already reporting on his experimental apparatus in the early 1960s. Levin's approach assumed that any Martian microorganisms, like those on Earth, would assimilate (eat) simple organic compounds, decompose them, and produce gases such as CO_2, methane, or hydrogen as end products. For this reason, a dilute aqueous solution of seven such organic compounds, radioactively labeled for detection purposes, was added to the incubation chamber containing the Mars soil sample. The experiment tested for the expected labeled release of the gas produced as any organisms ate the organics and breathed out the decomposition products.[194] The output was in the form of radioactive disintegrations, measured by a carbon-14 detector in counts per minute.

The second biology test, the "gas exchange" experiment, was developed by Vance Oyama of NASA Ames Research Center, a veteran of life detection experiments on Apollo lunar samples. The gas exchange experiment tested for life under two different conditions. In the first mode, it was assumed that any organism in the dry Martian environment would be stimulated to metabolic activity by the addition of slight water moisture, and would give off a gas that could be detected by chromatography in the area immediately above the sample. In the second, "wet nutrient" (or chicken soup) mode, a rich nutrient of 19 organic compounds was added as an additional stimulus to metabolic activity, the products to be detected in the same manner. In both cases, the liquid added did not come

193 These experiments were described in detail in a special issue of *Icarus,* 16 (February 1972), 1–228, as cited later. For more on the experimental strategy see *Origins of Life,* 7 (1976), 271–333. On the constraints imposed on the biology package see H. P. Klein, "General Constraints on the Viking Biology Investigation," *Origins of Life,* 7 (1976), 273–279. On the Space Science Board/National Academy of Sciences recommendation see *Planetary Exploration, 1968–1975. Report of a Study by the Space Science Board,* July 1968, cited in Soffen and Young, *Icarus,* 16 (February 1972), 1–16.

194 G. V. Levin, "Detection of Metabolically Produced Labeled Gas: The Viking Mars Lander," *Icarus,* 16 (February 1972), 153–166. For Levin's earlier work on the subject see Levin, A. H. Heim, J. R. Clendenning and M. F. Thompson, "Gulliver – A Quest for Life on Mars," *Science,* 138 (October 12, 1962), 114; G. V. Levin, A. H. Heim, M. F. Thompson, N. H. Horowitz, and D. R. Beem, "Gulliver – An Experiment for Extraterrestrial Life Detection and Analysis," *Life Sciences and Space Research,* 2 (Amsterdam, 1964), 124–133; and Levin and Heim, "Gulliver and Diogenes – Exobiological Antithesis," *Life Sciences and Space Research,* 3 (1965), 105.

into contact with the soil, but was added underneath the cell in which the soil "incubated." Water vapor gradually seeped up through the porous bottom of the incubation chamber, creating gradations of moisture through the soil. Experiments were also undertaken without the addition of any moisture.[195]

The "pyrolytic release" experiment (also called the "carbon assimilation" experiment) was headed by Norman Horowitz of Caltech. Horowitz, a member of the WESTEX group in 1959, had cooperated with Levin's project in the early 1960s, but after Mariner IV showed that liquid water could not exist on the planet, he split with Levin and became convinced that it was best to test for Martian organisms under conditions known to exist on Mars when the experiment was designed. Horowitz believed this:

> In every observable parameter – atmospheric pressure and composition, radiation flux, temperature regime, water abundance – the Martian environment is now perceived to be hostile to life to a degree unknown anywhere on Earth. Hostile though it may be, however, we can be certain that if life exists on Mars it is adapted to this extreme environment and maladjusted to extreme departures from it. It follows that we are not likely to succeed in detecting Martian metabolism, or growth, or any other vital function if, in the attempt, we deviate radically from actual Martian conditions – for example, by trying to culture Martian organisms in an aqueous medium. Rather, we have to devise experiments that use Martian materials under Martian conditions if we hope to elicit a positive response or, failing that, a meaningful negative one.

Thus, to the small sample of Martian soil, Horowitz proposed in his experiment to add only CO_2 and carbon monoxide, gases known to exist in the Martian atmosphere and now radioactively "tagged" for detection purposes. It was assumed that any organism on Mars would have developed the ability to assimilate these gases and convert them to organic matter. After 120 hours of incubation, the soil chamber was to be heated to 635°C to pyrolyze the organic matter and release the volatile organic

[195] V. I. Oyama, "The Gas Exchange Experiment for Life Detection: The Viking Mars Lander," *Icarus,* 16 (February 1972), 167–184. On the work on lunar life detection see further references therein. Other investigators were also involved in biological study of lunar samples. See, for example, J. W. Schopf, "Micropaleontological Studies of Lunar Samples," *Science,* 167 (January 30, 1970), 779–780, and other papers in the same issue of *Science.*

products, thus the name "pyrolytic release."[196] A radiation counter measured disintegrations per minute.

All three experiments sought to detect metabolic activities. Of these experiments, Oyama's wet nutrient mode was the most Earth-like approach in that it added rich terrestrial organics to stimulate any Martian organisms. Horowitz's was the most Mars-like, making few assumptions about Martian life except that it would be carbon based. Levin's, with his weak organic nutrient, fell in between. Levin and Oyama's experiments attempted to detect life by the decomposition of organics into gas during metabolism (a universal property of terrestrial organisms), while Horowitz sought to synthesize organic matter, which he would then pyrolyze in order to be able to detect. For detection purposes, both Levin and Horowitz used standard techniques of radioactive carbon-14 as a "tracer," a method that did not change the chemistry but provided a means of distinguishing atmospheric carbon from metabolized carbon. Oyama used the well-known method of gas chromatography for detection, as did Biemann (in conjunction with a mass spectrometer) for the organics experiment, which had nothing to do with metabolism. Since they were ignorant of the nature of Martian life, the fondest hope of all the experimenters was that at least one of the experiments – hopefully their own – would turn up something.

The integration of these experiments into a coherent biology package that would survive the extremes of interplanetary space and the Martian environment was no small task. Under contract to Martin Marietta (the lander prime contractor), the TRW Defense and Space Systems Group began development of the biology package in 1970. "A single instrument of the complexity required to conduct the three active biology experiments has never before been assembled for spaceflight," they wrote. This complexity was due to the many functions that had to be performed: "Small amounts of soil are transported, tiny amounts of gases and liquids are metered, gas flowrates are controlled, 750 degree C temperatures are reached, and soil temperatures are held below those of the surrounding environment by [a] thermoelectric cooling system." In addition, nuclear detection systems were needed, as well as a gas chromatograph and illumination by a Xenon arc lamp to simulate the Sun (minus the UV rays). A

[196] N. H. Horowitz, J. S. Hubbard, and G. L. Hobby, "The Carbon-Assimilation Experiment: The Viking Mars Lander," *Icarus,* 16 (February 1972), 147–152: 147. On Horowitz as a member of WESTEX see Cooper (footnote 201), 99. In 1959 Horowitz had chaired an "Astrobiology" session of the Lunar and Planetary Exploration Colloquium; see *Proceedings of the Lunar and Planetary Exploration Colloquium* (September 23–24, 1959), 2, no. 1, 2–14.

total of 4000 mechanical parts and 1800 electronic parts went into the Viking biology package, which was delivered in 1975 after 5 years of design, fabrication, and testing.[197]

Both Viking spacecraft were launched in summer 1975, and during the 348 million kilometer journey, more than one participant undoubtedly contemplated the possibly historic outcome: the detection of life on Mars would prove much more than a fitting birthday gift for a country that prided itself on technological prowess and curiosity; it would also be a culmination to an age-old question and a watershed in human history. The prospects just before landing were assessed in an article by Lederberg and Sagan in the journal *Icarus,* edited by Sagan. With an eye on future planetary programs, the authors emphasized that Viking's instruments were limited to detecting only certain types of organisms and that follow-up spacecraft would be needed, but that Viking represented a significant first step. As for the expectations of the biology team itself prior to the Viking landing, opinions were split: Horowitz was pessimistic, Levin was optimistic, and Oyama and Klein gave the lander a 50–50 chance of detecting Martian life.[198]

Summer 1976 finally brought the day that Lowell, Kuiper, and a host of scientific ghosts would have savored: the landing of two spacecraft on the surface of Mars to test for life in situ. They would not have been disappointed: Viking 1 landed successfully on the Chryse plain on July 20, and the first results of the biology experiments returned from Viking were exciting, to say the least. Although no visible life forms walked across the field of view of the camera,[199] once the soil samples were collected on July 28, the biology experiments quickly began to return major surprises. Levin's experiment evolved gas into the chamber after the nutrient was added; then the reaction tapered off. Horowitz's pyrolytic release test was also positive, and Oyama's gas exchange experiment evolved not only CO_2 but also oxygen, the latter a reaction never before seen in tests on terrestrial or lunar soils. Because of the speed and course of the latter reaction, Oyama's experiment was not believed to be biological in nature. In short, two of the three biology experiments gave

[197] On the technological challenges in construction of the instruments from the point of view of the engineers , see Fred S. Brown, "The Biology Instrument for the Viking Mars Mission," *Review of Scientific Instruments,* 49 (February 1978), 139–182. The instruments are also described in H. P. Klein, *Origins of Life,* 5 (1974), 431, and H. P. Klein et al., "The Viking Mission Search for Life on Mars," *Nature,* 262 (July 1, 1976), 24.

[198] Carl Sagan and Lederberg, "The Prospects for Life on Mars, a Pre-Viking Assessment," *Icarus,* 28 (June 1976), 291–300. For the experimental strategy see *Origins of Life,* 7 (1976), 271–333, for articles by Young, Klein, Hubbard, Levin, and Oyama.

[199] The camera results for biology are given in Levinthal, Jones, Fox, and Sagan, "Lander Imaging as a Detector of Life on Mars," *JGR,* 82 (September 30, 1977), 4468–4478: 4468.

"presumptive positive results" for biology, and the third gave evidence of an oxidizing material in the surface at the Viking site. There was only one problem: in another unexpected finding, Biemann's organic analysis showed no organic molecules present to the level of a few parts per billion, a result later called "probably the most surprising single discovery of the mission." As Klein has subsequently recounted, these first results caused the carefully laid out experimental strategy to be abandoned as the scientists attempted to discover whether chemical or biochemical reactions were taking place.[200]

Hour by hour and day by day the experimenters grappled with their unexpected data. The August 27, 1976, issue of *Science* reported preliminary results from the Viking orbiter and lander, but since the first surface samples had been collected only on July 28, no results from the biology experiments were yet available.[201] By October, however, the biology team reported in the same journal its preliminary results of the first month's operation. The watchword was caution as the joint team warned, by way of preface:

> All three of the experiments yielded data indicating that the surface material of Mars is chemically or biochemically quite active. Under normal circumstances, it would be premature to report biological experiments in progress before the data are amenable to ready interpretation. However, the unique nature of this investigation impels us to make this report, and we are fully cognizant of its preliminary nature.

Horowitz reported that a small amount of gas was indeed converted into organic (not necessarily living) material and that heat treatment of a duplicate sample prevented such conversion – a key control test. But, he concluded, "although these preliminary findings could be attributed to biological activity, several experiments remain to be done before such an interpretation can be considered likely." Levin reported that a substantial evolution of radioactive gas was registered, which also did not occur with

200 G. W. Snyder, "The Planet Mars as Seen at the End of the Viking Mission," *JGR*, 84 (December 30, 1979), 8508; H. P. Klein, "The Viking Biological Investigation: General Aspects," *JGR*, 82 (September 30, 1977), 4677–4680.

201 G. A. Soffen and C. W. Snyder, "The First Viking Mission to Mars," *Science*, 193 (August 27, 1976), 759–766. An appreciation of the hour-by-hour "instant science" can be obtained from the many journalistic accounts of the Viking biology experiments, especially Henry S. F. Cooper's accounts in *The New Yorker*, published in book form as *The Search for Life on Mars* (New York, 1980). While this fine-grained activity is a gold mine for those interested in how scientists reach conclusions (in this case, under the pressure of intense media scrutiny), we have concentrated here on the published record, which represents the scientific results published after considered judgment of the evidence, even if still incomplete.

a heat-treated sample. Oyama reported that a substantial amount of oxygen was detected in his first chromatogram only 2.8 hours after humidification, and subsequently significant increases in CO_2 were found. In their general conclusion taking into account all three experiments, as well as Biemann's, the authors stated that while the experiments gave clear evidence of chemical reactions, "the essential question is whether they are attributable to a biological system. We are unable at this time to give a clear answer to that question, partly because the planned experimental program is not yet completed and partly because of the inherent difficulty in defining complex living organisms which may have developed and evolved in an environment completely different from that of the planet Earth." In his overview, Soffen reported that "the biology experiment is indeterminate but has yielded some clues to the chemistry of the surface."[202]

Ten weeks later, as Mars was out of communication during conjunction, an entire issue of *Science* was devoted to the Viking results, and by now the biology experimenters were ready to give what they called "interim results." After describing his positive results, Horowitz was still unwilling to commit to a biological explanation, saying that further experimentation was needed to resolve ambiguities. Levin's evidence, in the form of plots of radioactivity supposedly caused by metabolic activity (Figure 3.13), also concluded that "available facts do not yet permit a conclusion regarding the existence of life on Mars." And Oyama, reporting in *Nature,* agreed. Soffen was still forced to conclude that "the tests revealed a surprisingly chemically active surface – very like oxidizing. All experiments yielded results, but these are subject to wide interpretation. No conclusions were reached concerning the existence of life on Mars."[203]

By eight and a half months after the first Viking had landed, 26 biological experiments had been carried out and the first relatively complete results were reported, along with the results of other Viking experiments, in the *Journal of Geophysical Research*. By then, shortly before the biological experiments were terminated in May 1977, Klein's considered judgment was that the positive result of Horowitz's pyrolytic release experiment was probably nonbiological in origin, while Levin's labeled release experiment remained ambiguous. Ironically, the gas exchange experiment of Oyama – the scientist most optimistic about Martian life

[202] H. P. Klein et al., "The Viking Biological Investigation: Preliminary Results," *Science,* 194 (October 1, 1976), 99–105; and Soffen, "Status of the Viking Mission," 57–58.

[203] N. H. Horowitz, G. L. Hobby, and J. S. Hubbard, "The Viking Carbon Assimilation Experiments: Interim Report," *Science,* 194 (December 17, 1976), 1321–1322; Gilbert V. Levin and P. A. Straat, "Viking Labeled Release Biology Experiment: Interim Results," ibid., 1322–1329; and G. A. Soffen, "Scientific Results of the Viking Missions," ibid., 1274–1276.

(after Vishniac's death) – showed no evidence at all of biological activity. Oyama and most of his colleagues concluded that the spontaneous evolution of oxygen was due to a chemical reaction involving "superoxides" such as hydrogen peroxide, perhaps by the effect of solar radiation on the small amount of water vapor in the upper atmosphere of Mars. "It's like the three bears," Klein later said. "Not too much water, not too little water, just the right amount of water in its atmosphere to produce something like this. This is one of the big mysteries, and any future missions to Mars have to find out what this stuff is."[204]

In the end, there was no complete consensus among the experimenters themselves. Writing for *Scientific American*, Horowitz concluded that although "it is not easy to point to a nonbiological explanation for the positive results" of his pyrolytic release experiment, "it appears that the findings of the pyrolytic-release experiment must also be interpreted nonbiologically," mainly because the reaction was less sensitive to heat than one expected from a biological process. But there was another factor in his decision:

> There are doubtless some who, unwilling to accept the notion of a lifeless Mars, will maintain that the interpretation I have given is unproved. They are right. It is impossible to prove that any of the reactions detected by the Viking instruments were not biological in origin. It is equally impossible to prove from any result of the Viking experiments that the rocks seen at the landing sites are not living organisms that happen to look like rocks. Once one abandons Occam's razor the field is open to every fantasy. Centuries of human experience warn us, however, that such an approach is not the way to discover the truth.[205]

Levin, however, did not agree; in the 1980s he continued to argue forcefully that a biological interpretation of his data was still possible.[206]

Clearly sensitive to their own assumptions, the Viking biologists continued to ponder the strategy of their experiments: "In arriving at any final

204 Steven J. Dick interview with H. P. Klein, September 15, 1992, p. 22. See also the special issue of *JGR*, 82, no. 28 (September 30, 1977), 3959–4681, especially V. Oyama and B. Berdahl, "The Viking Gas Exchange Experiment Results from Chryse and Utopia Surface Samples," 4669–4676, and H. P. Klein, "The Viking Biological Investigation: General Aspects," 4677–4680. This volume was reprinted as *Scientific Results of the Viking Project* (Washington, D.C., 1977). The papers of the other biology experiments are cited in Table 3.5.

205 N. H. Horowitz, "The Search for Life on Mars," *SciAm*, 237 (November 1977), 52–61: 61.

206 Gilbert V. Levin, "A Reappraisal of Life on Mars," *The NASA Mars Conference*, ed. Duke B. Reiber (San Diego, Calif., 1988), 187–208.

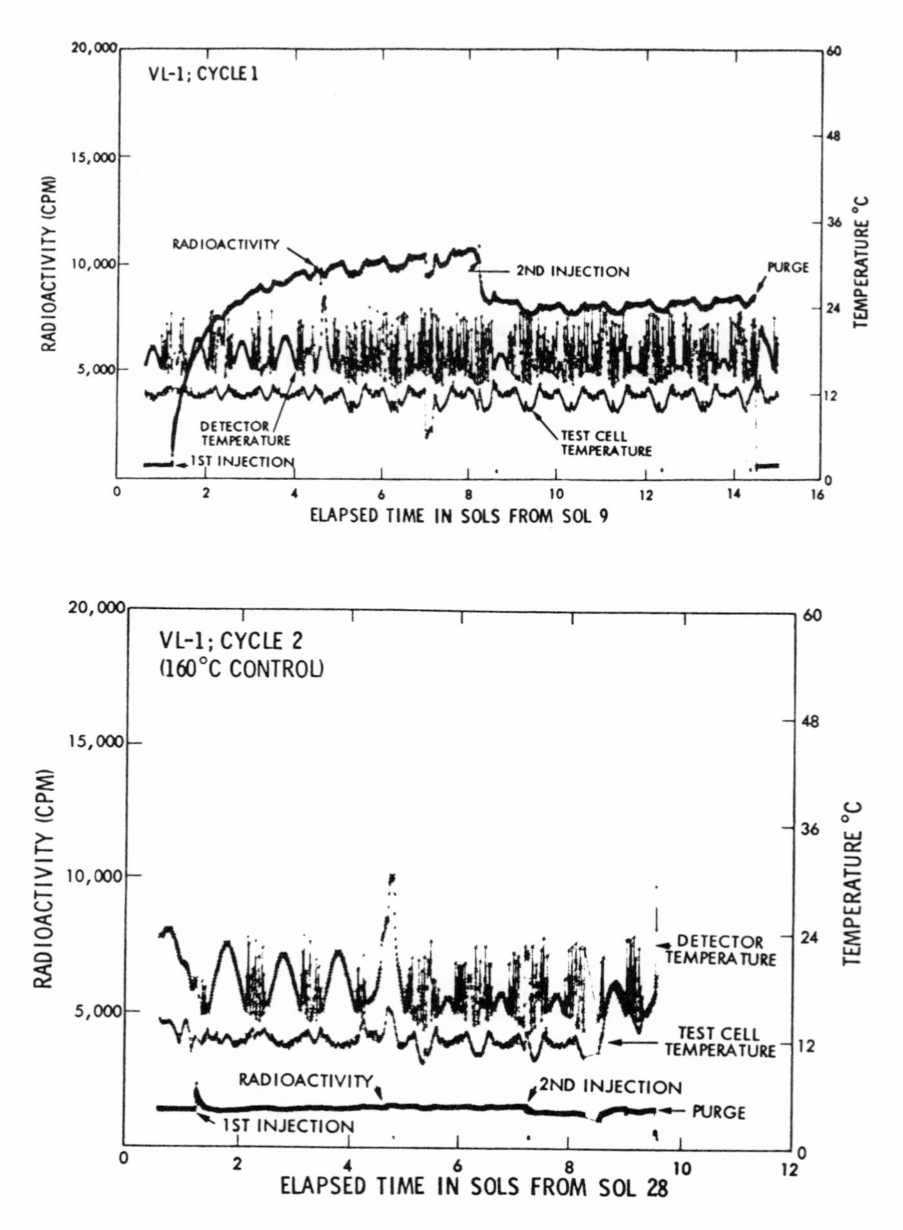

Figure 3.13. Data from the labeled release experiment on the Viking lander, which Levin interpreted as possible evidence of life on Mars. The upper graph

judgement on the fundamental question, Is there life on Mars?, we must carefully consider not only the actual experimental results that have been obtained but also the context within which these data were obtained. We must examine the assumptions on which each of the experimental techniques was based, the conditions under which the experiments were actually carried out, and the data themselves," Klein wrote after the mission. After a discussion of each of the experiments with this caveat in mind, Klein concluded his summary of Viking biology results with this astonishing remark: "Finally, we must not overlook the fact, in assessing the probabilities of life on Mars, that all of our experiments were conducted under conditions that deviated to varying extents from ambient Martian conditions, and while we have accumulated data, these and their underlying mechanisms may all be coincidental and not directly relevant to the issue of life on that planet."[207]

After 10 years of contemplation of the experiments conducted for some 10 months on the surface of Mars, Horowitz remained convinced that they not only proved the absence of life on Mars but also, by extension, "Since Mars offered by far the most promising habitat for extraterrestrial life in the solar system, it is now virtually certain that the earth is the only life-bearing planet in our region of the galaxy. We have awakened from a dream. We are alone, we and the other species, actually our relatives, with whom we share the earth."[208] On the other hand, "Most biologists feel that the results of this first set of metabolic experiments are indecisive," Soffen wrote in 1988, contemplating future Mars missions. "They believe that no life was detected, but that we cannot state for certain that we have exhausted the possibility . . . to dismiss biology once and for all on Mars with our meagre data may be premature."

Still, it is also fair to say that most scientists were much less optimistic about life on Mars in the aftermath of Viking. A National Academy of

[207] *JGR*, 82 (September 30, 1977), 4679.
[208] Horowitz, *To Utopia and Back*, 146. A summary was given at the NASA Mars Conference in 1986, cited earlier, "The Biological Question of Mars," 177–185.

Caption to Fig. 3.13 (*cont.*).
shows a high level of radioactive gas on addition of a radioactive nutrient to the Mars sample, compared to a sterile control sample below. A similar response was seen when the labeled release experiment tested biologically active soils on Earth. Most scientists opted for nonbiological explanations of the Viking data. With permission from G. Levin and P. Straat, "Viking Labeled Release Biology Experiment: Interim Results, *Science*, 194 (December 1976), 1323. Copyright 1976 American Association for the Advancement of Science.

Sciences report in the late 1980s asserted that Viking had lowered the possibilities of life on Mars, in sharp contrast to its opinion 25 years before, at the influential 1964 Summer Study. "I have devoted 20 years to thinking about this question," Soffen himself wrote in 1990. "In my role as the Viking project director, I began with an optimistic view of the chances of life on Mars. I now believe that it is very unlikely. But one doubt lingers: we have not explored the planet's polar regions."[209]

Only history will tell whether subsequent spacecraft will prove the Viking biology experiments a fluke, as Mariner 9 did its predecessors. But its results were impressive enough that most scientists shifted the focus of their biological Martian interests either to past Martian history or to different Martian environments such as rocks, polar caps, subsurface soil, or volcanic regions. Certainly the ancient river valleys still remained a major challenge, and it would be one of the striking ironies of history if, a century after Lowell's canal hypothesis, the "channels of Mars" gave evidence of fossil life. In presenting the latest results on the channels of Mars in a volume by that name, one scientist wrote that "Had this book been written only fifteen years earlier, its title would likely have relegated it to the realm of speculative philosophy, perhaps even to fictional accounts. That the channels of Mars are a matter of serious and intense scientific interest follows directly from the National Aeronautics and Space Administration program of planetary exploration."[210] "Channels" did not have the implications of "canals," but a century of canal bashing had not dulled the imaginations of scientists when it came to the implications of evidence for water in the early history of Mars.

As Viking project scientist Gerald Soffen concluded without exaggeration, the exploration of Mars by spacecraft had important implications for both astronomy and humanity: "Comparative planetology was conceived with Mariner and born with Viking. Our rendezvous with history was a major milestone in human affairs."[211] The Viking experiments, one of the great exploratory adventures of the twentieth century, had failed to find life of any kind on Mars. Most scientists accepted this conclusion, but some likened it to two spacecraft landing on two remote spots on Earth and concluding against any form of life. Neither Russian nor Ameri-

209 G. A. Soffen, "What Should We Look for When We Return to Mars?" in *Bioastronomy – The Next Steps* (Dordrecht, 1988), 3–13: 13; and Soffen, "Life in the New Solar System?" in *The New Solar System* (3d ed., Cambridge, 1990), 275–280; the NAS report is cited on page 278.

210 Victor R. Baker, *The Channels of Mars* (Austin, Tex., 1982, Preface). See also Lynn Rothschild, "Earth Analogs for Martian Life: Microbes in Evaporites, a New Model System for Life on Mars," *Icarus*, 88, (November 1990), 244–260.

211 *JGR*, 82 (September 1977), 3970.

can scientists were willing to exclude the possibility that future spacecraft might find evidence of past or present life on Mars.[212]

Although beyond Mars loomed the question of organic molecules on Jupiter or Saturn's moon Titan, Mars had been the last best hope for life in the solar system.[213] While the importance of Lowell's legacy is evident even in the Space Age, what also stands out is the extraordinary progress made during the twentieth century on the question of life on Mars, despite Lowellian preconceptions. Grave difficulties and problems of interpretation notwithstanding, the process of observation eliminated from Mars first intelligence, then vegetation, and finally organic molecules.

But long before the Viking denouement, eyes had turned beyond the solar system, to the myriad of stars and their potential planetary systems.

[212] Lee Dye, "Soviets to Press Search for Signs of Life on Mars," *Washington Post,* July 19, 1987 section A; C. P. McKay, "Exobiology and Future Mars Missions: The Search for Mars' Earliest Biosphere," *Advances in Space Research,* 6 (1986), 269–285.

[213] See Carl Sagan, "The Solar System beyond Mars: An Exobiological Survey," *Space Science Review,* 11 (1971), 827–866; C. F. Chyba and G. D. McDonald, "The Origin of Life in the Solar System: Current Issues," *Annual Review of Earth and Planetary Science,* 23 (1995), 215–249.

4

PLANETARY SYSTEMS: THE LIMITS OF THEORY

We begin to suspect that life is not the normal accompaniment of a sun, since planets capable of sustaining life are not the normal accompaniments of suns. Astronomy does not know whether or not life is important in the scheme of nature, but she begins to whisper that it must necessarily be somewhat rare.

James Jeans (1923)[1]

Millions of planetary systems must exist, and billions is the better word. Whatever the methods of origin, and doubtless more than one type of genesis has operated, planets may be the common heritage of all stars except those so situated that planetary materials would be swallowed up by greater masses or cast off through gravitational action.

Harlow Shapley (1958)[2]

Should we not come to the rescue of a cosmic phenomenon trying to reveal itself in a sea of errors?

Peter van de Kamp (1983)[3]

If life in the solar system was closely linked to planetary science, the question of life in other solar systems was no less closely linked to "planetary systems science," as the study of other solar systems was becoming known toward the end of the twentieth century.[4] The widespread existence of life beyond the Earth, by any standard definition of life, required that planetary systems be a common feature of the universe. The existence of such systems became an increasingly important research problem in astronomy during the twentieth century. The observation of planetary systems, however, was extremely difficult. Whereas Mars had posed prob-

[1] James Jeans, *The Nebular Hypothesis and Modern Cosmogony, being The Halley Lecture delivered on 23 May, 1922* (Oxford, 1923), 30.

[2] Harlow Shapley, *Of Stars and Men: The Human Response to the Expanding Universe* (Boston, 1958), Chapter 8, "The Fourth Adjustment."

[3] Peter van de Kamp, "The Planetary System of Barnard's Star," *VA*, 26 (1982), 141–157: 157.

[4] David C. Black and Mildred Shapley Matthews, eds., *Protostars and Planets II* (Tucson, Ariz., 1985), Preface, xviii. In *Protostars and Planets: Studies of Star Formation and of the Origin of the Solar System* (Tucson, Ariz., 1978), xviii, editor Tom Gehrels spoke of the development of a new discipline with the purpose of cross-fertilization of ideas on the the origin of stars and the origin of the solar system. In 1985 Black and Matthews wrote, "The elements of Gehrels' 'new discipline' are now beginning to emerge and their interrelationships are being defined. Pursuing his attempt to name this new discipline, it is perhaps appropriate to designate this area of overlap between planetary science and stellar astronomy/astrophysics *planetary systems science*" (Preface, xviii).

lems when observed only tens of millions of miles away, the nearest stars were a million times more distant. The direct observation of planetary companions to those stars was out of the question for most of the century and is still problematic at its end. Whereas in our solar system the crucial problem was to observe planetary surfaces and atmospheres, outside the solar system only the *effects* of supposed planets could hope to be observed, not the planets themselves. Even here the difficulties were so great that, as observation played the central role in the search for life in the solar system, theories of solar system formation assumed the dominant role for much of the century in the quest for other solar systems. By century's end, observation once again assumed an important – but still problematic – role as the arbiter of theory.

It is not our purpose in this chapter to give a detailed history of theories of the origin of our planetary system, long a fascinating problem in astronomy and a task that, to some extent, astronomers and historians of science have already accomplished.[5] We are interested not so much in the technical details of the theories as in the bearing these theories had on the idea of other planetary systems. That bearing was not always made explicit, but the link was never far below the surface; Sir Harold Spencer Jones stated it clearly in his *Life on Other Worlds* (1940): "if we can find out how the solar system came into being we shall possibly be able to judge what likelihood there is that other stars have families of planets."[6] We also examine in this chapter how the observational evidence developed over the course of the century, how theory and observation interacted, and to what extent nontechnical factors may have played a role in an extremely contentious debate on which – as all the participants knew – hinged the question of the widespread distribution of life in the universe. We shall see that the attitude toward planetary systems was skeptical in the first half of the twentieth century, that the period 1943–1958 formed a definite turning point, and that the latter half of the century was almost universally optimistic, not always with good reason.

[5] See especially Stephen G. Brush, "From Bump to Clump: Theories of the Origin of the Solar System 1900–1960," in *Space Science Comes of Age: Perspectives in the History of the Space Sciences,* eds. Paul Hanle and Von del Chamberlain (Washington, D.C., 1981), 78–100; Brush, "Theories of the Origin of the Solar System 1956–85," *RMP,* 62 (January 1990), 43–112; Aleksey E. Levin and Stephen G. Brush, eds., *The Origin of the Solar System: Soviet Research 1925–1991* (New York, 1995); and Stanley L. Jaki, *Planets and Planetarians: A History of Theories of the Origin of Planetary Systems* (Edinburgh, 1978). Among the astronomers reviewing this history were Sir Harold Jeffreys, "The Origin of the Solar System," *MNRAS,* 108 (1948), 94–103; R. Jastrow and A. G. W. Cameron, eds., *Origin of the Solar System* (New York, 1963); and H. Alfvén, "Origin of the Solar System," in *The Origin of the Solar System,* ed. S. F. Dermott (New York, 1978), 19–40.

[6] Harold Spencer Jones, *Life on Other Worlds* (New York, 1940), 247.

4.1 SKEPTICISM: OBSERVATIONAL HINTS AND STELLAR ENCOUNTERS

At the turn of the twentieth century, scientific opinion on the subject of planetary systems was far from unanimous. The late nineteenth century had generally favored the notion of abundant planetary systems, often for philosophical reasons: the British astronomers R. A. Proctor and J. E. Gore, for example, had both agreed that other stars should have planets because otherwise these stars would serve no purpose.[7] Others made the same claim based on physical considerations, especially the nebular hypothesis, which implied to many the universality of solar systems as a result of the condensation of planets from primordial rotating nebulae. Simple analogy with our solar system, combined with a nonanthropocentric viewpoint, yielded the same conclusion.

On the other hand, as we have seen in Chapter 2, A. R. Wallace argued that planetary systems are rare, based on a limited number of single stars around which planets could develop.[8] Nor was his opinion an aberration: many agreed with his conclusion, if not his reasoning, that solar systems were uncommon. The historian of astronomy Agnes Clerke wrote in 1905 that there are "strict limitations to the possible diffusion of planetary worlds." In particular, "We have become aware of incapacitating circumstances, by which a multitude of stars are precluded from maintaining retinues of subordinate globes." The large proportion of double stars discovered by spectroscopic means, she wrote, "makes it impossible any longer to regard the solar system as a pattern copied at large throughout the sidereal domain. We cannot, then, compare it with any other; the mechanism of which the earth forms part must, perforce, be studied in itself and by itself, and it may, for aught that appears, be the outcome of special and peculiar design."[9]

The undecided scientific mind was therefore quite open to any evidence – theoretical or observational – that would shed light on the subject of planetary systems. In the absence of direct observation of planets, prohibited by their relatively small size and by the glare of the stars they were supposed to surround, astronomers necessarily resorted to indirect methods. The most common claim for the observed effects of planets came in the debate over "unseen companions" of stars, as revealed by the gravitational perturbation of a star by one or more encircling planets. Advances in positional astronomy enabled Friedrich Wilhelm Bessel to claim in 1844,

[7] R. A. Proctor, *Other Worlds Than Ours* (New York, 1871), 256; J. E. Gore, *The Worlds of Space* (London, 1894), chapter 3.
[8] A. R. Wallace, *Man's Place in the Universe* (New York, 1904), 278–290.
[9] Agnes Clerke, *Modern Cosmogonies* (London, 1905), 234.

based on visual transit circle observations, that variations in the "proper motions" of the stars Sirius and Procyon indicated the presence of unseen companions. The effect was only a few seconds of arc, but when these companions (now known to be stars and not planets) were subsequently confirmed visually, astronomy had a method for planet detection if only delicate enough observations could be made.[10]

Aside from the change in position of a star due to gravitational perturbation by the planet (now called the "astrometric" method), two other planetary effects on a central star were possible. A change in the amount of light emanating from the star (the "photometric" method) would be caused simply by the slight diminution of light as a planet passed in front of the star as seen from Earth, an argument that had its roots in the postulation in 1782 of an eclipsing companion of Algol. And a change in the spectrum of the star (the "spectroscopic" method) would be caused by changes in the radial velocity of the star as a planet tugged it one way or the other in the course of its orbit, an effect first detected for a companion star by E. C. Pickering with the discovery of "spectroscopic binaries" in 1889. These three techniques – the astrometric, the photometric, and the spectroscopic – were all proven methods for the detection of unseen companions by the turn of the century and would maintain their validity as planet detection techniques at the end of the century, albeit with a much greater appreciation of the difficulties involved when searching for objects of low mass.

At the turn of the century, the idea of "unseen" or "dark" companions was therefore widespread; the question was whether they were stellar or planetary in nature. We know today that those early observations were clearly stellar, but in the context of the times this was far from clear. While today we are accustomed to thinking about Algol, for example, as an "eclipsing binary star," Henry Norris Russell remarked in 1943 that he well remembered the days early in the century when the variation in light of Algol "was ascribed with little or no protest to eclipses by an enormous planet."[11] While astronomers today talk with confidence about "spectroscopic binary stars," it was those stars that Newcomb erroneously interpreted as surrounded by planets when he wrote in 1902 that "the history of science offers no greater marvel than the discoveries of invisible planets moving round many of the stars which are now being made, and in which the Lick Observatory has recently taken the lead."[12]

[10] The visual confirmations were announced, respectively, by Alvan Clark in 1862 and J. M. Schaeberle in 1896. Edmond Halley had shown in 1718 that some of the "fixed" stars had a long-term motion of their own, known as "proper motion."

[11] Henry Norris Russell, "Physical Characteristics of Stellar Companions of Small Mass," *PASP*, 55 (April 1943), 79–86: 85.

[12] Simon Newcomb, *Astronomy for Everybody* (Garden City, N.Y., 1902), 333. On the work to which he was referring, see footnote 17.

While the changes of position caused by unseen companions were widely attributed to stars, they led to the suspicion that unseen companions could undoubtedly extend to planetary size. No one could fail to see the ambiguity when Lowell Observatory astronomer T. J. J. See announced with regard to the double star 70 Ophiuchi that "there is some dark body or other cause disturbing the regularity of its elliptical motion, but heretofore all efforts to see it with the telescope have been unsuccessful." And it was in conjunction with his discussion of double stars and unseen companions that See made the astonishing claim that

> Our observations [of double star systems] during 1896–97 have certainly disclosed stars more difficult than any which astronomers had seen before. Among these obscure objects about half a dozen are truly wonderful, in that they seem to be dark, almost black in color, and apparently are shining by a dull reflected light. It is unlikely that they will prove to be self-luminous. If they should turn out dark bodies in fact, shining only by the reflected light of the stars around which they revolve, we should have the first case of planets – dark bodies – noticed among the fixed stars.[13]

Although this was only an early example of See's penchant for controversial statements in many areas of science, this particular claim shows that planetary systems were very much under discussion at the beginning of the century.

See's early claim for the visual sighting of planets around another star was quickly dismissed by others; indeed, he himself realized that any such planets would have to be enormous, since planets as known to us "would be invisible even if the power of our telescopes were increased a hundredfold." Moreover, the intention of See's discussion and the thrust of his conclusion was not that planetary systems were common, but that the solar system was unique in its character, and therefore required a different explanation for its origin than did double star systems. The well-measured perturbations of some of those systems, See asserted, "lead us to suppose that there are many dark bodies in the heavens; but not even such bodies furnish us evidence of any other system similar to our own, as respects complexity and orderly arrangement . . . our planetary system is now shown to be absolutely unique among the thousands of known

[13] T. J. J. See, "Recent Discoveries Respecting the Origin of the Universe," *Atlantic Monthly,* 80 (1897), 484–492: 489, 491. See does not specify around which stars these "planets" were seen. See's work on 70 Ophiuchi, announced in *AJ* (November 1895 January, 1896), was the subject of a bitter dispute with F. R. Moulton (*AJ*, May 1899) and resulted in See's banishment from publishing further in that journal.

systems, and in the present state of our knowledge appears to be an exceptional formation."[14] On the other hand, in his response to See, the American science writer Garrett P. Serviss, while asserting that planets so large as to be visible from the Earth could not really be planets in the normal sense, believed that double stars were outnumbered 100 to 1 by single stars, which should have planetary companions.

> It is true that we do not know, by visual evidence that the single stars have planets, but we find planets attending the only representative of that class of stars that we are able to approach closely – the sun – and we know that the existence of those planets is no mere accident, but the result of the operation of physical laws which must hold good in every instance of nebula condensation.[15]

Here was the crux of the matter, clearly grasped by both See and Serviss, and passed down through the century: was the solar system a freak or a common occurrence in the universe? And was analogy from general physical principles sufficient to discover the truth, or was hard-nosed observation or theoretical insight necessary?

The resolution of this puzzle was not to come quickly from observation, even in the newborn era of large reflecting telescopes. In 1905 Newcomb could still write:

> evidence is continually increasing that dark and opaque worlds like ours exist and revolve around their suns as the earth on which we dwell revolves around its central luminary. Although the number of such globes yet discovered is not great, the circumstances under which they are found lead us to believe that the actual number may be as great as that of the visible stars which stud the sky. If so, the probabilities are that millions of them are essentially similar to our own globe.[16]

But that same year, W. W. Campbell (Lowell's opponent in the canals controversy) and H. D. Curtis, whose work on spectroscopic binaries at Lick Observatory Newcomb was drawing on for his statement, came to a different conclusion. Realizing that the velocity of our Sun through space varies about .03 km/sec because it is attended by planets, they

[14] Ibid., 490–491. See would later assert the abundance of planetary systems in volume 2 of his *Researches* (footnote 27), 232–233, 562–565, 710–714.

[15] Garrett P. Serviss, "Are There Planets Among the Stars?," *Appleton's Popular Science Monthly, 12* (December 1897), 171–177: 175.

[16] Simon Newcomb, "Life in the Universe," *Harper's Magazine,* 111 (1905), 404, reprinted in Donald Goldsmith, *The Quest for Extraterrestrial Life: A Book of Readings* (Mill Valley, Calif., 1980), 24–27: 24–25.

wrote that "An observer favorably situated in another system, provided with instruments enabling him to measure speeds with absolute accuracy, could detect this variation, and in time say that our sun is attended by planets. At present, terrestrial observers have not the power to measure such minute variations." While here was clearly a promising observational technique that might eventually be applied to planetary systems, Campbell and Curtis cautioned that their program of spectroscopic binaries might have quite different consequences for planetary systems:

> the star which seems not to be attended by dark companions may be the rare exception. There is the further possibility that the stars attended by massive companions, rather than by small planets, are in a decided majority; suggesting, at least, that our solar system may prove to be an extreme type of system, rather than a common or average type.[17]

The abundance of binary stars, which it was believed would preclude planetary systems from forming, was in fact one of the primary results of the spectroscopic binary technique and was perceived as a valid argument against planetary systems early in the century.[18]

In the absence of decisive observational evidence for planetary systems, one might expect that theories of solar system formation would play an especially important role, at least in determining the plausibility of such systems. This had indeed been the case for the nebular hypothesis, which we have noted favored abundant planetary systems because it seemed to be a universal process. But that hypothesis was under serious attack in 1900, and for the first two decades of the twentieth century the new theory, to the limited extent that it addressed the issue at all, gave conflicting indications about the possibility of other planetary systems. The abundance of planetary systems, however, would be a major topic of discussion in the 1920s and 1930s as the theory was refined.

The conception of the new theory of the origin of the solar system, therefore, was not related to the issue of other planetary systems. Rather, more practical considerations, followed by a realization of the technical weaknesses of the nebular theory, inspired what came to be known as the "Chamberlin–Moulton hypothesis," destined to displace for four decades the venerable nebular hypothesis. It is well established that when T. C. Chamberlin, chairman of the Geology Department at the University of Chicago, expressed misgivings in 1897 about the nebular hypothesis, the

[17] W. W. Campbell and H. D. Curtis, "First Catalogue of Spectroscopic Binaries," *Lick Observatory Bulletin*, no. 79 (1905), 145.

[18] Clerke, *Modern Cosmogonies*, 234.

impetus came from problems in his own field of geology.[19] In particular, he believed that the Earth could not have condensed from the hot fluid of a solar nebula and still hold an atmosphere in accordance with the kinetic theory of gases. In searching for weaknesses in the nebular hypothesis, Chamberlin sought and received support from F. R. Moulton, a graduate student in astronomy and therefore a generation younger than Chamberlin, also at the University of Chicago. In 1900 both Chamberlin and Moulton published their objections to the nebular hypothesis, without having any detailed idea of what would replace it.[20]

Looking to the heavens for some indication of solar systems in formation, however, both Chamberlin and Moulton drew attention to the spiral nebulae. The existence of spirals had been known since the mid-nineteenth century, but James Keeler, using the new Crossley reflector at Lick Observatory, had recently photographed them in much more detail and found them to be the most abundant form of nebula. Decades before it was realized that these objects were actually enormous galaxies outside our own, both papers of 1900 suggested that spiral nebulae might be solar systems in formation. "It seems a necessary inference from the results of the discussion that the solar nebula was heterogeneous to a degree not heretofore considered as being probable, and that it may have been in a state more like that exhibited in the remarkable photographs of spiral nebulae recently made by Professor Keeler," Moulton wrote. Moreover, the spiral nebulae did not exhibit the rings of Laplace, one of the technical objections to the nebular hypothesis. Keeler himself lent support to this idea just months before his untimely death.[21]

But what was the origin of the spiral nebulae? In the search for the answer to this question, Chamberlin and Moulton by 1901 adopted a concept that two decades later would have a profound effect on the belief in planetary systems – the close encounter or actual collision of stars in space. Even a close approach without collision, Chamberlin concluded, would cause an elongation, or "projection," of the rotating Sun, and "the

[19] Stephen G. Brush, "A Geologist among Astronomers: The Rise and Fall of the Chamberlin–Moulton Cosmogony," *JHA*, 9 (February and June 1978), 1–41, 77–104.

[20] T. C. Chamberlin, "An Attempt to Test the Nebular Hypothesis by the Relations of Masses and Momenta," *Journal of Geology*, 9 (January–February 1900), 58–73; F. R. Moulton, "An Attempt to Test the Nebular Hypothesis by an Appeal to the Laws of Dynamics," *ApJ*, 11 (March 1900), 103–130. A joint paper was published as "Certain Recent Attempts to Test the Nebular Hypothesis," *Science*, 12 (August 10, 1900), 201–208.

[21] On Keeler's statement see James E. Keeler, "The Crossley Reflector of the Lick Observatory," *ApJ* (June 1900), 325–349: 348. Chamberlin had written to Keeler as early as January 30, 1900, requesting photos of spiral nebulae, which Keeler promptly sent. For the context of Keeler's work on spirals see Donald Osterbrock, *James E. Keeler: Pioneer American Astrophysicist* (Cambridge, 1984), 314–318.

effects of explosive projection combined with concurrent rotation must obviously give rise to a spiral form."[22] Differences in the nature of the interacting bodies might even account for the observed variety of forms of spiral nebulae that Keeler had observed; the interacting bodies might be "alive" gaseous stars or an "extinct" cold star passing a gaseous star. By 1905 both spiral nebulae and the idea of close encounters of stars to produce them had become an integral part of a well-developed theory now referred to as the "planetesimal hypothesis." According to this theory (Figure 4.1), the Sun had been approached by another star, whose tidal force caused material to be ejected from the Sun. The passing intruder then caused the ejected material to form spiral arms. These arms contained knots of denser material that condensed into nuclei, which in turn grew into planets and satellites by the capture of planetesimals, cold particles in the nebula. So central was the role of spiral nebulae by this time that in his *Introduction to Astronomy* of 1906, Moulton even called the planetesimal hypothesis "The spiral nebula hypothesis."[23]

By contrast to the spirals, Chamberlin later wrote that the idea of stellar encounters was not an essential part of his theory, but that it was first proposed only as a way of explaining how the nebula might have originated in order "to develop the hypothesis as definitely and concretely as possible." Though it was only one possible explanation for the origin of the nebula, Chamberlin believed it a plausible one because the event seemed "very likely to have happened" and "the form of the nebula supposed to arise in this way is the most common form known, the

[22] T. C. Chamberlin, "On the Possible Function of Disruptive Approach in the Formation of Meteorites, Comets and Nebulae," *ApJ*, 14 (July 1901), 17–40: 34. The idea of collisions of celestial bodies to produce solar systems was common in the eighteenth century. See Jaki, *Planets and Planetarians,* chapter 4, "Convenient Collisions," 87–110. As Brush ("A Geologist," 17–18) points out, many astronomers in the late nineteenth and early twentieth centuries believed novae were caused by the collision of a star with other bodies, and the appearance of Nova Persei on February 22, 1901, affected Chamberlin's thinking in this regard. Moreover, Brush finds (19) that both Arrhenius (1901) and Lowell (1903) had proposed that the collision of two stars might form a nebula and then a solar system. The observation of solar prominences – giant eruptions of gas from the Sun – also provided graphic evidence of Chamberlin's idea. The idea that novae were caused by "a great swarm of meteorites or a body of planetary dimension" colliding with a star was still current in 1927, according to the textbook of Henry Norris Russell, Raymond S. Dugan, and John Q. Stewart, *Astronomy,* vol. 2 (Boston, 1927), 922, who credit H. von Seeliger and W. H. Pickering with the idea.

[23] T. C. Chamberlin, "Fundamental Problems of Geology," *Carnegie Institution Yearbook for 1904* (Washington, D. C., 1905), 195–258. This report to the Carnegie Institution, which heavily supported Chamberlin's work, is the first statement of the complete Chamberlin–Moulton hypothesis. See also F. R. Moulton, "On the Evolution of the Solar System," *ApJ*, 22 (October 1905), 165–181: 166–169; Moulton, *Introduction to Astronomy* (New York, 1906), 463–488; and Brush, ("A Geologist,"), 27. The term "planetesimal" was first used in early 1903 (Brush, 21).

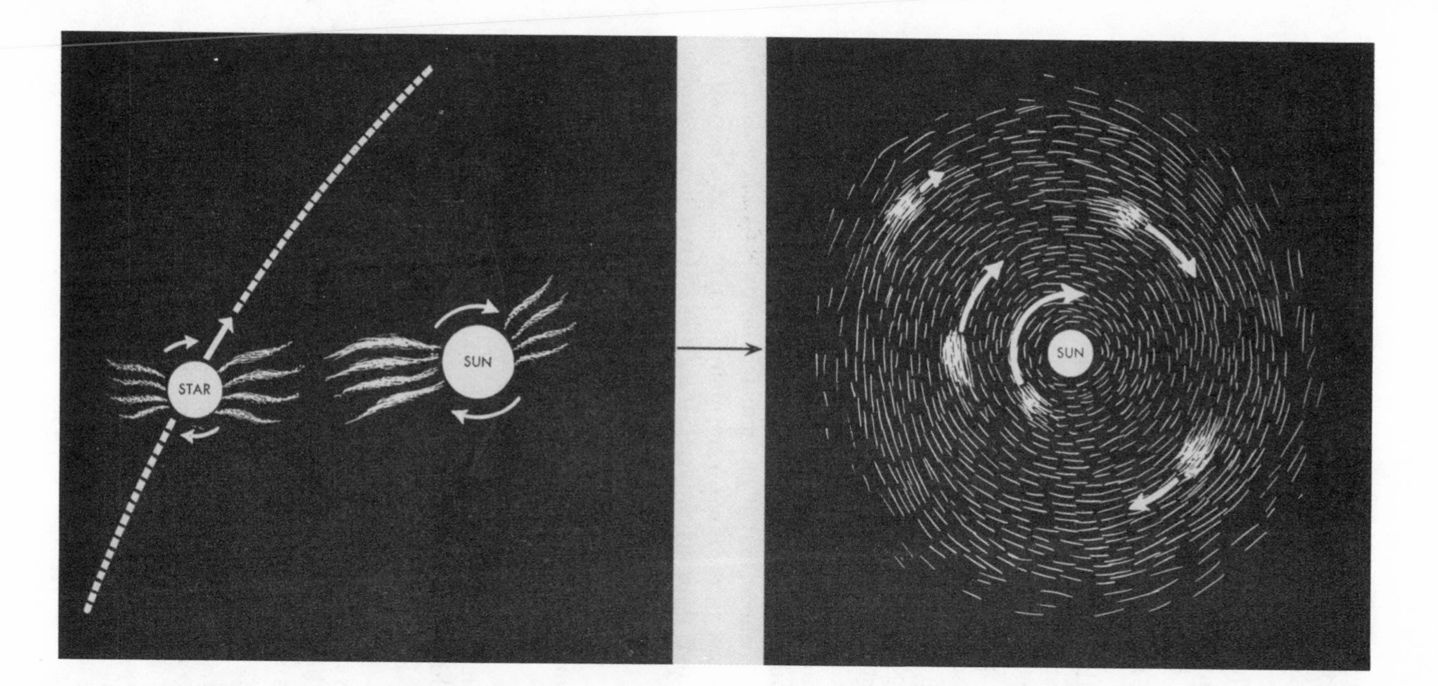

Figure 4.1. The Chamberlin–Moulton planetesimal hypothesis (1905), according to which a close encounter of another star with the Sun causes gases to erupt from both (left). These gases condense to form a large number of planetesimals that, in turn, accrete to form the planets (right). The spiral part of the theory, dropped a few years after it was proposed, is not depicted. (See also Figure 4.2.) From a review of theories of the origin of the Earth by astronomer Thornton Page in *Physics Today* (October 1948), at a time when abundant planetary systems were again being proposed.

spiral."[24] Although he emphasized that the planetesimal doctrine did not stand or fall on the merits or demerits of the stellar encounter hypothesis, the latter idea would persist for decades – long after the connection with spirals was severed.

This theory, breathtaking in the scope of events it sought to explain, had implications for many fields. "In 1900," Moulton recalled many years later, "all the world was indeed before those who had given up the Laplacian theory of the origin of the planets and all theories essentially related to it."[25] This was true for the abundance of planetary systems no less than for other fields, but on this subject the Chamberlin–Moulton theory gave contradictory indications. As long as planetary systems were associated with spiral nebulae, they would seem to exist in abundance, since they were actually observed in abundance, at least in their formative stages. The close encounter of two stars, on the other hand, seemed to imply rarity, since in the vast stretches of space such encounters were believed to be extremely rare occurrences. Curiously, rarity or abundance of planetary systems does not seem to have been an issue for Chamberlin or Moulton. In no case did Chamberlin or Moulton, during the period they were formulating their theory, refer to any implied rarity of planetary systems, much less attempt to calculate just how rare they would be. Chamberlin did believe that if an actual collision was necessary, such events were too rare to account for the many spiral nebulae observed. But if only a disruptive approach was necessary, this did not seem a problem to him.[26]

The spiral nebulae, however, did not long survive their link to planetary systems. T. J. J. See, though by this time commanding little influence in the astronomical community, argued in 1910 that the Chamberlin–Moulton theory "can easily be shown to be theoretically unsound, and it is emphatically contradicted by the observed distribution of the spiral nebulae." If spiral nebulae were caused by stellar encounters, he pointed out, they should be most abundant where the stars are densest because encounters would be most frequent there. In particular, one would expect the maximum frequency of spirals in the globular clusters, where no spirals at all were observed. Moreover, See emphasized that the spiral nebulae were thousands of times larger than our solar system, and thus unlikely to be associated with solar system formation at all, a point made by Agnes Clerke as early as 1903 and probably evident to many.[27] Never-

[24] Chamberlin, "Fundamental Problems of Geology," 210.
[25] F. R. Moulton, *Consider the Heavens* (Garden City, N.Y., 1935), 143.
[26] Brush, "A Geologist," 18.
[27] T. J. J. See, *Researches on the Evolution of the Stellar Systems,* vol. 2 (Lynn, Mass., 1910), 104–108. Agnes Clerke, *Problems in Astrophysics* (London, 1903), 445. Already

theless, it was with hopes of shedding light on the origin of the solar system that Percival Lowell put V. M. Slipher to work in 1909 measuring the radial velocities of spiral nebulae.[28] The astounding result that they were receding at great velocities was later interpreted by Hubble as evidence for the expanding universe. While Chamberlin realized by 1907 that spiral nebulae could not play the role he had hoped, as the spiral nebulae went on to play greater roles in the history of astronomy, he still could portray them as visible models of a process perhaps still invisible to the telescope.[29]

In his later writings, Chamberlin did appear to accept the rarity of planets: "Our planetary family had an aristocratic birth. It was no everyday affair," he wrote in 1928 long after the heat of battle over his theory.[30] The lack of attention to this matter is perhaps understandable in the case of Chamberlin, whose interest was more in the Earth than the stars. One might have expected Moulton, an astronomer, to show more interest in the number of planetary systems, but this was apparently not a question of burning concern to him. In a popular work of 1935, Moulton noted that a stellar collision might occur only every thousand million million years, but he argued that many physical and biological phenomena known to occur are based on exceptional events much rarer than stellar collisions. Although he thus seemed to believe that planetary systems were not uncommon, this was not the conclusion he drew. Rather, he concluded only that the rarity of such events did not preclude the stellar collision theory for the origin of our solar system.[31] He was more worried about the plausibility of his theory than about its implications for other planetary systems.

Thus, despite early observational hints of other planetary systems, their

in 1905 Chamberlin had stated that spirals had dimensions "that vastly transcend those of the solar system," and they could not be taken as precise examples of solar systems in formation (note 23, 214).

[28] For Lowell's view of spiral nebulae in relation to solar system formation, see his *The Evolution of Worlds* (New York, 1909), 24–25. For Lowell's instructions to Slipher and their connection to other solar systems, see William G. Hoyt, *Lowell and Mars* (Tucson, Ariz., 1976), 147–149. Both Lowell and Slipher had concluded by 1916 that the spiral nebulae had nothing to do with solar system formation but were extragalactic "island universes." See also Robert W. Smith, *The Expanding Universe, Astronomy's 'Great Debate', 1900–1931* (Cambridge, 1982).

[29] Brush, "A Geologist," 81–82. Even in 1928 (footnote 30, 135ff.), Chamberlin used spiral nebulae as a model for the smaller spirals he believed were associated with planetary formation, and still held out hope that such smaller ones might exist undetected. Moulton finally gave up on the spiral nebulae connection in 1939. On the role of spiral nebulae in later debates see Smith, *The Expanding Universe*.

[30] T. C. Chamberlin, *The Two Solar Families* (Chicago, 1928), 148.

[31] F. R. Moulton, *Consider the Heavens* (New York, 1935), 144–145.

existence had a more ambiguous status in the Chamberlin–Moulton hypothesis than in the previously accepted nebular hypothesis. To the extent that their rarity or abundance was an issue at all, it oscillated between the twin pillars of the planetesimal hypothesis: the spiral nebulae, which implied abundance, and stellar encounters, which implied rarity. With the gradual realization that spiral nebulae were too large to represent planetary systems in formation, the stellar encounter aspect of the theory was free to gain the upper hand – and with it the implication of the rarity of planetary systems.

This, in fact, is precisely what occurred, not in America but in Britain, where, in the tradition of Whewell and Wallace (though not of Proctor), the scientific community seemed more skeptically inclined toward planets and life. It was at the hand of the British mathematical physicist and astronomer James Jeans that the question of other solar systems would become closely linked with the rarity of planets and life in the universe. Jeans, a 1903 graduate of Trinity College, Cambridge, became well known early on with his *Dynamical Theory of Gases* (1904), which he tackled as a problem in molecular physics but which would later be applied to planetary atmospheres, among other subjects.[32] During 4 years at Princeton University (1905–1909), Jeans wrote textbooks on theoretical mechanics and the theory of electricity and magnetism, and after his return to England distinguished himself again with his support of the quantum theory, on which he wrote an influential Report in 1914. That year marked a milestone in Jeans's life.[33] Prior to 1914 he did important work on atomic theory and statistical mechanics. After 1914 he turned from the microscopic to the macroscopic, from atoms to astronomy, and specifically to cosmogony.

In his new work, Jeans made clear that he was following a line of research initiated by Laplace and C. Maclaurin and extended by E. A. Roche, Lord Kelvin, C. Jacobi, Henri Poincaré, and Sir George Darwin (fifth child of Charles), all distinguished mathematical physicists. His approach was therefore very different from that of Chamberlin. Jeans's strategy was first to derive theoretical models and then to see if they had any bearing on astronomical objects; in his words, "when a firm theoretical framework had been constructed, it seemed permissible and proper to try to fit the facts of observational astronomy into their places."[34] Chamberlin was motivated by a problem of nature, while Jeans's motivation

[32] On Jeans see E. A. Milne, *Sir James Jeans: A Biography* (Cambridge, 1952), and A. E. Woodruff, "Jeans," *DSB*, 7 (New York, 1973), 84–86.

[33] Milne, *Sir James Jeans*, 20.

[34] Jeans, *Problems of Cosmogony and Stellar Dynamics* (Cambridge, 1919), Preface, vi.

was a mathematical problem which he clearly saw had applications to nature. In particular, Darwin had written that progress in cosmogony depended on further research on rotating gaseous masses.[35]

Thus Jeans's attention was at first devoted to the stability of rotating bodies, on which subject he published two lengthy papers in 1915 and 1916.[36] This work he applied to cosmogony in 1916 with reference to tidally distorted masses – in other words, how a rotating astronomical body would be affected by tidal forces raised by another passing astronomical object, as would happen in a close stellar encounter. In a paper read before the Royal Astronomical Society in 1916 and published in the society's *Memoirs* the following year, Jeans dealt not only with the origin of solar systems, but also with binary star formation and spiral nebulae. In contrast to the binaries and spirals, Jeans concluded that the solar system might well have been formed from a tidally distorted mass, in particular by another star approaching our sun. Unlike the Chamberlin–Moulton hypothesis, however, Jeans's analysis showed that neither spiral nebulae nor planetesimals played a role in planet formation, and he thus emphasized that for solar systems "the origin which seems most probable is not that of the planetesimal hypothesis." Instead, his analysis showed that rather than the streams of gas torn from the Sun condensing into numerous small, cold planetesimals that, in turn, accreted to form the planets, a single cigar-shaped filament of hot gas would be ejected and condense directly into the planets (Figure 4.2). As the theory was later elaborated, he pointed out that the largest planets would form near the center where the filament was thickest, and the smaller ones at each end, giving the distribution of planets observed in our solar system.[37]

The central question in determining whether this mathematical conclusion could really occur in nature was the frequency of close stellar encounters. Again in contrast to Chamberlin and Moulton, it is clear at the outset of the paper that Jeans was already thinking in these more general terms, not only with regard to the origin of our solar system, but also in

[35] Ibid., Preface. In 1918 he set forth his objections to Laplace's nebular hypothesis as an explanation for the formation of the solar system in "The Present Position of the Nebular Hypothesis," *Scientia (Revista di Scienza),* xxiv (1918), 270–281. See also Jeans's Halley Lecture, *The Nebular Hypothesis and Modern Cosmogony* (Oxford, 1923).

[36] Jeans, "The Potential of Ellipsoidal Bodies and the Figures of Equilibrium of Rotating Liquids." *PTrA,* 215 (1915), 27–78, and "The Instability of the Pear-Shaped Figure of Equilibrium," *PTrA,* 217 (1916), 1–34.

[37] Jeans, "The Motion of Tidally-Distorted Masses, with Special Reference to Theories of Cosmogony," *MRAS,* 62 (1917–1923), 1–48. The paper was read before the Society on November 10, 1916. Despite the differences, in a stinging attack on Jeans and Jeffreys, Moulton used the occasion of his review of Chamberlin's *The Two Solar Families* to demonstrate the similarities of the tidal hypothesis to the planetesimal hypothesis; "The Planetesimal Hypothesis," *Science* (December 7, 1928), 3–31.

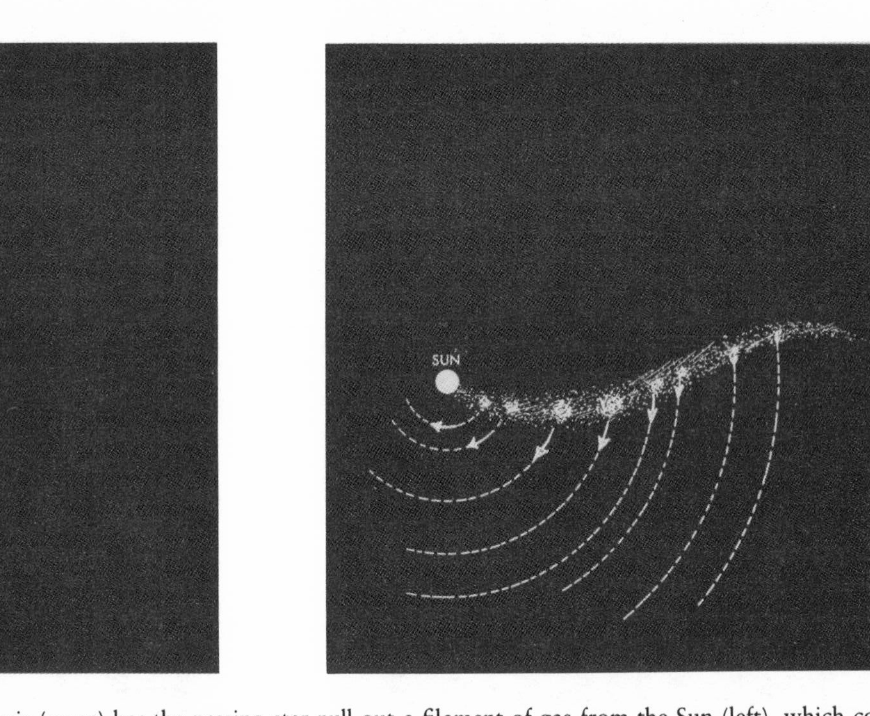

Figure 4.2. The Jeans–Jeffreys hypothesis (1917) has the passing star pull out a filament of gas from the Sun (left), which cools and condenses into planets, with the largest planets in the middle (right). From Thornton Page, *Physics Today* (October 1948).

connection with the frequency of planetary systems.[38] In his earliest statement on what would become a lifelong contentious issue, Jeans wrote:

> We have absolutely no knowledge as to whether systems similar to our solar system are common in space or not. It is quite possible, for aught we know to the contrary, that our system may have been produced by events of such an exceptional nature that there are only a very few systems similar to ours in existence. It may even be that our system is something quite unique in the whole of space. Thus an explanation of the formation of our system which implies that such systems are of very rare occurrence cannot be entirely dismissed, although naturally, as between two otherwise equally plausible explanations, we shall feel a preference for the one which does not postulate the occurrence of very rare or exceptional events.[39]

Jeans's analysis showed that the issue of abundance was very sensitive to the assumptions made about a variety of parameters, including the density of stars in the universe, the velocity of the stars in space, the age of the stars and of the universe, and the size and mass of the stars at the time of the encounter. All of these parameters were subject to change in the discussion that ensued over the next three decades. For now, using the best estimates known in 1916, Jeans found that at most 1 star in 4000 might have experienced a "non-transitory" encounter at the distance of Jupiter in a lifetime of 10 billion years, the upper limit that he placed on the age of the universe.[40] If the encounter distance were 100 times greater and the other parameters adjusted accordingly, one star in three might have experienced such an encounter, and "we may, without postulating anything very improbable, suppose our system to have experienced an encounter as close as this." However, in order to produce a solar system, certain criteria for velocity, size, and mass had to be met, and Jeans clearly did not think all these conditions would ensue at one time.[41] In the end he labeled these occurrences "somewhat improbable" and systems similar to our own "somewhat rare," but in general the entire process was not "impossible or

[38] Jeans, "Tidally-Distorted Masses," 1.

[39] Ibid., 46.

[40] Ibid., 42. This assumed stars with the present mass of the Sun, moving through space at 25 km/sec, and a density in space of 10^{-56} stars per unit volume. Of particular interest in Jeans's general treatment is the explicit parallels with the theory of gases; the motions of stars in space were treated similarly to the motions of molecules in a gas.

[41] Ibid., 46–47. In order to have produced a solar system, Jeans found, the encounter speeds at this distance would have had to be only about 4 km/sec, the size of the star encountered 2.5 times the radius of Neptune's orbit (considered a possibility in the Sun's early history), and the intruding star double the encountered star's mass.

very improbable." Given the number of parameters and their uncertainty, Jeans's waffling is not surprising. But he emphasized that no reasonable choice of parameters was likely to alter the result that only very few stars can have experienced nontransitory encounters. And most important, Jeans stressed, the theory violated no quantitative criterion.[42]

In his classic work *Problems of Cosmogony and Stellar Dynamics* (1919), Jeans discussed the problem in more detail and ended with results even more pessimistic: only one encounter in 30 billion years, a situation so improbable in the present universe as to cast doubt on the validity of the close encounter hypothesis. Pointing out that the parameters were not well known, Jeans concluded that while tidal breakup by a passing star was hardly a likely event, its improbability was not grounds for rejecting the tidal theory. In whatever case one adopted, the solar system seemed to be very exceptional, "and for aught we know may be unique."[43] In his 1923 lecture "The Nebular Hypothesis and Modern Cosmogony," Jeans carried his train of thought one step further, arguing that it was just possible, though not probable, that only the Earth could support life in the universe. "Astronomy does not know whether or not life is important in the scheme of nature, but she begins to whisper that life must necessarily be somewhat rare."[44]

In the hands of Jeans, this whisper soon grew to a crescendo. In both his technical and popular publications by the late 1920s, Jeans (seen at about this time in Figure 4.3) spread his view far and wide. The numbers varied somewhat, but always present was the basic scenario that the stars are sparsely scattered in space, close encounters exceedingly rare, and the conditions for life very exacting. "All this suggests," Jeans inevitably concluded, "that only an infinitesimally small corner of the universe can be in the least suited to form an abode of life."[45] In his popular works this view of the disruptive approach of stars was vividly drawn, and the rarity of such approaches and their ensuing solar systems was an integral part of this picture – clear even to the public.

Although this rarity was occasionally raised as a factor in the acceptance of the theory, to others it was not an important consideration,

[42] Ibid., 48. Jeans referred particularly to the Babinet criterion, which concerned the amount of angular momentum transferred from the primordial nebula to the planets in the nebular hypotheses.

[43] Jeans, *Problems*, 279, 290.

[44] Jeans, *Nebular Hypothesis*, 30.

[45] Jeans, "Life and the Universe," in *The Universe Around Us* (Cambridge, 1930; 1st ed., 1929), 331–336: 335. A similar view is given in *Astronomy and Cosmogony* (Cambridge, 1929; 1st ed., 1928), 420–421, "The Wider Aspects of Cosmogony," *Smithsonian Report for 1928*, 165–178, and *The Mysterious Universe* (New York, 1930), 4–5.

Figure 4.3 Walter S. Adams, James Jeans (center), and Edwin Hubble with a model of the 100-inch Mount Wilson telescope in April 1931. While this telescope helped solve the mysteries of Mars, it could not resolve the problem of planetary systems. Reproduced by permission of the Huntington Library, San Marino, California.

including Sir Harold Jeffreys, the other British scientist who had worked in the mathematical tradition and made original contributions to the tidal theory since 1917. When in 1931 the German astronomer Friedrich Nölke argued that not only angular momentum but also the rarity of planetary systems implied by the Jeans–Jeffreys tidal hypothesis was "evidently a defect," Jeffreys replied that he had already dealt with this argument in his book *The Earth,* "but as it has frequently been revived some further comment is needed." The most stinging aspect of this further comment was tellingly methodological:

> The argument is in fact a fallacious application of probability. The task of science is to coordinate observed data and predict observable ones. The observed datum here is the existence of one solar system. Other solar systems, whether they exist or

> not, are unobservable, and it is no task of science to explain the unknown and unknowable. Prof. Nölke's objection is based on the belief that a large fraction of the stars have attendant planetary systems, and this is a pure speculation.[46]

For two decades the Jeans–Jeffreys tidal theory was widely accepted, and when the beginning of the end came in 1935, it was once again because of problems with physical principles. This time it was the Americans' turn again, in the form of Henry Norris Russell, who had returned to Princeton (where he had been trained) the same year as Jeans, and by now was enjoying a reputation as the dean of American astronomers. Russell criticized the tidal hypothesis because it could not account for the present orbits of the planets; he could not see how a close stellar encounter would remove the planets so far from the Sun and give them most of the angular momentum of the system rather than the Sun, which was a thousand times more massive. He also could not see how the planets could condense out of the high-temperature matter ejected from the Sun, an objection given definitive form by Russell's student Lyman Spitzer 4 years later.[47] In their discussion the possibility of other planetary systems played no role, but their fatal objections left science without a workable theory of the origin of the solar system, and by association placed in limbo the idea that such systems were rare.

While the rarity of planetary systems in the first four decades of the twentieth century may seem pessimistic by reference to the predominant opinion today that planets and life are somehow good, this was not always the case during the period under discussion. On the one hand, the worldview that Jeans expressed in his many popular works was in some ways depressing, with its lonely picture of the flowering of life in only one corner of the universe or at best on remote islands, surrounded by torrential deluges of life-destroying radiation. But we have seen that Chamberlin, on the other hand, believed that the rarity of planetary systems conferred on Earth the status of "noble birth," a property he characterized late in his life as one of two "valuable assets" of the planetesimal hypothe-

[46] Nölke's own words were that under the Jeans–Jeffreys hypothesis "The development of our system must be held to be a rare exceptional case in cosmical evolution. This is evidently a defect." Nölke, *Gutenberg's Handbuch der Geophysik*, 1 (1931), 8–68, as translated in Jeffreys, "On the Origin of the Solar System," *MNRAS*, 92 (1932), 887–891: 888, reprinted in *Collected Papers of Sir Harold Jeffreys on Geophysics and Other Sciences*, vol. 5 (London, 1976), 73–77: 74. This is clearly a case where a predisposition to abundant planetary systems did affect one man's acceptance of a theory of solar system formation. On Jeffreys's reply see Jeffreys, *Collected Papers*, 74. On Jeans and Jeffrey's work see also Jaki, *Planets and Planetarians*, 202ff.

[47] Russell, *The Solar System and Its Origin* (New York, 1935); L. Spitzer, "The Dissipation of Planetary Filaments," *ApJ*, 90 (1939), 675–688; Brush, "A Geologist," 86–91.

sis, the other being that the rarity of disruptive approaches made the future of life on Earth more secure.[48] And Eddington, agreeing in 1928 with Jeans's conclusion that "The solar system is not the typical product of development of a star; it is not even a common variety of development; it is a freak," did not seem disappointed when he concluded a chapter on "Man's Place in the Universe" by saying:

> I do not think that the whole purpose of the Creation has been staked on the one planet where we live; and in the long run we cannot deem ourselves the only race that has been or will be gifted with the mystery of consciousness. But I feel inclined to claim that *at the present time* our race is supreme; and not one of the profusion of stars in their myriad clusters looks down on scenes comparable to those which are passing beneath the rays of the sun.[49]

Whether optimistic or pessimistic, the nineteenth century view of abundant planetary systems, which Eddington believed made it in the twentieth century "a presumption, bordering almost on impiety, to deny to them life of the same order of creation as ourselves," was thwarted for decades by the tidal theory of Jeans and Jeffreys. Far from the teleological view of Proctor or Gore, Eddington asked, "How many acorns are scattered for one that grows to an oak? And need she be more careful of her stars than of her acorns? If indeed she has no grander aim than to provide a home for her greatest experiment, Man, it would be just like her to scatter a million stars whereof one might haply [*sic*] achieve her purpose." To have provided a theoretical underpinning for this startlingly different worldview was no small part of the legacy of James Jeans.

But alas, this worldview had no more claim to objective truth than the nineteenth-century belief in abundant planetary life, for if the early observational claims for planetary systems at the turn of the century had yielded no definitive result, by 1940 neither had theory solved the problem – nor could it – especially with the departure of spiral nebulae as confirming evidence. The discredited nebular hypothesis had been superseded by the planetesimal hypothesis of Chamberlin and Moulton and then by the tidal theory of Jeans and Jeffreys, only to have Russell and Spitzer overturn the latter, leaving only the void. Reviewing the collisional and nebular hypothe-

48 Chamberlin, *Two Solar Families,* 148–149. Chamberlin made these statements in a chapter titled "Positive Assets and Residual Working Hypotheses." As we have seen, he did not claim the rarity of planetary systems as a "positive asset" to his theory at the time of its formulation.

49 A. S. Eddington, *The Nature of the Physical World* (New York, 1929; 1st ed. Cambridge, 1928), 178. This book was reprinted and reissued many times, at least as late as 1948.

ses in 1938, Lick Observatory Director Emeritus Robert G. Aitken still saw the development of planetary systems as an "exceptional event."[50] "Exceptional" did not mean unique to Aitken, who pointed out that even if only one star in a million had planets, there would still be 30,000 solar systems in the Milky Way Galaxy – and 2 million galaxies were within the range of current telescopes.

Aitken, however – an observational astronomer – held out no promise for an observational conclusion. There was, in his view, "no hope" of a visual sighting. And he had the full weight of his distinguished career in double star astronomy behind him when he pointed out that the gravitational perturbations of Sirius and Procyon were caused by bodies believed to be one-third as massive as the primary stars, while our solar system constituted in total only 1/745th of the mass of the Sun. In a pointed analogy, Aitken correctly deduced that our solar system would therefore be much too small to be detected by the gravitational perturbation method from even a nearby star with current techniques. This makes all the more surprising the events that were to follow only 5 years later, and that – justifiably or not – would tranform skepticism into optimism.

4.2 TURNING POINT: 1943–1958

The 15 years between 1943 and 1958 saw a remarkable turning point in the fortunes of planetary systems. A knowledge of precisely how this turnabout occurred is important to an understanding of the subject for the remainder of the century. It had begun with Russell's criticism of the Jeans–Jeffreys tidal theory, but it was fueled by the revival of a modified nebular hypothesis, developments in fields as diverse as double star astronomy, supernovae, the measurement of stellar rotation periods, and geochemistry and – most surprising of all – by insistent claims that planetary systems, or their effects, had been actually observed. Moreover, broader events in the field of cosmology also conspired toward change, events that Jeans himself could not ignore.

The implications of the revolution in cosmology of the 1920s and 1930s – a greatly enlarged Galaxy, the existence of innumerable "island universes" full of stars, a universe expanding in space and expanded in time – are evident in Jeans's review of the subject of life on other worlds published in 1941. Having given a dim view of the chances of life on Mars and Venus, Jeans turned to the realm of the stars and the origin of planetary systems. He pointed out that under present conditions in the

[50] R. G. Aitken, "Is the Solar System Unique?" *ASP Leaflet No. 112* (June 1938), 98–106: 105.

universe the frequency of stellar encounters would be only 1 in 10^{18} years, so that for stars 2 billion years old, 1 in 500 million might have planets. So far this was his old argument. But in a sign of the times, he went on to say that though this seemed like a small fraction, in a universe with 10 billion galaxies, each with 100 billion stars, this minute fraction still represented 2 million million stars that might have planetary systems! Statistics – and the new cosmology – had caught up with Jeans, even if only 2000 of these systems might be located in our own galaxy. Straining the definition of "rare," Jeans was forced to conclude that "although planetary systems may be rare in space, their total number is far from insignificant."[51]

By the following year, however, Jeans's view had undergone a much more radical change. In a letter to *Nature* of June 20, 1942, reacting to recent claims of serious dynamical problems arising for the tidal theory, which assumed that the Sun was approximately its present size at the time of encounter, Jeans asserted that the Sun was most likely comparable in size to the present orbit of Uranus or Neptune when an encounter took place. In a last-ditch effort to save the tidal theory from dynamical objections, Jeans was forced to increase greatly the size of the Sun at the time of supposed planetary formation, a concession that greatly increased its cross section and, by analogy, the cross sections of other suns. Not only did this address the dynamical objections in Jeans's opinion, it also led to another conclusion: that the total chance of planet formation was now one in six with such a size for the Sun. Thus, "there is no longer any need to strain the probabilities to account for the existence of the planets." And the final conclusion is one hardly expected from Jeans: "A far larger proportion of the stars than we have hitherto imagined must be accompanied by planets; life may be incomparably more abundant in the universe than we have thought."[52] The whole exercise demonstrated the fragility of the argument and the dangers of using equations whose parameters were not well determined. For 25 years Jeans had epitomized the concept of the rarity of life in the universe. Now in the last years of his life he recanted, and his death in 1946 left no substantial heirs to his theory.

Jeans's turnabout was just the beginning, and the cracks opening in the

[51] Sir James Jeans, "Is There Life on Other Worlds?" *Science,* 95 (June 12, 1942), 589; reprinted in Goldsmith, *The Quest,* 81–83: 83.

[52] Jeans, "Origin of the Solar System," *Nature,* 149 (July 17, 1942), 695. In its early form (1919), Jeans's tidal hypothesis had proposed a Sun the size of Neptune's orbit, but Jeffreys had modified this for astrophysical reasons to require the Sun to have its present size. In response to Jeans's letter to *Nature,* Jeffreys restated his reasons for believing that the Sun must have had approximately its present size at the time of any encounter. Jeffreys, "Origin of the Solar System," *Nature,* 153 (January 29, 1944), 140, reprinted in *Collected Papers,* 91.

tidal theory in 1941–1942 were to become a breach through which the floodwaters of change would rush in the following year. In that year, strong and independent observational claims were made for the existence of two planetary systems around nearby stars. In that year also, a new theory would begin to be developed that would eventually sweep away the tidal theory. And late in that year, the American Astronomical Society sponsored the first modest "Symposium on Dwarf Stars and Planet-Like Companions" at its annual meeting. Many astronomers were quick to draw general conclusions, especially in light of the new observations; as Henry Norris Russell wrote in 1943, "On the basis of this new [observational] evidence, it therefore appears probable that among the stars at large there may be a very large number which are attended by bodies as small as the planets of our own system. This is a radical change – indeed practically a reversal – of the view which was generally held a decade or two ago."[53]

The surprising observational claims for planetary systems in 1943 resulted from the astrometric method, and arose quite naturally out of ongoing astronomical research programs for measuring stellar positions and motions. Since the beginning of the century, great advances had been made in techniques for the determination of stellar positions, particularly in the form of photographic plates taken with long-focus refracting telescopes. In the United States, Frank Schlesinger at Yale pioneered long-focus photographic techniques in attempts to measure stellar parallax – those small apparent motions of stars due to the Earth's annual motion around the Sun.[54] Of course, Bessel had first *visually* measured stellar parallax in 1838, but the new photographic techniques allowed much smaller parallaxes to be determined, and much more accurately. Several observatories around the world specialized in this work, determining parallaxes year after year for a given list of stars – valuable work because the measure of stellar parallax was a measure of stellar distance. Another astrometric tradition, begun by the Danish astronomer Ejnar Hertzsprung, used similar photographic techniques for double star orbits. These methods could detect extremely small shifts in position; the largest parallax was less than a second of arc, and most were much smaller. In the tricky business of angular measurement in astronomy, a process beset by random and systematic errors, this was its most exacting test – and the search for planetary systems would go a step beyond that (Table 4.1).

In order to understand the role of this astrometric technique destined

[53] Russell, "Anthropocentrism's Demise," *SciAm* (July 1943), 18–19: 19.

[54] Frank Schlesinger, "Photographic Determinations of Stellar Parallax Made with the Yerkes Refractor," *ApJ*, 32 (December 1910), 372–387, and 33 (January and March 1911), 8–27, 161–173.

Table 4.1 *Planetary companions in the context of angular measurement in astronomy*

Phenomenon	Date of first measurement	Method	Angle (arcseconds)
Apparent diameter of Moon		Visual	1800
Apparent diameter of Mars		Visual	4 to 25
Proper motion of stars	1718	Visual	Largest 10/year most <.1/year
Aberration of light	1728	Visual	40/year
Nutation of Earth	1748	Visual	18/year
Stellar parallax	1838 1910	Visual, photographic	Largest .75
Perturbations in proper motions by *stellar* companions	1844 1936	Visual, photographic parallax	Periodic with few arcseconds to few *tenths* of arcsecond amplitude
Perturbations in proper motions by *planetary* companions	1938 1963	Photographic, photographic parallax	Periodic with few *hundredths* of arcsecond amplitude
Perturbations in double star orbits by *planetary* companions	1897 1943	Visual, photographic	Periodic with few *hundredths* of arcsecond amplitude

to dominate twentieth-century observational claims of planetary systems, we need to focus on the career of Peter van de Kamp, the towering figure and common thread in much of what would happen in the field after the 1930s. A Dutch-born student of Hertzsprung, van de Kamp (Figure 4.4) was still in high school in 1916, the year James Jeans first proposed his close encounter theory of the origin of solar systems. But 10 years later, with degrees in astronomy from Utrecht, Gröningen, and the University of California, he was working in the field of astrometry at the University of Virginia's Leander McCormick Observatory. Ten years after that, by 1937, he was director of the Sproul Observatory at Swarthmore College, and it was in this position, after a quarter century of observations, that he would make the startling announcement in 1963 that he had discovered a planet around another star.

Figure 4.4. Peter van de Kamp, pioneering planet hunter of the century, began his search in 1937 at Sproul Observatory of Swarthmore College and announced a planetary companion to Barnard's star in 1963. He is shown here with the 24-inch Sproul refractor used to make the observations. From van de Kamp, "The Planetary System of Barnard's Star," *Vistas in Astronomy*, 26 (1982), 146. Copyright 1982 with permission from Elsevier Science Ltd., The Boulevard, Langford Lane, Kedlington OX5 1GB, UK.

Van de Kamp had come to the United States in 1923 to help carry out a program, suggested by J. C. Kapteyn, of determining proper motions of stars using photographic techniques with long-focus refractors.[55] It was thus precisely this tradition of stellar positions out of which grew van de Kamp's work, for the same principle is employed in the search for planetary companions. Although in the 1920s and 1930s a planet search was not a specific goal (the goal was merely to measure parallaxes and calculate proper motions), the detection of perturbations in proper motion was a by-product of this parallax method, which was often repeated year after year for the same star.

It was in 1936 that the first perturbation in the proper motion of a star based on photographic observations had been announced at the Leander McCormick Observatory, one of the observatories specializing in parallax observations. Van de Kamp was working there under Observatory Director S. A. Mitchell at the time, but the discoverer was van de Kamp's cousin, Dirk Reuyl, who had preceded him by a few years in his immigration from the Netherlands to the United States. Many years later, van de Kamp well recalled the day when Reuyl made this discovery. Reuyl feared an error in his measurements, but van de Kamp and Mitchell realized that it was just possible that the measurements were correct. Closer scrutiny showed that they were, and this proved to van de Kamp the usefulness of the method of photographic astrometry in the search for low-mass companions. The next year, 1937, he began his parallax program at Sproul for the detection of low-mass companions by looking for perturbations in the motions of stars.[56]

Reuyl's unseen companion was clearly stellar in mass, but it opened up the possibility of detecting what cautious astronomers euphemistically called "substellar masses" and bolder astronomers termed "planets." It also put emphasis on the very interesting theoretical question of just how small a star could be and still be a star or, to put it another way, how large a planet could be and still be a planet. Thus the search for low-mass stellar companions was an important astronomical question in its own right, but the sought-after prize was clearly planetary systems.

In 1938, the year after van de Kamp began his search, the Leiden

[55] Steven J. Dick, interview with van de Kamp, May 24, 1988, 4.

[56] Ibid., 12, 17. The discovery of a companion to Ross 614 was announced by Dirk Reuyl, "Variable Proper Motion of Ross 614," *AJ*, 45 (July 23, 1936), 133–135. Visual and photographic confirmation was made by Walter Baade in the 1950s. On the Sproul program see van de Kamp, "The Astrometric Study of Unseen Companions to Nearby Stars," *PASP*, 55 (October 1943), 263, and "The Astrometric Study of Unseen Companions of Single Stars," *AJ*, 51 (June 1944), 7–11.

astronomer Erik Holmberg made both claims: that from parallax plates he had found perturbations in the motions of several stars and that perturbations in one of them, Procyon, were small enough to indicate that it had a companion of planetary size. "If, according to the mass–luminosity relation, the mass of the principal star is put equal to 0.1 [solar masses]," Holmberg wrote, "the probable mass of the companion will be only 0.0018 [solar masses], i.e. about two times the mass of Jupiter." His claim, however, was not widely believed or cited for several reasons: only two series of parallax measurements had been made; there was uncertainty in the mass of Procyon itself; and (perhaps most important) the Jeans universe still held sway. Holmberg urged more observations of Procyon and other stars in order "to decide if the frequency of planetary systems in space is so exceedingly small as appears from the theory of J. H. Jeans."[57] Although later proven spurious, Holmberg's was the first claim for planets based on photographic evidence.

Another claim 5 years later caused a much greater stir. Working under van de Kamp at the Sproul Observatory, Kaj Strand, a student of Hertzsprung and later scientific director of the U.S. Naval Observatory, was pursuing the other line of photographic astrometry: double stars, a field pioneered by Hertzsprung. Visual claims for perturbations in double star orbits had been made in the late nineteenth century by Seeliger, but in 1943 Strand announced that he had discovered photographically perturbations indicating a *planetary* companion to the star 61 Cygni, famous as one of the first stars to have its parallax measured – by Bessel in 1838. Using photographic observations from the Potsdam, Lick, and Sproul observatories covering the years 1914–1918 and 1935–1942, Strand announced in no uncertain terms:

> The only solution which will satisfy the observed motions gives the remarkably small mass of 1/60 that of the sun or 16 times that of Jupiter. With a mass considerably smaller than the smallest known stellar mass (Kruger 60B = 0.14 [solar masses]), the dark companion must have an intrinsic luminosity so extremely low that we may consider it a planet rather than a star. Thus planetary motion has been found outside the solar system.[58]

[57] Erik Holmberg, "Invisible Companions of Parallax Stars Revealed by Means of Modern Trigonometric Parallax Observations," *Meddelande fran Lunds Astronomiska Observatorium*, ser. 2, no. 92 (Lund, 1938), 4–23: 6, 14–15. See also the diagram of perturbations in this artricle. Holmberg cited Jeans, *Astronomy and Cosmogony* (1928), for the rarity of planetary systems.

[58] Kaj Strand, "61 Cygni as a Triple System," *PASP*, 55 (February 1943), 29–32.

Almost simultaneously, Reuyl and Holmberg, based primarily on observations made at the Leander McCormick Observatory, announced that they had discovered a planetary companion around the star 70 Ophiuchi, the very star that had gotten T. J. J. See into so much trouble at the turn of the century.[59] Though they spoke of only a "third body" and not a "planet," the deduced mass for the third body was between 0.008 and 0.012 solar masses (compared to 0.016 for Strand's claimed planet). That this was in the planetary mass range escaped no one.

The reaction to these discoveries was considerable. Immediately on publication of the results, Russell sat down and wrote an excited account that appeared in the June issue of *Scientific American*. The following month, in an article entitled "Anthropocentrism's Demise," Russell put the results in a broader context for the same magazine, arguing that the new evidence required a reversal of the previous opinion regarding the rarity of planetary systems. And during the same period, he wrote another article examining from a theoretical viewpoint the physical characteristics of stellar companions of small mass, concluding regarding a body such as claimed by the new observations that "It is well within the bounds of accepted usage to call the new body a planet."[60] A few months later, a discussion erupted in the pages of *Nature* on the significance of the discovery for theories of the origin of solar systems, in which Jeans himself participated, arguing that with his modification of the previous year, the new discoveries did not affect the status of the tidal theory.[61] But there is no doubt that these observations had now pushed Jeans to the opposite extreme of his earlier view of the rarity of planetary systems. Though he published no more on the subject, in a BBC broadcast in the final year of his life now preserved only in his archives, Jeans declared to the listening audience, "Our sun is surrounded with planets, and a fair proportion of these other stars are likely to be so too. For two of the nearer stars in our galaxy have recently been found to have planetary bodies revolving round them, and there is no reason why the more remote stars should be different. The

59 Dirk Reuyl and E. Holmberg, "On the Existence of a Third Component in the System 70 Ophiuchi," *ApJ*, 97 (January 1943), 41–45. Strand was indirectly involved in this claim since he had determined the latest orbit for 70 Ophiuchi (*Leiden Annals*, 18, part II, 1937) . Strand, however, had stated that his observations did not support the theory of a third body.

60 Russell, "Planet Companions," *SciAm* (June 1943), 260–261; Russell, "Anthropocentrism's Demise"; and Russell, "Physical Characteristics of Stellar Companions of Small Mass," *PASP*, 55 (April 1943), 79–86.

61 A. Hunter, "Non-Solar Planets," *Nature*, 152 (July 17, 1943), 66–67; H. K. Sen, "Non-Solar Planetary Systems," *Nature*, 152 (November 20, 1943), 600; and Jeans, "Non-Solar Planetary Systems," *Nature*, 152 (December 18, 1943), 721.

total number of planets in the whole of space seems to be millions of millions of millions at the lowest."[62] Only 5 years before, one could not have believed it was Jeans speaking.

The observational discoveries were thus widely cited and applauded, even making the pages of *Time* magazine. Fuel was occasionally added to the fire, for example by van de Kamp's announcement in 1944 of a low-mass companion to Barnard's star and another one around Lalande 21185. Although he concluded that these were probably stellar rather than planetary in mass, they added to the debate over low-mass companions and 20 years later would be the subject of more sensational announcements.[63] In 1957 Strand claimed confirmation of his own 1943 discovery of a planet around 61 Cygni, and in 1960 A. N. Deich at the Pulkovo Observatory in Leningrad claimed independent verification, a claim reasserted in 1977 and 1978. Although these claims of Strand, Reuyl, and Holmberg are now generally considered erroneous, they could not be finally disproven for many years; in the meantime, they played a central role in reversing opinions about the frequency of planetary systems.[64]

The increased interest in the subject of low-mass companions generated by these discoveries, as well as its origin in ongoing astrometric programs, is evident in a modest symposium on "Dwarf Stars and Planet-Like Companions" held in late 1943 under the auspices of the American Astronomical Society. Six participants, including Strand, Russell, and van de Kamp, gave papers.[65] Representing the two photographic astrometric approaches to the subject, van de Kamp reviewed work on unseen companions of single stars and Strand did the same for

[62] Jeans, "Man and the Universe," BBC broadcast to Sweden, November 10, 1945, James Jeans Papers, Royal Society Archives, London.

[63] Van de Kamp, "A Photographic Study of the Proper Motion of Nearby Stars," *Proceedings of the American Philosophical Society,* 88 (November 1944), 372–374. For the *Time* coverage see "Dark Companions," *Time*, 42 (August 2, 1943), 46–47, where H. N. Russell's reaction is also given. *Time* reported, "The first clear proof of the existence of a planet outside the solar system has now been accepted by astronomers," but cited only Russell.

[64] Strand, "The Orbital Motion of 61 Cygni," abstract in *AJ*, 62 (February 1957), 35; A. N. Deich, *Izvestiya Glavnaia Astronomisches Observatoria Pulkovo,* 166 (22, no. 1) (1960); 138; Deich and O. N. Orlova, "Invisible Companions of the Binary Star 61 Cygni," *Soviet Astronomy,* 21 (March–April, 1977), 182–88; Deich, "New Data on Unseen Companions of 61 Cygni," *Soviet Astronomy Letters,* 4 (January–February 1978), 50–52. On the lack of confirmation of the companion of 61 Cygni, see F. J. Josties, "The Hertzsprung Multiple Exposure Technique and Its Application to 61 Cygni," *Current Techniques in Double and Multiple Star Research,* IAU Colloquium 62, eds. Robert S. Harrington and Otto G. Franz (Flagstaff, Ariz., 1983), 16–27.

[65] Symposium on Dwarf Stars and Planet-Like Companions, *AJ*, 51 (June 1944), 1–17. The papers were delivered at the 71st meeting of the American Astronomical Society in Cincinnati, November 5 and 6, 1943. Other speakers included Everett Yowell, William J. Luyten, A. N. Vyssotsky, and Emma Williams.

unseen companions of double stars, while Russell elaborated on his previous theoretical discussion of the physical characteristics of low-mass companions. An important offshoot of the latter was the question of the difference between stars and planets; in other words, when did a low-mass star become a planet? Russell concluded that an object less than 1/20th the size of our Sun would have a surface temperature of about 700 K and would be invisible even under the best circumstances. Van de Kamp the following year adopted Russell's value of 1/20 (0.05) of the Sun's mass "as a conventional borderline between visible stars and the *per se* invisible bodies which we shall designate by the general term *planet*."[66] An agreement on definitions was perhaps a fitting conclusion to the flurry of activity of 1943–1944. But everyone realized that confirmation and further progress in the search for planetary systems would be slow and difficult.

As these observational developments were occurring, news came from Europe that the theoretical vacuum left by Russell's criticism of the tidal hypothesis was beginning to be filled, apparently independently of any knowledge of the observational results announced in the United States. In a Germany engulfed in war, the influential physicist and future philosopher Carl Friedrich von Weizsäcker (1912–) had been working at the University of Strassburg, and in 1944 he published what the first American reviewers called "a significant new paper on the origin of the solar system."[67] A modified version of the nebular hypothesis, it surmounted the traditional problems of that theory by supposing that the greater part of the primordial solar nebula (being composed, in his view, of hydrogen and helium) dissipated into space, carrying with it the angular momentum that had troubled previous nebular theories. The Sun was thus left with little angular momentum, and that of the planets was ascribed to the motion of the original nebula. The planets condensed from the part of the nebula that remained, which included solid particles afloat in a sea of hydrogen and helium. Instead of the accretion process of the Chamberlin–Moulton planetesimal hypothesis, however, Weizsäcker proposed that the interaction of certain "allowed" streams of gas resulted in eddies or vortices (*Wirbeln*) that, in turn, yielded planetary orbits roughly analogous to quantized orbits in the Bohr atom (Weizsäcker too was a physicist, no less than Jeans). This, in his view, might account for the Bode–Titius law of planetary distances, long the

[66] Van de Kamp, "Stars or Planets?" *S&T*, 4 (December 1944), 5–7, 22, and Russell, "Physical Characteristics," 79–86, and "Notes on White Dwarfs and Small Companions," *AJ*, 51 (June 1944), 13–17: 17.

[67] G. Gamow and J. A. Hynek, "A New Theory by C. F. Von Weizsäcker of the Origin of the Planetary System," *ApJ*, 101 (January–May 1945), 249–254.

subject of speculation. The mechanism of formation Weizsäcker discussed mathematically in great detail, a fact which undoubtedly accounts for its serious reception.[68]

A measure of the originality of Weizsäcker's theory is that it attracted the attention of scientists of the caliber of Chandrasekhar, George Gamow, and many others. It was first made known in the United States in 1945 in a review by Gamow and J. Allen Hynek, who concluded that the theory was probably erroneous in details but perhaps correct in some of its qualitative ideas. Among those ideas were the larger mass of the primordial solar nebula; the escape of much of this mass into space before the formation process began, resolving the angular momentum problem; and the physical explanation of the Bode–Titius law of planetary distances. Most important of all, they believed, Weizsäcker "has directed a fresh stream of thought into the long-stagnant pool of theories of planetary origin."[69]

Perhaps sensitized to the issue of the frequency of planetary systems by the emphasis it had been given in Jeans's tidal theory, many astronomers immediately raised this issue, if only in a cursory way. Gamow and Hynek themselves pointed out that "if the planets were formed in this manner, other stars have planetary systems in which similar Bode relations obtain. In any event, if the Weizsäcker theory holds, planetary systems of a wide variety of types must be the rule rather than the exception." With an eye toward confirmation, they also wondered if the process of solar system formation might be going on in the galaxy at present. They concluded that even if it was, a nebular cloud around a star would probably not be visible optically or spectroscopically.[70] Similarly, in a 1948 review of theories of the origin of the Earth, physicist Thornton L. Page noted of Weizsäcker's theory, "One of the interesting consequences is that the formation of planets should be an extremely common occurrence. Possibly in the process of formation of every star the conditions would be correct to form planets. Thus we might expect billions, if not hundreds of billions of planets in our galaxy, the strong likelihood that life has developed on a million or more of these, the high probability that there are other civilizations of mankind."[71] This was only the beginning of attempts to make numerical estimates, however uncertain.

Other scientists equally influenced by the work of Weizsäcker, however, did not even mention the implications of his theory for other plane-

[68] Von Weizsäcker, "Uber die Entstehung des Planetensystems," *Zeitschrift fur Astrophysik,* 22 (1944), 319–355; Gamow and Hynek, "A New Theory," 249–250.

[69] Gamow and Hynek, "A New Theory," 254.

[70] Ibid., 250, 253.

[71] Thornton Page, "The Origin of the Earth," *Physics Today* (October 1948), 12–24: 21.

tary systems, although they presumably believed that similar processes would be in operation elsewhere. Yerkes theoretical astronomer S. Chandrasekhar, for example, seemed content merely to follow up on the mathematical plausibility of Weizsäcker's theory.[72] The same was true for D. Ter Haar, who failed to broach the subject even in a semipopular treatment, though he did comment on the aesthetic appeal of a return to the nebular hypothesis because it "starts from probably the simplest possible hypothesis – a sun, surrounded by a gaseous envelope."[73]

One astronomer greatly influenced by Weizsäcker's theory, and destined to make his own contribution to the gathering planetary systems debate, was Yerkes astronomer Gerard P. Kuiper, whose claim of lichenlike plant life on Mars we examined in the previous chapter. Though he concluded early on that Weizsäcker's theory had to be abandoned in its details, like Gamow and Hynek he noted that it had the great merit of "making a fresh start with this difficult problem and of introducing new concepts capable of theoretical analysis."[74] Kuiper first came in the 1930s to the problem of the solar system's origin from the field of observational double star astronomy, and it was in this field that he would make his own contribution to the expanding belief in the abundance of planetary systems. As he recalled near the end of his life:

> At the beginning of my career I was asked to review a book on the origin of the Solar System. The analytical part of this book impressed me greatly. The second, synthetic part was entirely disappointing. After the review was written, I continued for many months to struggle with this problem and had to conclude that the state of Astronomy did not permit its solution. I was nevertheless fascinated by it, and had become aware of at least part of the extensive and difficult literature written in search for solutions. I then determined to find a closely-related problem, that with finite effort would probably lend itself to a solution.[75]

72 Chandrasekhar, "On a New Theory of Weizsäcker on the Origin of the Solar System," *RMP,* 18 (January 1946), 94–102.

73 Ter Haar, "Recent Theories about the Origin of the Solar System," *Science,* 107 (April 23, 1948), 405–411, a summary of an earlier work, and "Further Studies on the Origin of the Solar System," *ApJ,* 111 (January 1950), 179.

74 Gerard P. Kuiper, "On the Origin of the Solar System," chapter 8 in *Astrophysics: A Topical Symposium Commemorating the Fiftieth Anniversary of the Yerkes Observatory and a Half Century of Progress in Astrophysics,* ed. J. A. Hynek (New York, 1951), 357–424: 366. On Kuiper, see Chapter 3, footnote 113.

75 Gerard P. Kuiper, "Discourse Following Award of Kepler Gold Medal at A.A.A.S. Meeting, Franklin Institute, Philadelphia, December 28, 1971," *Communications of the Lu-*

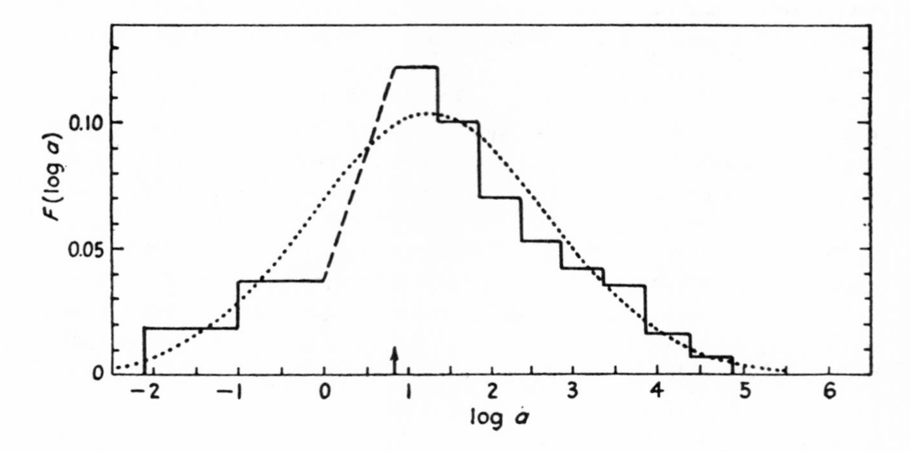

Figure 4.5. Kuiper's frequency curve of separations in binary stars, showing how the average of these separations approximates the size of our solar system. The arrow shows the mean position of the planets, and the median separation of the binaries is at $a = 10^{1.3}$ or 20 astronomical units. First published in 1935, it was not used as an argument for the abundance of planetary systems until 1951. From Kuiper (1951), as cited in footnote 74.

This related problem capable of solution with finite effort was found in the study of double stars. As early as 1935, Kuiper's work on double stars showed that their median separations were about 20 astronomical units, close to the distance of the major planets from the Sun (Figure 4.5).[76] But it was not until the late 1940s – after the difficulties with the collision hypothesis were well known – that Kuiper connected this result to planetary systems. By 1951 Kuiper was speculating on this basis that "it almost looks as though the solar system is a degenerate double star, in which the second mass did not condense into a single star but was spread out – and formed the planets and comets." On the basis of data showing that 10 percent of binary stars had components with a mass less than 1/10th that of the primary, Kuiper estimated that "planetary systems might be one or two orders of magnitude less frequent than the 10 percent just mentioned" – in other words, 1 in 100 or 1 in 1000. "Probably some 10^9 planetary systems

nar and Planetary Laboratory, 9, 403–407: 404. The book in question was Russell's *The Solar System and Its Origin,* published in 1935, 2 years after Kuiper obtained his Ph.D. For a similar statement that solar system problems led Kuiper into a study of double star statistics see Kuiper, "The Formation of the Planets," *JRASC*, 50 (March–April 1956), 51–68.

[76] Kuiper, "Problems of Double-Star Astronomy," *PASP*, 47 (April 1935), 121–150: 128, 138–139.

occur in the galaxy alone."[77] This was orders of magnitude larger than any collision theory predicted, a result that Kuiper noted some found difficult to believe. Referring to the lingering predominance of the theory of Chamberlin, Moulton, Jeans, and Jeffreys, Kuiper recalled, "I announced this result on 4 September, 1949 at a regular Sunday broadcast of the University of Chicago Roundtable. I still remember the skepticism of my astronomical colleagues; so strong was astronomical tradition."[78]

Kuiper also addressed the question of whether planetary systems were forming at present, and if so, whether observable celestial objects might be identified with such formation. Spiral nebulae were by this time out of the question as candidates, but he found two other classes of objects promising: a class of young stars known as "T Tauri stars" that often had surrounding nebulae and "Bok globules," dark, circular patches sometimes seen against a bright, nebulous background.[79] Kuiper thus found the empirical study of planetary system formation a possibility in 1956.

Kuiper clearly also believed that this conclusion was aesthetically pleasing: "It is perhaps satisfying that the present developments indicate that the process of planetary formation is but a special case of the almost universal process of binary-star formation," he wrote in 1951. A few years later he wrote in the same vein, "The great frequency of planetary systems here predicted appears comforting to this writer when contemplating our own system – which apparently is no freak of nature any more than the sun itself."[80] This idea of planetary systems as failed double stars Kuiper kept to the end of his life. "Planetary systems clearly had to originate as the low-mass extremity of the almost universal process of double-star formation," he wrote in 1971. "Indeed, the median separation in double stars was just of the dimension of the system of the massive Jovian planets, about 10 AU. A basis had thus been found for estimating the *frequency* of planetary systems in our galaxy."[81] The frequency first estimated in 1949 Kuiper also held to the end of his life. One of his last papers, in press at the time of his death in 1973, still put the frequency of

[77] Kuiper, "On the Origin," 364–365 and 416–417.

[78] Kuiper, "Discourse," 405. The broadcast was printed in Round Table Pamphlet no. 598 (University of Chicago Press), as cited in Kuiper, "On the Origin of the Solar System, I" *Celestial Mechanics*, 9 (May 1974), 328.

[79] Kuiper, "Formation of the Planets," 164. The T Tauri stars had been discovered in 1945 by Alfred Joy. On their history and possible relation to planetary systems see Catherine L. Imhoff, "In Search of the T Tauri Stars: An Historical Perspective," *Astronomical Quarterly*, 1 (1977), 213–238. Bok globules are named after astronomer Bart Bok, who focused attention on them in published papers beginning in 1947.

[80] Kuiper, *Astrophysics*, 416 and "Formation of the Planets," 167. Citing Chamberlin (1928), Kuiper noted that some had derived comfort from the rarity of planetary systems.

[81] Kuiper, "Discourse," 404.

planetary systems in the Galaxy at 1 in 100 or 1 in 1000, yielding a total of 10^9 planetary systems in the Galaxy.[82]

Kuiper's colleague at the University of Chicago, Harold P. Urey, knew of the former's work when he wrote his pioneering *The Planets: Their Origin and Development,* which attacked the problem of the planets and their origin from yet another point of view: geochemistry. Urey's interest was piqued in about 1949, when he gave a course on "Chemistry in Nature" at a University of Chicago summer session. Like T. C. Chamberlin, he came to the subject from a problem in the Earth's formative history: during his course preparations, he was surprised to read that the Earth's temperature was believed to be rising, not falling. "This led on to a consideration of the curious fractionation of elements which must have occurred during the formation of the earth. One fascinating subject after another came to my attention, and for two years I have thought about questions relating to the origin of the earth for an appreciable portion of my waking hours, and have found the subject one of the most interesting that has ever occupied me," he wrote in the Preface to that work.[83] Reviewing theories of solar system origin, Urey pointed out that collisional hypotheses yielded only one stellar collision every 3 billion years, a scheme in which our Sun would not be an average star at all, but Urey proclaimed the objections to this theory fatal and promised evidence "regarding both the star and the planet with life on it as average."[84]

Urey firmly believed that the origin of the solar system had to be understood in the context of stellar evolution. He cited Struve's just published *Stellar Evolution* for the theory that "the solar system is only one orderly possibility as the initial dust clouds of varying masses and angular momenta and very probably of somewhat different chemical composition evolved into stars."[85] Urey noted astronomer Fred Whipple's attempt to explain the solar system based on visible masses of gas and dust, and he cited Kuiper's results on double stars. Moreover, he cited the recent results of van de Kamp and Strand regarding small, possibly dark companions surrounding stars near the Sun. Since these were so close to the Sun, "it seems very likely that all stars are multiple, if we include as stars objects of the size of the asteroids, and that solar systems are only special cases of multiple stars."[86] The results of 1943 were still exerting their effect a decade later. On the theoretical side, Urey also cited Weizsäcker on the evolution of the solar system. Urey assumed that star and planet formation

82 Kuiper, "On the Origin . . . I," Paper II was not published because of Kuiper's death.
83 Urey, *The Planets: Their Origin and Development* (New Haven, Conn., 1952), ix.
84 Ibid., 3. For the objections to the collision theory, Urey cited Russell's *Solar System*.
85 Ibid., 10.
86 Ibid., 6.

were normal processes of evolution in the universe; however, he attempted no estimate of the number of planetary systems in the galaxy. Due in part to differences in their theories of the origin of the solar system, especially the cold-accretion versus molten concepts of the formation of the Moon, Urey and Kuiper would become involved in bitter disputes.[87]

Aside from observations of stellar perturbations, the revival of the nebular hypothesis, double star astronomy, and the field of geochemistry, support for planetary systems during this period came unexpectedly from yet another quarter – the study of stellar rotation. It had been known since the time of Galileo that the Sun rotates on its axis, and Frank Schlesinger was the first to actually observe the rotation of stars in 1909. Otto Struve, however, was the pioneer in this field, utilizing stellar rotation as early as 1923 in developing his thesis to explain the broadening of spectral lines. His earliest published results in 1929 and 1930 established the procedure and the fact of widespread stellar rotation.[88] In his 1930 paper, Struve had also noted a sharp slowdown in stellar rotation of the F spectral type, an effect subsequently pinned down more precisely to the F5 spectral type.[89] The cause of this effect was unknown, but already in 1930 Struve proposed that it might be related to the breakup of the star into components, forming a binary star.[90] By 1945 Struve was speculating that the observed formation of a shell around some stars "represents the process of rotational breakup, and we are thus probably witnessing by means of our spectrographs the actual formation of the rings of Laplace," referring to the nebular hypothesis. But he apparently did not think that such shells would necessarily form solar systems, noting that if Laplace knew what we know now about the spectra of these stellar shells, "He would hardly have suggested that the gaseous rings would condense into planets."[91]

87 On the Urey–Kuiper dispute see Ronald E. Doel, *Solar System Astronomy in America: Communities, Patronage, and Interdisciplinary Research, 1920–1960* (Cambridge, 1996), chapter 4; also Stephen Brush, "Nickel for Your Thoughts: Urey and the Origin of the Moon," *Science*, 217 (1982), 891.

88 G. Shajn and O. Struve, "On the Rotation of Stars," *MNRAS*, 89 (1929), 222–239, and Struve, "On the Axial Rotation of Stars," *ApJ*, 72 (July 1930), 1–18. On Struve see K. Krisciunas, "Otto Struve (1897–1963)," *BMNAS* (1992), 351–387, including a select bibliography of his works; Z. K. Sokolovskaya, "Otto Struve," *DSB*, 13 (New York, 1976), 115–120; and T. G. Cowling, *Biographical Memoirs of Fellows of the Royal Society*, 10 (1964), 283–304.

89 Struve, *Stellar Evolution* (Princeton, N.J., 1950), 130.

90 Struve, "Axial Relation," note 88, 3. The fission theory to which Struve refers is that of G. H. Darwin and James Jeans.

91 Struve, "The Cosmogonical Significance of Stellar Rotation," *PA*, 53 (May 1945), 201–218, and (June 1945), 259--276: 275–276. Struve first presented this paper in February 1942; it may have been revised, but nowhere does it mention Weizsäcker. That Struve was still under the spell of collision theories is evident in his explanation of the origin of

Struve, however, kept turning the problem over in his mind, and by late 1946 was more favorably inclined to the idea that stellar rotation was revealing the possible formation of planetary systems. He asked:

> Can it be that a large fraction of those late-type stars which would normally possess rapid rotations have, by some as yet unknown process, produced planetary systems and in this manner relieved themselves of a large fraction of their angular momenta? After all, the sun is also a late-type star; and computation shows that if the planets were combined with the sun the velocity of rotation of the latter would be much more rapid than it is at present. It is only fair that I should warn the reader to accept this speculation with reserve. It tells us nothing about the origin of the solar system. It is like a fleeting glance of a cool spring in the heat of the desert. Perhaps it is only a mirage.[92]

Struve gradually converted this mirage into reality. In his classic *Stellar Evolution* (1950), he proposed that condensation from an interstellar cloud produced a star of the early spectral type with a very high angular momentum and a large mass. As it condensed further, it shed more mass, and (after intermediate stages which included various forms of observed double stars) eventually shed its angular momentum by "the formation of a planetary system with a single star having approximately the mass of the sun and devoid of appreciable angular momentum, surrounded by one or more planets at considerable distances, which contain a considerable fraction of the original momentum." Summarizing his ideas of stellar evolution in a figure that included a planetary system as one of its possible outcomes, Struve nevertheless stressed that "The formation of planets is not an essential feature of our hypothesis. We have introduced it only because the common envelope of a close double star represents the sort of medium required for their origin, and because the sun is a typical, cool, slowly rotating dwarf."[93]

By 1952, however, Struve was not only explicitly attributing the F5 rotational discontinuity to the formation of planetary systems, he was also suggesting a means for their detection.[94] More than 20 years after he

stellar rotations as due to close stellar encounters or to the infall of interstellar or nebular matter, 272–273.

92 Otto Struve, "Stellar Rotation," *S&T,* 6 (November 1946), 3.

93 Struve, *Stellar Evolution,* 235–239. Thus Struve in 1950 did not seem to be quite ready to proclaim the abundance of planetary systems, as he did only 2 years later, perhaps under the influence of the theories of solar system formation of Hoyle and Kuiper.

94 Struve, "Proposal for a Project of High-Precision Stellar Radial Velocity Work," *Observatory,* 72 (October 1952), 199-200. Frank Drake also recalls being impressed by

clearly accepted axial rotation of stars and the discontinuity in rotation between the early and late stars, Struve firmly invoked planetary systems as the explanation. This explanation, in turn, became an important argument in the solar system theory of British theorist Fred Hoyle and others,[95] and the importance of the stellar rotation argument was spread especially by Struve's student Su-Shu Huang.[96]

Finally, as retired Harvard Observatory Director Harlow Shapley made clear in his popular work *Of Stars and Men: Human Response to an Expanding Universe* (1958), the new cosmology was a continual force in the background favoring abundant planetary systems. That cosmology, he argued, required us to believe that we are not the only life in the universe, and he pointed in particular to three developments. First, the discovery that the nebulae are actually galaxies of stars meant that we have at hand "more than one hundred million million million sources of light and warmth for whatever planets accompany these radiant stars." Second, the expanding universe implied that "a few thousand million years ago . . . the average density in the unexpanded universe must have been so great that collisions of stars and gravitational disruptions of both planets and stars were inevitably frequent . . . at that time countless millions of other planetary systems must have developed, for our sun is of a very common stellar type. . . . Millions of planetary systems must exist, and billions is the better word." If the nebular hypothesis operated – and Shapley stated even at this late date that he believed there is more than one mode of planet formation – then this also implied the existence of a large number of planetary systems. Finally, biochemistry now indicated to Shapley that "whenever the physics, chemistry and climates are right on a planet's surface, life will emerge, persist and evolve."[97] In Shapley's view, the Earth and its life are "on the outer fringe of one galaxy in a universe of millions of galaxies. Man becomes peripheral among the

Struve's discussion of planetary systems during a lecture at Cornell in 1951. David Swift, *SETI Pioneers* (Tucson, Ariz., 1990), 382–383.

95 Hoyle, *Frontiers of Astronomy* (New York, 1955), 83, 104–105; and *The Nature of the Universe*, 2d ed. (New York, 1960), 32, 81, 90. It is of interest that the first edition of *The Nature of the Universe* (1950), 26 and 101, did not yet use the stellar rotation argument in connection with planetary systems, whereas by 1955 it was Hoyle's most important argument in this regard. Hoyle's 1950 estimate for planetary systems was based on his theory that supernovae played a role in solar system formation, and was therefore based on the rate of occurrence of supernovae in our Galaxy. Among others accepting the stellar rotation argument was M. H. Briggs, "The Detection of Planets at Interstellar Distances," *JBIS*, 17 (March–April 1959), 59–60.

96 Su-Shu Huang, "A Nuclear Accretion Theory of Star Formation," *PASP*, 69 (October 1957), 427–430: 428, and "Occurrence of Planetary Systems in the Universe as a Problem in Stellar Astronomy," *VA*, 11 (1969), 217–263.

97 Shapley, *Of Stars and Men* (Boston, 1958), 108–114. See also chapter 4, "An Inquiry Concerning Other Worlds," 53–75.

billions of stars in his own Milky Way; and according to the revelations of paleontology and geochemistry he is also exposed as a recent, and perhaps an ephemeral manifestation in the unrolling of cosmic time."[98] This view, poetically expressed by one who had helped to build it, is basically the modern conception of the universe.

Several significant characteristics emerge from this crucial period in the history of planetary systems. First, in at least two independent cases, observations made in the 1930s were not interpreted until the early 1950s as bearing on the abundance of planetary systems. Although Kuiper published his statistical data on separation of binaries in 1935, it was not until 1951 that he used them as support for planetary systems. And though Struve wrote his classic paper on stellar rotation in 1930, and although the case for a sharp discontinuity in rotational speeds was becoming increasingly clear during the 1930s, not until about 1952 did he propose that the slower rotational velocities of late-type stars were due to angular momentum having been lost to planetary systems. In both cases, this was after the revived nebular hypothesis was beginning to be interpreted as indicating numerous planetary systems. During this transition period from pessimism to optimism, therefore, observation did not so much affect theory – although it may have spurred the acceptance of the theory – as theories of planetary system formation influenced the interpretation of observations already made.

Second, none of the many independent arguments for planetary systems were made with anything approaching deductive rigor. Kuiper's binary star separations, while suggestive of his hypothesis that solar systems were failed double stars, fell far short of proof. While the slowdown in stellar rotation examined by Struve might have been the result of the transfer of angular momentum to planetary systems, that transfer might equally have been made to the interstellar medium or other places. Hoyle's supernovae theory, while it explained the abundance of heavy elements, was removed by even more steps of inference from his conclusion that the abundance of planetary systems was linked to the frequency of supernovae. And Weizsäcker's theory merely reverted to the preplanetesimal hypothesis argument that there seemed to be no reason why the process should not be common throughout the universe. For all these reasons, great weight was given to the more direct observation of planetary systems without the

[98] *Of Stars and Men* (Boston, 1958), 108, in a chapter adapted from an article in *The American Scholar* (Autumn 1956). The view is in marked contrast to Shapley's skeptical view of extraterrestrial life in the 1920s. Shapley first returned to the subject with a more favorable attitude in "On Climate and Life," in *Climatic Change: Evidence, Causes and Effects* (Cambridge, Mass., 1953), the result of a conference sponsored by the American Academy of Arts and Sciences in May 1952. The article in slightly different form was originally printed in *The Atlantic Monthly* earlier in 1953.

Table 4.2. *Estimates of frequency of planetary systems, 1920-1961*

Author	Argument	No. of planetary systems in galaxy	No. of habitable planets in galaxy
Jeans (1919, 1923)	Tidal theory	Unique	1
Shapley (1923)	Tidal theory	"Unlikely"	"Uncommon"
Russell (1926)	Tidal theory	"Infrequent"	"Speculation"
Jeans (1941)	No. of stars	10^2	—
Jeans (1942)	> Diameter of Sun	one in six stars	Abundant
Russell (1943)	Observation of companions	Very large	$>10^3$
Page (1948)	Weizsacker	$>10^9$	$>10^6$
Hoyle (1950)	Supernovae	10^7	10^6
Kuiper (1951)	Binary star statistics	10^9	—
Hoyle (1955)	Stellar rotation	10^{11}	—
Shapley (1958)	Nebular hypothesis	10^6 - 10^9	—
Huang (1959)	Stellar rotation	10^9	10^9
Hoyle (1960)	Stellar rotation	10^{11}	10^9
Struve (1961)	Stellar rotation	$>10^9$	—

Source: Steven J. Dick, in Heidmann and Klein (reference 122 in this chapter), p. 359, by permission of Springer-Verlag.

intermediary of these theories – and the observations themselves were fraught with difficulty.

Nevertheless, whether accepting the arguments of Kuiper, Struve, Hoyle, and Weizsäcker, independently or in concert, one thing was clear: during this 15 year period, planetary systems were returned to their status as a normal outcome of stellar evolution, a comeback evident in Table 4.2. That general idea – opposed by Chamberlin, Moulton, Jeans, and Jeffreys – supported an abundance of planetary systems no matter which of the ascending theories of this period one chose. Moreover, it was once again in consonance with the Copernican principle that our Earth and our solar system were nothing special in the universe. Even Jeans at the outset of his astronomical labors had written that although we have no a priori knowledge of whether solar systems are common or not, "as between two otherwise equally plausible explanations, we shall feel a preference for the one which does not postulate the occurrence of very rare or exceptional events."[99] Forty years after he had written that statement – his own theory now discounted – it was fully operative as a working principle of many astronomers.

However doubtful the force of the arguments, there is no doubt of the result of this period that was so tumultuous in science, as well as in world affairs. In 1940 the Astronomer Royal Sir Harold Spencer Jones, echoing Jeans, Eddington, and the conventional wisdom espoused by most astronomers, had written in his volume *Life on Other Worlds,* "life is not widespread in the universe . . . not more than a small proportion of the stars are likely to have any planets at all. With the usual prodigality of Nature, the stars are scattered far and wide, but only the favoured few have planets that are capable of supporting life." In the second edition of 1952, citing Weizsäcker, he wrote, "On this important problem of the origin of the solar system and of planetary systems in general, there has been a marked change in outlook in the last few years from that of twenty years ago. Astronomers then felt pretty confident that the solar system was something very exceptional; now it appears much more probable that the formation of a planetary system may occur as one of the normal courses of stellar evolution."[100] A turning point had been passed, and there would be no going back for the rest of the century.

4.3 OPTIMISM: OBSERVATION TO THE RESCUE?

It is some measure of the difficulty of the search for planetary systems that the Space Age did not bring immediate advances in the problem. Unlike solar system studies, where planetary spacecraft brought immediate and revolutionary progress in our knowledge of the planets, no such prospect was in store for planetary systems. It is true that increased knowledge of our own planetary system provided voluminous data for the refinement of theories of the origin of the solar system, which by the usual gross analogies could be applied to other solar systems. But, although substantial, these refinements changed little the fortunes of planetary systems. Perhaps the largest impact of the Space Age on planetary systems science was the infusion of funds from space agencies such as NASA, which displayed an

[99] Jeans, "Tidally-Distorted Masses."
[100] Sir Harold Spencer Jones, *Life on Other Worlds* (London, 1940), 293, and (2d ed., London, 1952), 243.

interest in both the observational and theoretical aspects of the subject almost from the beginning, but with delayed results.

We should therefore not be surprised that, while most astronomers in the second half of the twentieth century were optimistic about other planetary systems, observational proof of their existence throughout the 1970s remained entirely dependent on the old astrometric technique. That technique, the results of which remained elusive in many cases, created a public and scientific sensation with the announcement in the 1960s of the detection of several planetary systems. The promise and limitations of this technique, and the difficulties of tackling a problem at the limits of science, may best be seen in the famous case of Barnard's star. The central figure in the case is Peter van de Kamp, whom we last saw beginning his search for low-mass companions at the Sproul Observatory of Swarthmore College in Pennsylvania in 1937, in the wake of Dirk Reuyl's discovery the previous year of perturbations caused by a dark *stellar* companion to the star Ross 614. Such is the long-term nature of the problem of determining perturbations in stellar motions that only now, 25 years later, was van de Kamp beginning to announce results with planetary companions. He did so in an era when scientists were increasingly taking the issue of extraterrestrial life seriously. But even without that momentous issue hanging in the balance, it is unlikely that the conflict and controversy that Barnard's star generated throughout the 1970s could have been avoided.

The interesting history of Barnard's star began in 1916, when Yerkes astronomer E. E. Barnard announced its discovery – curiously, in light of subsequent history, during a meeting of the American Astronomical Society at the Sproul Observatory. As Barnard emphasized, it was a star of 9.5 magnitude, peculiar for its enormous proper motion of about 10.3 arcseconds per year. This meant that it was a close star (the closest known star after the Alpha Centauri system), and it was immediately placed on observational programs, including the parallax program at Sproul in 1916–1919. As we saw in the previous section, in 1938 van de Kamp had placed it back on the Sproul parallax program with his arrival as director in 1937, and by 1944 he announced a low-mass companion stellar in nature.[101] Over the next 20 years, as Hoyle, Kuiper, Struve, and others were predicting an abundance of planetary systems based on their own

101 The initial aims of the Sproul program are described in a lecture van de Kamp presented before the Swarthmore Chapter of the Society of the Sigma Xi on February 15, 1938, published as "The Invisible Universe," *Sigma Xi Quarterly*, 26, no. 2 (June 1938), 103–09, 118. Van de Kamp pointed out that the largest known planetary mass (Jupiter) was 1/1000th the mass of the Sun, and the smallest known stellar mass (Kruger 60 B) was 1/7th the mass of the Sun. The purpose of his study was to "throw light on intermediate masses."

work, and as theory once again made plausible abundant planetary systems, van de Kamp patiently collected data on Barnard's star and other nearby stars.[102]

There is no doubt that van de Kamp was sensitive to the question of whether low-mass companions were stars or planets, at least since his 1944 article on the subject, an article undoubtedly stimulated by Strand's observational claims of 1943 and the symposium on "Dwarf Stars and Planetary Companions" also held in that year.[103] In a progress report on "Planetary Companions of Stars" in 1956, van de Kamp pointed out that while numerous unseen objects had been detected over the last two decades with masses 0.05 of the Sun's or greater, it was "extremely likely" that these objects were stars. "There are tentative indications of unseen companion objects with about .01 solar masses *or more*, and these *may* be planetary companions. However, definitive interpretation can hardly be reached at present, partly due to limitations of accuracy," he wrote in that year. In particular, van de Kamp labeled the 1938 claims of Holmberg "spurious" and "illusory"; he pointed out that the 1943 claims of Reuyl and Holmberg for a planetary companion of 70 Ophiuchi had not been confirmed by Strand as of 1952; and he held out hope only for Strand's claim in 1943 for a companion of 61 Cygni of .016 solar masses, well into the planetary range, as confirmed by two other observers.[104] As for his own program, now almost two decades old, van de Kamp claimed only that Lalande 21185 was probably a star of low luminosity, and that no satisfactory explanation existed for some perturbations seen in the motion of Barnard's star. Almost two decades after the start of his observational program at Sproul, no one could accuse van de Kamp of rushing to judgment on planetary companions!

All of this was to change in the 1960s, a decade of uproar in this field of astronomy, as in so many other activities of American life. First, in 1960 Sarah Lippincott, van de Kamp's colleague at Sproul, announced that the companion of Lalande 21185 had a mass of only .01 that of the Sun. Although this was at the upper edge of the planetary range (recall

[102] Van de Kamp's lack of attention in his published papers to theories of solar system formation supports the view that theory had little, if any, influence on his work. On the other hand, once van de Kamp announced the discovery of a planetary companion in 1963, he wrote that the results "make us more receptive to the theoretical evidence that such objects may exist." "The Discovery of Planetary Companions of Stars," *Yale Scientific Magazine*, 38 (December 1963), 6–8: 8.

[103] P. van de Kamp, "Stars or Planets?," *S&T*, 4 (December 1944), 5–7, 22, reprinted in *The Origin of the Solar System*, eds. Thornton Page and Lou Williams Page (New York, 1966), 176–180. On the symposium, see note 65.

[104] Van de Kamp, "Planetary Companions of Stars," *VA*, 2 (1956), 1040–1048. The two observers cited were Deutch [a variant spelling of Deich, as cited in note 64] and Nowacki for work they published in 1951.

that the 1943 claims for planetary companions to 61 Cygni and 70 Ophiuchi gave them about this same mass), Lippincott's technical article in the *Astronomical Journal* made no mention of the word "planet," and perhaps for this reason her announcement did not raise much of a stir.[105] But van de Kamp's 1963 article with the mundane title "Astrometric Study of Barnard's Star from Plates Taken with the 24-inch Sproul Refractor" created a sensation. In it he announced the discovery of a companion to Barnard's star with a mass of only .0015 the mass of the Sun, only 1.6 times the size of Jupiter, which he specifically characterized as a planet. He further found the distance of the planet from Barnard's star to be similar to that of Jupiter from the Sun, and its surface temperature about 60 K compared to 120 K for Jupiter.

Van de Kamp's claim was based on 25 years of photographic observations, using three types of photographic emulsions and 50 observers, yielding 2413 plates. To the extent that the public was aware of such details, they might have been persuaded by this alone that such an intensive scientific effort must have yielded a definitive result. But they would not have been aware of the subtleties of the technique. Three reference stars were chosen as the background stars against which to measure the position of Barnard's star itself. In coming to his conclusions, van de Kamp had to take into account all known effects and causes of error. This included "magnitude equation," for the reference stars were unavoidably of different brightnesses, so that their diameters varied on the photographic plate. He had to account for "color equation," because the reference stars were of different spectral types, and the objective lens had been adjusted during the course of the observations. A "personal equation" entered because the plates had to be measured on a machine, and each person making the measurements had his or her own errors of measurement. Van de Kamp even had to take into account the change in proper motion of Barnard's star due to the fact that it was moving closer to the Sun, and thus the apparent angular movement was gradually increasing – by 1/1000th of a second of arc per year.

Having taken into account all these effects as best he could, van de Kamp found a perturbation in the motion of Barnard's star with a period of about 24 years (Figure 4.6). In order to come up with an actual mass for the companion, he further had to carry out a "dynamical interpretation" calculation, using the mass of Barnard's star. By Kepler's law, once

[105] Sarah L. Lippincott, "Astrometric Analysis of Lalande 21185," *AJ*, 65 (September 1960), 445–448, with a minor erratum in *AJ* , 67 (October 1962), 570. The dividing line between "star" and "planet" was still open to discussion, but Lippincott could well have argued – as others had before her – that she had discovered a planet. In 1963 S. S. Kumar found that a body smaller than .07 of the mass of the Sun could not support thermonuclear reactions and would contract to become a black dwarf. Kumar, "The Structure of Stars of Very Low Mass," *ApJ*, 137 (May 15, 1963), 1121–1125.

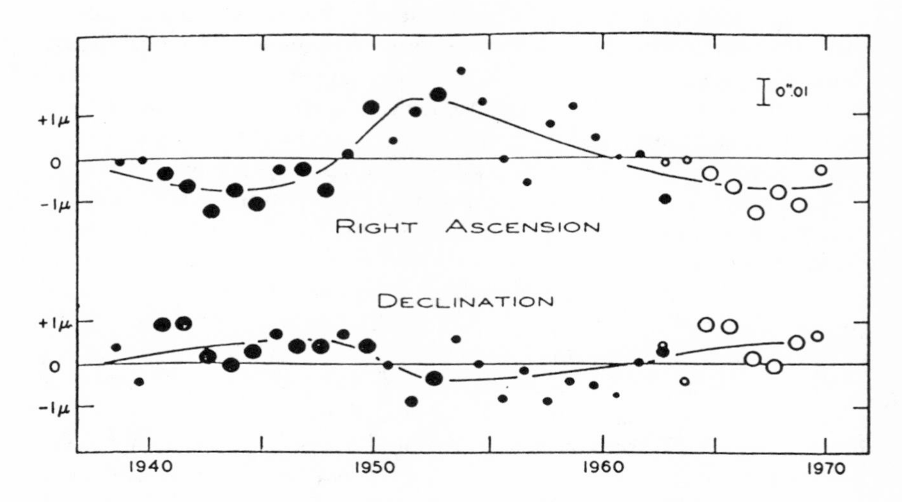

Figure 4.6. One of four methods of planet detection (see also Figures 4.7, 4.8, and 4.9). Van de Kamp's data for Barnard's star (1963) represent the classical astrometric method showing minute gravitational perturbations of a few hundredths of an arcsecond over a period of decades; plots using the spectroscopic radial velocity method look similar but need not cover such a long period of time. Used with permission from Elsevier Science Ltd.

this mass and the period of the orbiting body were known, the mass of the latter could be calculated. It was here that van de Kamp finally came to the figure of .0015 times the mass of the Sun for his new planet: "The orbital analysis leads, therefore, to a perturbing mass of only 1.6 time the mass of Jupiter. We shall interpret this result as a companion of Barnard's star, which therefore appears to be a planet, i. e., an object of such a low mass that it would not create energy by the conventional nuclear conversion of hydrogen into helium."[106]

As with the announcements 20 years before, the reaction to van de Kamp's result was swift. From *Time* to popular science magazines and more sober scientific journals, countless reports of van de Kamp's results hailed the discovery of another planetary system. Independent verification of the result, on the other hand, was more difficult, since the observations were very specialized and required decades to reach a result – van

[106] Van de Kamp, "Astrometric Study of Barnard's Star from Plates Taken with the 24-Inch Sproul Refractor," *AJ*, 68 (September 1963), 515–521: 521. Popularizations of his discovery by van de Kamp in the same year are "Barnard's Star as an Astrometric Binary," *S&T*, 26 (July 1963), 8–9, and "The Discovery of Planetary Companions of Stars," *Yale Scientific Magazine* 38 (December 1963), 6–8.

de Kamp had been at it for a quarter century. It is not surprising, therefore, that van de Kamp himself was the first to reinforce his own result. In 1969, with 5 more years of photographic measures of Barnard's star, he reiterated his claim that a planetary companion to that star existed, with a slightly revised mass of 1.7 times that of Jupiter.[107] In the same year, van de Kamp proposed an alternative analysis of his data which held out the possibility that two planets orbited Barnard's star, with periods of 26 and 12 years and masses of 1.1 and .8 times that of Jupiter.[108]

But trouble was around the corner, and the 1970s saw serious questions raised about van de Kamp's momentous result. In 1973, John Hershey, one of van de Kamp's own student's, found that changes made to the Sproul telescope, in particular a change of lens cell in 1949, caused jumps in the data at that point and may have affected the results for Barnard's star.[109] In the same year, George Gatewood of the Allegheny Observatory in Pittsburgh and Heinrich Eichhorn of the University of South Florida announced that they could not independently confirm the existence of a companion to Barnard's star. Utilizing 241 photographic plates from two observatories spanning the period 1916–1971, they presented in detail the customary explanation of their reference system (which, significantly, included 19 reference stars rather than van de Kamp's 3), their photographic plate measurement technique, and their method of "reduction" or analysis of the data obtained. In the end, they were forced to conclude "with disappointment" that no perturbation existed in the motion of Barnard's star. Attempting to explain their result, they pointed to the disadvantages of van de Kamp's reduction technique, known as the "method of dependences," over their "plate overlap method"; to the changes in the optical system of van de Kamp's telescope over the extended period of time of his study; and to the fact that his claimed perturbation was just "on the verge of significance." With regard to the last, they cautioned that in the past, astrometric studies had suggested "the reality of actually unreal things," notably van Maanen's rotation of external galaxies and the claimed measurement of parallax before Bessel. A similar analysis by Gatewood published the following year gave the same null result for Lippincott's 1960

107 Van de Kamp, "Parallax, Proper Motion, Acceleration, and Orbital Motion of Barnard's Star," *AJ*, 74 (March 1969), 238–240. "The Prevalence of Planets and the Probability of Life," *Time*, 84 (September 25, 1964), 49–50.

108 Van de Kamp, "Alternate Dynamical Analysis of Barnard's Star," *AJ*, 74 (August 1969), 757–759. These results were reported in "The Mysterious Companions of Barnard's Star," *Time*, 93 (April 25, 1969), 93–94, and in "Two Planet Solar System," *SciAm*, 220 (June 1969), 58.

109 John Hershey, "Astrometric Analysis of the Field of AC +65 6955 from Plates Taken with the Sproul 24-Inch Refractor," *AJ*, 78 (June 1973), 421. These results were consistent with earlier indications reported by Lippincott in *Publications of the Leander McCormick Observatory*, 16 (1971), 59.

claim of a planetary companion around Lalande 21185. Two other studies of van de Kamp's data in 1973 were more favorable to his claim of one or more planetary companions, but a decade after the first announcement by the van de Kamp group, planetary systems were once again in trouble.[110]

Though Gatewood and Eichhorn acknowledged the assistance of van de Kamp in refereeing their paper for the *Astronomical Journal,* the latter understandably did not take lightly this negation of the main result of his work of 25 years. In order to take the objections into account, especially the finding that changes in the instrument might have affected positional measurements, van de Kamp remeasured his plates on a new machine and included only material from 1950 on. He confirmed the existence of the shorter-period planet, with a mass now .4 that of Jupiter, while the second planet was "less well determined." In 1977, 60 years after the discovery of Barnard's star, van de Kamp took the occasion to reassert his belief in the reality of its planetary companions. In addition to the now familiar scientific defense, the article concluded with a Rembrandt etching on the appearance of Christ to Thomas, with the caption "Blessed are they that have not seen, and yet have believed," suggesting a religious invocation of faith undoubtedly seen by some as not readily transferable to the scientific realm. Van de Kamp's last paper based on new data, published in 1982, again supported the conclusion of two planets around Barnard's star, a conclusion he had not relinquished in 1988.[111]

The Barnard's star episode was only the most notorious of several claims made for planetary companions by the mid-1970s, all subject to the same limitations of technique and inference. Although the assault on Barnard's star continued to receive the greatest attention, in the field of astrometric perturbations it was not unique and thus could not be written off as a fluke. In an extensive 1975 review of the subject of unseen astrometric companions, van de Kamp could list 17 "well-established perturbations" of stars by unseen companions, including Barnard's star

[110] George Gatewood and Heinrich Eichhorn, "An Unsuccessful Search for a Planetary Companion of Barnard's Star (BD +4 3561)," *AJ*, 78 (October 1973), 769–776; Gatewood, "An Astrometric Study of Lalande 21185," *AJ*, 79 (January 1974), 52–53. The two further studies were a mathematical analysis in Oliver G. Jensen and Tadeusz Ulrych, "An Analysis of the Perturbations on Barnard's Star," *AJ*, 78 (December 1973), 1104, and David C. Black and Graham Suffolk, "Concerning the Planetary System of Barnard's Star," *Icarus*, 19 (July 1973), 353–357.

[111] Van de Kamp, "Astrometric Study of Barnard's Star from Plates taken with the Sproul 61-cm Refractor," *AJ*, 80 (August 1975), 658–661; van de Kamp, "Barnard's Star 1916–1976: A Sexagesimal Report," *VA*, 20 (1977), 501–521. Also see "Dark Companions of Stars, Astrometric Commentary on the Lower End of the Main Sequence," *Space Science Reviews*, 43 (1986), 211–327, reprinted in 1986 under the same title; van de Kamp, "The Planetary System of Barnard's Star," *Vistas in Astronomy*, 26 (1982), 141–157. Steven J. Dick, interview with van de Kamp, 1988.

and 3 others with possible planetary companions.[112] Another 14 stars, including the famous 61 Cygni, were listed with "perturbations of provisional, suspected, or uncertain nature."

Not everyone was convinced that even those stars showing well-established perturbations necessarily harbored planets, for this depended on theoretical ideas about the cutoff point for stable hydrogen burning in stars. Shortly after van de Kamp's 1963 announcement regarding Barnard's star, CalTech geochemist Harrison Brown had supported the idea of numerous planets by an extension of the "luminosity function" (the distribution of the stars with their visual magnitudes) to low masses. If his assumptions were correct, Brown argued, "then we must conclude that planetary systems are far more abundant than we have so far suspected. Virtually every main-sequence star should have a planetary system associated with it."[113] At about the same time, however, S. S. Kumar, a University of Virginia astronomer, began to argue that all of the objects claimed as planets were probably very low mass "degenerate" objects that he termed "black dwarfs." These objects, below the .07 solar mass limit believed to be necessary for sustained nuclear burning, Kumar believed would shine faintly while contracting, but without passing through any stellar evolution, and were not planets. "The Galaxy probably contains billions of black dwarfs of very low mass, either as single objects or as members of multiple systems. At the present time we do not know if it also contains a large number of solar systems similar to our own. It seems quite clear that the number of solar systems in the Galaxy cannot be as large as previously estimated by many workers. Consequently, the number of Earth-like planets has also been overestimated." By 1971 Kumar estimated that the total number of solar systems in the Galaxy was less than 1 million out of several hundred billion stars, making it very unlikely that van de Kamp or anyone else had detected a planet around the nearby stars they had studied so assiduously over the decades.[114]

No independent confirmation of the astrometrically determined planetary companions came during the century, though in one case a claim was

[112] Van de Kamp, "Unseen Astrometric Companions of Stars," *Annual Reviews of Astronomy and Astrophysics*, 13 (1975), 295–333: 312–313. The three other possible planetary companions were Epsilon Eridani, BD +43 4305, and BD + 68 946, all within 5 parsecs of the Sun (ibid., 331). Other reviews of the subject include George Gatewood, "On the Astrometric Detection of Neighboring Planetary Systems," *Icarus*, 27 (January 1976), 1–12, and *Icarus*, 41 (February 1980), 205–231.

[113] Harrison Brown, "Planetary Systems Associated with Main-Sequence Stars," *Science*, 145 (September 11, 1964), 1177-1181.

[114] S. S. Kumar, "On Planets and Black Dwarfs," *Icarus*, 6 (January 1967), 136–137; "Planetary Systems," in W. C. Saslaw and K. C. Jacobs, eds., *The Emerging Universe* (Charlottesville, Va., 1971), 25–34. See also his articles in *Zeitschrift fur Astrophysik*, 58, 248 (1964), and *Annals of the New York Academy of Sciences*, 163 (1969), 94.

made of independent confirmation by a new technique, only to be retracted a year later. This case involved the U.S. Naval Observatory, which had built a 61-inch astrometric reflector, considered the state-of-the-art instrument in the parallax field. In 1983 Robert S. Harrington (yet another van de Kamp student) and Varkey V. Kallarakal announced that after 10 years of observation, the very-low-luminosity stars Van Biesbroeck 8 and Van Biesbroeck 10 showed variations in their proper motions, indicating that "they each apparently have an unseen companion, either or both of which could have masses of a few milli-Suns." This was, of course, in the planetary mass range, and astronomers at the University of Arizona soon used a new technique known as "speckle interferometry" in an attempt to confirm the object. Confirmation was reported in 1985, with the conclusion that "These observations may constitute the first direct detection of an extrasolar planet," a conclusion the authors believed supported by both astrometric and astrophysical considerations. By late 1986, however, it became clear that the existence of the object could not be confirmed elsewhere.[115] Their reputations at stake, the original speckle team attempted to reproduce their observation, an attempt in which they failed. Finally, to complete the circle, Harrington at the Naval Observatory reported that newer astrometric data did not confirm the perturbation that had set off the whole episode.

Even as this experience pointed up the difficulties still inherent in the astrometric search for planets, new techniques were on the ascendant. As seen in Table 4.3, planetary detection in the 1980s belonged not to astrometry but to those new (or newly refined) techniques ranging from ground-based "charge-coupled device" cameras, infrared speckle interferometry, spectroscopy, and even radio pulsar signals to observations from spacecraft high above the Earth's obscuring atmosphere.[116] The turn-

[115] The original observations were reported in Robert S. Harrington and V. V. Kallarakal, "Astrometry of the Low-Luminosity Stars VB8 and VB10," *AJ*, 88 (July 1983), 1038–1039. The speckle observations are reported in D. W. McCarthy, Jr., Ronald G. Probst, and F. J. Low, "Infrared Detection of a Close Cool Companion to Van Biesbroeck 8," *ApJ*, 290 (March 1, 1985), L9–L13. Lack of confirmation is given in Michael F. Skrutskie, William J. Forrest, and Mark A. Shure, *Astrophysics of Brown Dwarfs*, eds. M. C. Kafatos, R. S. Harrington and S. P. Maran (Cambridge, 1986), 82; in *ApJ Letters*, 312 (January 15, 1987), L55; and in C. Perrier and J. M. Mariotti, *ApJ Letters*, 312 (January 1, 1987), L27. In the debate over whether the object was a planet or a brown dwarf star, scientific journals tended to report it as a brown dwarf, as in *Physics Today*, "VB 8B. The First Brown Dwarf outside the Solar System" (January 1986), S6, and *SciAm*, "When is a Planet?" (April 1985), 70; others reported it as a planet (" 'Planet' Detected beyond the Solar System," *Science News* [December 15, 1984), 373]; and some debated the point ("VB 8B: Brown Dwarf or Planet?" *S&T*, 69 [February 1985], 126).

[116] For reviews of proliferating planetary detection techniques in the 1980s, see David Black, "In Search of Other Planetary Systems," *Space Science Reviews*, 25 (1980), 35;

Table 4.3. *Observational milestones in the search for planetary systems*

Author	Star	Method
See (1897)	Six stars	visual
Holmberg (1938)	Procyon	Astrometric (parallax)
Strand (1943)	61 Cygni	Astrometric (double star)
Reuyl & Holmberg (1943)	70 Ophiuchi	Astrometric (double star)
Lippincott (1960)	Lalande 21185	Astrometric (parallax)
Van de Kamp (1963)	Barnard's star	Astrometric (parallax)
Harrington (1983)	VB8B	Astrometric (parallax)
Aumann et al. (1984)	Vega, 40 more	IRAS, infrared excess
Smith and Terrile (1984)	Beta Pictoris	CCD camera image of circumstellar dust
McCarthy et al. (1985)	VB8B	Infrared speckle interferometry
Beckwith (1985)	R Monocerotis HL Tauri	Infrared speckle interferometry
Zuckerman and Becklin (1987)	Giclas 29-38	Infrared excess
Forrest et al. (1988)	Gliese 569	Infrared direct image: brown dwarf?
Campbell et. al (1988)	9/18 stars	Spectroscopic 1-10 Jup. mass
Latham et al. (1988)	HD 114762	Spectroscopic: brown dwarf?
Lyne and Shemar (1991)	Pulsar PSR1829 -10	Radio pulsar timing
Wolszczan and Frail (1992)	Pulsar PSR1257+12	Radio pulsar timing
O'Dell et al. (1992)	15 stars in Orion nebula	Hubble Space Telescope extended dust disks
O'Dell et al. (1994)	56 of 110 stars in Orion nebula	Preplanetary disks

ing point came during the years 1983–1984 with the announcement of both ground-based and spacecraft results, fanning hope that the question of the abundance of planetary systems would soon be resolved observationally. A half century after the beginning of van de Kamp's sustained efforts, astrometry would never regain its dominance in the field of planet detection.

The new era in planet hunting began 25 years after the dawn of the Space Age with unexpected observations from a spacecraft not specifically designed to search for planetary systems, but rather to survey space for objects at wavelengths just beyond visible light and largely beyond Earth-bound detectors – the infrared region of the spectrum. Beginning in the fall of 1983, a team of scientists centered at NASA's Jet Propulsion Laboratory (JPL) reported that observations made with the Infrared Astronomical Satellite (IRAS), launched in January 1983, indicated the presence of a cool cloud of solid particles around the bright star Vega in the constellation Lyra. The discovery, one of the first to come from the spacecraft, was made serendipitously while the telescope's detectors were being calibrated. The published paper, entitled "Discovery of a Shell around Alpha Lyrae," focused on the nature of the shell, which the authors concluded was composed of solid particles at least 1 millimeter in radius, at a distance of 85 astronomical units (about twice the size of our solar system), heated by the central Sun Vega. The material was believed to be the remnant of the cloud out of which Vega had formed. These results, the authors wrote, "provide the first direct evidence outside of the solar system for the growth of large particles from the residual of the prenatal cloud of gas and dust." Although the scientific paper stopped short of calling this a "solar system," a news release from JPL at the time of the discovery the previous August noted that "the material could be a solar system at a different stage of development than our own," and this was the interpretation emphasized by the press. By early 1984 similar "circumstellar shells" or "protoplanetary disks" had been found around six more stars, and by July, with a total of 40 such stars, the shells were being reported as "a widespread phenomenon." The discoverers in general were careful to emphasize that planets had not been found; instead, "the presumption is that these rings will eventually condense into solar systems like our own; if so, that makes the Vega phenomenon the first semidirect evidence that planets are indeed common in the universe."[117]

Jill Tarter, David Black, and John Billingham, "Review of Methodology and Technology Available for the Detection of Extrasolar Planetary Systems," *JBIS*, 39 (1986), 418.

117 H. H. Aumann et al., "Discovery of a Shell around Alpha Lyrae,", *ApJ*, 278 (March 1, 1984), L23–L27: L23; front page of The *Washington Post* for August 10, 1983: "Satellite Discovers Possible Second Solar System"; "Infrared Evidence for Protoplanetary Rings around Seven Stars," *Physics Today* (May 1984), 17–20; "Protoplanetary Systems," *Science*, 225 (July 6, 1984), 39.

This evidence of circumstellar shells was found in the infrared region invisible to the human eye, but by late 1984 one of the circumstellar disks found by IRAS had actually been photographed by a ground-based optical telescope. Employing the new charge-coupled device technology in conjunction with the 2.5-meter telescope at Las Campanas Observatory in Chile, these observations of the star known as "Beta Pictoris" produced one of the most famous images in the astronomy of the 1980s (Figure 4.7), one the public could appreciate more than the infrared excess detected by IRAS. After explaining the details of their observations (which involved a coronograph to block out the glare of the star) and the interpretation of their results, the authors concluded, "Because the circumstellar material is in the form of a highly flattened disk rather than a spherical shell, it is presumed to be associated with planet formation. It seems likely that the system is relatively young and that planet formation either is occurring now around Beta Pictoris or has recently been completed." Once again, however, these results were open to interpretation; though several years later the Hubble Space Telescope confirmed complex activity in the encirlcing gas disk, it remained far from certain whether planets had formed, or would form, from this material.[118]

With the announcement a few months later of the ground-based detection of VB8B by infrared speckle interferometry, planet hunting was at a fever pitch by the mid-1980s. One of the most intriguing results was the claim of indirect detection of actual planets by the spectroscopic method, the fulfillment of an idea foreseen by W. W. Campbell and H. D. Curtis at Lick Observatory in 1905 and the subject of Struve's 1952 proposal. Techniques for determining by spectroscopic means the variations in radial velocity caused by tugs of *stellar* companions had been continually refined over the decades, to the level of several hundred meters per second. But only in the 1970s did breakthroughs occur that would reveal such variations at the 10 meter per second level necessary to detect the effect of *planets* tugging on stars, and not until the 1980s would any results relevant to planetary systems be announced. Building on a technique first proposed by Roger Griffin in 1973, Bruce Campbell of the University of British Columbia refined the spectroscopic method using an absorption cell of hydrogen fluoride, with an eye toward the search for extrasolar planets. By 1983 Campbell and his team concluded that the technique was stable enough to be used in the search for substellar com-

[118] Bradford A. Smith and Richard J. Terrile, "A Circumstellar Disk around Beta Pictoris," *Science*, 226 (December 21, 1984), 1421–1424. Beta Pictoris is about 56 light years from Earth, with an estimated age of 1 billion years. On the Space Telescope results, obtained January 12, 1991, see Space Telescope Science Institute *Observer*, 1, no. 3 (1991), 1–5.

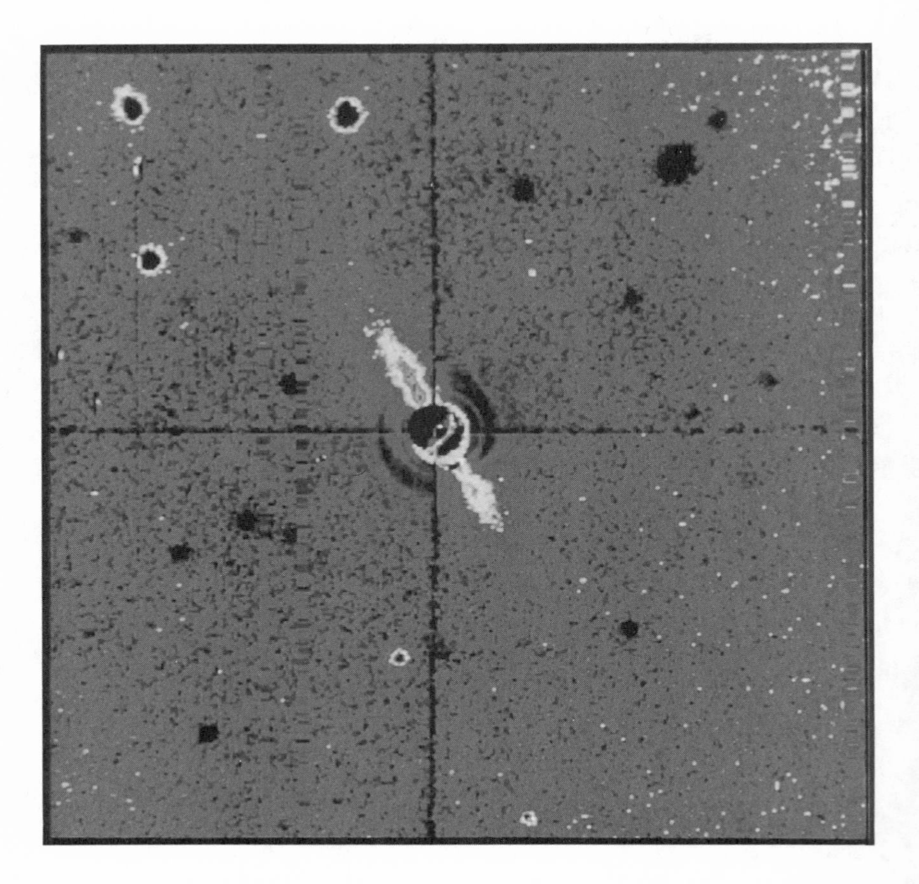

Figure 4.7. The charge-coupled device (CCD) image of a disk around Beta Pictoris (1984). Courtesy of the Jet Propulsion Laboratory.

panions, and in 1988 they announced that after 6 years of observations, 7 of 16 stars examined showed evidence of "long term low level variations" in radial velocity, at the level of 25–60 meters per second. These, he concluded, were probably planets with 1 to 9 Jupiter masses, which "probably represent the tip of the planetary mass spectrum."[119]

[119] The hydrogen fluoride was used to generate a reference spectrum, thus eliminating systematic errors. The technique is described in Bruce Campbell and G. A. H. Walker, "Precision Radial Velocities with an Absorption Cell," *PASP*, 91 (August 1979), 540–545, and "Precision Radial Velocities," *PASP*, 95 (September 1983), 577–585; results were announced in Bruce Campbell, G. A. H. Walker, and S. Yang, "A Search for Planetary Mass Companions to Nearby Stars," in *Bioastronomy – the Next Steps*, ed.

At the same time, an important role was still to be played by radial velocity surveys at precisions of several hundred meters per second, capable of detecting companions as small as 30 Jupiter masses with periods as long as 10 years. The two major efforts in this respect were made by David Latham and colleagues at the Harvard–Smithsonian Center for Astrophysics and by the Geneva Observatory, which used CORAVEL instruments. By 1992 these surveys had produced several "brown dwarf" candidates too small for sustained nuclear burning and tantalizing hints of possible planets, rekindling the old controversy about the difference between stars and planets. Particularly tantalizing was the companion to the star known as HD 114762, which Latham and his colleagues reported could have a mass as small as .001 of the Sun. Not only was this value smaller than the .08 "traditional dividing line between brown dwarfs and stable hydrogen-burning stars," it was also smaller than the "proposed dividing line" between a brown dwarf and a planetary companion. "Thus the unseen companion of HD 114762 is a good candidate to be a brown dwarf or even a giant planet," they concluded, allowing that there was less than a 1 percent chance (depending on the unknown orbital inclination) that this companion could be massive enough to burn hydrogen stably.[120] Although the method was incapable of detecting Earth-like planets, the Harvard team argued that any star with a Jupiter-size planet was likely to harbor a terrestrial planet as well, reasoning incorporated into the star-selection procedure of NASA's Search for Extraterrestrial Intelligence (SETI).[121]

By 1990, in a remarkable parallel to the fate of van de Kamp's as-

G. Marx (Dordrecht, 1988), 83–90: 83; and Campbell, Walker, and Yang, "A Search for Substellar Companions to Solar-Type Stars," *ApJ*, 331 (August 15, 1988), 902–921. Other high-precision radial velocity programs (with accuracies of 10 to 50 meters/sec and various spectroscopic techniques) were underway at the Lunar and Planetary Laboratory of the University of Arizona, the McDonald Observatory in Texas, the Lick Observatory, and New Zealand. On W. W. Campbell and Curtis see "First Catalogue of Spectroscopic Binaries"; on Struve's proposal, see "Proposal for a Project of High-Precision Stellar Radial Velocity Work."

120 D. W. Latham, T. Mazeh, R. P. Stefanik, M. Mayor, and G. Burki, "The Unseen Companion of HD 114762: A Probable Brown Dwarf," *Nature*, 339 (May 4, 1989), 38–40. See also M. Major in A. G. D. Philip and D. W. Latham, eds., *Stellar Radial Velocities* (Schenectady, N.Y., 1985), 1; Latham in ibid., 21; and Latham, in H. McAlister and W. Hartkopf, eds., *Complementary Approaches to Binary and Multiple Star Research* (San Francisco, 1992), 110. On brown dwarfs and their relation to stars and planets, see R. S. Harrington, ed., *Astrophysics of Brown Dwarfs* (New York, 1986), and J. Liebert and R. G. Probst, *Annual Review of Astronomy and Astrophysics*, 25 (1987), 473–519. Note the refinement of the .05 solar mass dividing line between stars and planets accepted by Russell and van de Kamp in 1944. Additional claims include B. Zuckerman and E. E. Becklin, "Excess Infrared Radiation from a White Dwarf – an Orbiting Brown Dwarf?," *Nature*, 330 (November 12, 1987), 138–140.

121 David R. Soderblom and D. W. Latham, "Target Selection Strategy for NASA's SETI/MOP," in Seth Shostak, ed., *Third Decennial US–USSR Conference on SETI* (San Francisco, 1993), 231–248.

trometric claims, cracks began to appear in the Canadian assertion that 50 percent of solar-type stars had planetary systems. Campbell himself now showed that at least some of the radial velocity variations were due to astrophysics on the surface of the stars – motions in the stellar material itself – rather than to planetary systems.[122] While planetary systems were still claimed for the other stars, this retrenchment cast doubt on them as well. In the last decade of the twentieth century, it was not clear whether these spectroscopic observations would eventually go the way of 61 Cygni, 70 Ophiuchi, and Barnard's star or whether they were revealing the tip of a planetary iceberg.

While both the classical and new techniques of planet detection remained problematic as the century entered its last decade,[123] a stunning discovery in a seemingly unrelated branch of astronomy turned up what finally appeared to be the first confirmed planetary system, giving renewed hope to planet hunters everywhere. In a bizarre twist to the planet detection story, this planetary system was found in 1992 circling a pulsar, one of a class of extremely dense, rapidly rotating neutron stars believed to be born from supernova explosions – long past the evolutionary stage that might harbor conditions for life.[124] The technique still depended on the gravitational tug of a planet on the star, but with a considerable difference

[122] B. Campbell, S. Yang, A. W. Irwin, and G. A. H. Walker, "Towards an Estimate of the Fraction of Stars with Planets from Velocities of High Precision," in J. Heidmann and M. J. Klein, *Bioastronomy: The Search for Extraterrestrial Life* (Berlin, 1991), 19–20.

[123] The photometric method mentioned earlier in the chapter proved so elusive that no results were announced. For discussions of the technique see A. T. Lawton, "Photometric Observation of Planets at Interstellar Distances," *Spaceflight*, 12 (September 1970), 365–373; F. Rosenblatt, "A Two-Color Photometric Method for Detection of Extra-Solar Planetary Systems," *Icarus*, 14 (February 1971), 71, and W. J. Borucki et al., "A Photometric Approach to Detecting Earth-Sized Planets," in Marx, *Bioastronomy*, 107–116.

[124] The first report of a pulsar planet of about 10 Earth masses was made by British astronomers M. Bailes, G. G. Lyne, and S.L. Shemar, "A Planet Orbiting the Neutron Star PSR1829–10," *Nature*, 352 (July 25, 1991), 311–313. This claim was retracted 6 months later in Lyne and Bailes, "No Planet Orbiting PSR1829-10," *Nature*, 355 (January 16, 1992), 213, when they reported that they had failed to account for the eccentricity of the Earth's orbit. The week before this retraction, however, an announcement was made by an American team using the 305-meter Arecibo radio telescope of two or more planets around another pulsar (spinning once every 6.2 milliseconds): A. Wolszczan and D. A. Frail, "A Planetary System around the Millisecond Pulsar PSR1257 + 12," *Nature*, 355 (January 9, 1992), 145–147. This claim was confirmed by D. Backer, S. Sallmen, and R. Foster, "Pulsar's Double Period Confirmed," *Nature*, 358 (July 2, 1992), 24–25. It was further confirmed and elaborated in A. Wolszcan, "Confirmation of Earth-Mass Planets Orbiting the Millisecond Pulsar PSR B1257 + 12," *Science*, 264 (April 22, 1994), 538–542; see also the commentary on 506–507. The theoreticians were not far behind in explaining how pulsar planets might have formed: I. R. Stevens, M. J. Rees and P. Podsiadlowski, "Neutron Stars and Planet-Mass Companions," *MNRAS*, 254 (February 1992), 19p–22p (Short Communications.) The latter argued that the pulsar destroyed its companion star, and planets were formed from the debris.

from the classical method: because pulsars emitted extremely accurate clocklike pulses at radio wavelengths, the tug could now be measured by timing pulse differences amounting to only a few thousandths of a second as the pulsar was tugged first one way, then another, by the circling planet – making this the most sensitive (if unexpected) planet detection technique known (Figure 4.8). Of the three planets discovered around the pulsar, two were three times the mass of the Earth and one was the mass of the Earth's Moon. While some pointed out that these were hardly the kind of planets of interest for life, others argued that if planets could form around pulsars, they could form anywhere. "The unexpected evidence for planetary-mass companions to pulsars suggests that accumulation processes in circumstellar disks may lead to sizeable bodies in very different phases of stellar evolution. This evidence greatly strengthens the case for the existence of planetary systems similar to the solar system," a group of NASA-sponsored planet hunters wrote in 1992 in contemplating their own program.[125] While one might argue about the implications, the claim of pulsar planets, almost all astronomers agreed, was "completely irrefutable." Had Struve still been alive, he would have marveled that planets had seemingly been found, not around those slowly rotating stars that he believed in the 1950s had given up their angular momentum to the planets, but around the most rapidly rotating objects in the universe.

During the last decade of the century, the planet search pressed forward on several fronts, including the Hubble Space Telescope's detection of protoplanetary disks around 56 of 110 young stars in the Orion Nebula (Figure 4.9). But most significantly, planet hunting began to achieve a central focus as NASA contemplated a systematic, comprehensive program of its own. Although the American space agency had supported limited study from the 1970s,[126] the "call to action," as radio astronomer Bernard F. Burke called it, came in 1985 when NASA's Solar System Exploration Division established a Planetary Astronomy Committee whose scope included not only the study of our solar system, but also the search for planetary systems. In the same year, the Space Science Board of

[125] Bernard F. Burke and R. A. Brown, eds., *TOPS: Toward Other Planetary Systems, A Report by the Solar System Exploration Division* (Washington, D.C., 1992), 16.

[126] NASA studied the problem in connection with its SETI program by sponsoring two workshops on "Extrasolar Planetary Detection" in 1976, chaired by Jesse L. Greenstein. See J. L. Greenstein and David Black, "Detection of Other Planetary Systems," in P. Morrison, J. Billingham, and J. Wolfe, eds., *The Search for Extraterrestrial Intelligence*, NASA SP-419 (Washington, D.C., 1977), 53–62; the workshop's agenda and members are listed on 269–274. This was followed by David Black, ed., *Project Orion*, NASA SP-436 (Washington, D.C., 1979), and by Black and W. Brunk, *An Assessment of Ground-Based Techniques for Detecting Other Planetary Systems*, vol. 1: *An Overview* and vol. 2: *Position Papers*, NASA CP-2124 (Washington, D.C., 1980). Project Orion was a design study for a dedicated ground-based system for detection of planets.

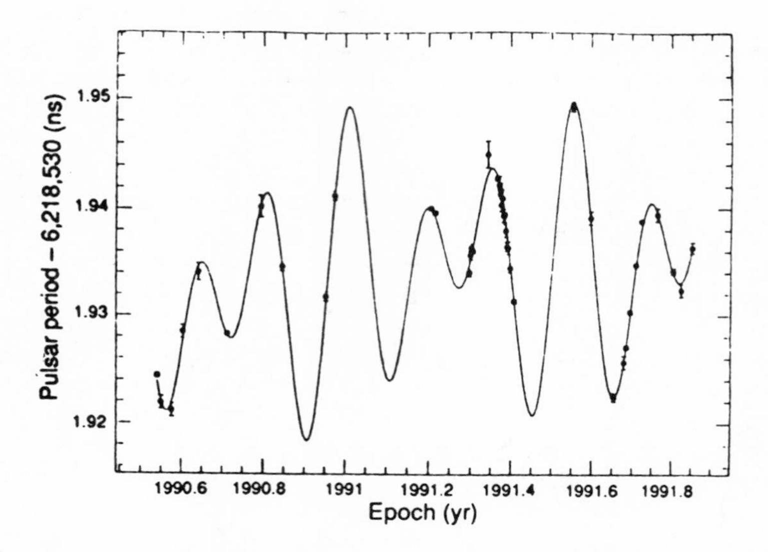

Figure 4.8. Data indicating pulsar planets (1992), the strongest evidence to date of planetary systems, is believed in this case to be irrefutably confirmed. The points represent observations, and the solid line indicates changes in period predicted by a two-planet model of this pulsar system. The vertical axis is in nanoseconeds (billionths of a second), so that the period variations of the pulsar are only .03 billionths of a second, or ±15 picoseconds (thousandths of a billionth of a second). Note that only 1 year was needed for the pulsar method compared to decades for the astrometric method. Used with permission from A. Wolszczan and D. A. Frail, "A Planetary System around the Millisecond Pulsar PSR1257 + 12," *Nature*, 355 (January 9, 1992), 145–147, copyright 1992 Macmillan Magazines Limited.

the National Academy of Science directed its Committee on Planetary and Lunar Exploration (COMPLEX) to widen its scope to planetary systems. By 1988 they had decided that preparations for the search should begin in earnest, and accordingly the Solar System Exploration Division created a Science Working Group "to formulate a strategy for the discovery and study of other planetary systems." During a workshop in Houston in early 1990 a three-phase program known as "Toward Other Planetary Systems (TOPS)" was formulated, to begin with ground-based observations, followed by a space-based effort and study of the discovered planetary systems. In 1993 the program was renamed "Astronomical Studies of Extrasolar Planetary Systems (ASEPS)."[127]

[127] Burke and Brown, *TOPS;* on ASEPS see *Science*, 264 (April 22, 1994), 506. See also the reports of its predecessor committees, Planetary Astronomy Committee of the Solar

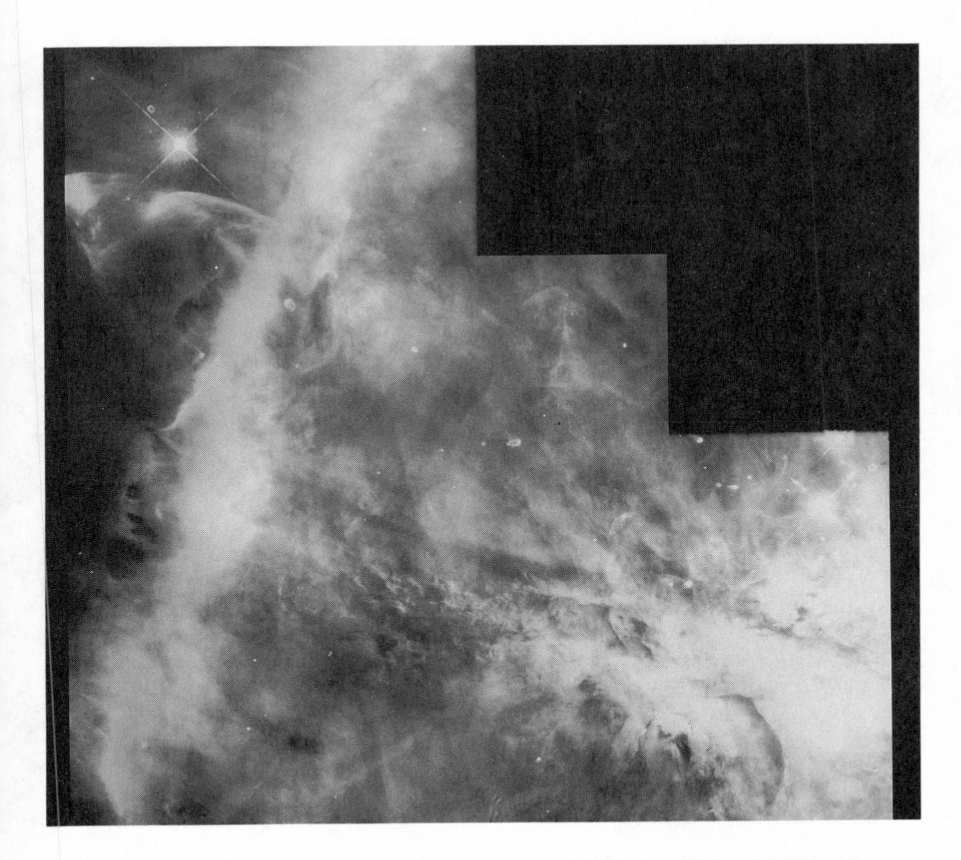

Figure 4.9. Another new technology for planet detection: The Hubble Space Telescope's image of protoplanetary systems in the Orion nebula (1994). By permission of NASA, the Space Telescope Science Institute, and C. R. O'Dell.

These reports discussed the full range of planet detection problems in a practical and programmatic way, defining the common understanding of such terms as "planet," "brown dwarf," and "star" and quantifying them in graphic form (Figure 4.10). While the participants contemplated mainly

System Exploration Division, *Other Worlds from Earth: The Future of Planetary Astronomy* (Washington, D.C., 1989), and Committee on Planetary and Lunar Exploration, Space Studies Board, *Strategy for the Detection and Study of Other Planetary Systems and Extrasolar Planetary Materials: 1990–2000* (Washington, D.C., 1990). These three reports taken together constitute the best strategic thinking on the subject of planetary detection during the 1990s, while the three *Protostars and Planets* volumes contain detailed papers. For a brief time during 1992–1993 NASA's SETI was a part of the TOPS program, until SETI was terminated by Congress in October 1993.

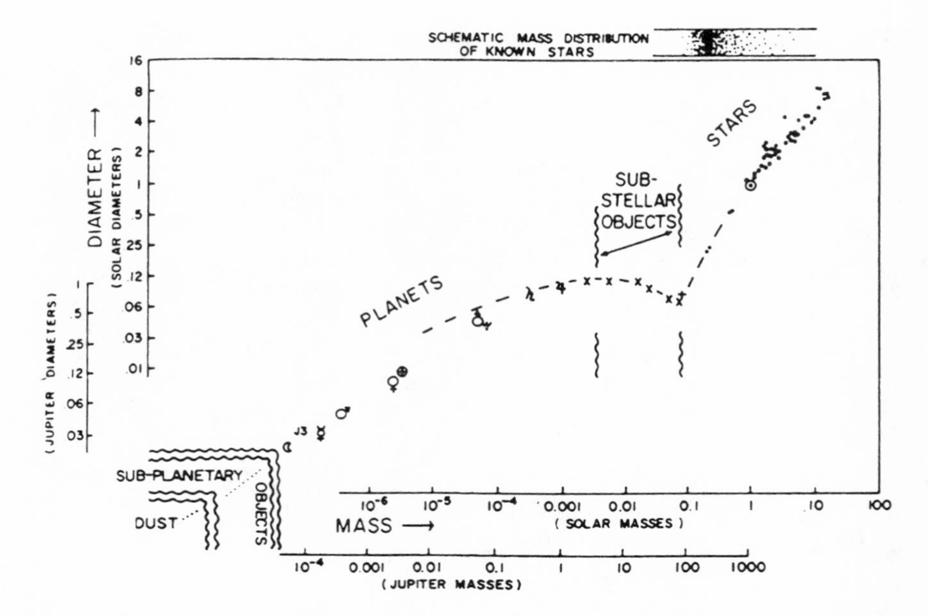

Figure 4.10. Diagram of diameter versus mass showing classes of objects in the search for planetary systems, ranging from dust (lower left) to planets, brown dwarfs, and stars (upper right). Reprinted with permission from *Strategy for the Detection and Study of Other Planetary Systems and Extrasolar Planetary Materials*. Copyright 1990 by the National Academy of Sciences. Courtesy of the National Academy Press, Washington, D.C.

an indirect search in the early stages of the project, with the advance of space technology even the direct imaging of extrasolar planets was not out of the question.[128] If funded, the whole effort would be comparable to the

[128] Such imaging was one of the early justifications for the Hubble Space Telescope. For an optimistic assessment regarding the latter, see William A. Baum, "The Search for Planets in Other Solar Systems through Use of the Space Telescope," in Michael D. Papagiannis, ed., *Strategies for the Search for Life in the Universe* (Dordrecht, 1980), 163–166, and Jane L. Russell, "Prospects for Space Telescope in the Search for Other Planetary Systems," in M. D. Papagiannis, ed. *The Search for Extraterrestrial Life: Recent Developments* (Dordrecht, 1985), 75–84. For a pessimistic assessment, see R. A. Brown, "Systematic Aspects of Direct Extrasolar Planet Detection," in Marx, *Bioastronomy,* 117–123. For an early suggestion see Lyman Spitzer, Jr., "The Beginnings and Future of Space Astronomy," *American Scientist,* 50 (September 1962), 473–484. Other proposals for space telescopes dedicated to planet detection indicated that it would be feasible. See Richard J. Terrile, "Direct Imaging of Extra-Solar Planetary Systems with a Low-Scattered Light Telescope," in Marx, *Bioastronomy,* 125–130, and E. H. Levy et al., "Discovery and Study of Planetary Systems Using Astrometry from Space," in Marx, *Bioastronomy,* 131–136.

Viking effort to search for life on Mars; indeed, the involvement of both the Space Science Board and NASA found precedent in the Viking preparations a quarter century earlier. With these activities, it was clear that the search for planets beyond the solar system was no longer a solitary occupation, as it had been with van de Kamp, but was developing into the discipline of planetary systems science proclaimed by its participants.

By century's end, then, the effort was intensifying but the desired knowledge about the frequency of planetary systems was still lacking. By the mid-1980s, the scientists who gathered to review progress in the field of planetary systems found that there had been consensus for several decades that the modified nebular hypothesis for the formation of the solar system is "but one example of a phenomenon that occurs frequently in nature." But theory was intrinsically unable to validate this claim. Even in particular cases, such as the relation of binary stars to solar systems, theory remained inconclusive. If binary star systems and planetary systems were the results of mutually exclusive evolutionary processes, then a determination of the frequency of binaries would, by elimination, give some idea of the number of single stars around which solar systems might form. However, some claimed that planets – even habitable planets – might exist in binary star systems.[129] Both general and particular theories depended on observation for their validation.

While it seemed that in the last four decades of the twentieth century observation might come to the rescue of a theory unable to resolve definitively the question of the abundance of planetary systems, in the end all techniques of observation proved difficult and elusive. The astrometric technique encompassed many steps, from observation and measurement of the photographic plates to reduction techniques. The spectroscopic technique, though a breakthrough of impressive proportions for astrophysics in general, was still subject to analytical problems when it came to the search for planetary effects. The photometric technique never developed adequately to the extent needed for the search for planets. Direct imaging of planets remained at the tantalizing edge of possibility, even from outer space. And circumstellar shells, while lending credence to solar systems in formation, did not prove the actual existence of planets on which life might eventually develop. Ironically, confirmation came

[129] Black and Matthews, *Protostars and Planets II,* xvii; H. Abt and S. G. Levy, "Multiplicity among Solar Type Stars," *ApJ Suppl.*, 30 (1976), 273–306; R. S. Harrington, "Planetary Orbits in Multiple Star Systems," in John Billingham, ed., *Life in the Universe* (Cambridge, Mass., 1981), 119–122, and Harrington, "Planetary Orbits in Binary Stars," *AJ*, 82 (September 1977), 753–756. H. Abt, "The Companions of Sunlike Stars," *SciAm*, 236 (April 1977), 96–104 concluded that virtually all stars belong to multiple systems, with the "component" at the low-mass end taking the form of planetary systems. This is similar to Kuiper's "failed binary star" scenario.

only in the case of pulsar planets, on which only the most imaginative science fiction writers could envision life.

The whole problem of planetary systems science at the end of the twentieth century could therefore be summarized as follows: The existence of planets around even the closest stars could not be confirmed directly by observation or indirectly by observation of their effects. Nor could circumstellar shells that were directly observed or indirectly inferred be conclusively linked to planetary systems. And though in the last half of the twentieth century a consensus had been reached on a theory of the origin of planetary systems, theory in itself could never ensure the existence of planetary systems without observational confirmation.[130] And the tantalizing detection of exotic pulsar planets hardly proved the existence of conventional planets – and conventional planets were the holy grail on the way to extraterrestrial life.

As in planetary science, indefinite results invited a broader role for preconceived notions. Planet searchers, driven by considerations of stellar evolution, a revived nebular hypothesis, and the Copernican principle of mediocrity and simple analogy, certainly expected other planets to exist and fervently hoped to find them. But the search for planetary systems should not be seen in the same class as Lowell's canals. Unlike the canals, stellar perturbations by planetary bodies were not a result of direct visual perception, but rather of a whole series of steps, from the detector, to the measurement process, to statistical analysis of the data. The high stakes and fervent hopes notwithstanding, in the end the search for planets outside the solar system is another example of science attempting to function at its outermost limits.

In the end, regarding the quest for planetary systems, one is struck by the parallel to early attempts to determine stellar parallax – attempts that were, of course, eventually successful and the results of which now determine the basic stellar distance scale. This early quest to measure parallax, John Herschel wrote, was a goal that seemed within reach of the astronomer, "only to elude his seizure when apparently just within his grasp, continually hovering just beyond the limits of his distinct apprehension, and so leading him on in hopeless, endless, and exhausting pursuit." Today, 150 years later, the same can be said of the search for planetary systems. Instead of parallax, astronomers looked for *variations* in proper

[130] Consensus did not imply unanimity. Aside from differences of opinion over details of the current model of the nebular hypothesis, a modified collision hypothesis survived. See John R. Dormand and Michael M. Woolfson, *The Origin of the Solar System: The Capture Theory* (New York, 1989). Should observation fail in the end to find solar systems, it is possible that a collisional hypothesis will come back into vogue.

motions using the parallax technique; instead of a few *tenths* of a second of arc, they attempted to extract from their instruments a few *hundreds* of a second of arc. Instead of stellar companions causing velocity variations in the primary star at the level of *kilometers per second*, they now searched for planets at *meter per second* variations. And where once tenths of a magnitude pushed the limits of photometry, now differences in magnitude at the level of hundredths were required. What may seem at times like wishful thinking in fact has a long precedent in the history of science, where difficult observations are more often the rule than the exception.

Throughout all of the debate, there should be no doubt of the ultimate aim of the field now known as "planetary systems science." Although the search for extrasolar planets arose naturally as part of ongoing research programs in astronomy, in the twentieth century this search became an important component of what may be viewed as a new cosmological worldview.[131] This cosmology assumed that planetary systems were common, that life had developed on many of those planets, and that intelligence may have evolved to the point where we can communicate by radio waves. The point was to test the validity of this cosmology. As one NASA astronomer wrote in 1987:

> There are many reasons to search for planets, all stemming from a basic curiosity about possible cosmic neighbors. Both the theory of star formation and the Copernican Principle suggest that planets exist throughout the Universe. We wish to test this speculation and learn whether the Solar System is indeed unique, rare, common or typical. The search for planets is the first step in the search for extra-terrestrial intelligence, as the surface of a planet is probably the only viable location for the origin and evolution of life.[132]

The same incentive is explicitly stated in the report of the scientists working to take NASA "Toward Other Planetary Systems." Seen in this light, the search for planetary systems has taken on a vigorous life of its own, the first and most crucial of many tests of this new cosmology, conferring on it a role of even more significance to the history of science.

[131] Steven J. Dick, "The Concept of Extraterrestrial Intelligence – An Emerging Cosmology?" *The Planetary Report*, 9 (March–April, 1989), 13–17, and Dick, "The Biophysical Cosmology: The Place of SETI in the History of Science," International SETI Conference, Toronto, October 1988.

[132] J. D. Scargle, "Planetary Detection Techniques: An Overview," in Marx, *Bioastronomy*, 79–82: 79.

5

EXTRATERRESTRIALS IN LITERATURE AND THE ARTS: THE ROLE OF IMAGINATION

With infinite complacency men went to and fro over this little globe about their affairs, dreaming themselves the highest creatures in the whole vast universe, and serene in their assurance of their empire over matter. . . . Yet across the gulf of space minds that are to our minds as ours are to those of the beasts that perish, intellects vast and cool and unsympathetic, regarded this earth with envious eyes, and slowly and surely drew up their plans against us.

H. G. Wells (1897)[1]

Space . . . the final frontier. These are the voyages of the starship Enterprise. Its five-year mission to explore strange new worlds, to seek out new life and new civilizations – to boldly go where no man has gone before.

"Star Trek" (1966)[2]

It is, I admit, mere imagination: but how often is imagination the mother of truth?

Sherlock Holmes[3]

As flies to wanton boys are we to th' gods.

Shakespeare (1605)[4]

One of the most surprising developments in the saga of extraterrestrial life in the twentieth century was the spread of the idea from science to the realms of literature and the arts. Despite the popular appeal of the idea in the past, no one could have foreseen that this subject, of all the rich variety of scientific subjects, would be transformed into one of the universal themes of literature. That this is precisely what happened is some measure of how deeply felt and firmly ingrained was the alien concept in the human mind. And with this important step, it became even more entrenched, for now the concept evoked not only an intellectual but also an emotional response.

During the twentieth century, the concept of the alien became a leitmotif of that young genre known as "science fiction," developed successively in

[1] H. G. Wells, *The War of the Worlds* (serialization, 1897; book, 1898), in *Seven Science Fiction Novels of H. G. Wells* (New York, 1934), 309.

[2] Opening lines of each episode of the *Star Trek* television series, beginning September 8, 1966. Paramount Television/NBC, created by Gene Roddenberry.

[3] Sherlock Holmes, "The Valley of Fear," part 1, chapter 6, as quoted in A. G. Cairns-Smith, *Seven Clues to the Origin of Life* (Cambridge, 1986), 107.

[4] Shakespeare, *King Lear,* Act 4 scene 1, 32–37.

literature and film and culminating in some of the most popular movies of all time, including *2001: A Space Odyssey* (1968), *Close Encounters of the Third Kind* (1977), and *ET: The Extraterrestrial* (1982). For vast segments of the public, the idea of extraterrestrial life entered their lives not as science or philosophy, but as a fictional account that gave readers, and eventually viewers of film, an emotional experience of the alien, unfettered by bothersome scientific technicalities. Science stimulated the literary imagination, and literature and film stimulated, in turn, the popular imagination, to such an extent that a history of the alien in science fiction is necessary to a full understanding of why extraterrestrials came to have such a grasp on the popular mind. Moreover, a surprising number of scientists were influenced by what we might call "alien literature" and even produced some of the best of it themselves, forming an increasingly symbiotic relation with science in the second half of the twentieth century.

In this chapter we document the origin, development, and coming of age of the alien theme in science fiction, with an eye toward its role in establishing the alien in popular culture as well as in the scientific imagination. So pervasive did the theme become that our approach cannot be encyclopedic.[5] Instead, it will be sufficient to show from a few exemplars the broad patterns in the development of the alien theme, and the interaction of the alien in literature with science and the extraterrestrial life debate.

5.1 THE INVENTION OF THE ALIEN: VERNE, WELLS, AND LASSWITZ

It is a remarkable fact of history that only in the last third of the nineteenth century did extraterrestrials enter the realm of literature. Though Lucian's armies of the Sun and the Moon clashed in the second century, though Kepler's remarkable imagination placed thick-skinned Selenites on the Moon, and though Voltaire had mile-high Sirians in his *Micromegas*, the true extraterrestrial alien, replete with its own physical and mental characteristics, is a relatively recent invention. Why should this be when the concept of extraterrestrials is so old? It is the thesis of this section that the birth of the alien in literature is closely tied to late-nineteenth-century science, especially evolutionary theory, astronomy,

5 Among the best treatments of the alien in science fiction is Brian Stableford's entry "Aliens" in John Clute and Peter Nicholls, *The Encyclopedia of Science Fiction* (New York, 1993). Because this is much revised from the first edition of 1979 and more accessible, I will used the 1993 edition (hereafter cited as *SFE*) unless otherwise specified. See also Mark Rose, *Alien Encounters: Anatomy of Science Fiction* (Cambridge, Mass., 1971).

and the plurality-of-worlds tradition. We shall see these influences repeated in three traditions – the French, German, and British – while the imaginative science of the American astronomer Percival Lowell fueled these traditions after their birth.

The history of the alien in literature may be approached in many ways: the history of science fiction, the history of the extraterrestrial life debate, the history of science, and even the history of religion. To contrast only two of these, it is immediately evident that if the alien came late in the history of science, it came remarkable early in the history of science fiction.[6] Though precursors to science fiction literature are well known, notably Mary Shelley's *Frankenstein* (1818), three figures are most often associated with its birth: Jules Verne in France, H. G. Wells in England, and, for those who probe a bit deeper, Kurd Lasswitz in Germany. If one would expect in Proctor's England that an H. G. Wells would arise to send Martians to invade the Earth, one would expect no less in Flammarion's France and even in a Germany still reeling from Kant's philosophy. It is curious, then, that while Wells and Lasswitz follow this expectation, Jules Verne does not, a circumstance that gives us a clue to the invention of the alien in literature.

We may begin in France, for it is here that the remarkable imaginations of Camille Flammarion and Jules Verne flourished, and here that the no less remarkable but now more obscure works of J. H. Rosny the Elder were penned. As we have seen in Chapter 1, in 1862, at the age of only 20, Flammarion's *La pluralité des mondes habités* (Plurality of Inhabited Worlds) appeared, a factual account of the possibility of life on other worlds that was to go through 33 editions by 1880, as well as numerous translations and reprintings well into the twentieth century. This work was fully in the plurality-of-worlds tradition, as was his *Les mondes imaginaire et les mondes réels* (Real and Imaginary Worlds), published 2 years later. But Flammarion soon went beyond the bounds of science when in 1872 he wrote his *Recits de l'infini*, translated in 1874 as *Stories*

[6] On the history of science fiction see Brian W. Aldiss, *Billion Year Spree* (New York, 1973), and its revision by Brian W. Aldiss and David Wingrove, *Trillion Year Spree: The History of Science Fiction* (New York, 1986); James E. Gunn, *Alternate Worlds: The Illustrated History of Science Fiction* (New York, 1975); and Lester del Rey, *The World of Science Fiction: The History of a Subculture* (New York, 1979). Encyclopedic treatments include Brian Ash, ed., *The Visual Encyclopedia of Science Fiction* (New York, 1977); James Gunn, ed., *The New Encyclopedia of Science Fiction* (New York, 1988); and the indispensable volume edited by John Clute and Peter Nicholls, *The Encyclopedia of Science Fiction* (New York, 1993). Another indispensable volume that I have used extensively is Neil Barron, ed., *Anatomy of Wonder: A Critical Guide to Science Fiction* (New York, 1987), a complete guide to science fiction and its historical and critical studies. An anthology of science fiction works with historical commentary is Gunn's four-volume *The Road to Science Fiction* (New York, 1977–1982).

of Infinity, which included three tales treating other worlds: "Lumen," "The History of a Comet," and "In Infinity," in which a disembodied spirit travels throughout the universe, reincarnating on other worlds in forms that might have resulted from evolution under alien conditions. Flammarion's "boldest scientific romance," *La fin du monde* (translated as Omega: The Last Days of the World) (French, 1893–1894, English, 1897), has been seen as a precursor to the British philosopher Olaf Stapledon's *Last and First Men* (1930), a seminal work of science fiction that we shall discuss in the next section. In both fact and fiction, then, the alien pervaded Flammarion's work, and through him spread into the intellectual life and popular culture of France.[7]

Jules Verne (1828–1905) was not as single-minded as Flammarion, nor did he have a scientific background. He was 34 when he launched his career with *Cinq semaines en balloon* (Five Weeks in a Balloon) in 1863, 1 year after Flammarion's *Pluralité*. There followed in rapid succession the *Voyage au centre de la terre* (Journey to the Center of the Earth, 1864), *De la terre a la lune* (From the Earth to the Moon, 1865), *Autor de la lune* (Around the Moon, 1870), and *Vingt mille lieues sous le mers* (Twenty Thousand Leagues under the Sea, 1870), among others. Like Flammarion, Verne lived in Paris and was familiar with the young astronomer's work. Yet, throughout a writing career that lasted until his death in 1905, Verne never wrote a novel that made significant use of aliens.[8]

Verne's brief discussion of extraterrestrials in his novel *Around the Moon*, however, provides a clue as to why he never wrote an alien novel. The question of life on the Moon, one of the characters says, "requires a double solution. Is the moon habitable? Has the moon ever been inhabited?" Verne decides that the Moon is not habitable because of "her surrounding atmosphere certainly very much reduced, her seas for the most part dried up, her insufficient supply of water, restricted vegetation, sudden alternations of cold and heat, her days and nights of 354 hours." Moreover, Verne's character adds decisively as they approach the Moon, viewed from 500 yards there is no sign of motion and thus no sign of life.[9]

7 Surprisingly, Flammarion's *Pluralité* never appeared in English. *Stories of Infinity* was revised and published in English as *Lumen* (1887), with yet more material added in 1897. On Flammarion's science fiction writings, see *SFE*, 432. The earlier edition (p. 224) makes the comparison of *Fin du monde* with Stapledon.

8 For Flammarion's influence on Verne see Michael Crowe, *The Extraterrestrial Life Debate, 1750–1900* (Cambridge, 1986), 384, citing Kenneth Allot, *Jules Verne* (London,1940), 108, 152, 208. On Verne see also Peter Costello, *Jules Verne: Inventor of Science Fiction* (New York, 1978), and Jean Jules-Verne, *Jules Verne: A Biography* (New York, 1976).

9 Gunn, *Road to Science Fiction*, vol. 1, 283.

Verne concludes, however, that the Moon was inhabited in the past, "for nature does not expend herself in vain; and a world so wonderfully formed for habitation must necessarily be inhabited." And inhabited "by a human race organized like our own; that she has produced animals anatomically formed like the terrestrial animals; but I add that these races, human or animal, have had their day, and are now forever extinct!" The Moon, Verne had his character say, had grown older more quickly than on Earth, its rarified atmosphere now rendering it uninhabitable, as some day the Earth would be.[10]

Here we have our clue to Verne's relative neglect of extraterrestrials in his stories. Though it has been said that Verne wanted to do for science and geography what Dumas had done for history – to excite through the power of literature an emotional response to otherwise mundane facts of everyday existence – by comparison to Flammarion and others, his imagination was constrained by science. Although willing to grant that the evolution of worlds would yield an inhabited Moon in the past, Verne required evidence for a Moon presently inhabited. With his fertile imagination the French author would venture around the world and into space, but he never went beyond the Moon to Mars or took up Flammarion's passion for the alien.

Verne's imagination paled by contrast to that of his younger countryman J. H. Rosny the Elder (1856–1940). Though now obscure, Rosny was the one rival Verne had to contend with in France, and it is generally considered that Verne's decline in the 1880s and 1890s coincides with the rise of Rosny's remarkable works. Undoubtedly influenced by Flammarion, his very first work, "Les xipehuz" (The Shapes, 1887) dealt fictionally with wholly alien themes. Not for Rosny a discussion of life on the Moon; rather, he tackled the nature of life itself, though with little literary merit at this early stage in his career. In "The Shapes" mysterious translucent creatures appear on Earth, threatening the existence of humanity. Unlike anything in terrestrial experience or likely to emerge from terrestrial evolution, the creatures are living minerals unable to communicate with humans. Although they are eventually destroyed, Rosny succeeds in vividly conveying the idea that life in the universe may be completely unlike life on Earth. These themes of strange life, though detached from their extraterrestrial associations, are continued in "Un autre monde" (Another World, 1895) and *La mort de la terre* (The Death of the Earth, 1910), the latter dealing with ferromagnetic life.[11]

[10] Ibid., 284–285.

[11] On Rosny see *SFE*, 1029–1030, and J. P. Vernier, "The SF of J. H. Rosny the Elder," *Science Fiction Studies*, 2 (July 1975), 156–163. English translations of the works mentioned here are found in J. H. Rosny Aîné, *The Xipehuz and the Death of the Earth,*

In his conception of life and intelligence totally incompatible with terrestrial life, Rosny was almost a half century ahead of his time, and one wonders how the alien theme might have developed had not the world become almost obsessed with the myth of a Martian civilization. As we saw in Chapter 3, this idea had begun with Schiaparelli's observations in 1877, but it captured the popular and literary imagination only with the publication of Lowell's *Mars* in 1895.[12] It was in November of that year that the German philosopher and historian Kurd Lasswitz (1848–1910) began his novel *Auf Zwei Planeten* (On Two Planets, 1897), in which the Martians traveled to Earth and accidentally came face to face with its inhabitants. A teacher of mathematics, physics, and philosophy at Gotha, a town southwest of Berlin, Lasswitz (Figure 5.1) was steeped in the traditions of science and German idealism. Best known for his biography of the physicist and philosopher Gustav Fechner (which appeared as he was writing his Martian novel) and his *Geschichte der Atomistic von Mittelalter bis Newton* (History of Atomism from the Middle Ages to Newton, 1889–1890), he is often given the title "father of German science fiction," holding the place in Germany that H. G. Wells does in England and that Verne and Rosny hold in France. Beginning as early as 1878, Lasswitz began to examine the basis of fiction about science, arguing that it satisfied a basic human need and was a legitimate form of art.[13]

What led Lasswitz to write his novel is nowhere explicitly stated, but we can make a plausible case that it was a combination of his interest in

translated from the original 1888 and 1912 French editions by George Edgar Slusser (New York, 1978), and "Another World," translated by Damon Knight in Damon Knight, ed., *A Century of Science Fiction* (New York, 1962), 270–297. Knight calls Rosny "a giant of French letters" and says that "Rosny, not Verne, is considered the father of French science fiction."

[12] Although Martian stories existed prior to Lowell's *Mars* – most notably in British historian and novelist Percy Greg's *Across the Zodiac* (1880), in his discussion of the Martian theme Mark R. Hillegas calls them "almost totally lacking in vitality, and the Martian myth [of life] is hardly developed in them." See Hillegas, "Martians and Mythmakers: 1877–1938," in Ray B. Browne, Larry N. Landrum, and William K. Bottorff, eds., *Challenges in American Culture* (Bowling Green, Ohio, 1970), 150–177: 156. See also William B. Johnson and Thomas D. Clareson, "The Interplay of Science and Fiction: The Canals of Mars," *Extrapolation*, 5 (1963–164), 37–48.

[13] On Lasswitz see William B. Fischer, *The Empire Strikes Out: Kurd Lasswitz, Hans Dominik, and the Development of German Science Fiction* (Bowling Green, Ohio, 1984); Fischer, "German Theories of Science Fiction: Jean Paul, Kurd Lasswitz, and After," *Science Fiction Studies* (1976), 254–265; Franz Rottensteiner, "Kurd Lasswitz: A German Pioneer of Science Fiction," in Thomas D. Clareson, ed., *SF: The Other Side of Realism* (Bowling Green, Ohio, 1971), 289–306; Edwin M. J. Kretzmann, "German Technological Utopias of the Pre-War Period," *Annals of Science*, 3 (1938), 417–430; and Karl S. Guthke, *The Last Frontier* (Ithaca, N.Y., 1990), 382–392. Fischer's volume is by far the most detailed treatment of Lasswitz, while Kretzmann places him in the context of technological utopias and Guthke compares him with H. G. Wells.

Figure 5.1. A pioneer in alien literature: Kurd Lasswitz.

extraterrestrials and his desire to see the improvement of terrestrial society. It is unlikely that Lowell provided the initial stimulus for Lasswitz's novel, but the latter may have known of Schiaparelli's work and perhaps of Flammarion's *La planète Mars* (1892), and internal evidence indicates that he must have learned of Lowell's *Mars* as he wrote the novel. Whatever the proximate causes, it is clear that Lasswitz was affected by the earlier plurality-of-worlds tradition. In an article on extraterrestrial life written near the end of his life, he stated that ever since science had made the Earth a planet and the stars suns,

> we have not been able to lift our gaze to the starry firmament without thinking, along with Giordano Bruno, that even on those inaccessible worlds there may exist living, feeling, thinking creatures. It must seem absolutely nonsensical indeed that in the infinity of the cosmos our Earth should have remained the only supporter of intelligent beings [*Vernuftwesen*]. The

> rational order of the universe [*Weltvernunft*] demands that there should necessarily even be infinite gradations of intelligent beings inhabiting such worlds.[14]

But what set Lasswitz to writing his novel was more than a belief in extraterrestrials; it was a belief that they could be used to illuminate some of his strongly felt ideas about society and the important role of science and technology in it. A better society, he felt, was not only within reach but had already been achieved in some part of the universe: "We do dream of a higher civilization [*Kultur*], but we would also like to come to know it as something more than the hope for a distant future," he wrote years after finishing his novel. "We tell ourselves that what the future can sometime bring about on Earth must even now, in view of the infinitude of time and space, have already become a reality somewhere." In this endeavor, the general philosophical influences on Lasswitz were Kant and, more directly, Fechner, from both of whom he took the idea of free moral will that pervades the novel.[15]

On Two Planets portrays Martians who have arrived in advance of a larger Martian expedition, established a solar-powered space station "hovering" above the Earth's North Pole, and constructed a base at the Pole itself, where they are discovered by a German balloon expedition. The Mars they left behind was straight out of Lowell: In order to win new areas for cultivation, the Martians

> crossed the entire desert area with a network of broad, straight canals and in this manner they distributed over the entire planet, when the snows melted at the beginning of summer in each hemisphere, the water that had accumulated in the form of snow at the poles. On both shores of the canals, vegetation was plentiful. The network of canals crossing the entire desert area created an extremely fertile belt of vegetation, about a hundred kilometers wide, which contained an uninterrupted chain of thriving settlements.[16]

14 Fischer, "German Theories of Science Fiction," 63, translating from one of Lasswitz's last writings, "Unser Recht auf Bewohner anderer Welter" (Our Claims on Inhabitants of Other Worlds), which appeared in the *Frankfurter Zeitung* 1 day before his death in 1910. It was reprinted in a posthumous volume of his works *Empfundenes und Erkanntes* (Things Felt and Known) (Leipzig, 1919), 163–185.

15 Fischer, "German Theories of Science Fiction," 63 translating from "Unser Recht. . . ." On the role of Kantian philosophy in the novel, see Fischer, 146ff., and Hillegas in the English translation (footnote 16 below), 379ff. Lasswitz was eulogized in *Kantstudien* (1911).

16 *Two Planets*, trans. Hans H. Rudnick (Carbondale, Ill., 1971), 41–42, 65. The translation includes an Epigraph by Wernher von Braun and an Afterword by Mark Hillegas. This first English edition is based on the 1948 German edition abridged by Erich

It was not desperation that brought the Martians to Earth but curiosity: with their more transparent atmosphere their telescopes had been able to detect cities on Earth, and after repeated attempts at communication by signal, they had discovered how to control gravitation and travel to Earth.

Lasswitz's Martians, having mastered technology in order to survive the scarcity of water on their planet, were advanced in every way over their terrestrial counterparts: "Mathematics and natural science had reached a climax in their development which looms before us humans as a distant ideal. . . . The conditions of Mars favored the development of culture and civilization to a much higher degree than the conditions of earth." Although the Martians were humanoid and capable of interbreeding, they "had large heads, very light, nearly white hair; shining, powerful, piercing eyes." More important, they had reached a higher ethical level than their terrestrial counterparts. Free from instincts and desires, they had found contentment in following their "immortal philosopher," Imm.

In biological, social, and technological aspects, Lasswitz thus adopted an evolutionary universe, showing the influence of Kant, Fechner, and Lowell.[17] Employing both the Earth and Mars as settings, he uses the situation to explore the relation between two cultures and the problems of the improvement of society. When the superior Martians place the Earth under a benign protectorate, the Earthlings find their human dignity at risk and eventually rebel. When they are finally able to seize the polar base and space station, a treaty is negotiated and communication between the two civilizations is limited to light signals. Equality between the two civilizations is achieved, and as the story ends, there is hope that the new planetary order may lead to the long-sought Utopia.

Although the lack of an English translation before 1971 prevented its having much effect in England or America, the influence of *Auf zwei Planeten* in Germany is witnessed by its sales of almost 100,000 copies and Wernher von Braun's statement at the beginning of the English edition that "I shall never forget how I devoured this novel with curiosity and excitement as a young man." Within 10 years of its publication in 1897, *Auf zwei Planeten* had been translated into Swedish, Norwegian, Danish, Dutch, Spanish, Italian, Czech, Polish, and Hungarian.[18]

Lasswitz, with incidental additions from the Frankfurt (1969) German edition, but is overall much shorter than the original German novel (Leipzig, 1897).

[17] Ibid., 41–42, for the Lowellian influence. The name "Imm" may be taken from Immanuel Kant.

[18] See also Hillegas, Afterword to the English edition of *Two Planets,* 375. The novel was banned in Germany from 1933 to 1935, but by that time it had sold more than 70,000

As Karl Guthke has pointed out, if Lasswitz's story left a legacy of hope, H. G. Wells's Martian fantasy left a legacy of fear. *The War of the Worlds*, serialized in 1897 just as Lasswitz's novel was emerging in Germany and first published in book form in 1898, is striking in its originality: Wells not only led the first invasion from outer space but did so in fine literary style. The influence and appeal of the work are apparent not only in its many editions and in the well-known effect of Orson Welles's radio play adaptation in 1938, but also in innumerable imitations that have followed in the century since. As Arthur C. Clarke has written, the Menace from Space was virtually unknown before Wells, but it has become all too common since.[19]

How do we explain the origin of this prototype of alien literature at this time and place? Wells (Figure 5.2) was a man of his times in many ways, not least in his championship of Darwin's theory of evolution.[20] As Fechner was Lasswitz's great influence, so the biologist T. H. Huxley was Wells's. Huxley, the nineteenth century's greatest champion of Darwinian evolution, taught Wells from about 1883 to 1886 at the Normal School of Science (later known as the Royal College of Science) in London. "I believed then that he was the greatest man I was likely to meet," Wells wrote in 1901, recalling his student days, "and I believe that all the more firmly today."[21] Huxley's evolutionary viewpoint is pervasive in Wells's early writings in science and science fiction, and it forms the broad background to the advanced beings in *The War of the Worlds*.

Moreover, it is clear that Wells was familiar with the plurality-of-

copies. Willy Ley discussed the novel's role as background to the space age in *Rockets, Missiles, and Men in Space* (New York, 1969), 65–69.

19 Arthur C. Clarke, Introduction to *The Invisible Man* and *The War of the Worlds* (New York, 1962), xv. *War of the Worlds* was serialized in *Pearson's Magazine* (April–December, 1897) in England and in *Cosmopolitan* in the United States. The serialization from *Pearson's*, with original illustrations, is reprinted in *The Collector's Book of Science Fiction by H. G. Wells* (Secaucus, N.J., 1978). The book (London, 1898, and New York, 1898) has gone through many editions, including the important "Atlantic Edition" (1924), but the edition cited in this chapter for reasons of wider accessibility will be the Dover reprint of the novel found in *Seven Science Fiction Novels of H. G. Wells*, 307–453.

20 The literature on Wells and his work is extensive. Among biographies see David C. Smith, *H. G. Wells: Desperately Mortal* (New Haven, Conn., 1986); Lovat Dickson, *H. G. Wells: His Turbulent Life and Times* (New York, 1969); and Anthony West, *H. G. Wells: Aspects of a Life* (New York, 1984). The last author was Wells's son. Wells's own story of his life is *Experiment in Autobiography* (New York, 1934).

21 Mark R. Hillegas, *The Future as Nightmare: H. G. Wells and the Anti-Utopians* (New York, 1967), 19, where Huxley's influence is further discussed. Hillegas is quoting Geoffrey West, *H. G. Wells: A Sketch for a Portrait* (New York, 1930), 106. For another possible Huxley influence see "The Extinction of Man," *Pall Mall Gazette* (September 25, 1894), reprinted (according to Hillegas, *The Future as Nightmare*, 182, n. 14) in *Certain Personal Matters* (London, 1897), 173.

Figure 5.2. H. G. Wells. Courtesy of The Rare Book and Special Collections Library, University of Illinois at Urbana-Champaign.

worlds tradition. He knew that the debate was at least as old as the seventeenth century, as his prefatory quotation from Kepler indicates: "But who shall dwell in these worlds if they be inhabited? . . . Are we or they the Lords of the World? . . . And how are all things made for man?"[22] As a widely read man of his times, he could not fail to have known of Richard Proctor's *Other Worlds Than Ours*, which by then had gone through at least seven English editions. He may have known of Huxley's own belief on the subject, as well as Flammarion's *Pluralité*, though no English edition had appeared. Perhaps most important, it was no accident that Wells's invaders came from the Mars of Percival Lowell, whose Martian canal controversy had by now reached England.

We know in particular that Wells was interested in life beyond the Earth as early as 1888, when he spoke before the Debate Society at the Royal College of Science on the subject "Are the Planets Habitable?" In June 1894 he wrote "The Living Things That May Be," a review of J. E. Gore's *The Worlds of Space*, in which he faulted Gore for lack of imagination and suggested the possibility of silicon life. In December of the same

[22] Wells, *The War of the Worlds*, 307, citing Kepler, as quoted in Robert Burton's *The Anatomy of Melancholy* (1621).

year he followed up on this theme in "Another Basis for Life." "Very attractive is the question of whether life extends beyond the limits of this little planet of ours," Wells wrote, citing the recent work of astronomer Sir Robert Ball on conditions for life in space. In particular, Wells pointed to the possibility of silicon life, based on the recent work of Emerson Reynolds. Finally, in April 1896, 1 year before *War of the Worlds* began to appear serially, his essay "Intelligence on Mars" appeared, in which he argued that if Martians exist, they would be very unlike us.[23]

In 1893 Wells began to turn out short stories in earnest, and as early as 1894 they showed a penchant for strange lifeforms, although all on Earth. "Aepyornis Island" (February 1894) incorporated a large extinct bird; "The Flowering of the Strange Orchid" (April 1894) featured a man-attacking plant; and "The Stolen Bacillus (June 1894) featured an anarchist trying to poison the water of London with bacteria, while "In the Avu Observatory" Wells penned his first story in an astronomical setting, again including a large and unusual bird. By 1897 Wells's fascination with the extraterrestrial is apparent, not only in *War of the Worlds* but also in "The Crystal Egg" and "The Star," both of which include Martians. By this time he had also written three major novels: *The Time Machine*, *The Wonderful Visit,* and *The Island of Dr. Moreau*; in 1897 he would also pen *The Invisible Man.* All of these foreshadow in different ways elements that would appear in *War of the Worlds*, especially the evolutionary viewpoint. His son Anthony West believed that the seed of the idea is found in *The Time Machine,* where Wells spoke of humanity evolving into "something inhuman, unsympathetic, and overwhelmingly powerful."[24]

All of this was in the background, but Wells himself tells us the circumstances that led him to set pen to paper in 1897: "The book was begotten by a remark of my brother Frank. We were walking together through some particularly peaceful Surrey scenery. 'Suppose some beings from another planet were to drop out of the sky suddenly,' said he, 'and begin laying about them here!' Perhaps we had been talking of the discovery of Tasmania by the Europeans – a very frightful disaster for the native Tasmanians! I forget. But that was the point of departure." We cannot say whether or not Wells's Martian novel would have been written without the offhand remark of his brother, but once the seed was planted, all the background we have detailed came into play. In any case, Wells also com-

[23] On the talk before the debate society see Bernard Bergozni, *The Early H. G. Wells: A Study of the Scientific Romances* (Toronto, 1961), 123, citing "Report of a Debate," *Science Schools Journal* (November 1888), 58. Bergozni is one of the best sources for background to *War of the Worlds*. On the review of Gore see Robert M. Philmus and David Y. Hughes, eds., *H. G. Wells: Early Writings in Science and Science Fiction* (Berkeley, Los Angeles, and London, 1975), 236, and on Ball, ibid., 144–147.

[24] Anthony West, *H. G. Wells: Aspects of a Life* (New York, 1984), 233.

mented that in his short stories he wished to comment on the false securities of everyday life in tranquil England, in a way he regarded as realistic but others considered imaginative gymnastics. "The technical interest of a story like *The War of the Worlds* lies in the attempt to keep everything within the bounds of possibility. And the value of the story to me lies in this, that from first to last there is nothing in it that is impossible."[25]

The invention of the alien unfolded with the plot of the story, a story simple in concept but ingenious in execution, full of scientific ideas gleaned from evolutionary biology and astronomy. It is not just that Martians invade the Earth, blindly laying waste to all in their path. Wells invests his Martians with a certain plausibility; reason informs their every action, and their very presence betrays our smallness in an enormous universe. The chilling opening lines clearly indicate our place in the chain of being:

> No one would have believed in the last years of the nineteenth century that this world was being watched keenly and closely by intelligences greater than man's and yet as mortal as his own; that as men busied themselves about their various concerns they were scrutinised and studied, perhaps almost as narrowly as a man with a microscope might scrutinise the transient creatures that swarm and multiply in a drop of water.

These "minds that are to our minds as ours are to those of the beasts that perish" were not a haphazard creation of Wells's imagination. If the nebular hypothesis has any truth, Wells pointed out, Mars must be older than our world. Its smaller size by comparison with the Earth would have cooled it faster to a temperature at which life could begin. It has a much more attenuated atmosphere, lower temperature, a surface only one-third ocean, and huge snow caps. The planet is in its last stages of exhaustion, and so its inhabitants look across space at the Earth, "a morning star of hope . . . green with vegetation," crowded only "with what they regard as inferior animals": creatures as alien and lowly as are the monkeys and lemurs to us. And so the plot is set, for "to carry warfare sunward is, indeed, their only escape from the destruction that generation after generation creeps upon them."

A central element of the story was triggered by an article Wells had seen in *Nature*, the British scientific journal, reporting the observation of a strange light seen on Mars. And so in the story, during the opposition of 1894 – the period during which Lowell began his scrutiny of the Martian

[25] Bergozni, *The Early H. G. Wells*, 124–125, cited from *Strand Magazine*, 59, 154. See also the Introduction to the 1924 edition. The first English edition of Wells's book was dedicated "To my brother Frank Wells, This Rendering of His Idea."

surface – a light is seen on the surface of Mars, followed by an outbreak of incandescent gas, signs that Wells attributes to the "huge gun" by which the Martians later sent the spacecraft to Earth. More signs were seen during the next two oppositions, and finally the spectroscope analyzed a mass of hydrogen gas moving toward the Earth with enormous velocity. The local astronomer scoffed at the very idea of Martians, but thought a meteorite shower was striking Mars or a volcanic explosion. Organic evolution had probably not taken the same course on two adjacent planets, he argued; "the chances of anything man-like on Mars are a million to one."[26]

One by one during the course of the story the missiles land on the Earth. They are believed at first to be meteorites, but eventually two creatures emerge from the first missile. As the reader continues, Wells's description is clearly meant to instill horror, for the tentacles that emerged first do not indicate something humanlike. And then, with the appropriate metaphor of the creatures emerging from the cylinder, we have the birth of the monstrous alien:

> A big greyish rounded bulk, the size, perhaps of a bear, was rising slowly and painfully out of the cylinder. As it bulged up and caught the light, it glistened like wet leather. Two large dark-coloured eyes were regarding me steadfastly. The mass that framed them, the head of the thing, it was rounded, and had, one might say, a face. There was a mouth under the eyes, the lipless brim of which quivered and panted, and dropped saliva. The whole creature heaved and pulsated convulsively. A lank tentactlar appendage gripped the edge of the cylinder, another swayed in the air.

This was only the first of Wells's many detailed descriptions of the alien – desperate, unfriendly, monstrous. That the alien appearance was more than skin deep Wells makes clear in his occasional descriptions of Martian physiology. The Martians were heads, with no digestive organs, and communicated by telepathy. They did not sleep, were sexless, and reproduced by budding, a direction in which Wells sees evolution might carry humanity. For mobility in the heavier gravity of Earth, they constructed machines. Ironically, the Martians were most like humans in their penchant for violence. When a deputation is sent with a flag to signal the good intentions of humanity, it is destroyed by the Martian Heat Ray. The Martians in their machines move toward London, wreaking destruction, and victory seems to be theirs. But in the end bacteria

[26] Wells, *The War of the Worlds* , 313. In the story, 10 flashes were observed on 10 nights.

against which their bodies are defenseless destroy them, and the Martians lie slain "after all man's devices had failed, by the humblest things that God, in his wisdom, has put upon this earth."[27]

The immense success of *The War of the Worlds* launched not only Wells's career but also the career of the alien. Reprintings, variations on the theme, and imaginative art began immediately (Figure 5.3) and have continued to the present. Adaptations have sometimes changed the setting, as in the 1938 Orson Welles American radio version. But always the effect has been powerful, and the imitations, variations, and elaborations on the story have echoed down the century. As late as 1908 Wells cited "the work of my friend, Mr. Percival Lowell," as testimony that *The War of the Worlds* was not too farfetched. Pronouncing Lowell's case for canals created by intelligent Martians "very convincing," Wells discussed the forms that such Martians might take.[28] Although Lowell's claims were soon discredited, Mars remained a favorite setting for the alien through many decades. And after that, there were other settings around other stars; the possibilities were almost unlimited.

Despite the success of *The War of the Worlds*, only twice more in his long career did Wells return to his extraterrestrial theme in fiction. *The First Men in the Moon* (1901) featured interplanetary travel, with humans meeting extraterrestrials in their own environment instead of as invaders of Earth.[29] And in *Star Begotten: A Biological Fantasia* (1937), Wells speculated about cosmic rays from Mars mutating the human spirit. But both of these novels pale by comparison to the originality and influence of *War of the Worlds*. By the time *Star Begotten* was written, Wells had already accomplished much else in his long career, including proposals for a world order foreshadowed in his 1901 discussion of lunar society. For Wells, the invention of the alien and the elaboration of what would become several of its standard themes were only one contribution on the way to understanding life on Earth.[30]

27 Ibid., 444. On Martian physiology see 407ff.

28 H. G. Wells, "The Things That Live on Mars," *Cosmopolitan Magazine* (March 1908), 335–343, also notable for its illustrations. Wells particularly cites Lowell's *Mars and Its Canals* (1907).

29 The story, in which Wells has his characters travel to the Moon to discover successively plants, animals, and intelligence, is particularly notable not only for the contemporary science that Wells incorporates, but especially for its discussion of communication with extraterrestrials, both face-to-face and at long distance. For mentions of Francis Galton, Nikola Tesla, and Marconi (discussed in Section 8.1) see *The First Men in the Moon* in *Seven Science Fiction Novels*, 525, 559, 587.

30 Many imitations of Wells appeared in the following years, including several in the United States. See Garrett P. Serviss's "sequel" to Wells, *Edison's Conquest of Mars* (1898 in the *New York Journal*; 1947 as a book), in which the Earthlings invade Mars. See also Mark Wicks, *To Mars Via the Moon* (London, 1910), among others.

Figure 5.3. Cover from *Amazing Stories* of August 1927 illustrating H. G. Wells's *War of the Worlds*. Copyright 1927 by Experimenter Publishing Company.

The invention of the alien, then, occurred independently in different forms in France, Germany, and England, in the latter two cases spurred on by Lowellian claims in the United States. But in all cases the evolutionary world view and the plurality-of-worlds tradition played an important role in the birth of the alien. In France, Lamarckian and Darwinian evolution formed the background to the alien fiction of Flammarion, who had himself been France's chief champion of the plurality of worlds since 1862. The same background influenced Rosny, who undoubtedly knew Flammarion's work on plurality of worlds. In Britain a direct line exists from Darwin to Wells via Huxley, amid the Proctorian plurality-of-worlds tradition. And in Germany, Lasswitz was influenced by the Kantian tradition. The alien thus emerged at the beginning of the nineteenth century from a confluence of science and philosophy, combined with the usefulness of the alien in exploring social as well as biological evolution.

5.2 THE DEVELOPMENT AND USES OF THE ALIEN: BURROUGHS TO BRADBURY

Once invented, the alien took on a life of its own. Though it might have died after the initial bursts of creativity of Flammarion, Rosny, Lasswitz, and Wells, once unleashed the alien proved a deep-seated and multipurpose concept. Its potential as a character in space adventure spanning the universe was obvious. More subtle were its uses in expanding areas of human thought like philosophy and religion, including such traditional problems as good and evil and the uniqueness or triviality of humanity. And while such areas were being explored, the nature of the alien and interspecies communication were subthemes of almost endless potential. Set adrift in the post-Darwinian era that had placed humanity at the pinnacle of terrestrial creation, the alien showed how provincial was the view of life on Earth, giving new meaning to Shakespeare's line "As flies to wanton boys are we to th' gods."

In this section we concentrate on the development and uses of the alien during the first half of the twentieth century, the period bracketed by two major – but vastly different – American writers of alien literature: Edgar Rice Burroughs and the early Ray Bradbury. Although characterized by no simple linear development, the alien evolved during this time from a rather predictable character in the "space opera" adventures of Burroughs and others to the much more subtle, impressionistic, and almost ethereal creatures in Bradbury's *Martian Chronicles* (1950). In the interim, we shall see the nature of alien morphology and behavior explored in several representative writers of the 1920s and 1930s. And we shall explore the uses of the alien in philosophy and religion through the work of two British writers: Olaf Stapledon and C. S. Lewis, who at the same time made their own contributions to defining the possibilities of alien morphology and psychology. While we will resist the temptation to conclude that the alien steadily evolved from the simple to the complex, the progress in its career at the hands of talented and imaginative writers will be evident.

The decades immediately following its invention clearly demonstrate that alien literature needed not be profound to be popular and influential. The alien could merely play the role of protagonist or antagonist in swashbuckling space adventures, with no serious attempt to illuminate its nature or the nature of humanity by comparison. That this lack of profundity did not translate into lack of popular impact (indeed, the relation may be inverse) may be seen from the case of Edgar Rice Burroughs (1875–1950), who took up Lowell's Martian theme in the United States several years before Lowell's death in 1916. It may well be true (as Arthur C. Clarke has said) that if one is no longer a teenager, it is too late to read Burroughs; it is no less true that for many during that impressionable age (including Bradbury, Clarke, and Carl Sagan), Burroughs's adventures under the moons of Barsoom (Mars) had a lasting effect on the imagination.[31] Although perhaps best known as the inventor of Tarzan (1912), between 1912 and 1948 Burroughs also penned 10 volumes of the exploits of John Carter, battling or befriending various Martian life forms in the context of Martian culture. *A Princess of Mars*, published in novel form in 1917, was the first of the series, with *The Chessman of Mars* (1922) and *The Swords of Mars* (1936) considered among the best.[32] If nothing else, fanned by the imaginations of Lowell and then Burroughs, Mars was firmly entrenched in the United States as an adventurous and mysterious place in the solar system – a place one might want to visit some day.

We need not belabor the work of Burroughs; a summary of his first

[31] On Burroughs (who was a failed salesman when he began writing at age 36) see *SFE*, 177–178. Clarke's comment is in the Introduction to the 1962 paperback edition of H. G. Wells's *The Invisible Man* and *The War of the Worlds*, xviii. For Burroughs's influence on Bradbury, Clarke, Sagan, and others see Ray Bradbury, Foreword to Neil McAleer's biography of Clarke (footnote 62). A comparison of the Mars of Lowell and Burroughs is given in R. D. Mullen, "The Undisciplined Imagination: Edgar Rice Burroughs and Lowellian Mars," in Thomas D. Clareson, ed., *SF: The Other Side of Realism* (Bowling Green, Ohio, 1971), 229–247. Although Mullen shows that Burroughs played fast and loose with Lowell's views, Lowell is nevertheless the undisputed source of Burroughs's ideas.

[32] Less well known is Burroughs's Venus series (1934–1946). It is notable that except for a few stories written in the last decade of his life, Burroughs's stories were set in the solar system.

Mars novel suffices to characterize all the rest. In *The Princess of Mars* John Carter is whisked to Mars by magical means; Burroughs did not bother with an examination of antigravity propulsion mechanisms such as Lasswitz had described. On Mars, the alien setting is established immediately on arrival as Carter watches Martians hatching from giant eggs before he is captured by six-limbed green giants who ride huge "thoats." That Mars had achieved no utopia is immediately evident as the green giants capture their bitter rival Dejah Thoris, Princess of Helium. Carter's task is to rescue the princess, with whom he falls in love. This he does with all the elements of romance and adventure that Burroughs can summon, in a Lowellian setting that is not entirely faithful to Lowell but provides the exotic sense that certainly adds to the story. Although Burroughs's Martians were no terrestrials – they had differently colored skin, variations in anatomy, and oviparous women, among other exotic characteristics – they were driven by essentially terrestrial passions. There is therefore little serious examination of either the alien nature or the alien psychology and morality. All is subordinated to the story, which at times borders on fantasy. None of this mattered to Burroughs's impressionable young fans. Action was the watchword of Burroughs's writing, and his readers loved it. None of the novelty of pure and simple space adventure of this type seemed to have worn off by the end of the century in similar stories such as *Star Wars*, also replete with aliens. Sophisticated or not, through such stories the alien gained a place in young minds, inspiring some of them to develop careers in science or science fiction writing themselves.

Despite numerous examples of such adventurous alien "space opera" over the following decades,[33] the alien phenomenon would hardly deserve serious attention if it had been used only for adventure. More serious questions abounded, among them the basic nature of the alien being. Lasswitz and Wells had already presented the two extreme possibilities – the extraterrestrial essentially good, though capable of retaliation, and the monstrous alien intent only on invasion. Although the invading monster carried the day throughout the 1930s, perhaps because of its dramatic story potential, a great deal of ground remained to be investigated between these two extremes.[34] The works of three authors of the 1920s and 1930s demonstrate the possibilities inherent in the alien,

[33] For example, science is subordinated to adventure in the interstellar exploits of E. E. "Doc" Smith (1890–1965), *The Skylark of Space* (1928; modern edition New York, 1975), and its many sequels, as well as in the conflict between civilizations in his Lensman series beginning in 1934 and running throughout the 1950s. On Space Opera see *SFE,* 1138–1140.

[34] Peter Nicholls, "Aliens," in *SFE,* 15–19.

both in physical characterization and in plot: David Lindsay's *A Voyage to Arcturus* (1920), Stanley G. Weinbaum's "A Martian Odyssey" (1934), and John W. Campbell's "Who Goes There?" (1938).

Unlike Lasswitz, Wells, and Burroughs, the Scottish-born writer David Lindsay (1878–1945) placed his aliens on the distant planet Tormance, orbiting Branchspell, the larger of two stars in the double star system known on Earth as "Arcturus." But more than setting separated Lindsay from his predecessors in alien literature; his influences were the music of Beethoven, the novels of fellow Scottish author George MacDonald, and the philosophical works of Schopenhauer and Nietzsche. Although he wished to attend university, duty to his family and his Scottish Calvinist upbringing led him unhappily to accept work as a clerk at Lloyd's of London. Only in 1919, having moved to Cornwall with the support of his wife, did Lindsay begin a writing career that lasted until 1932, but that never gained him fame or money in his lifetime.[35]

A Voyage to Arcturus, the first and best of Lindsay's novels, went further than most in his imaginative characterization of alien lifeforms, and thus in giving the reader a feel for the truly alien. This is carried out in part by additional senses that Lindsay gives not only his aliens but also the Earthling (Maskull), who on landing on Tormance finds himself sprouting new organs that extend his perceptions of his new environment. With these continually developing new senses, Maskull interacts with the inhabitants of Tormance, whose skin is opalescent, its hue "continually changing with every thought and emotion," who communicate by telepathy, and whose habits (such as producing from the mouth a brightly colored crystal as an "overflowing of beauty") are completely foreign to terrestrials. The reader is subjected to a continually changing pageant of life forms, for life on the young planet is "energetic and lawless, and not sedate and imitative. Nature is still fluid – not yet rigid – and matter is plastic. The will forks and sports incessantly, and thus no two creatures are alike." This applies to plant and animal life alike; the alien world of Tormance is vividly conveyed and sustained in numerous passages such as the following:

> The jale-colored blossoms of a crystal bush were emitting mental waves, which with his breve [one of the new sense organs] he could clearly distinguish. They cried out silently, "To me To me!". While he looked a flying worm guided itself

[35] On Lindsay and his works see J. Clute,"Lindsay," in *SFE*, 723; J. B. Pick, E. H. Visiak, and Colin Wilson, *The Strange Genius of David Lindsay* (London, 1970); and Gary K. Wolfe, *David Lindsay* (Mercer Island, Wash., 1982). George MacDonald had written *Lilith* (1895; modern edition Grand Rapids, Mich., 1981), a search for self set in two universes, with which Lindsay was undoubtedly familiar.

> through the air to one of these blossoms and began to suck its nectar. The floral cry immediately ceased.[36]

But these physical descriptions are only the setting for a plot that surpasses adventure and is at once philosophical, theological, and moral. The events surrounding Arcturus were not, anthropologist Loren Eiseley said in his Introduction to the 1963 edition of Lindsay's novel, simply "a superficial tale of odd beings with odd organs on a planet remote from our own. This is not a common story of adventure. Rather, it is a story of the most dangerous journey in the world, the journey into the self and beyond the self." Lindsay's journey is almost Dantean but in an extraterrestrial setting, a modern-day exploration of self that nevertheless shows few traces of modern astronomy. Tormance was, of course, imagined, since no planets were known to exist around any star, but this did not matter; modern science was not Lindsay's concern. His concern was the search for a deeper beauty and reality than most people could commonly perceive, a reality he called "Muspel." In encountering the many lifeforms in his journey across Tormance, Maskull is encountering beings who have, for different reasons, not yet found the deeper reality, a vision he himself finally sees in the climax of Arcturus. By clearly using alien lifeforms to expand the sense of the possible and explore deep questions, Lindsay placed himself squarely in a nascent tradition of alien literature that went not only beyond Mars but beyond invasion and adventure.[37]

By contrast, much of the rest of alien literature seems tame, if not parochial, by the narrowness of its themes. But the continuing potential of new themes in the old Martian setting is clear in Stanley G. Weinbaum (1902–1935), who graduated with a degree in chemical engineering at the University of Wisconsin in the same class as Charles Lindberg and made good use of his scientific background in his fiction. Remarking on this scientific base, science fiction author and critic Sam Moskowitz wrote that "the only reason why Weinbaum was successful at all is the high degree of scientific authenticity he imparted to his otherworldly creations, rescuing them from the realm of the fairy tale." For this reason, "Many devotees of science fiction sincerely believe that the true beginning of modern science fiction, with its emphasis on polished writing, otherworldly psychology, philosophy and stronger characterization began with Stanley G. Weinbaum," wrote Moskowitz. "Certainly few authors in this branch of the literature have exercised a more obvious and

[36] David Lindsay, *A Voyage to Arcturus* (New York, 1963), 46, 63, 65.

[37] Loren Eiseley, Introduction to *A Voyage to Arcturus* (New York, 1963), viii. For a more sophisticated analysis, see Galad Elflandsson, "David Lindsay and the Quest for Muspel-Fire," introducing the Citadel Press edition of *A Voyage to Arcturus* (Secaucus, N.J., n.d.).

pervasive influence on the attitudes of his contemporaries," he added. The loss to science fiction was considerable, then, when Weinbaum died of throat cancer at the age of 33, 15 months after the publication of his first science fiction story.[38]

Though Weinbaum placed his first published science fiction story on Mars, his portrayal of the alien is generally considered a breakthrough: the Martians were not simply transported terrestrial monsters but truly alien beings, intent on living their own lives rather than invasion; the ideas of Flammarion and Rosny were combined with the Martian myth to produce a totally alien world. Weinbaum's story featured what Moskowitz called "the most delightful zoology of life forms since L. Frank Baum wrote *The Wizard of Oz*": Tweel, the bird-like creature who befriends a visiting Earth expedition and humorously attempts communication; a silicon creature producing bricks as waste material; the "dream beast," a foreshadowing of Bradbury's work; and the people of the mud cities along the canals, more alien than any of the others. Here was a theme for the future: the study of alien mentality and motives and their reflection on the creature we call human. In the story the chemist explains one lifeform as follows: "The Beast was made of silica! There must have been pure silicon in the sand, and it lived on that. Get it? We, and Tweel, and those plants out there, and even the biopeds are carbon life; this thing lived by a different set of chemical reactions. It was silicon life!" Scientists had broached the subject before, but to incorporate such a creature as an active character in a story plot was something else. The imagination could go on and on:

> there the thing was, alive and yet not alive, moving every ten minutes, and then only to remove a brick. Those bricks were its waste matter. . . . We're carbon, and our waste is carbon dioxide, and this thing is silicon, and its waste is silicon dioxide – silica. But silica is a solid, hence the bricks. And it builds itself in, and when it is covered, it moves over to a fresh place to start over. No wonder it creaked! A living creature half a million years old![39]

One has the feeling, reading such literature, that the horizons of life have been expanded, that the scope for new literary themes is endless, and that

38 On Weinbaum see *SFE*, 1307–1308, and Sam Moskowitz, "The Wonder of Weinbaum," Introduction to Stanley G. Weinbaum, *A Martian Odyssey* (New York, 1966).

39 Weinbaum, "A Martian Odyssey," in *A Martian Odyssey and Other Great Science Fiction Stories* (New York, 1966), 28. The story first appeared in *Wonder Stories* (July 1934). Weinbaum's "The Lotus Eaters" (1935), set on Venus, gives the worldview of an intelligent plant.

even the scientist might have his or her interest piqued by literary explorations extending beyond scientific claims.

As if to contrast the spectrum of alien possibilities yet again, 4 years after "A Martian Odyssey" appeared in *Wonder Stories*, John W. Campbell, Jr. (1910–1971), destined as both writer and editor to bring science fiction into the mainstream, used the alien as the central focus for a horror thriller published in *Astounding Science Fiction*. Having attended MIT and graduated from Duke with a degree in physics in 1932, Campbell was another for whom science was not enough. It had to be supplemented by the imagination, and this he had done through science fiction beginning in 1930 with galactic epics incorporating his knowledge of science.[40] "Who Goes There?" voted by the Science Fiction Writers of America as the greatest novella of all time, showed that the monstrous alien was not the the sole preserve of unsophisticated writers (though there were many), and could in fact be used as the main character in a psychological thriller as great as any played out in a more mundane setting. Appearing in 1938, exactly 40 years after *The War of the Worlds* was published in novel form, it amply demonstrated that original use of the monstrous alien had not been exhausted. "Who Goes There?" the climax to Campbell's writing career, featured an alien discovered at an Antarctic base camp and unfrozen after 20 million years. Not only does the alien come to life after being brought back to the camp for examination, it has the ability to insinuate itself into the bodies of humans while imitating their behavior. As the camp members gradually realize this, the classic mutual suspicions arise, and medical tests are devised to determine whose body has been invaded. Finally, all traces of the monster are exterminated, except for his last manifestation: "Like a blue rubber ball, a Thing bounced up. One of its four tentacle-like arms looped out like a striking snake. In a seven-tentacled hand a six-inch pencil of winking, shining metal glinted and swung upward to face them. Its line-thin lips twitched back from snake-fangs in a grin of hate, red eyes blazing." This monster too is finally killed, in the kind of scene that had lost none of its appeal when reworked in the movie *Alien* 40 years later.[41] When it is found that the alien had mastered atomic power, and had constructed an antigravity device with which it could have traveled over the Earth at will, Campbell gives reader a sense of the fragility of life on Earth in the face of powers that may be much more highly evolved: "By the grace of

[40] On Campbell see *SFE*, 187–188, and Lester del Rey, "The Three Careers of John W. Campbell," in Lester del Rey, ed., *The Best of John W. Campbell* (New York, 1976), and its "Postscriptum" by Mrs. Campbell.

[41] "Who Goes There?" in *The Best of John W. Campbell*, 290–353: 349. The story was filmed as *The Thing* in 1951.

God . . . and the margin of half an hour, we keep our world, and the planets of the system." Though "Who Goes There?" was one of Campbell's last stories, his influence was only beginning; as the editor of *Astounding Stories* beginning in 1937, Campbell would go on to a role of great influence in science fiction, shaping the alien all the while.

If Lindsay, Weinbaum, and Campbell extended our understanding of the physical possibilities of the alien and its uses, others would go beyond the physical to focus on its mental and moral capacities – as Lindsay had already shown was possible. No one did this with such flair and imagination as the British philosopher Olaf Stapledon (1886–1950). A graduate of Oxford in history and of Liverpool University in philosophy (1925), Stapledon spent most of his life near Liverpool, though not in academic positions. He began writing essays as early as 1908, but only in 1930, at the age of 44, did he take up fiction, in which aliens immediately played a role. Though we meet many aliens in Stapledon's work, his aliens are not an end in themselves, nor is adventure the dominant theme: "Political, religious, and philosophical ideas, rather than gadgets or adventure are central to his works," wrote one of his admirers.[42] Breathtaking in their scope, Stapledon's novels have served as a source of ideas for many subsequent science fiction writers.

Three of Stapledon's works especially embrace the alien theme: *Last and First Men* (1930), *Star Maker* (1937), and *The Flames* (1947). The first covers a vast expanse of time – some 2 billion years – the second a vast expanse of both space and time. *Last and First Men* – which may show the influence of biologist J. B. S. Haldane's *Possible Worlds* (1927) – tells the story of 18 species of humanity, some of them naturally evolved, others designed or bred, over a 2-billion-year span.[43] Four of the species live out their lives on Earth before the Moon spirals in and destroys it; 4 more species live on a Venus terraformed for the purpose; and the last 10 species live on Neptune, before a supernova finally destroys the solar system and humanity with it. The many crises and triumphs of these 18 species of humanity – all of which carry the message of Stapledon's philosophical ideas – are not our concern here. But in two cases, alien beings did play a prominent role.

In Stapledon's scenario, it was with the Second Men, a mere 10 million years in the future, that humanity had its first encounter with

[42] C. C. Smith, "Olaf Stapledon's Dispassionate Objectivity," in Thomas Clareson, ed., *Voices for the Future: Essays on Major Science Fiction Writers*, vol. 1 (Bowling Green, Ohio, 1976), 44. On Stapledon see *SFE*, 1151–1153; P. A. McCarthy, *Olaf Stapledon* (Boston, 1982); and Leslie A. Fiedler, *Olaf Stapledon: A Man Divided* (New York, 1983).

[43] On Haldane's influence see *SFE*, 533.

extraterrestrial intelligence in the form of Martians. Like Wells 33 years earlier, Stapledon too had his Martian invasion, but it was only a prelude, comprising 2 of the 16 chapters of the book and only a moment in the span of the novel. Because Stapledon's Mars was also Lowell's dying planet, terrestrial water and vegetation were given as the Martians' rationale for invading Earth. But Stapledon's loftier reason for a Martian invasion was clear: it was intended to put humanity in perspective as a biological, spiritual, and philosophical entity. Thus, one day the Martians literally floated in from outer space, beings "fantastically different yet fundamentally similar." Martian life had developed the ability of "maintaining vital organization as a single conscious individual without continuity of living matter. Thus the typical Martian organism was a cloudlet, a group of free-moving members dominated by a 'group-mind.' But in one species individuality came to inhere, for certain purposes, not in distinct cloudlets only, but in a great fluid system of cloudlets. Such was the single minded Martian host which invaded the Earth."[44] So different from terrestrials were the Martian cloudlets that at first they did not recognize us as intelligence, nor we they. Many invasions occurred over 50,000 years as humanity devised different weapons against the cloudlets. In the end, humans finally developed a virus that killed the Martians – and almost all the humans.

Millions of years later, driven from Earth by the approach of the Moon, Stapledon's Fifth Men escaped to Venus, a hot world almost completely covered by a shallow ocean. Here they found marine intelligence and other forms of life, "some sessile, others free-swimming, some microscopic, others as large as whales," all of them powered not by photosynthesis or chemical reactions, but by "the controlled disintegration of radioactive atoms."[45] All of these lifeforms were reluctantly destroyed in the process of terraforming the planet for the needs of humanity. In due time Venus is abandoned for Neptune, and in the end it appears that humanity will be destroyed, except for one last attempt to seed its cells among the stars – an idea Stapledon undoubtedly took from the panspermia hypothesis.

In the total expanse of 2 billion years of this story, extraterrestrials played a relatively minor role, except for the Mars and Venus episodes. Although Stapledon believed life was found throughout the universe, according to the hypothesis of British astronomer James Jeans current at the time, it was rare:

[44] Stapledon, *Last and First Men* and *Star Maker* (New York, 1968), 116–117.

[45] Stapledon, *Last and First Men,* 187–191. Stapledon here adopted Arrhenius's view of Venus, discussed in chapter 3.

> Not before there are stars, and not after the stars are chilled, can there be life. And then, rarely. In our own galaxy there have occurred hitherto some twenty thousand worlds that have conceived life. And of these a few score have attained or surpassed the mentality of the First Men. . . . There are the millions of other galaxies, for instance the Andromedan island. We have some reason to surmise that in that favoured universe mind may have attained to insight and power incomparably greater than our own. [Beyond planets] we have evidence that in a few of the younger stars there is life, and even intelligence. How it persists in an incandescent environment we know not, nor whether it is perhaps the life of the star as a whole, as a single organism, or the life of many flame-like inhabitants of the star . . . .

Even a few burned-out stars, Stapledon conjectured, may harbor life.[46]

Seven years later, in Stapledon's *Star Maker,* all the expanse of the previous novel was reduced to a few paragraphs, a "long human story, most passionate and tragic in the living . . . but an unimportant, a seemingly barren and negligible effort, lasting only for a few moments in the life of the galaxy." Unlike *Last and First Men*, *Star Maker* is replete with alien biologies that play an essential part in the story, but again the aliens themselves are not the main point. The purpose of the story is "not only to explore the depths of the physical universe, but to discover what part life and mind were actually playing among the stars." Ultimately, the aliens join the terrestrial protagonist in their search for the highest intelligence in the universe – Star Maker. And not only that: the joint purpose of the aliens and the terrestrial "was not merely scientific observation, but also the need to effect some kind of mental and spiritual traffic with other worlds, for mutual enrichment and community."[47] The communication problem is solved by an enhanced form of telepathy in which the pilgrim "inhabits" the bodies of alien individuals.

Even for such a grandiose scheme, science provided the background to Stapledon's story. In his preface the author expressed his gratitude to certain professors, but especially to J. D. Bernal's book *The World, the Flesh and the Devil* for the idea of artificial planets. In addition, it is clear that Stapledon had embraced the cosmology of his time, especially the

[46] Ibid., 230–231. At the end of *Star Maker*, in "A Note on Magnitude," p. 435, Stapledon cites W. J. Luyten's *The Pageant of the Stars* for his view of the size of the universe. Stapledon used this idea of life in stars later for his novel *The Flames* (1947). The astronomer Harlow Shapley would later take up the idea briefly.

[47] Ibid., 379, 263, 271.

rarity of planetary systems. But Stapledon's pilgrim discovers that although planets are comparatively rare, the universe is so large that many planets still exist: "It was indeed a stirring experience to see spread out before us a whole 'universe,' containing a billion stars and perhaps thousands of inhabited worlds; and to know that each tiny fleck in the black sky was itself another such 'universe,' and that millions more of them were invisible only because of their extreme remoteness." The view that humanity is unique proves "ludicrously false."[48]

In his description of the different species of life and their search for Star Maker, the process of evolution was everywhere evident. The humanoid races inhabited planets of much the same size and nature as the Earth and were molded by evolution to much the same form. To be sure, great variety existed, including such variations as feathered, penguinlike men and human echinoderms with a different method of reproduction. Among nonhumanoids, evolution was still the dominant guiding principle, with "the physical and mental form of the conscious beings . . . an expression of the character of the planet on which they live." Among the nonhumanoid races, Stapledon described nautiloids, an intelligent fishlike race, symbiotic races, composite organisms forming a single personality, the Martian composite cloudlets described in *Last and First Men*, and plant men. Although all had their own peculiar cultural problems, all were similar to humans mentally in their involvement in spiritual crisis. In the latter regard, Stapledon made clear that not all races had triumphantly achieved what he called "a more awakened state"; for every one that did, hundreds of thousands of civilizations ended in disaster. Stapledon indeed saw a great chain of being throughout the universe, with Star Maker at the apex.[49]

All the while, Stapledon did not forget his purpose: the search for the spirit of the universe, what humans on Earth formerly called God. Nor did he hesitate to draw poignant conclusions; midway in their journey, having seen the pain and suffering in the universe, the frequent destruction of civilization for every one triumph, "it was becoming clear to us that if the cosmos had any lord at all, he was not that spirit but some other, whose purpose in creating the endless fountain of worlds was not fatherly toward the beings that he had made, but alien, inhuman, dark."[50]

In the final stages of his journey, Stapledon became part of a "communal mind" capable of contacting worlds far beyond the mentality of

[48] Ibid., 266, 303. Stapledon's universe was also expanding, as had just been proven by Hubble and others.
[49] Ibid., 360, 311, 334, 345.
[50] Ibid., 317.

humanity. In this state he undertook a comparison of civilizations obviously designed to place our own terrestrial civilization in perspective, describing enterprises ranging from eugenic manipulation of species to "sub-atomic energy" and interstellar travel.[51] Toward the end of his story, stars began to explode in our galaxy and the story moved to other galaxies, where more exotic life was encountered. Even stars were described as intelligent entities, giving stellar evolution a new meaning. Independent organisms of fire were encountered, ranging from cloud size to Earth size. Overall, Stapledon found three grades of intelligence inhabiting our universe: planetary, stellar, and nebular, all the product of Star Maker.[52]

The immense span of time and space that Stapledon described – our own universe – was, in his view, only one creation in a vast series, an early mature work of Star Maker, but one that by comparison with later universes was juvenile. "Musical" creatures with complex patterns and rhythms; universes with no time; worlds with antigravity rather than gravity, and reverse evolution were among Stapledon's concepts that provided an inexhaustible source of ideas to science fiction writers. In many ways Stapledon's work reflected Arnold Toynbee's just-published *A Study of History*, expanding the historian's discussion of civilizations beyond the Earth.[53]

In the end, Stapledon succeeded in making his reader feel the smallness of humanity in the face of the vast universe, in which ultimate meaning was to be found. A measure of his success is found in the fact that it is difficult to overestimate the influence of his work, especially in Britain. J. B. S. Haldane, a prominent biologist in the physiology and origins of life, knew Stapledon's work, as did C. S. Lewis, Arthur C. Clarke, and numerous others who went on to expound their own visions of meaning through literature.[54] Like Stapledon, their vision of life in the universe served as a vehicle for philosophical exploration, making use of contemporary knowledge of the physical universe, extrapolated to the biological universe in the search for a Cosmic Spirit.

If Stapledon's novels were a search for cosmic truth, the work of C. S. Lewis (1898–1963) was a defense of Christianity, a system that believed it had already found the truth, but for all that could still make use of the

[51] In his discussion of eugenics (p. 350), Stapledon perhaps shows the influence of the geneticist H. J. Muller. For interplanetary and interstellar travel, see 353; on subatomic power, p. 354; on artificial planets for interstellar travel (showing the influence of J. D. Bernal), see 355 and 364ff.

[52] Ibid., 385ff.

[53] Ibid., 415ff. Toynbee's vols. 1–3 were published in 1933, and volumes 3–6 in 1939, so influence is possible. See Stapledon, *Star Maker*, 348–350, for the utopian phase of civilization, and p. 362 for the term "chrysalis," also used by Toynbee.

[54] On Haldane see Smith, "Olaf Stapledon's Dispassionate Objectivity," 46.

alien. Lewis, a literature professor at Oxford since 1925, would go on to achieve fame for his Christian apologetics targeted at both adults (*The Screwtape Letters*, 1942) and children (*The Chronicles of Narnia*, 1950–1956) while capping his career as a Professor of Medieval and Renaissance Literature at Cambridge University after 1954. At Oxford he was closely associated with his fellow Anglican, Charles Williams, and Roman Catholic fantasy writer J. R. R. Tolkien, both of whom also composed works defending the Christian religion. It was in these surroundings that Lewis wrote a "Cosmic Trilogy" from 1938 to 1944, making heavy use of aliens and showing the influence of Stapledon's Mars and Venus episodes from *Last and First Men* and a more localized version of Star Maker. But it was Lindsay's *Voyage to Arcturus*, which Lewis read in 1936, that had the most immediate impact. "From Lindsay," he wrote, "I first learned what other planets in fiction are really good for; for spiritual adventures. Only they can satisfy the craving which sends our imaginations off the earth . . . my debt to him is very great."[55] Although he disagreed with Lindsay's philosophy, *Arcturus* was the springboard for what became the Cosmic Trilogy.

In *Out of the Silent Planet* (1938), a physicist and his colleague force a philologist to travel from Earth (Thulcandra) to Mars (Malacandra) to turn him over to the highest authority in return for gold. The philologist escapes, describes the many alien cultures he finds, and meets with the authority (Oyarsa) to discuss Earthly matters. The novel is a struggle between the spiritual and the scientific, probably based in part on the differences in worldview of Lewis (represented by the philologist) and J. B. S. Haldane, the Cambridge biologist whose *Possible Worlds* (1927) also influenced Lewis. The novel, which has been called "powerfully imagined" with scientific content that is "intermittently absurd," features a variety of Martians, in the form of sorns, hrossa, and pfifltriggi, and the spiritlike eldil, the latter with a physiology unlike that of humans or Martians. Although the philologist had feared Wellsian creatures, when actually sighted, the sorns "appealed away from the Wellsian fantasies to an earlier, almost an infantile, complex of fears. Giants – ogres – ghosts – skeletons: those were its key words. Spooks on stilts, he said to himself; surrealistic bogy-men with their long faces."[56] The giraffelike hrossa (capable of lan-

[55] George Sayer, *Jack: C. S. Lewis and His Times* (San Francisco, 1988), 152–153; see also 28 for the influence of Wells. Of the many works about Lewis, see especially A. N. Wilson, *C. S. Lewis: A Biography* (New York, 1990); Brian Sibley, *Shadowlands: The Story of C. S. Lewis and Joy Davidman* (1985), which was a stage play, a television series, and a movie; Lewis's autobiographical *Surprised by Joy* (London, 1955); and Mark R. Hillegas, *Shadows of Imagination: The Fantasies of C. S. Lewis, J. R. R. Tolkien and Charles Williams* (Carbondale, Ill., 1969). On the possible influence of Lindsay see *SFE*, 723.

[56] *Out of the Silent Planet* (New York, 1965), 47. Lewis makes several references to Wells in the novel, including allusions to *The First Men on the Moon* (70). On the "powerfully

guage) and the froglike pfifltriggi were no less vividly drawn, but it was the eldil (with bodies made of light) and the great eldil, Oyarsa, the ruler of Mars, toward which the plot led. Oyarsa "does not die . . . and he does not breed. . . . His body is not like ours, nor yours, it is hard to see and the light goes through it," one of the sorns explains. This creature tells the visitors that Earth is known as "The Silent Planet" for a reason: "Thulcandra [Earth] is the world we do not know. It alone is outside the heaven, and no message comes from it." Before life came to Earth, its Oyarsa became "bent," and he is forced to remain on Earth, where good must struggle with his evil. This enforced banishment prevents any taint of evil on other worlds. Through similar allegories, Lewis paints a vivid picture of the planet in which humanity is in a spiritual struggle from which it must extract itself – precisely the same message conveyed in Lewis's nonfiction.

In *Perelandra* (1943) Lewis multiplied his aliens and continued his parable, this time on Venus [Perelandra], a preparation for the struggle between good and evil that takes place in *That Hideous Strength* (1944). Ever more antiscience, and in particular anti-Wells, *Perelandra* and *That Hideous Strength* drew the wrath of Haldane, among others.[57] But taken as a whole, the novels received critical acclaim and a vast readership denied to the much more ambitious work of Lindsay, undoubtedly because of Lewis's more immediate appeal to Christian beliefs. All the while, even though the richly imagined aliens were in a sense only props for a story, the public interest in aliens was undoubtedly more than a little piqued. The years in which the trilogy was composed (1938–1944) span a crucial period of turnabout of belief in abundant planetary systems and thus extraterrestrial life.

It is some measure of the continuing allure of Mars that by midcentury the planet that had inspired Burroughs's adventures of John Carter in 1912 (and indeed continued to inspire the last of his Martian novels in 1948) could still give birth to the more sophisticated aliens in the work of Ray Bradbury (1920–). It has been said of Bradbury that he "brought the traditional image of Mars to a kind of impressionistic perfection" in *The Martian Chronicles* (1950), a collection of linked stories that had appeared in the science fiction magazines of the late 1940s. It has also been said that Bradbury's ancestors are not Verne and Wells, conditioned by a scientific world view, but Stapledon and Lord Dunsany.[58] Incorporat-

imagined" statement, see Peter Nicholls, "Lewis," in *SFE*, 716–717. Griffin, 139–140, discusses the possible models for Lewis's characters, including Haldane.

57 Haldane's review of Lewis's trilogy is in "Auld Hornie, F.R.S.," *Modern Quarterly* (Autumn 1946), Lewis's "A Reply to Professor Haldane" is reprinted in C. S. Lewis, *Of Other Worlds: Essays and Stories*, Walter Hooper, ed. (San Diego, Calif., 1966), 74–85.

58 Clifton Fadiman, in Prefatory Note to Bradbury, *The Martian Chronicles* (New York, 1970); Brian Stableford, "Mars," in *SFE,* 778. Fadiman's brief essay was written in

ing elements of poetry and fantasy, *The Martian Chronicles* tells of human colonists who undertake a wave of expeditions to a still-canaled Mars, only to come literally face to face with their own past. Instead of Martians, they inexplicably find their hometowns populated by their own dead relatives, brought back to life. But in reality they have found the Martians, who use not atomic weapons but "telepathy, hypnosis, memory and imagination" to first lure and then kill the invaders from Earth. It is pure Bradbury that the "hard" science and technology of minutely described spaceships and weapons are absent, replaced with memory and imagination. Eventually the Earthlings win out, for chicken pox destroys the Martians' already dying civilization. Continuing waves of crass, greedy settlers proceed to despoil the planet Mars, as they had the Earth, in contrast to the "graceful, beautiful and philosophical people" who had created the Martian civilization. Finally, Earth destroys itself in atomic war and the Martian settlers rush back to Earth. Only a single family, their sensibilities about what is important in life renewed, returns to begin civilization anew on Mars.[59]

Once again, the alien itself was not the main point of Bradbury's story. It was a method for reflecting on what terrestrial civilization had become. The Martians had found a balance between religion, art, and science, rather than letting science "crush the aesthetic and the beautiful." When the last family returns to Mars, the father tells his children, "Life on Earth never settled down to anything very good. Science ran too far ahead of us too quickly, and the people got lost in a mechanical wilderness, like children making over pretty things, gadgets, helicopters, rockets; emphasizing the wrong items, emphasizing machines instead of how to run the machines."[60] Bradbury used science fiction to criticize science and technology rather than to glorify it – the diametrical opposite of the goal of Wells and others. In this process an essential role was played by the alien, the detached and superior representative of life, even as it was dying, showing how civilization might develop along completely different principles.

If in the first half of the century science had come no closer to learning the truth about aliens, between Burroughs and Bradbury progress had

1958; the 1970 volume used here is the 21st printing of the 1958 Doubleday edition, giving some indication of the popularity of this book. On Bradbury see W. F. Nolan, *The Ray Bradbury Companion: A Life and Career History, Photolog, and Comprehensive Checklist of Writings* (1975), and its 1991 supplement by Donn Albright, *Bradbury Bits & Pieces: The Ray Bradbury Bibliography: 1974–1988*. Also see G. E. Slusser, *The Bradbury Chronicles* (San Bernardino, Calif., 1977); M. H. Greenberg and J. D. Olander, eds., *Ray Bradbury* (1980); and W. F. Toupence, *Ray Bradbury* (San Bernardino, Calif., 1989).

59 Bradbury, *The Martian Chronicles*, 46, 51, 54.

60 Ibid., 67, 179–180.

surely been made in defining the possibilities. Readers became familiar with the alien as a character in space opera, but more than that, aliens had helped humans to explore traditional themes from a new and less parochial perspective. One sees in cosmic and theological alien literature a pattern of search for a higher truth and wisdom, whether in Lindsay's Muspel, Stapledon's Star Maker, or Lewis's Oyarsa. The character of the alien could range from the ridiculous to the sublime; perhaps because of that very flexibility, its career would accelerate in the second half of the century as the alien reached new heights of recognition in popular culture.

5.3 THE ALIEN COMES OF AGE: CLARKE, ET, AND BEYOND

Two developments characterize the coming of age of the alien in the second half of the twentieth century. The further elaboration of old alien themes and the invention of new ones, of course, continued. But what was really new was the increasingly intimate relationship between science and science fiction, and the adaptation of these themes at an accelerating pace to the visually stunning and emotionally intense media of film and television. The visual media brought the alien one step closer to the hearts of the masses, while the marriage to science verified the alien theme as a plausible reality, not only because scientists themselves more frequently used fiction to speculate about alien contact, but also because alien science fiction influenced many who actually became involved in scientific programs to search for extraterrestrials. In other words, the role of scientist as author of alien literature in itself lent credibility (though not verification) to the subject that it had not possessed in the first half of the century.

In this section we examine how the alien in literature became increasingly wedded to science in several ways: first, in the Stapledonian tradition of cosmic perspective, as exemplified by Arthur C. Clarke; second, in the exploration of the nature of the alien in the tradition of the 1920s and 1930s pioneers, exemplified in the scientifically informed work of Hal Clement, Fred Hoyle, and Stanislaw Lem; and third, in the theme of contact with extraterrestrials, which now became more dominant as unidentified flying objects and radio searches for extraterrestrial intelligence brought the concept of contact to the fore. Even space opera was combined with serious scientific speculation about the relationship among intelligent species in the universe, as exemplified in the work of David Brin. Finally, we examine the importance of the alien in television and film. Along with the beginning of the Space Age, the visual media undoubtedly hastened science fiction's acceptance by the the masses beyond the wildest dreams of its pioneers in the first half of the century. Perhaps more than purely scientific progress, these developments in science fiction go a long way toward explaining the pervasiveness of belief in extraterrestrials in popular culture by the end of the twentieth century.[61]

The British author Arthur C. Clarke is prototypical of one whose preoccupation with the theme of alien encounter is used to place humanity in perspective while steeped in an adherence to the scientific tradition. At about the same time that Bradbury was discovering science fiction in the pulp magazines in America, Clarke (1917–) came to London as a civil servant in 1936. Here he worked as a radar officer in the Royal Air Force during World War II, predicted the development of communications satellites (1945), obtained a degree from Kings College (London) in physics and mathematics (1948), and served as president of the British Interplanetary Society (1946–1947, 1950–1953).[62] His nonfiction books *Interplanetary Flight: An Introduction to Astronautics* (1950), and *The Exploration of Space* (1951) made him well known as an advocate of space travel long before he was known for his science fiction; although already in the 1930s he had begun work on several versions of what would become his first novel, *Against the Fall of Night* did not appear until 1953. The most important influence on this work was none other than Olaf Stapledon's *Last and First Men,* which Clarke discovered in his local library shortly after its publication: "With its multimillion-year vistas, and its role call of great but doomed civilizations, the book produced an overwhelming impact upon me," Clarke recalled years later.[63]

Like Stapledon, his British successor thought in terms of eons. His first novel, expanded in 1956 as *The City and the Stars*, described the far future on Earth in which technology had kept the utopian city of Diaspar running smoothly, but completely isolated from the rest of the universe for a billion years. The story's protagonist, considering his civilization in a state of stagnation rather than utopia, travels across the galaxy, finds an alien

[61] This trend was sometimes reversed: science fiction authors occasionally wrote nonfiction books about extraterrestrial life. See Isaac Asimov, *Is Anyone There?* (New York, 1967), and *Extraterrestrial Civilizations* (see footnote 81); see also Poul Anderson, *Is There Life on Other Worlds?* (New York, 1963).

[62] On Clarke see Neil McAleer, *Odyssey: The Authorised Biography of Arthur C. Clarke* (Chicago, 1992); Clarke, *Ascent to Orbit: A Scientific Autobiography* (New York, 1984); Clarke, *Astounding Days: A Science Fictional Autobiography* (New York, 1989); Clarke, *The View from Serendip* (New York, 1977); and Thomas D. Clareson, "The Cosmic Loneliness of Arthur C. Clarke," in Thomas D. Clareson, ed., *Voices for the Future: Essays on Major Science Fiction Writers* (Bowling Green, Ohio, 1976).

[63] Arthur C. Clarke, Introduction to *The Lion of Comarre and Against the Fall of Night* (*AFN*) (New York, 1968). See also McAleer, 19. *AFN* was first published in *Startling Stories*, having been turned down twice by John Campbell. Before *AFN* was published as a novel, Clarke had written on the alien theme in "The Sentinel" (1948), "Encounter at Dawn" (1950), and "The Possessed" (1951). All three are published in Clarke's collection of stories *The Nine Billion Names of God* (New York, 1974).

intelligence, and initiates a cultural Renaissance. Although most of the novel is spent with the protagonist (with the unpretentious name of Alvin) describing the city and trying to determine how to leave it, his first glimpse of the stars is an epiphany that leads on in the final quarter of the novel to the stars. There he finds humanity's past – a past in which humanity had explored the Galaxy. In that age of exploration "everywhere he found cultures he could understand but could not match, and here and there he encountered minds which would soon have passed altogether beyond his comprehension." As he returns to Earth to brood about his fate, humanity began a great experiment of mind and genetics, spanning millions of years and involving other galactic civilizations, to build a pure mentality, a disembodied intelligence that would search for "a true picture of the Universe" unencumbered by physical limitations. This it did, at first resulting in disaster in a destructive Mad Mind and eventually culminating in a pure mentality that possessed knowledge, but not wisdom. This entity was left to roam the universe, while humanity, desiring peace and stability, retreated to Diaspar. It was this pure mentality that Alvin discovered, and from which the knowledge and inspiration is gained to once again strike out into the mysteries of the universe. Unlike Lasswitz, whose Earth found utopia only by keeping the Martians at a distance, Clarke found it only by seeking extraterrestrial civilizations, though in a much broader perspective than our own solar system. This search was to set a pattern in much of Clarke's science fiction, for Clarke, like his character Alvin, was an explorer. Indeed, in this regard the British author revealed the driving force behind his work when he noted in his novel that "all explorers are seeking something they have lost. It is seldom that they find it, and more seldom still that the attainment brings them greater happiness than the quest."[64]

Meanwhile, after a series of short stories dealing with the alien theme, in 1953 Clarke had published *Childhood's End*, which quickly became a classic of alien literature.[65] This time the Earth itself is visited by alien "Overlords" possessing a bodily form resembling the devil, who attempt to incorporate humanity into the scheme of universal sentience. The novel is set in the immediate future as humanity is about to journey to the Moon. The Overlords are benevolent aliens who impose an end to war and world government on the Earth. Amid some resistance, the first contact theme is played out: the eradication of poverty and ignorance is set against negative aspects that include a decline in the creative arts and in science brought on by the Overlords' vastly superior knowledge. Yet the Overlords, they reveal later, are only agents of an even more superior

64 Arthur C. Clarke, *The City and the Stars* (New York, 1956), 174–175, 178.
65 Arthur C. Clarke, *Childhood's End* (New York, 1953).

intelligence, and their ultimate goal is to usher the human race (by way of its children) into the Overmind, something the Overlords themselves cannot achieve. As in *The City and the Stars*, the concern of the novel is with the ultimate destiny of humanity, which, though different in detail, is still intimately connected to extraterrestrials. Both novels involve a religious vision: in *City*, the creation of the disembodied intelligence was recognized as "a conception common among many of Earth's ancient religious faiths," and it was noted as strange that an idea that "had no rational origin should finally become one of the greatest goals of science"; in *Childhood's End* the children give up their individual souls to unite with the universal soul that is the Overmind. Fifteen years after *Childhood's End*, Clarke would still be seeking an extraterrestrial destiny, via yet another route, in the mystical denoument of *2001: A Space Odyssey*.

Driven by a belief that extraterrestrials gave a true perspective on humanity, Clarke would make the alien theme a pervasive one throughout his career. "The idea that *we* are the only intelligent creatures in a cosmos of a hundred million galaxies is so preposterous that there are very few astronomers today who would take it seriously," he wrote in 1972. "It is safest to assume, therefore, that *They* are out there and to consider the manner in which this fact may impinge upon human society."[66] Although Clarke treated many themes in his long career, this real-life conviction about extraterrestrials was the dominant thread running through in his science fiction. By the early 1950s, this theme was not only an echo from Stapledon, but also an idea reinforced by science with its claims for abundant planetary systems.

While Clarke excelled in the Stapledonian search for meaning in the cosmic context, others explored the potential strangeness of the alien in ways that went beyond Lindsay, Weinbaum, and their contemporaries in the first half of the century. In the United States, Hal Clement (1922–), who obtained degrees in astronomy, chemistry, and education and worked as a high school teacher, set his highly original novel *Mission of Gravity* (1954) on a rapidly spinning, discus-shaped planet (Mesklin), whose gravity varied from 3 times (at the equator) to 700 times (at the pole) that of Earth. The quintessential "hard" science fiction writer, Clement elaborated his story around the natives who, having necessarily adapted to their world, were only 15 inches long and 2 inches in diameter and operated on a vastly different time scale than humans. When a human spaceship crash

[66] "When the Aliens Come," in Arthur C. Clarke, *Report on Planet Three and Other Speculations* (New York, 1972), 89–102. For further statements of Clarke's belief in extraterrestrials see "Life in Space," in Arthur C. Clarke, *The View from Serendip*, 143–156. This is followed by a reprinting of Clarke's review of two books about UFOs: "Last (?) Words on UFO's," 157–160.

lands on Mesklin, they undertake the dangerous journey to the poles to retrieve data required by the humans for an antigravity machine. In return, the Mesklinites desire knowledge, which they hope to gain from both the journey and the grateful humans. While Clement's plot is itself original, the characterization of the alien on a planet of extremely high gravity is an early example of how the scientific treatment of the alien theme could lead to the detailed construction of truly nonterrestrial intelligences.[67]

Clarke's British contemporary, the eminent astronomer Fred Hoyle (1915–), explored a wholly different kind of alien than Clement's, though with similarities to the disembodied intelligence of Clarke's *The City and the Stars*. Known as a brilliant maverick in the astronomical community, Hoyle was the chief champion of the steady-state theory of cosmology, which most consider today to be superseded by the big-bang theory. As we have seen in Chapter 4, in the early 1950s he was also among the first astronomers to champion abundant solar systems and the implied possibilities of life. This was undoubtedly the background to his first novel, *The Black Cloud* (1957), which demonstrates that while science could inform fiction, it need not keep a tether on imagination. The alien intelligence in this book takes on the form of a cloud of interstellar matter, some half billion years old, with which astronomers are eventually able to communicate. The cloud, which initially caused chaos and threatened destruction of the Earth, is convinced to retreat, showing the potentially benevolent character of intelligence. But not before Hoyle has the chance to explore the nature of both intelligence and communication. Planetary intelligence, the cloud reveals, is unusual because the gravitational force limits the size of its beings and the scope of their neurological activity, and the comparative lack of chemical food leads to the "tooth and claw existence" characteristic of Darwinian survival of the fittest. When radio communication is established with the cloud, the detailed comparisons of neurological structure between an intelligent cloud and a human are discussed, giving the sense of just how unique – and inferior – humanity might be. Significantly, Hoyle was a scientist for whom extraterrestrials played an important role in his worldview; he later wrote much more on extraterrestrial life, both in fiction and in nonfiction.[68]

67 Hal Clement, *Mission of Gravity* (New York, 1954), was followed by *Close to Critical* (New York, 1964) and *Starlight* (New York, 1971), set on the same world. Clement is a pseudonym for Harry Clement Stubbs. On Clement see Donald M. Hassler, *Hal Clement* (Mercer Island, Wash., 1982); Gordon Benson, Jr., *Hal Clement, Scientist with a Mission: A Working Bibliography* (San Bernardino, Calif., 1989); and *SFE*, 233–234. See also Clement's essay "Hard Sciences and Tough Technologies" in Reginald Bretnor, *The Craft of Science Fiction* (New York, 1976).

68 Fred Hoyle, *The Black Cloud* (New York, 1957). Most of the discussion of the nature of the cloud and communication with it are found in chapter 10. In addition, there is much

An even stranger exploration of alien nature emerged from the non-Anglo-Saxon world in the form of Stanislaw Lem's *Solaris* (1961), in which humans attempt to communicate with a living ocean on the planet Solaris. Lem (1921–), a Polish physician who wrote on the history and methodology of science (particularly cybernetics) and ran into trouble with Soviet Lysenkoism, began writing science fiction about 1950 as Bradbury's *Martian Chronicles* was being published in America. Although Lem was not affected by the Martian mania in the West, he was concerned with many of the same problems of human identity and purpose, and he made use of the human mind to conjure living beings in a way reminiscent of Bradbury.[69] But in *Solaris* he was able to play out these concerns in an alien setting unlike anything produced in the West. When the planet Solaris is found to contain an ocean that is in some sense alive, the space station Solaris hovers a few hundred meters above the surface of the ocean for the purpose of examination. Countless attempts in the past to establish contact had been frustrated; now much was learned about the ocean, but in the end scientists were convinced only "that they were confronted with a monstrous entity endowed with reason, a protoplasmic ocean-brain enveloping the entire planet and idling its time away in extravagant theoretical cognitation [*sic*] about the nature of the universe. Our instruments had intercepted minute random fragments of a prodigious and everlasting monologue unfolding in the depths of this colossal brain, which was inevitably beyond our understanding."[70]

Communication of a sort finally emerges from the scientists' subconscious minds as the ocean synthesizes out of each scientist's past a "Phantom" living person, constructed of neutrinos. There are parables here

of interest regarding the political reaction to such a discovery. On Hoyle see *SFE*, 590–591; Hoyle, *The Small World of Fred Hoyle: An Autobiography* (London, 1986); and Hoyle, *Home Is Where the Wind Blows: Chapters from a Cosmologist's Life* (Mill Valley, Calif., 1994). His subsequent novels with an alien theme include *Ossian's Ride* (1959) and (with John Elliot) *A for Andromeda* (New York, 1962). On Hoyle's controversial theory of the origin of life from outer space, developed in the late 1970s and 1980s, see chapter 7.

[69] On Lem see his "Reflections on My Life" in *Microworlds: Writings on Science Fiction and Fantasy*, ed. Franz Rottensteiner (San Diego, Calif., 1984), 1–30, reprinted from *The New Yorker*, January 30, 1984; Darko Suvin, "Lem," in *SFE*, 710–712; Suvin, "The Open-Ended Parables of Stanislaw Lem and *Solaris*," Afterword to the English translation of *Solaris* (New York, 1970). See also the special issue of *Science Fiction Studies*, 13, pt. 3 (1986). Lem had read much of the science fiction in the West in the 1950s, including the works of Bradbury, Asimov, Clarke, and Campbell, according to Suvin, *Solaris*, 214–215.

[70] Lem, *Solaris*, 28. Lem's fictional elaboration was based to some extent on his nonfiction writings. In particular, Suvin points out that in his *Summa Technologiae* (1964), Lem had defined Reason or Intelligence as a "second-degree homeostatic regulator able to counteract the perturbations of its environment by action based on historically acquired knowledge," as the Ocean Solaris, "Open-Ended Parables," 218.

about human communication and the meaning of life, but with regard to alien life one important question raised (rarely in other science fiction) is whether the very nature of the alien will prevent human communication with it: Lem portrays the attempts at contact as being "like wandering about in a library where all the books are written in an indecipherable language. The only thing that's familiar is the color of the bindings!" Beyond the issue of comprehensibility, there is the question (also submerged in most science fiction) of whether contact *should* be made. "We think of ourselves as the Knights of the Holy Contact," Lem has one his characters say. "This is another lie. We are only seeking Man. We have no need of other worlds. . . . We are searching for an ideal image of our own world: we go in quest of a planet, of a civilization superior to our own but developed on the basis of a prototype of our primeval past." And again, "Solaristics is the space era's equivalent of religion: faith disguised as science. Contact, the stated aim of Solaristics, is no less vague and obscure than the communion of the saints, or the second coming of the Messiah. Exploration is a liturgy using the language of methodology; the drudgery of the Solarists is carried out only in the expectation of fulfillment, of an Annunciation, for there are not and cannot be any bridges between Solaris and Earth." While the ultimate purpose of Lem's novel is to use the cosmos to learn about humans, it may also be read at a different level as an argument against attempting contact before humans understand themselves: "Man has gone out to explore other worlds and other civilizations without having explored his own labyrinth of dark passages and secret chambers, and without finding what lies behind doorways that he himself has sealed."[71] Unlike the search for meaning of Lindsay, Stapledon, and Clarke, Lem's search failed, imparting the message that our fate may lie not in the stars but in ourselves. Lem's novels, including *Solaris* (the first to be translated into English), were widely read in both the East and West, and *Solaris* was filmed in 1971.[72]

Despite Lem's warnings of alien incomprehensibility and human infancy, numerous science fiction novels were inspired by real-life efforts to

[71] Ibid., 167, 180, 165.

[72] At least some Russian science fiction writers seem to have taken Lem's advice to "know thyself" before attempting to understand aliens, though this does not prevent them from using aliens to do so. This is the case in the novel *Roadside Picnic* (1972; English translation by Antonina Bouis, New York, 1977) by perhaps the most famous Russian science fiction writers, the brothers Arkady and Boris Strugatsky. Patrick L. McGuire, "Russian SF," in Barron, *Anatomy of Wonder*, 441–473, comments (444) that in the decades following the Russian Revolution in 1917, "Soviet space stories might depict human flights into space or, less frequently, visits by extraterrestrials to Earth, but invaders from other planets, so common in American SF of the period, seem to have been virtually unknown." This is an interesting commentary on the cultural and political effects on literature.

detect extraterrestrial intelligence by radio telescopes, beginning with Frank Drake's Project Ozma in 1960.[73] Although already explored in fiction by Hoyle's *A for Andromeda* (1962) and Harrison Brown's *The Cassiopeia Affair* (1968), the theme found perhaps its classic expression in James Gunn's *The Listeners* (1972).[74] In Gunn's novel, the frustrations of a long search are vividly portrayed against a backdrop of "computer runs" which relate the history of the idea of other worlds. Gunn treats not only the institutional workings and frustrations, but also the possible implications of contact, with his portrayal of Jeremiah, the leader of a religious cult that denies the existence of extraterrestrial intelligence.[75]

All of these themes – cosmic perspective in the context of extraterrestrials, original explorations of the nature of the alien, and contact with extraterrestrial intelligence – would continue to be elaborated in ever more subtle form throughout the century. In the tradition of the cosmic perspective, subsequent authors did not often reach the metaphysical heights of Stapledon and Clarke (except for Clarke himself), but there was ample scope for examining narrower themes in the sense of C. S. Lewis's focus on Christianity. In this line, James Blish's *A Case of Conscience* (1958) portrayed a Jesuit priest and biologist who must deal with the implications of the discovery of intelligent reptilian inhabitants on the planet Lithia.[76] Imaginative portrayals of the alien in the tradition of Clement also continued, notably with Robert Forward, whose *Dragon's Egg* (1980) and *Starquake!* (1985) described life on a neutron star, an extremely dense object where a generation passes in 37 minutes.[77] David Brin intelligently explored even the idea of life in the Sun in *Sundiver* (1980). And the idea of radio contact was explored further by one of the pioneers in the field, Carl Sagan, whose novel *Contact* (1985) is grounded

[73] The problem of contact with other civilizations was inherent in many previous stories. In particular, Murray Leinster's "First Contact" (1945) summed up the major problem: the aliens may be good or evil, but since we cannot trust them, we must take an attitude of hostility. "First Contact" has been reprinted many times, including in the anthology *Contact*, ed. Noel Keyes (New York, 1963), 11–37. Dissenting from Leinster's view was the Russian paleontologist Ivan Yefremov's "Cor Serpentis" (Heart of the Serpent, 1959), English trans. in *More Soviet Science Fiction* (New York, 1962).

[74] James E. Gunn, *The Listeners* (New York, 1972). As we saw in Chapter 4, in 1964 Brown had argued for numerous planetary systems. Among other treatments of contact are Lem, *His Master's Voice* (San Diego, Calif., 1969).

[75] Alien contact by means other than radio signals was also explored. An abandoned spaceship entering the solar system is explored in Clarke's final alien novel, *Rendevous with Rama* (New York, 1973) and its sequels. A Bracewell probe whose purpose is to search for life is explored in Michael McCollum's *LifeProbe* (New York, 1983).

[76] James Blish, *A Case of Conscience* (New York, 1958). Blish graduated from Rutgers in microbiology and from Columbia in zoology.

[77] The Forward cites Frank Drake, "Life on a Neutron Star," *Astronomy*, 1, no. 5 (December 1973), 4–8.

in a scientific search that discovers, as in Arthur C. Clarke's novels, that humanity's true destiny is among the stars.[78]

At the same time, a more sophisticated space opera in the vein of Burroughs has never lost its charm or its readership. In this tradition are David Brin's *Startide Rising* (1983) and *The Uplift War* (1987). The continued acceptance of this tradition in the field of science fiction is evidenced by the fact that these novels captured the field's highest honor, the Hugo Award.

Although we have seen only the tip of the alien iceberg by examining some of the classics of alien literature,[79] it is equally important to note that the alien was not a major theme for every writer of science fiction. The most notable example in the United States is Isaac Asimov (1920–1992). Though Asimov is often compared to Arthur C. Clarke for popularity and for other reasons, the contrast in their attitude toward extraterrestrials is striking: both the universe of Asimov's Foundation series and that of his robot series (eventually related to each other) are devoid of aliens. Similarly, though Asimov had featured aliens in a few of his early short stories, his other mainstream novels also lack aliens.[80] Only a relatively late work, *The Gods Themselves* (1972), is notable for its alien beings. Asimov recalled, "I rarely had sex in my stories and I rarely had extraterrestrial creatures in them, either, and I knew there were not lacking those who thought that I did not include them because I lacked the imagination for it. I determined, therefore, to work up the best extraterrestrials that had ever been seen for the second part of my novel." The result, in one critic's words, was "among the most fascinating and believable aliens yet imagined in science fiction," but this does not lessen the fact that for even some of the most widely read science fiction authors, extraterrestrials were not an essential part of their world view or a criterion for public acceptance.[81]

[78] On a civilization revitalized by a message from the stars, see also M. P. Kube-McDowell, *Emprise* (New York, 1985) and its sequel *Enigma* (New York, 1986). For related novels, see Barron, *Anatomy of Wonder.*

[79] Some indication of the pervasiveness of the alien theme in the second half of the twentieth century is given by the anthologies of stories dealing with the alien theme, beginning at least as early as Leo Marrgulies and Oscar J. Friend, ed., *From Off This World* (1949), and continuing at an accelerated pace to the present.

[80] Two early short stories using aliens are "Blind Alley," *Astounding* (March 1945), and "No Connection," *Astounding* (June 1948), both found in *The Early Asimov* (2 vols., Garden City, N.Y., 1972). Much has been written about Asimov and his work, but the best source remains his own autobiography, *In Memory Yet Green* (New York, 1979) and *In Joy Still Felt* (New York, 1980).

[81] James Gunn, *Isaac Asimov: The Foundations of Science Fiction* (Oxford, 1982), 198. Gunn discusses the novel in chapter 7. For Asimov's attitude toward extraterrestrials see Asimov, *Extraterrestrial Civilizations* (New York, 1979). Asimov had also tackled the theme of extraterrestrial life in a series of nonfiction articles in the early 1960s, including

Moreover, it is well to remember that the alien theme was, of course, only a small part of the total scope of science fiction. Themes of the relation of humanity to the universe could be carried out without the alien, as could the other themes discussed earlier. Furthermore, one could argue that "New Wave" science fiction beginning in the 1960s, by eschewing traditional themes, also tended to avoid the well-worn subject of the alien more than conventional science fiction. Nevertheless, the amount and scope of alien literature in the twentieth century are truly impressive, as is its appeal to the public. Nowhere is this appeal more evident than in the success of science fiction in the visual arts, a success that seemed to have no bounds in the second half of the century.

For all of its imagination and innovation, science fiction – and the alien with it – would have remained the province of a limited readership had it not been for the expansion of the genre to the visual media. The alien boom in cinema had begun almost exactly at the turn of the half century, for reasons that probably have little to do with science and more to do with the onset of the cold war and the paranoia about communists.[82] Whatever the reason, cinema seems to some extent to have repeated the evolution of the alien in science fiction literature, from the unsophisticated aliens met in the early 1950s to the maturity of the mysterious, existential, and yet unseen aliens in Arthur C. Clarke and Stanley Kubrick's *2001: A Space Odyssey*.[83]

The alien boom in cinema is generally considered to have begun with *The Thing* (1951), based on John W. Campbell's "Who Goes There?" (discussed earlier). Called "a superior example of sf cinema" and "by far the most influential of the films that sparked off the sf/monster movie

two in The *New York Times Magazine* (November 29, 1964, and May 23, 1965). All four are reprinted in Isaac Asimov, *Is Anyone There?* (New York, 1967), 189–216.

[82] As suggested by Peter Nicholls, *SFE* (1979), 116. On the alien in cinema see Vivian Sobchak, *The Limits of Infinity: The American Science Fiction Film 1950–1975* (South Brunswick, N.J., 1980), expanded as *Screening Space: The American Science Fiction Film* (New York, 1987). Also, in a less academic mode, see John Brosnan, *Future Tense, the Cinema of Science Fiction* (New York, 1978), and Philip Strick, *Science Fiction Movies* (London, 1976). Philippe Hardy, ed., *The Aurum Film Encyclopedia: Science Fiction* (1991) is a comprehensive list, description, and analysis of science fiction cinema, while Clute and Nicholls, *SFE*, has many relevant entries, including Nicholls's "Cinema," 218–226. For more references see Neil Barron, "Science Fiction on Film and Television," in Barron, *Anatomy of Wonder*, 672–686.

[83] There were, of course, even cruder (though still influential) precursors – the serial films *Flash Gordon* (1936) and *Buck Rogers* (1939), both emerging from comic strips and later adapted for television. On the evolution of these series in comic strips, film, and television beginning in the late 1920s, see Gunn, ed., *The New Encyclopedia of Science Fiction*, 72–73 and 174–176. For a chronological listing of science fiction on television see Brosnan, *Future Tense*, 291–304.

boom of the 1950s," it featured James Arness as a humanoid vegetable who arrives on an unidentified flying object (UFO) and terrorizes an Arctic base. In the same year appeared the morally more sophisticated film *The Day the Earth Stood Still,* in which the alien Klaatu and his robot attempt to stem human violence. The year 1953 saw several notable alien films, including a vastly altered *War of the Worlds*, now set in contemporary California instead of 1890s England. *It Came from Outer Space* (1953), based on "The Meteor" by Ray Bradbury, portrayed aliens whose spaceship crash lands, who are able to change shape, and who duplicate humans to assist in the repair of their spaceship. *This Island Earth* (1955) featured a scientist who is taken to another planet in the hope that he can help keep its atomic shield functioning against attacks from enemies – a clear analogy to the cold war in full swing on Earth. Although not generally considered a great film, it has been called "an excellent bad film." With *Invasion of the Body Snatchers* (1956), which has been labeled "a subtle and sophisticated movie . . . possibly the most discussed B-grade movie in the history of American film"; *Earth vs the Flying Saucers*, which capitalized on the UFO mania spreading across the world and climaxed with a battle over Washington, D.C.; and, in the same year, *Forbidden Planet,* with its aliens both alive and vanished, the alien was firmly entrenched in the cinema.[84] The dawn of the space age the following year could only heighten the interest – and the potential realism – of such movies.

The 1960s saw the alien in film begin to mature in quite remarkable fashion. While alien films continued, the *Star Trek* television series, which ran initially from 1966 to 1969, became something of a popular culture phenomenon. *Star Trek,* which reminded its viewers at the beginning of each episode that its mission was "to seek out new life and new civilizations," carried out many of its alien episodes in imaginative and memorable fashion. Created by Gene Roddenberry, the television series was so popular that it spawned a following of "Trekkies," some of whom were undoubtedly fascinated with space travel in general, but others of whom were much taken with the treatment of the alien theme. It also spawned a new television series, *Star Trek: The Next Generation* (1987), seven feature films from 1979 to 1994, and well-attended Star Trek conventions.[85]

Perhaps even more significant for the alien theme during the tumultu-

[84] See the appropriate *SFE* entries for these films, as well as the entries on "Cinema" and "Invasion." These films were often remade, sometimes better and more faithful to the original, as in *The Thing* (1982) and *Invasion of the Body Snatchers* (1978), sometimes as a television series, as in *War of the Worlds* (1988–1990), and sometimes followed by novelization rather than preceded by it, as in *Forbidden Planet.*

[85] On the history of *Star Trek* and its spinoffs see Allan Asherman, *The Star Trek Compendium* (New York, 1986). For a briefer summary see *SFE,* 1156–1158.

ous decade of the 1960s, which witnessed assassinations on the one hand and the Apollo Moon landing on the other, Arthur C. Clarke's vision of cosmic perspective reached the cinema with *2001: A Space Odyssey* (1968), making spectacular but restrained use of the alien theme. Based on his short story "The Sentinel," written in 1948, his mystical treatment, in the hands of producer/director Stanley Kubrick, raised the science fiction film to a new level. Five years and 10 million dollars in the making, it produced mixed reviews but quickly became a classic discussed by many, if not understood by all.[86] It is significant that what made *2001* so popular and influential was not the book but the movie; only after the success of the movie did Clarke become the most popular science fiction writer in the world. Although the sequel novels and film, *2010: Odyssey Two* (1982; film, 1984) and *2061* (1988), were less successful, the impact of *2001* itself would be felt throughout the century.

During the 1970s, two spectacularly successful movies featured aliens. In *Star Wars* (1977) the variegated aliens were part of space adventure, expanded from Burroughs's Mars to galactic space, as others had often done before in science fiction, but which few had accomplished on film. As the freedom fighters tackled the evil Galactic Empire, aliens played a less than cerebral role. The continuing popular success of its sequels, *The Empire Strikes Back* (1980) and *The Return of the Jedi* (1983), with their renderings of Yoda, Ewoks, and Jabba the Hutt, demonstrate that space opera was no less popular at the end of the century than when Burroughs invented the genre.

The other film, *Close Encounters of the Third Kind* (1977), played on a more down-to-Earth subject, the continuing fascination of the public (if not scientists) with UFOs. Directed by Steven Spielberg, the movie took its name from the categorization of UFOs by astronomer J. Allen Hynek (who made a cameo appearance), the "third kind" being physical contact (see Chapter 6). The movie vividly evoked both the anxiety and the sense of wonder that extraterrestrials can generate, culminating in the final scene, in which humans and aliens communicate by means of musical tones.[87]

[86] For the reviews, ranging from The *New York Times* to *The New Republic*, see Jerome Agel, ed., *The Making of Kubrick's 2001* (New York, 1970), 207–284. Because Kubrick had intended to open the film with opinions about extraterrestrial life, interviews were undertaken with experts from a variety of backgrounds. Though these interviews were never used in the movie, transcripts of them are found in this volume on pages 27–57. Arthur C. Clarke, *The Lost Worlds of 2001* (New York, 1972), is Clarke's account of his collaboration with Kubrick, as well as transcripts of various versions of the script. The novel *2001: A Space Odyssey* (New York, 1968) was written for the movie, rather than the other way around.

[87] The novelization is S. Spielberg, *Close Encounters of the Third Kind* (New York, 1977).

Finally, to come full circle from our discussion of the works of Lasswitz and Wells in 1897, the last quarter of the century saw cinema explore the two opposite extremes of alien nature: the good and the evil. *Alien* (1979), followed by *Aliens* (1986) and *Alien III* (1992), evoked a vivid horror as the human visitors to a planet are attacked by alien creatures, one of which reproduces inside a crewman's body and escapes to terrorize the crew. By stark contrast, *ET: The Extraterrestrial* (1982), depicted a lovable alien left behind when his spacecraft hurriedly departed. Befriended by children even as he is hunted by suspicious government scientists, ET learns some of Earth's customs through his friends but becomes ill, dies, is resurrected, and, with the help of the children, escapes to the spaceship that returns for him. The fact that *ET* became the most commercially successful science fiction movie of the century until *Jurassic Park* a decade later demonstrated that the public preferred benevolent aliens to horrific ones; the *Alien* series, although popular, did not approach the success of *ET*. Popularity, however, had nothing to do with reality, and the fact remained that while *ET* and *Alien* starkly portrayed alternative alien natures, scientists had no idea which form might actually predominate.

By the end of the century the alien, barely invented 100 years before, had come to assume a central role in popular culture and scientific imagination.[88] We may trace, during the course of that century, several families of alien literature, drawing on a variety of traditions. The monstrous alien, beginning with Wells, extending through John W. Campbell and on to *Aliens*, is certainly one of the strongest and most vividly portrayed in film. The swashbuckling alien adventures prototyped by Burroughs extend through numerous authors in a direct line to *Star Wars*. More difficult to portray in film was the use of the alien in cosmic perspective, from the spiritual journey of Lindsay to the vast expanse of space and time in Stapledon, and thence from Lewis's limited Christian journey to Clarke's more cosmic vision; nevertheless, *2001: A Space Odyssey* succeeded in capturing on film even this alien genre. Finally, the hope for utopia via extraterrestrials, so optimistically treated in Lasswitz at the end of the nineteenth century, still found its adherents in authors like Clarke, a product of the British empire, but also drew in pessimists like Lem, who had experienced firsthand the result of the Russian empire.

During the century, science and science fiction increasingly complemented each other: speculative science fiction provided the perfect outlet

[88] Wayne Barlowe, Ian Summers, and Beth Meacham, *Barlowe's Guide to Extraterrestrials* (New York, 1979) is an entertaining book that visualizes the aliens in some of the classic science fiction works.

for scientists who wished to go beyond science.[89] Not only did scientists exercise their imaginations in science fiction, science fiction also inspired them to tackle questions in the real world.[90] Many of the pioneers in exobiology and SETI grew up on science fiction and were led to their careers by its imaginative lure. Having nurtured science fiction, science now received in return some of the rewards of imagination. This fact alone gives science fiction a prominent place in the twentieth century's fascination with extraterrestrial life.

A final truth must also be recognized. Despite all the advances of the twentieth century in uncovering the secrets of astronomy, terrestrial biology, and evolution, science can as yet add nothing to the question of the physical, mental, and moral nature of intelligence beyond the Earth. At best, science may shed a pale light on the question of the possible physical forms of the alien, but it can say nothing about its mental evolution – much less about whether good or evil or some compromise of the two rules such intelligences as might exist in the universe. For that, the speculations of science fiction – from the monstrous form conjured in the work of Wells, Campbell, and *Alien* to the enlightened beings of Lasswitz, Lindsay, and *ET* – are as valid as anything science can suggest. With all its science, this most crucial question of all remains for the twentieth century in the realm of the imagination.

89 This is explicitly acknowledged, for example, in Robert L. Forward, "About the Author," *Dragon's Egg* (New York, 1980), 309.

90 On the influence of science fiction on SETI scientists, see David Swift, *SETI Pioneers* (Tucson, Ariz., 1990), index under "science fiction."

6

THE UFO CONTROVERSY AND THE EXTRATERRESTRIAL HYPOTHESIS

Horatio: *O day and night, but this is wondrous strange!*
Hamlet: *And therefore as a stranger give it welcome.*
There are more things in heaven and earth, Horatio,
Than are dreamt of in your philosophy.

Shakespeare[1]

What we have here is a signal-to-noise ratio problem: There is indeed a fantastic amount of noise, represented by the many misidentifications of familiar objects seen under unusual or surprising circumstances – balloons, birds, satellites, meteors, aircraft, stars – yet, in all scientific honesty, one is led to ask whether there might not indeed be a signal somewhere in the noise.

J. Allen Hynek[2]

How often have I said to you that when you have eliminated the impossible, whatever remains, however improbable, must be the truth.

Donald Menzel quoting Sherlock Holmes[3]

No one would have believed in the last years of the nineteeth century, even as H. G. Wells was writing his *War of the Worlds*, that the twentieth century would witness on a sporadic but large scale a strange new phenomenon in the skies of planet Earth; that this phenomenon, eventually known as "flying saucers" or "unidentified flying objects (UFOs)," would become the subject of raging controversy among scientists and the public alike; and that, for better or worse, this controversy would become intimately associated with the debate over the existence of extraterrestrial intelligence. That this is precisely what happened is surely one of the most remarkable tales of the twentieth century. More remarkably still, as one of the principal scientists involved would point out, the UFO controversy was unlike other novel scientific ideas such as evolution, in which scientists tried to impose new ideas on the masses. Rather, like the nineteenth-century report of stones falling from the sky (now known as meteorites), it was the masses who tried to impose a novel idea on a mostly incredu-

[1] *Hamlet,* Act 1, scene 5, 164–167.

[2] J. Allen Hynek, testimony before U.S. Congress, House Committee on Armed Services, *Unidentified Flying Objects: Hearing by Committee on Armed Services of the House of Representatives,* 89th Congress, 2d session, April 5, 1966, 5991–6075: 6007.

[3] Donald Menzel, *Flying Saucers* (Cambridge, 1953), 51, citing Sir Arthur Conan Doyle, "The Sign of the Four," as part of Menzel's approach to analyzing UFOs.

lous community of scientists.[4] Between public gullibility and scientific close-mindedness, between perception and reality, lies an important chapter in the history of the extraterrestrial life debate and a story of the limits of science under the most trying circumstances.

UFOs are the subject of an enormous literature, some of it of more interest to the social sciences of psychology and sociology than to physical science. It is not our purpose to review this literature or to analyze whether UFOs really exist.[5] Nor is it our intention to write the history of the UFO debate.[6] Rather, we wish to examine the interaction of the UFO controversy with the extraterrestrial life debate. This immediately narrows our focus to what historically is known as the "extraterrestrial hypothesis" of UFOs – the idea that this mysterious aerial phenomenon represents spaceships from alien civilizations. Even here, we are not interested for present purposes so much in the details of specific cases as in the scientific and public reaction to the extraterrestrial hypothesis; how this reaction reflects on the scientific and public attitudes toward the existence of life beyond the Earth; and the nature of the arguments scientists brought to bear in reaching their conclusions. Although UFOs have been a worldwide phenomenon, for practical purposes of available documentation we focus chiefly on the United States, which has been the major center of UFO activity and study. In this chapter, then, we examine the rise of the extraterrestrial hypothesis beginning in 1947, how it peaked in scientific attention in the late 1960s with activity centered on the U.S. Air Force–sponsored "Condon Report," and the subsequent decline in the scientific community of extraterrestrials as a serious hypothesis for the elusive objects seen in the sky.

6.1 THE RISE OF THE EXTRATERRESTRIAL HYPOTHESIS

The rise of the extraterrestrial hypothesis of UFOs to 1965 is characterized by a central role for the media, a schizophrenic attitude by the U.S. Air Force, and (with two outstanding exceptions) the lack of substantial scien-

[4] J. Allen Hynek, in Foreword to David M. Jacobs, *The UFO Controversy in America* (Bloomington, Indiana, 1975), x–xi.

[5] Bibliographies of UFO literature include Lynn Catoe, *UFOs and Related Subjects: An Annotated Bibliography* (Washington, D.C., 1969), compiled by the Library of Congress for the U.S. Air Force in support of the Condon study (see footnote 77), and Richard Michael Rasmussen, *The UFO Literature: A Comprehensive Annotated Bibliography of Works in English* (Jefferson, N.C., and London, 1985);

[6] A detailed history of the UFO debate, largely limited to the United States, is Jacobs, *The UFO Controversy in America*. As the present volume went to press, Curtis Peebles, *Watch the Skies! Chronicle of the Flying Saucer Myth* (Washington, D.C., and London, 1994), was published.

tific participation. While much of the media – with their usual mixed motives of truth and profit – continually pushed the extraterrestrial hypothesis on a receptive public, and while the Air Force was understandably preoccupied with national security aspects of the issue, scientists in many ways abdicated their role as critical analyzers of an unexplained phenomenon. In part this was due to the reluctance of scientists to engage in a controversial issue that would do nothing to advance their careers and might do their reputations great harm. In part the fleeting nature of the phenomenon and the concomitant lack of hard data did not lend itself to easy scientific analysis, at least not by way of the normal scientific methodology. And perhaps most significantly, the concept of extraterrestrial intelligence – the favored explanation for UFOs among the media and the public – was not yet a part of the collective scientific consciousness, as it certainly would be in later years. For all these reasons, the UFO controversy to 1965 assumes the surreal character of mostly nonscientific individuals and organizations thrashing about in the midst of a potentially significant phenomenon, resulting in sporadic reaction rather than systematic study. And yet, for all that, it was unique in eliciting public and scientific attitudes toward extraterrestrial life that otherwise would never have been expressed.

One of the peculiarities of the UFO phenomenon is its wavelike character in terms of the numbers of sightings, at least in the United States. As documented by historian David Jacobs, the first wave of "mystery airships" came in 1896–1897 – before the Wright Brothers succeeded in heavier-than-air flight – followed by a half-century lull until the beginning of the modern debate in 1947, and then succeeding waves in 1952, 1965–1967, and 1973. This wavelike character of the phenomenon in the United States, against the steady background of yearly reports, may be seen in Table 6.1 for the years 1950–1968. In what follows, we shall be alert to any correlation of these peaks with concurrent developments, especially in science.

It is notable that the idea – one hesitates to use the more formal word "hypothesis" at this early stage – that the UFO phenomenon might be due to extraterrestrial spaceships was mentioned even in the first wave. The 1896–1897 sightings, after all, appeared to be airships, and although the favored explanation was secret experimental aircraft controlled by individuals or governments, the other alternative was that the supposed aircraft came from beyond the Earth. The few astronomers who offered an opinion agreed for the most part that there was a more mundane astronomical explanation, such as the planet Venus or a bright star; the idea of extraterrestrial spaceships, possibly from Mars, came from the popular side rather than the scientific. Even Percival Lowell, who had

Table 6.1. *Number of UFO reports received each month by Project Blue Book, 1950-68**

	J	F	M	A	M	J	J	A	S	O	N	D	Total
1950	15	13	41	17	8	9	21	21	19	17	14	15	210
1951	25	18	13	6	5	6	10	18	16	24	16	12	169
1952	15	17	23	82	79	148	536	326	124	61	50	42	1501
1953	67	91	70	24	25	32	41	35	22	37	35	29	509
1954	36	20	34	34	34	51	60	43	48	51	46	30	487
1955	30	34	41	33	54	48	63	68	57	55	32	25	545
1956	43	46	44	39	46	43	72	123	71	53	56	34	670
1957	27	29	39	39	38	35	70	70	59	103	361	136	1006
1958	61	41	47	57	40	36	63	84	65	53	33	37	627
1959	34	33	34	26	29	34	40	37	40	47	26	10	390
1960	23	23	25	39	40	44	59	60	106	54	33	51	557
1961	47	61	49	31	60	45	71	63	62	41	40	21	591
1962	26	24	21	48	44	36	65	52	57	44	34	23	474
1963	17	17	30	26	23	64	43	52	43	39	22	22	399
1964	19	26	20	43	83	42	110	85	41	26	51	15	562
1965	45	35	43	36	41	33	135	262	104	70	55	28	887
1966	38	18	158	143	99	92	93	104	67	126	82	40	1060
1967	81	115	165	112	63	77	75	44	69	58	54	24	937
1968	18	20	38	34	12	25	52	41	29				

* Sum of reports received from Air Force bases and those received directly from the public.
Source: E. U. Condon, *Scientific Study of Unidentified Flying Objects* (New York, 1969), Section V, Table I.

originated his theory of Martian canals in 1894 and was just beginning his long battle for acceptance of that idea, does not appear to have offered an extraterrestrial hypothesis for UFOs. But that did not prevent the press from doing so, probably under the influence of Lowell's writings. The *St. Louis Post Dispatch*, suggesting that "these may be visitors from Mars, fearful, at the last, of invading the planet they have been seeking," suggested that a message of peace and an invitation to land be sent, though it did not say how. Both the *Houston Post* and the *Washington Times* suggested links of the phenomenon to visitors from the planet Mars, while a Tennessee paper rejected the extraterrestrial hypothesis on the grounds that anything like human life could not be sustained during the voyage.[7] But like the 1896–1897 wave itself, the extraterrestrial hypothesis for UFOs faded into history for a half-century. Although the reasons for this have never been analyzed, it is remarkable that while the idea of Martian intelligence rose to full fury until 1910, the reports of UFO-like phenomena were confined to a 2-year period at the end of the nineteenth century. Because what did not happen in history may often be as important as what did, this clue deserves more attention in any study of the UFO phenomenon. Lowell's ideas of Martians offered the perfect scientific excuse for a continued UFO wave, and yet the public, while enamored of Lowell's vision, did not transport the Martians to terrestrial skies – at least not until the 1938 Orson Welles broadcast. And Lowell's silence indicates that even he placed limits on what was considered good scientific evidence for extraterrestrials.

Whatever the reason, the modern era of UFOs began only in 1947, and it was not long before the extraterrestrial hypothesis was put forth as a possible explanation. This time it would be the opening shot of a sustained hypothesis. Although mysterious sightings had been sporadically reported earlier,[8] on June 24, 1947, Kenneth Arnold, flying his private plane near Mount Rainier in Washington State, reported nine disc-shaped objects flying in formation at speeds he estimated to be over 1000 miles per hour. Arnold, a reputable businessman and deputy U.S. marshal, was taken seriously, and his description of the objects as flying "like a saucer if you skipped it across the water" led the newspapers to coin the term "flying saucer."[9] His report precipitated more than 850 sightings during the year from others. A variety of explanations were offered, but according to a Gallup poll taken at the time, very few people immediately sought an extraterrestrial explanation. Although within months 90 percent of the population had heard of flying saucers, most thought they

[7] Jacobs, *The UFO Controversy in America,* 29, citing the *St Louis Post Dispatch*, April 11, 1897, 4, and April 14, 1897, 7; the *Houston Post*, April 22, 1897, 9; the *Washington Times* and the *Memphis Commercial Appeal,* cited in the *Cincinnati Commercial-Tribune*, April 22, 1897, 4. It is notable that H. G. Wells's *War of the Worlds* began serial publication in England in April of this year, and in the United States in *Cosmopolitan.*

[8] For examples see Charles Fort (1874–1932), *The Book of the Damned* (New York, 1919), later referred to by Harvard astronomer Donald Menzel, as "the best reference to ancient saucers."

[9] Arnold's detailed descriptions of the sighting were published in his 16-page booklet "The Flying Saucer as I Saw It," (1950) and in a more substantial book with Ray Palmer *The Coming of the Saucers* (Amherst, Wisc., 1952).

were illusions, hoaxes, secret weapons, or other earthly phenomena.[10] Once again, it was not the public that seemed anxious to court extraterrestrials, and although astronomers were just beginning to hint that life might be common among the distant stars, they were certainly not prepared to associate this idea with a UFO phenomenon which implied that such intelligence had actually reached the Earth. The media, on the other hand, were more than willing to exploit extraterrestrials. In particular, it is of interest that Arnold's book *The Coming of the Saucers* (1952) was coauthored and published by Ray Palmer (1910–1977), the managing editor of the science fiction magazine *Amazing Stories*, who was promoting the extraterrestrial hypothesis even before the Arnold sightings. In the July 1946 issue of his magazine he had written, "If you don't think space ships visit the earth regularly . . . then the files of Charles Fort and your editor's own files are something you should see. . . . And if you think responsible parties in world governments are ignorant of the fact of space ships visiting earth, you just don't think the way we do."[11] Although Palmer's magazine reached only a specialized group, we see in his hostile attitude toward government and the inability or unwillingness to separate science fiction from science fact characteristics that would mark the extraterrestrial hypothesis of UFOs for many years. And the increasing interest in science fiction, which heavily emphasized extraterrestrials (see Figure 6.1 and the previous chapter), should not be underestimated as a contributor in itself to the rise of the extraterrestrial hypothesis.

The media and science fiction magazines were one thing, but ironically in view of subsequent history, it was only when the U.S. Air Force decided to investigate the flying saucer reports that the extraterrestrial hypothesis was recognized at an official level. During 1947 the Air Force, charged with the security of the skies for the United States, collected 147 flying saucer reports at its Technical Intelligence Division of the Air Materiel Command at Wright Field in Dayton, Ohio. On December 30, 1947, the order was given to begin a project to study the phenomenon, and an incident on January 7, 1948, reinforced the propriety of Air Force participation. On that date a large number of people (including the base commander) spotted a UFO in proximity to Godman Air Force Base, near Louisville and Fort Knox, Kentucky. When three F-51 planes, led by

[10] Jacobs, *The UFO Controversy in America,* 41, citing an August 1947 Gallup poll in George H. Gallup, *The Gallup Poll: Public Opinion 1935–48* (New York, 1972), 666.

[11] Quoted in Margaret Sachs, *The UFO Encyclopedia* (New York, 1980), 238, which also gives additional information on Palmer. Palmer went on to edit other magazines also featuring flying saucer articles, including one renamed *Flying Saucers from Other Worlds* (1957). At this time, Palmer "decided to concentrate all his energies on promoting UFOs and the occult," according to Peter Nicholls, ed., *The Science Fiction Encyclopedia* (New York, 1979), 446.

Figure 6.1. Cover from Hugo Gernsback's *Science Wonder Stories* for November 1929 depicting powerful aliens in a saucerlike spacecraft, foreshadowing the beginning of the modern UFO controversy two decades later. Copyright 1929 by Gernsback Publications, Inc.

Captain Thomas Mantell, went to check out the reports, Mantell's plane crashed after reporting that he was at an altitude of 22,000 feet. Although investigators concluded that he had blacked out from lack of oxygen, speculation persisted that Mantell had been shot down by an extraterrestrial spacecraft. While it is now believed that Mantell was chasing a Skyhook balloon outfitted with a camera (later used for secret reconnaissance over Iron Curtain countries), the more colorful and exciting extraterrestrial rumors were hard to squash. It was only the beginning of many hard lessons associated with the UFO phenomenon. The Air Force would investigate this and a growing number of UFO reports through Project Sign, set up January 22, 1948; Project Grudge, set up December 16, 1948; and finally Project Blue Book, set up in March 1952 and continuing for 17 years.[12]

The extraterrestrial hypothesis first officially emerged in Project Sign, where an "Estimate of the Situation" in late 1948 concluded that the UFOs were of extraterrestrial origin. But General Hoyt S. Vandenburg disagreed, and the report was returned, declassified, and burned.[13] For the time being, the extraterrestrial hypothesis lost ground in the Air Force, but Project Sign's final report still left open the possibility that the UFO phenomenon might be something extraordinary and extraterrestrial:

> The possibility that some of the incidents may represent technical developments far in advance of knowledge available to engineers and scientists of this country has been considered. No facts are available to personnel at this Command that will permit an objective assessment of this possibility. All information so far presented on the possible existence of space ships from another planet or of aircraft propelled by an advanced type of atomic power plant have been largely conjecture.[14]

In addition to this open-ended statement, James E. Lipp of the Rand Corporation contributed a significant Appendix dedicated to "the likelihood of a visit from other worlds as an engineering problem." Citing Harvard astronomer Fred Whipple's 1941 book on the solar system, Lipp

[12] For an early firsthand account of Air Force involvement by the man who headed its investigations from 1951 to 1953, see Edward J. Ruppelt, *The Report on Unidentified Flying Objects* (New York, 1956).

[13] Jacobs, *The UFO Controversy in America,* 46–47. To my knowledge, no copy of this document remains.

[14] The final report of Project Sign is U.S. Air Force, "Unidentified Aerial Objects: Project Sign," February 1949, no. F-TR-2274-IA, Air Force Archives, Maxwell Air Force Base, Montgomery, Alabama. Parts of the Project Sign report are reprinted in Brad Steiger, ed., *Project Blue Book* (New York, 1976), 170–216. The passage here quoted is on page 172. Although the editor is a UFO enthusiast, this volume is useful for its reprinting of primary documents.

pointed out that the solar system is unlikely to be inhabited by higher forms of life outside Earth, by his odds at least 1000 to 1. Assuming that Mars or Venus were inhabited by intelligence, Lipp could not see why an advanced race would travel this far and not make contact unless it had been intimidated by the five atomic bomb explosions that had taken place by the spring of 1947, when the Arnold UFO sightings occurred. This left the possibility of the flying saucers originating from other solar systems, and here Lipp reflected the emerging idea (documented in Chapter 4) that planets were a normal part of stellar evolution rather than rare events. Following an interesting discussion of the likelihood of intelligence developing, he concluded that space travelers from neighboring stars were much more likely than spaceships from Mars. But this would require propulsion systems as yet unconceived on Earth. Even if it were possible, Lipp argued, the distances were enormous and "a super race (unless they occur frequently) would not be likely to stumble upon Planet III of Sol, a fifth-magnitude star in the rarefied outskirts of the Galaxy." Finally, Lipp argued that the observed maneuvers and appearance of reported UFOs were inconsistent with spaceships. Altogether, Lipp presented a balanced and interesting report, but in the end he concluded that although visits from space were possible, they were "very improbable," and the actions attributed to flying saucers in 1947 and 1948 "seem inconsistent with the requirements for space travel."[15]

When Project Grudge replaced Project Sign at the end of 1948, it had a less open-minded strategy. In the words of historian Jacobs, "New staff people replaced many of the old personnel who had leaned toward the extraterrestrial hypothesis. In the future, Sign personnel would assume that all UFO reports were misidentifications, hoaxes, or hallucinations." Project Grudge shifted the focus from explaining an unusual phenomenon in the atmosphere as something real to explaining it as illusion. A *Saturday Evening Post* article gave the new Air Force philosophy, backed up by Nobel Prize–winning chemist Irving Langmuir, a Project Sign consultant, whose advice to the Air Force on UFOs was to "Forget it!"[16]

[15] The Lipp Appendix "D" is in the form of a letter, "J. E. Lipp to Brigadier General Putt, Project Sign, no. F-TR-2274-IA." It appears on pages 202–214 of Steiger, *Project Blue Book,* and is reprinted in the Condon Report (see footnote 77) as its Appendix D, 844–852. The Whipple volume cited is *Earth, Moon and Planets* (Philadelphia, 1941). Lipp's argument about the unlikelihood of space travelers stumbling on the planet Earth is the opposite of the "where are they?" argument so prominent in the context of SETI in the 1970s and 1980s.

[16] Jacobs, *The UFO Controversy in America,* 50–51, citing Ruppelt, *The Report,* and interview with Hynek. The *Post* article is Sidney Shallett, "What You Can Believe about Flying Saucers (Part I)," *Saturday Evening Post*, April 30, 1949, 20; and Part II, 186.

Project Grudge did, however, take one step that would be of profound importance to UFO history. It hired J. Allen Hynek, an astronomy professor at nearby Ohio State University, to examine possible astronomical explanations for UFOs. Hynek (1910–1986) had come to Ohio State immediately after graduating with a Ph.D. from the University of Chicago in 1935. In 1956 he would go on to become the associate director of the Smithsonian Astrophysical Observatory, and 4 years later he became chairman of the Astronomy Department at Northwestern University. The year 1949 marked the beginning of his lifelong association with the UFO problem, culminating with his founding of the Center for UFO Studies in 1973.[17]

In his capacity as the author of many articles and books, and as a consultant to the Air Force on UFOs, Hynek would make a long voyage from skepticism to openness toward the extraterrestrial hypothesis. For now, in 1949, he was entirely skeptical. "Before I began my association with the air force," Hynek later recalled, "I had joined my scientific colleagues in many a hearty guffaw at the 'psychological postwar craze' for flying saucers that seemed to be sweeping the country and at the naivete and gullibility of our fellow human beings who were being taken in by such obvious 'nonsense.' It was thus almost in a sense of sport that I accepted the invitation to have a look at the flying saucer reports."[18] While Hynek argued that Mantell had chased the planet Venus, the fact remained that from his viewpoint in 1949, 33 percent of all sightings could not be explained by astronomical or other physical phenomena. The Final Report of Project Grudge, issued in August 1949, attributed those sightings to psychological causes.[19] With that conclusion Project Grudge was officially terminated, although some activity continued over the next 2 years.

It was thus not the Air Force that disseminated the extraterrestrial hypothesis, even if it raised the discussion of that hypothesis to an official level for the first time. Rather, this role was taken up by Donald Keyhoe (1897–1988), a retired Marine Corps major and pilot who had managed the public tours of Admiral Byrd and Charles Lindberg following their accomplishments and had written on Lindbergh, war, and aviation. Al-

Langmuir was interested in the characteristics of what he called "pathological science" and described his experiences with Project Sign in a talk in 1953, published in "Pathological Science," *Physics Today* (October 1989), 36–48.

17 On Hynek see Karl G. Henize, "J. Allen Hynek, 1910–86," *S&T,* 72 (August 1986), 117.

18 J. Allen Hynek, *The UFO Experience, A Scientific Inquiry* (New York, 1974), 2.

19 A summary was released to the press in December. Hynek's "Final Report" for Project Grudge, dated April 30, 1949, and other Project Grudge material are reprinted in Steiger *Project Blue Book,* 217–259.

though Keyhoe was not a journalist, as a former chief of information at the Department of Commerce he knew what media exposure meant and how to package an idea successfully for public consumption. Keyhoe became interested in the UFO question in 1949, and after receiving little cooperation from the military, surmised that the flying saucers really were from outer space. This he made the thesis of a 1950 article, "The Flying Saucers Are Real," which has been called "one of the most widely read and discussed articles in publishing history."[20]

With the expansion of his article into several books, Keyhoe became the chief early exponent of the extraterrestrial hypothesis.[21] Throughout his books, however, Keyhoe was peculiarly selective in his arguments for the extraterrestrial hypothesis. On the one hand, he did not hesitate to cite the most recent favorable opinions of astrophysicist Carl F. von Weizsäcker, based on the revival of the nebular hypothesis, that "billions upon billions of stars may each have their own planets revolving about them. It is possible that these planets would have animal and plant life on them similar to the earth's." On the other hand, he cited Percival Lowell's idea of Martian canals as indicative of a dying race trying to conserve its planet's water, an idea that, if true, would have lent credence to the extraterrestrial hypothesis because of Mars's proximity to the Earth. But when Keyhoe wrote this in the early 1950s, Lowell's idea had been rejected by most scientists for more than 40 years.[22] Moreover, Keyhoe tended to explain anything mysterious by the extraterrestrial hypothesis. Radio signals of unknown origin might be attributable to navigation beacons by spacefaring aliens. Such explanations reveal Keyhoe's major weakness: throughout his book, he had little idea of what constituted good scientific evidence or how to make sound scientific inferences from the evidence. Neither, unfortunately, did the public.

There things might have rested – a far-out and passing idea used to

[20] Donald E. Keyhoe, "The Flying Saucers Are Real," *True Magazine* (January 1950). On the popularity of this article, see Jacobs, *The UFO Controversy in America*, 57. Another article by Navy Commander R. B. McLaughlin followed in March and also championed the extraterrestrial hypothesis. Robert B. McLaughlin, "How Scientists Tracked a Flying Saucer," *True* (March 1950), 28.

[21] Donald E. Keyhoe, *The Flying Saucers Are Real* (New York, 1950), followed by *Flying Saucers from Outer Space* (New York, 1953); *The Flying Saucer Conspiracy* (New York, 1955); *Flying Saucers – Top Secret* (New York, 1960); and *Aliens from Space: The Real Story of Unidentified Flying Objects* (New York, 1973). Menzel later wrote that he first believed Keyhoe was just an opportunist trying to make money, but concluded that he sincerely believed he was writing the truth. Menzel to Condon, December 15, 1966, Menzel Papers, American Philosophical Society, Philadelphia, Box 3.

[22] For the von Weizsäcker quote see Keyhoe, *Flying Saucers from Outer Space*, 151, and, on Lowell, 159. Ruppelt, *The Report*, 309, noted that while Keyhoe's details about the sightings studied by the Air Force are correct, his interpretations are "into the wild blue yonder."

explain a few isolated instances – had not nature once again intervened to keep events rolling. In September 1951 an Air Force pilot reported a UFO over Fort Monmouth, New Jersey, and the Air Force decided to revitalize its UFO investigation, naming Captain Edward J. Ruppelt to head the new study. But this was only the beginning. Within months the United States was enveloped in its first full-blown modern "wave" of flying saucers, and in March 1952 the Air Force responded by upgrading its investigation, now code named "Project Blue Book."

The position of the Air Force, which at about this time renamed flying saucers "unidentified flying objects," was that "the possibility of the existence of interplanetary craft has never been denied by the Air Force, but UFO reports offer absolutely no authentic evidence that such interplanetary spacecraft do exist."[23] By year's end the number of reports topped out at just over 1500, with 536 reports in the month of July alone. In that month a series of sensational sightings over the Washington, D.C., area, including radar sightings from National Airport, brought widespread interest among the public and concern in the government. Once again, while the Air Force investigated, the scientists dawdled, and the press took the lead in the search for explanations. This was to prove a fatal lapse on the part of the scientists, whose investigation at this point might have changed the subsequent sensationalist history.[24]

As matters stood, though, the press was left to investigate, and an article by H. B. Darrach, Jr., and Robert Ginna in *Life* magazine asking "Have We Visitors from Space?" soon rivaled Keyhoe's 1950 article for popular influence. Among the propositions the article considered "firmly shaped by the evidence" were that "disks, cylinders and similar objects of geometrical form, luminous quality and solid nature for several years have been, and may be now, actually present in the atmosphere of the earth," as well as green globes of fire; that "these objects cannot be explained by present science as natural phenomena – but solely as artificial devices, created and operated by a high intelligence"; and that "no power plant known or projected on earth could account for the performance of these devices."[25] In short, in the opinion of the authors, the extraterrestrial hypothesis was the only plausible explana-

[23] Ruppelt, *The Report,* 315. On the renaming of flying saucers Ruppelt says "Obviously the term 'flying saucer' is misleading when applied to objects of every conceivable shape and performance. For this reason the military prefers the more general, if less colorful name: unidentified flying objects, UFO [pronounced Yoo-foe] for short," 13.

[24] The Air Force statement is in Ruppelt, *The Report,* 315; the numbers are from Table 6.1. The Washington radar sightings were later explained by a Civil Aeronautics Board investigation, to the satisfaction of some, as caused by temperature inversions.

[25] H. Bradford Darrach and Robert Ginna, "Have We Visitors from Space?" *Life*, April 7, 1952, 80–96: 80–81.

tion left. This theme the *Life* authors brought home with a description of 10 incidents, including one involving "one of the U.S.'s top astronomers" (unnamed), another reported by astronomer Clyde Tombaugh (discoverer of Pluto), and several more involving Air Force officers or airline pilots. Although the article claimed to have undertaken a "careful review of the facts with some of the world's ablest physicists, astronomers and experts on guided missiles," the only experts cited were Walther Riedel, former research director at Peenemünde, and a certain Dr. Biot, a mathematical physicist. The former stated, "I am completely convinced they have an out-of-world basis," and the latter concluded that "The least improbable explanation is that these things are artificial and controlled. My opinion for some time has been that they have an extraterrestrial origin." Other scientists, including Tombaugh, do not seem to have offered an explanation.[26]

Clearly, leaving a subject requiring scientific analysis to reporters was a dangerous game. On the other hand, tackling such a controversial subject could also be dangerous to the career of any scientist, especially one not well established. Perhaps for these reasons – and because Hynek, as an Air Force consultant, seemed unwilling or unable to write for public consumption – one prominent astronomer, and only one, did step forward in the early 1950s as a scientific spokesman about the UFO phenomenon. Donald Menzel (1901–1976), who became acting director of Harvard Observatory in 1952 and served as director from 1954 to 1966, was one of the pioneering astrophysicists in the United States. He had received his Ph.D. from Princeton in 1924 under Henry Norris Russell. Among his first accomplishments was the analysis of the observations of Coblentz and Lampland for the temperature of Mars in 1924 (see Chapter 3). From there he went on to a distinguished career in solar physics and more general problems of astrophysics at both Lick (1926–1932) and Harvard observatories (1933–1971).[27] Menzel clearly enjoyed mystery, challenge, and imagination, and in his early days had published regularly (under assumed names) in science fiction magazines. Moreover, during World War II he had studied the effect of atmospheric irregulari-

[26] Tombaugh reported several UFO sightings during his career, including a geometrically arranged group of six to eight rectangular lights over Las Cruces, New Mexico, on August 20, 1949. See Sachs, *The UFO Encyclopedia,* 322. Also see *Science Digest* (August 1975) interview.

[27] On Menzel see the obituaries by Owen Gingerich, *Physics Today* (May 1977), 96, 98; Zdenek Kopal, *Nature,* 267 (May 12, 1977); and especially Leo Goldberg, *S&T,* 53 (April 1977), 244–251. Menzel's papers on UFOs, covering the period 1952–1976, comprise about 12,000 items (12 linear feet) in the library of the American Philosophical Society. In addition, a photocopy of his manuscript 669-page autobiography was deposited in 1979.

ties on radar waves, and therefore felt himself particularly suited for analyzing radar sightings of UFOs such as occurred in 1952 over Washington, D.C. In many ways an ideal scientist to study UFOs, Menzel had a reputation that was above reproach, and association with the subject of UFOs was unlikely to harm his well-established career. As the reports of UFOs grew after 1947, Menzel recalled, that, as a devotee of science fiction and as one interested in space travel, he had studied them with an eye toward natural explanations, and early on concluded "with a slight feeling of disappointment" that the flying saucers were not extraterrestrial but were mundane objects.[28] That Menzel maintained this position unchanged throughout his life left him, in the eyes of many, a biased observer, but for the history of science he provides an excellent case study of the scientific approach – or at least one scientist's approach – to the subject of UFOs.

As early as April 1952 – just at the time he was becoming acting director of Harvard Observatory – Menzel had presented his ideas on UFOs at a conference with Air Force officials in Washington. From then on, in articles, interviews, or books, he preferred radar ghosts or mirages, and sometimes astronomical objects themselves, as explanations for the majority of UFO reports.[29] In his approach to UFOs, Menzel proposed to combine the skill of the detective with the logic of the scientist. He would follow Sherlock Holmes's advice that "when you have eliminated the impossible, whatever remains, however improbable, must be the truth." While this alone might lead to adoption of the hypothesis of alien spacecraft, the scientific side demanded that "we can always find an infinite number of hypotheses that will explain a given set of observational data. The scientist must choose the simplest."[30] A scientist, Menzel went on, looks for explanations but does not arbitrarily invent forces that make explanation unnecessary. Extraterrestrial beings, in Menzel's view, were a complex hypothesis to be considered only if simpler, natural explanations failed. He believed natural explanations could account for the observed phenomena.

At the same time, Menzel clearly separated the extraterrestrial hypothesis from belief in extraterrestrials. While he "saw no strong reason against the idea that planets inhabited by super-beings should not exist in great

[28] Donald H. Menzel and Lyle G. Boyd, *The World of Flying Saucers: A Scientific Examination of a Major Myth of the Space Age* (New York, 1963), Preface.

[29] Donald Menzel, "The Truth about Flying Saucers," *Look* (June 17, 1952), 35–39; "Those Flying Saucers: An Astronomer's Explanation," interview with Donald Menzel, *Time*, June 9, 1952, 54–56; *Flying Saucers* (Cambridge, 1953).

[30] Menzel, *Flying Saucers,* 51. For the scientific maxim of the simplest hypothesis, Menzel cites the French scientist Henri Poincaré, *Science and Hypothesis* (New York, 1952).

abundance" outside the solar system, he was unwilling to associate the possible abundance of life in the universe with the UFO phenomenon and appealed to common sense: "If these flying saucers represent potential interplanetary tourist trade, where is everybody? Can you imagine us Americans, for example, traveling millions or billions – or perhaps even millions of billions – of miles through space, without making some attempt to communicate with what are obviously friendly people, just as we reach our destination?"[31] Despite his belief in the possibility of intelligent life beyond the earth, Menzel completely opposed the extraterrestrial hypothesis of UFOs during his career in three books and several articles. Even as scientific ideas of extraterrestrial life became more common, this was a distinction that the public for the most part refused to make.

While Menzel was the most outspoken astronomer on the subject of UFOs at this time, thanks to Hynek we do know at least the attitudes of astronomers toward the subject at this crucial turning point in 1952. As part of his work as an Air Force consultant on UFOs, during the summer of 1952 Hynek conducted a poll of 45 astronomers about their opinions on UFOs. He found that five of them had themselves seen UFOs that they could not explain. Most of the astronomers he found "neither hostile nor overly interested" until Hynek explained that many of the reported cases were genuinely unexplained. They were happy to have more information but feared ridicule and harm to their scientific reputations by talking about the subject.[32] In one of the few independent statements on UFOs, astronomer Otto Struve held the evidence for the extraterrestrial hypothesis of flying saucers to be "completely negative," agreeing that they were physical phenomena either along the lines of Menzel's natural explanations or produced artificially by humans. Flying saucers were not an astronomical problem, Struve maintained, but he used the occasion to give one of the earliest statistical arguments for the abundance of inhabited planets in the Galaxy and concluded that "to us it should be a source of inspiration that science now favors the belief that within our galaxy there must be thousands of planets that now support life that is not too

[31] Menzel, *Flying Saucers,* 274. It is notable that Harlow Shapley, Menzel's predecessor as director of the Harvard Observatory, changed his mind in favor of extraterrestrial life in 1952, just at about the time the UFO wave (and Menzel's interest) began (see our Chapter 4, footnote 98). For Menzel's later views on extraterrestrial life see "Is There or Isn't There Life in the Universe?" *Graduate Journal,* 7, no. 1 (1965), 195–219, and Harvard Reprint no. 713; "Life in the Universe," chapter 19 of *Astronomy* (New York, 1970), 274–278; and with Fred L. Whipple and Gerard De Vaucouleurs, "Extraterrestrial Life," chapter 18 of *Survey of the Universe* (Englewood Cliffs, N.J., 1970), 412–427.

[32] Hynek, "Special Report on Conferences with Astronomers on Unidentified Aerial Objects to Air Technical Intelligence Center, Wright–Patterson Air Force Base," August 6, 1952, reprinted in Steiger, *Project Blue Book,* 268–285.

different from the kind we observe on the earth."[33] Like Menzel, Struve totally separated the extraterrestrial hypothesis of UFOs from the probability of intelligence in the universe.

The Central Intelligence Agency, however, seems not to have been satisfied with the explanations of Menzel or the investigation of the Air Force. In December 1952, fearing that UFO reports could be used to clog communications channels during times of national emergency, the agency secretly appointed a panel of five scientists "to evaluate any possible threat to national security posed by Unidentified Flying Objects ('Flying Objects') and to make recommendations thereof." The panel included H. P. Robertson (chair), a Caltech professor known especially for his work in cosmology and relativity theory; Samuel A. Goudsmit, a codiscoverer of electron spin, on the physics staff of Brookhaven National Laboratory; Luis Alvarez of the University of California, Berkeley, a high-energy physicist who had worked with Robert Oppenheimer on the atomic bomb project; Lloyd Berkner, an expert in ionospheric physics and the president of Associated Universities Incorporated, the organization that operated Brookhaven; and Thornton Page, a professor of astronomy at Wesleyan University in Connecticut. All were respected scientists who had worked on government projects during World War II. Had the public known about the assemblage of this group, no one could have complained about its credentials.

The Robertson panel met for 3 days in mid-January 1953 and reviewed some 75 case histories of sightings selected as best documented by the Air Technical Intelligence Center (ATIC, the successor of the Intelligence Division) at Wright-Patterson Air Force Base. The panel considered the extraterrestrial hypothesis, and noted that "none of the members of the Panel were loath to accept that this earth might be visited by extraterrestrial intelligent beings of some sort, some day. What they did not find was any evidence that related the objects sighted to space travelers."[34] Although the committee listened to Blue Book staffer Dewey Fournet's claim that an extraterrestrial origin was the only explanation for many UFO cases, astronomer Page pointed out that the existence of intelligence in the solar system beyond the Earth was "extremely unlikely." Although Page had suggested in 1948 that billions of planets might exist with life in our own galaxy, and although Berkner would 8 years later play an important role

[33] Otto Struve, "What I Don't Know About Flying Saucers," *The Griffith Observer* (December 1952), 138–140. I thank Kevin Krisciunas for bringing this article to my attention. For the context of Struve's career, including his proposal to search for planetary systems, see chapter 4.

[34] Robertson Panel, "Report of Meetings of Scientific Advisory Panel on Unidentified Flying Objects (Robertson Panel), 14–18 January, 1953," reprinted in the *Condon Report* (see footnote 77), 905–921: 910–911.

in supporting the first radio search for extraterrestrial intelligence at the National Radio Astronomy Observatory, their contribution to the conclusions of the Robertson report shows that scientists continued to dissociate the idea of extraterrestrial life from the extraterrestrial hypothesis of UFOs. Thus the panel believed that all sightings could be explained by prosaic causes. The Robertson Report concluded that while UFOs were no physical threat to national security, the continued clogging of communications channels by UFO reports was. The panel suggested that the UFO phenomenon be stripped of any aura of mystery, and that the public be reassured that UFOs represented nothing inimical – an example of Franklin Roosevelt's assertion at his first presidential inaugural address that we had "nothing to fear but fear itself." But because the report was classified as secret, the conclusions of the few scientists who officially considered the UFO question remained unknown to the public until they were partly declassified in 1966.[35]

Thus, although the Robertson committee carried out its responsibility in determining that UFOs were not a threat to national security, it left open the question of what exactly they were. From the scientific point of view, Menzel's explanation that UFOs were mirages remained the sole conclusion of science on the subject. Even if not all scientists were convinced, they remained silent. But J. Allen Hynek, the one astronomer aside from Menzel who knew most about the data, had been growing increasingly skeptical of Menzel's explanations. Finally breaking his silence on the subject, in 1953 Hynek wrote that "in the absence of any universal hypotheses for the phenomena which stimulated these [UFO] reports, it becomes a matter of scientific obligation and responsibility to examine the reported phenomena seriously, despite their seemingly fanciful character." Hynek cautioned that "stones from the sky" were once ridiculed also, and that there were cases in which scientists should "beware the ready explanation." He was not proposing to accept the extraterrestrial hypothesis: "We don't have space ships that disregard physical laws. But, do we have a natural phenomenon?"[36] Hynek was beginning to think so but could not quite bring himself to speak out. And he must

[35] Ibid., 918–919. The convening of a panel had first been disclosed by Ruppelt in 1956, causing the Air Force to release a summary of the report in April 1958. This is reprinted in Philip J. Klass, *UFOs: The Public Deceived* (New York, 1983), 7. *Saturday Review* science editor John Lear first revealed the CIA involvement in "The Disputed CIA Document on UFOs," *Saturday Review* (September 3, 1966), 45–50. For the circumstances surrounding the full release of the report, and much else regarding CIA involvement in the UFO study, see Klass, *UFOs*, 5–50. Klass concludes that the CIA at no time had anything to hide about the nature of UFOs.

[36] J. Allen Hynek, "Unusual Aerial Phenomena," *Journal of the Optical Society of America*, 43 (April 1953), 311–114.

have been chagrined when astronomer Morris K. Jessup of the University of Michigan did. Jessup, at the end of a career at the margins of astronomy that included double star work, wrote favorably about the extraterrestrial hypothesis in *The Case for the UFO* (1955) but applied very little critical thinking to reach his conclusions.[37] Hynek must have realized that he could have done better, but remained silent, understandably tied to his job in mainstream astronomy in academia.

Again, in the absence of further serious scientific study, the claims of unqualified individuals received more attention than they warranted. Such was the claim of George Adamski that he had made actual contact with alien spacecraft and their occupants, and had even journeyed on their spacecraft. Adamski (1891–1965), founder of a cult called the Royal Order of Tibet, lived on the southern slope of Mount Palomar, leading some to believe that he was associated with the great observatory at its summit (an illusion Adamski furthered by referring to himself as "Professor Adamski"). In reality, Adamski worked as a handyman at a hamburger stand in the vicinity. But his book *Flying Saucers Have Landed* (1953), as well as others that followed, were probably more widely known to the public than Menzel's sole attempt to discredit the UFOs. They brought new scientific disrepute to the subject of UFOs, undoubtedly reinforcing scientists' belief that they should stay out of the controversy.[38] Scientists seemed as unwilling to distinguish a potentially credible UFO phenomenon from Adamski's claims as the public was to separate scientific belief in extraterrestrials from UFOs.

With scientific study seemingly more remote than ever, certain members of the public took matters into their own hands. One result was the founding of private UFO organizations to collect and analyze reports – as if the public, fed up with scientists inattention to stones falling from the sky, had set up its own organization to study the phenomenon. With little or no scientific guidance, such organizations were subject to wild varia-

[37] Morris K. Jessup, *The Case for the UFO* (New York, 1955); *UFOs and the Bible* (New York, 1956); and *The Expanding Case for the UFO* (London, 1957). Jessup was an electrical engineer most noted for his work with astronomers Richard Rossiter and Henry F. Donner at the Lamont–Hussey Observatory, the Southern Hemisphere station of the University of Michigan Observatory. Here Jessup discovered some 500 double stars, published in the *Annals of the Michigan Observatory* and today still considered good observations. I am indebted to Prof. Rudi Lindner of the University of Michigan for information on Jessup. For more on Jessup, who committed suicide in 1959, see Jacques Vallee, *Revelations: Alien Contact and Human Deception* (New York, 1991), 198–204.

[38] George Adamski, *Flying Saucers Have Landed* (New York, 1953); *Inside the Spaceships* (London, 1955); and *Flying Saucers Farewell* (London, 1961). The last also appeared as *Behind the Flying Saucer Mystery* (New York, 1967). For more on Adamski, see Klass, *UFOs Explained* (New York, 1974), 290–293, and Hynek and Vallee, *The Edge of Reality: A Progress Report on Unidentified Flying Objects* (Chicago, 1975), 177–181. Klass discusses other "famous contactees" in the same chapter.

tions in the quality of their study, yet they filled a vacuum the scientists refused to occupy. In 1952 Coral and Leslie Lorenzen founded the Aerial Phenomena Research Organization (APRO) to study further the UFO phenomenon in the belief that the extraterrestrial hypothesis was plausible and that the extraterrestrials might be engaged in systematic study of the Earth. And in 1956 the National Investigations Committee on Aerial Phenomena (NICAP) was founded, and was under the direction of Keyhoe until 1969. Unlike APRO, NICAP gained some prominent members who "gave it the prestige and national outlook that no other UFO organization had."[39]

At the same time, the Air Force continued its study, as detailed in Edward J. Ruppelt's *Report on Unidentified Flying Objects* (1956). According to Donald Keyhoe, the Air Force had never accepted Menzel's explanations except for a small proportion of the sightings. "These explanations were known to the Project, and carefully considered, long before Menzel published his theories. They explain only a small percent of the sightings. . . . At the request of ATIC, prominent scientists analyzed Menzel's claims. None of them accepted his answers."[40] The continuing Air Force investigation, and Ruppelt's own opinion that many unexplained UFO sightings remained, gave the subject respectability at the dawn of the Space Age, even as science remained publicly aloof from the subject.

Even the dawn of the Space Age, however, did not inspire immediate scientific reaction to UFOs. Although Hermann Oberth, the father of German rocketry, held that UFOs were "definitely interplanetary vehicles," Frederick C. Durant III, president of the International Astronautical Federation (a federation of the world's space societies), reported in 1954 that "none of the delegates representing the rocket and space flight societies of all the countries involved had strong feelings on the subject of saucers. Their attitude was essentially the same as [that of] professional members of the American Rocket Society in this country. In other words, there appear to be no confirmed saucer fans in the hierarchy of the professional societies."[41] Perhaps more telling than any scientific opinion is the fact that the opening of the Space Age in 1957 itself coincided with a considerable UFO wave (Table 6.1), with the peaks of the wave, as reported to ATIC, occurring in October and November, just after the launch of Sputnik. Whether or not the Space Age precipitated this wave is

39 Jacobs, *The UFO Controversy in America,* 148. On the history of APRO and NICAP see ibid., 84, 132–157, and passim.
40 Keyhoe, *Flying Saucers from Outer Space,* 14, citing the Air Force; see also ibid., 20–21.
41 Ruppelt, *The Report,* 316–317, 311. Durant had been the recording secretary for the Robertson Panel, and may have authored its report.

an important historical question; it is surely a significant but ambiguous piece of data that as more people briefly turned their attention to the skies, the UFO phenomenon also increased in number. Believers in a "real" UFO phenomenon could interpret this effect as more observers seeing what was always there if only one would look carefully enough, while skeptics could interpret the increase as public inexperience or gullibility about the heavens.

Despite scientific uninterest, public fascination with the extraterrestrial hypothesis continued. There were undoubtedly many reasons for this, some of them psychological. Writing in 1959, psychologist Carl Jung argued that UFOs were a response to the lack of security and world peace. He also made a telling observation about media responsibility for UFOs. Based on his experience in 1958 of the press reporting that he was a "saucer believer" and its refusal to correct the erroneous statement, Jung believed that

> one must draw the conclusion that news affirming the existence of UFOs is welcome, but that skepticism seems to be undesirable. To believe that UFOs are real suits the general opinion, whereas disbelief is to be discouraged. This creates the impression that there is a tendency all over the world to believe in saucers and to want them to be real, unconsciously helped along by a press that otherwise has no sympathy with the phenomenon."[42]

Other psychological interpretations of the UFO phenomenon were more closely related to the extraterrestrial hypothesis. As Ruppelt had perceptively said in his Air Force study in 1956, the "will to see" may have

> deeper roots, almost religious implications, for some people. Consciously or unconsciously, they want UFO's to be real and to come from outer space. These individuals, frightened perhaps by threats of atomic destruction, or lesser fears – who knows what – act as if nothing that men can do can save the earth. Instead, they seek salvation from from outer space, on the forlorn premise that flying saucer men, by their very existence, are wiser and more advanced than we. Some people may reason that a race of men capable of interplanetary travel have lived well into, or through, an atomic age. They have survived and they can tell us their secret of survival. . . . To

[42] Carl Jung, *Flying Saucers: A Modern Myth of Things Seen in the Sky* , trans. by R. F. C. Hall (New York, 1959), x.

such people a searchlight on a cloud or a bright star is an interplanetary spaceship.[43]

Coming from a military man steeped in the atomic age, this may well have come closer to a true psychological explanation than Jung's more academic explanation.

Those believing that the extraterrestrial hypothesis was more than a psychological yearning, however, were finally supported by one young astronomer more serious than Jessup, though as yet no match for Menzel's credentials. The French astronomer and computer scientist Jacques F. Vallee (1939–) received his B.S. in mathematics from the Sorbonne in 1959, his M.S. in astrophysics from Lille University in 1961, and his Ph.D. in computer science from Northwestern University in 1967, where he came into contact with Hynek. While still at Lille University, he published two science fiction novels, testimony to a fertile imagination. Impressed by the 1954 wave of UFO sightings in France and influenced by Aimé Michel, the leading UFO researcher in that country, Vallee took up the subject of UFOs and by 1965 offered his own appraisal of the subject in *Anatomy of a Phenomenon*.[44] Vallee challenged the scientific method of Menzel and others, labeling it "the Harvard Syndrome" and finding that "the present prevailing attitude is too limited, that there is more to be found in this phenomenon than is claimed, because the samples of data they study are too small and the techniques they employ too narrow." Vallee opposed "the system which distorts a set of unknown phenomena until it is recognizable by ordinary standards," arguing that while this is the normal method of science, it is inadequate for revealing unknown phenomena. Rather, he argued, investigation of individual cases (selected by criteria better than publicity) should be combined with general analysis in which classes of phenomena are given attention, avoiding the ad hoc explaining away of individual cases.[45] Vallee believed it was wrong to invoke a new natural phenomenon to explain UFOs if the extraterrestrial hypothesis was plausible; extraterrestrials seemed to be his working hypothesis for the UFO phenomenon, and much of his book therefore aimed at showing the plausibility of advanced civilizations, interstellar travel, and UFOs as alien spaceships. "It is scientifically permissible to work under the general hypothesis

[43] Ruppelt, *The Report*, 17.

[44] Jacques Vallee, *Anatomy of a Phenomenon: Unidentified Flying Objects in Space* (Chicago, 1965). On Vallee see Jacques Vallee, *Forbidden Science: Journals 1957–1969* (Berkeley, Calif., 1992). The connection with Michel serves as a reminder that UFOs were an international phenomenon; see Aimé Michel, *The Truth about Flying Saucers* (New York, 1956); *Flying Saucers and the Straight Line Mystery* (New York, 1958). Michel (1919–) served as a clearinghouse for UFO reports in France.

[45] Vallee, *Anatomy of a Phenomenon*, 88–90, 104–106.

that UFO's are material objects, not excluding the possibility of their being nonhuman vehicles," Vallee wrote. While he did not claim to have a final solution, he left little doubt of the direction in which he leaned: "Through UFO activity, although no physical evidence has yet been found, some of us believe the contours of an amazingly complex intelligent life beyond the earth can already be discerned."[46] Vallee would persist in his UFO career, joining Hynek and becoming one of the leaders in the field. As we shall see, the views of both would eventually evolve beyond a "nuts and bolts" alien spaceship interpretation of UFOs, in a synergistic effort that symbolized the promise and the perils of a scientific approach to UFOs. But Vallee's book was an early indication that even when scientists took up the UFO problem, not only a phenomenon was at issue, but also the scientific method used to analyze it.

By contrast, Menzel had not changed his mind in the slightest and was not about to change his method. In his second book on the subject in 1963, he still argued that all UFOs were explainable in natural terms, despite professing an open mind on the question of extraterrestrial life.

> The creative scientist, eternally curious, keeps an open mind toward strange phenomena and novel ideas, knowing that we have only begun to understand the universe we live in. He remembers, too, that Biot's discovery that meteorites were 'stones from the sky' was at first greeted with disbelief, and he hopes never to be guilty of similar obtuseness. But an open mind does not mean credulity or a suspension of the logical faculties that are man's most valuable asset.

Although visits to and from other worlds may occur in the future, Menzel concluded, "No evidence yet found indicates that such visits have begun. No fact so far determined suggests that a single unidentified flying object has originated outside our own planet."[47] Nevertheless, over the next few years, that hypothesis would finally receive more serious scientific scrutiny

6.2 THE PEAK OF THE EXTRATERRESTRIAL HYPOTHESIS, 1965–1969

During the last half of the 1960s the extraterrestrial hypothesis of UFOs was the subject of more attention from mainstream science than at any

[46] Ibid., 194–196. The following year, Vallee and his wife published a book that emphasized the global nature of the UFO phenomenon; see Jacques Vallee and Janine Vallee, *Challenge to Science: The UFO Enigma* (Chicago, 1966).

[47] Donald Menzel and Lyle G. Boyd, *The World of Flying Saucers. A Scientific Examination of a Major Myth of the Space Age* (New York, 1963), 288–289.

time before or since. Whereas for 17 years since 1947 the Air Force had been the primary investigator of the UFO phenomenon and the media the primary purveyor of the extraterrestrial hypothesis, from 1965 to 1969 at least a few in the U.S. scientific community finally took the subject seriously, in part because of pressure from the U.S. Congress. Whereas before 1965 one could not speak of a spectrum of scientific opinion on UFOs because only a few scientists, such as Hynek and Menzel, had ventured an independent opinion, by 1970 that spectrum was well populated. If the media and the Air Force dominated the period to 1965, in the last half of the 1960s it was the turn of the scientists.

Amid the political and intellectual turmoil that characterized the 1960s, the broader scientific and technical landscape had also changed. Extraterrestrials were the subject of renewed interest; in 1960 they had even been the object of a radio telescope search, to much acclaim by the public and the press (see Chapter 8). In the early 1960s, claims of detection of other planetary systems made extraterrestrials seem more likely. And in 1966 McGowan and Ordway's *Intelligence in the Universe* and Shklovskii and Sagan's *Intelligent Life in the Universe* were published, the latter destined to become the bible of scientific thought on extraterrestrial life. Both books were the vanguard of a plethora of literature on the subject. By 1965, too, news of space travel was a reality for much of the populace; the United States was halfway to President John F. Kennedy's goal of landing a man on the moon, and the first Mars and Venus probes had reached into the solar system. Space was becoming a real place in the popular mind, and it made sense that if earthlings were traveling in space, aliens might too. All of this background was combined, often through loose thinking by association, in such popular books as Frank Edwards's *Flying Saucers – Serious Business* (1966), which sold in excess of 1 million copies and was still in print in the last decade of the century. With its large section on the possibility of life in outer space, Edwards's book and others became best-sellers in part because they favored the extraterrestrial explanation for UFOs, a sure bet to catch the public imagination.[48] For those who thought more critically and appreciated the immense distances likely to exist between civilizations, the link between extraterrestrial intelligence and UFOs was not so obvious, though the immensity of time argued that advanced civilizations had billions of years to colonize the universe and therefore might well be in Earth's vicinity. Argument without critical observation could go on forever, but positive evidence was the cru-

[48] Frank Edwards, *Flying Saucers – Serious Business* (New York, 1966). An even more sensationalist approach advocating alien spacecraft is Brad Steiger's *Strangers from the Skies* (New York, 1966).

cial factor for any explanation of the UFO phenomenon. And that is where the scientists could potentially be of most help.

The factor precipitating renewed interest in the extraterrestrial hypothesis in the 1960s was a new wave of UFOs, which filled the skies from July 1965 to mid-1967. Instead of the usual 30 or so reports per month, hundreds were reported; the ATIC received 887 in 1965 alone.[49] In part because of the changed scientific climate, this time the reaction was different from the response to the 1953 wave. No longer was it possible for a 1-week panel like the Robertson committee to undertake a limited study and pronounce that all was well. Instead, the trajectory of events led from the Air Force to congressional hearings; a 2-year university study headed by the physicist Edward U. Condon, precipitating a mix of scientific reactions; and a symposium on the subject sponsored by the prestigious American Association for the Advancement of Science (AAAS). These events began with increased public interest in UFOs and in the Air Force's previous investigations, prompting Hynek to propose to the Air Force a panel of civilian scientists to investigate UFOs.[50] As a result, the Air Force set up the Ad Hoc Committee to Review Project Blue Book in the fall of 1965, composed mostly of members of the Air Force Scientific Advisory Board. Chaired by Brian O'Brien, a member of that board, the group met on February 3, 1966. It found that of 10,000 sightings reported and classified over the previous 19 years, "there appears to be no verified and fully satisfactory evidence of any case that is clearly outside the framework of presently known science and technology." Hedging its bet, the committee recommended strengthening of the Air Force program by contracts to universities to study the problem.[51]

The O'Brien recommendation would soon be carried out, but not before other events intervened. On March 20 and 21, 1966, more than 100 witnesses reported UFOs in the cities of Dexter and Hillsdale, Michigan. The Air Force sent Hynek to investigate; his tentative explanation was that swamp gas had spontaneously ignited, causing a faint glow.[52] This

49 Table 6.1, this volume, and Jacobs, *The UFO Controversy in America,* 193–195.

50 Jacobs, *The UFO Controversy in America,* 197–198. That the Air Force considered this a public relations problem is evident in the fact that Hynek's suggestion was acted on by its Director of Information, Major General E. B. LeBailly, who wrote a "Memorandum for Military Director, Scientific Advisory Board," dated September 28, 1965, requesting that such a panel be convened. The memorandum is reprinted in the *Condon Report* (see footnote 7), 816–817.

51 O'Brien, "Special Report of the USAF Scientific Advisory Board Ad Hoc Committee to Review Project 'Blue Book' " (March 1966). Reprinted in the *Condon Report* (see footnote 77), Appendix A, 811–818, and discussed by Condon, 541ff. Other members of the committee included Carl Sagan, Jesse Orlansky, Launor Carter, Willis Ware, and Richard Porter.

52 Hynek, "Are Flying Saucers Real?" *Saturday Evening Post*, December 17, 1966, 20.

explanation was believed by virtually no one; Hynek was ridiculed, and the press had a field day over the issue of the inadequacy of scientific explanations of UFOs. Hynek later recalled that this was a watershed experience for him: while he had private misgivings as early as 1953 about Blue Book policies "it wasn't until after the 'swamp gas' incident that I said, 'I've had it! This is the last time I'm going to try to pull a chestnut out of the fire for the Air Force.' "[53]

Hynek's transformation was not the only result of the Michigan sightings; his explanation was considered so inadequate that Michigan congressmen Weston Vivian and Gerald Ford called for congressional hearings. On April 5, 1966, the House Armed Services Committee held open hearings on the subject of UFOs. Air Force Secretary Harold Brown, Blue Book head Hector Quintanilla, Jr., and Hynek were invited to testify. Secretary Brown brought with him statistics showing that of the 10,147 UFO sightings reported to the Air Force between 1947 and 1965, the Air Force had identified 9501 as natural phenomena wrongly interpreted. The remaining 646, Brown asserted – carefully choosing his words – were "those in which the information available does not provide an adequate basis for analysis, or for which the information suggests an hypothesis but the object or phenomenon explaining it cannot be proven to have been here or taken place at that time."[54] When asked by Committee Chairman L. Mendel Rivers of South Carolina whether UFOs have come from beyond the Earth, Secretary Brown answered, "To the best of my knowledge, no one in the Air Force and no one in the executive branch has expressed such a belief. Nor have I ever heard a Member of Congress make such a statement. I know of no one of scientific standing or executive standing, or with a detailed knowledge of this, in our organization who believes that they come from extraterrestrial sources," making clear that he was referring to flying craft and not meteorites. Hynek, reading a statement that he called "a little daring," supported further scientific investigation and specifically called for a civilian panel of physical and social scientists to examine whether UFOs were a significant scientific problem. He recalled that in the history of science

> all too often it has happened that matters of great value to science were overlooked because the new phenomenon simply

[53] J. Allen Hynek and Jacques Vallee, eds., *The Edge of Reality* (Chicago, 1975), 199–200.

[54] U. S. Congress, House Committee on Armed Services, *Unidentified Flying Objects: Hearing by Committee on Armed Services of the House of Representatives,* 89th Congress, 2d session, April 5, 1966, 5991–6075: 5992. Gerald Ford's letter to Rivers requesting hearings is reproduced on pages 6046–6047 of the hearings; the following 20 pages are journalistic material on UFO sightings inserted into the record. Thus journalism continued to substitute for scientific study.

> did not fit the accepted scientific outlook of the time. Thus, the evidence of fossils for biological evolution was overlooked; X-rays were overlooked, meteorites were overlooked as astronomers steadfastly refused to accept stories of stones which fell from the sky.[55]

Out of the hearings came a decision to implement the O'Brien Committee's recommendation for further scientific investigation. In April the Air Force moved to implement such an investigation; in May it disclosed its plan; in October, after a difficult search, it named the University of Colorado as the site for the study and Edward U. Condon as its head.

But why Condon? In a letter to astronomer Nicholas U. Mayall, Menzel revealed the reasoning behind the decision. He noted that Harvard was ruled out because of his own anti-UFO stance, the University of Arizona because of McDonald's pro-UFO stance, and Northwestern because of Hynek's long association with the Air Force. "The agency of the Air Force responsible for placing the grant did seek my advice," Menzel continued. "My number one choice was Ed Condon, whose independence cannot be questioned, and who most certainly is a competent scientist. He also is a responsible citizen and will undertake the assignment seriously."[56] In many ways Condon (1902–1974) was an ideal choice for the study. A reputable scientist who also had brief experience in journalism, Condon had no previously expressed opinion on UFOs but had been through the mill on a number of controversial issues. He had received his Ph.D. in physics from the University of California, Berkeley, in 1926. After a brief period in Germany working on quantum mechanics under Max Born and Arnold Sommerfeld, at Bell Telephone Laboratories writing popular science, and as a lecturer at Columbia University, Condon joined the physics faculty at Princeton. Except for brief stints at Stanford and the University of Minnesota, he remained at Princeton until 1937. During that time he published two books, *Quantum Mechanics* (1928) and *The Theory of Atomic Spectra* (1935), and numerous articles on quantum mechanics. He joined Westinghouse Electric in 1937, undertaking war-related work, as did many of his scientific colleagues. At the urging of the Secretary of Commerce, in 1945 President Harry S. Truman appointed Condon director of the National Bureau of Standards. Among his many accomplishments in revitalizing the bureau after the war, Condon established major new facilities for it in Boulder, Colorado. When Condon became one of the accused during the infamous House

[55] Ibid., 6007.
[56] Menzel to Mayall, November 1, 1966, Condon Papers, Donald H. Menzel correspondence file, Box O-5 (see footnote 57). On McDonald see p. 293.

Un-American Activities Committee hearings, in 1951 he left the Bureau for Corning Laboratories. About 1955 he became chairman of the Physics Department at Washington University at St. Louis, and later came to Boulder as a professor and fellow of the Joint Institute for Laboratory Astrophysics.[57]

It was in this context, during an "incautious moment," as one of his biographers has said, that Condon accepted the task of heading the Air Force investigation of UFOs. By his own account, Condon was not eager: "I did not want to do the UFO study," he wrote several years later to Carl Sagan, "but was talked into it in August 1966 by staff of the U.S. Air Force Office of Scientific Research, largely on the basis of appeals to duty to do a needed public service." By another account, during a losing campaign to become a regent of Washington University, Condon had stressed the need for more federal research money, and the $300,000 the Air Force was offering (eventually supplemented with another $200,000) was not to be taken lightly.[58] Undoubtedly, both the scientific challenge of the problem and the federal money were factors in Condon's acceptance of the task.

Even before the Condon study began, the new wave of UFOs brought some scientists out of the closet, and Condon knew he would have to deal with them. Particularly important was the work of James McDonald, an atmospheric physicist at the Department of Meteorology and the Institute of Atmospheric Physics at the University of Arizona in Tucson. A member of the National Academy of Sciences, McDonald had been casually interested in UFOs for many years. A particular sighting in Tucson in March 1966, as well as the Michigan sightings during the same month, led him to study the subject much more closely; by mid-1966 he had visited Hynek and Blue Book officials, and had come to believe that the extraterrestrial hypothesis was the best explanation for UFOs.[59] In 1967 he

[57] On Condon see the obituary by Philip M. Morse in *Biographical Memoirs of the National Academy of Sciences* (Washington, D.C., 1976), vol. 48, 125–151; Lewis M. Branscomb, "Edward Uhler Condon," *Physics Today*, 27 (June 1974), 68–70; Churchill Eisenhart, "Edward Uhler Condon," *The Technical News Bulletin of the National Bureau of Standards*, Dimensions, 58 (1974), 151. See also Condon, "Reminiscences of a Life In and Out of Quantum Mechanics," *Proceedings of the 7th International Symposium on Atomic, Molecular, Solid State Theory and Quantum Biology,* ed. Per-Olov Lowdin (New York, 1973), and Jessica Wang, "Science Security and the Cold War: The Case of E. U. Condon," *Isis,* 83 (June 1992), 238–269. The Condon Papers, including approximately 22 linear feet devoted to UFOs, are deposited in the Library of the American Philosophical Society in Philadelphia.

[58] Condon to Carl Sagan, October 6, 1969, Condon Papers, Box O-6. The second explanation is from David R. Saunders and R. Roger Harkins, *UFOs? Yes! Where the Condon Committee Went Wrong* (New York, 1968), 46. See also Boffey (footnote 72) for Condon's reluctance to take on the task.

[59] In a note to Gerard Kuiper dated June 13, 1966, McDonald stated, "My trip to Washington, New York, Project Blue Book, and Dearborn Observatory leaves me with the

presented a detailed lecture in Washington, D.C., in which he argued that UFOs were "the greatest scientific problem of our times," possibly explainable only by the extraterrestrial hypothesis. As the Condon Committee was undertaking its work, McDonald's ideas were being publicized in the national press.[60]

McDonald's activity is particularly important historically not only because of his claims as a reputable scientist, but also because of the reaction of his colleague at the University of Arizona, astronomer Gerard P. Kuiper. Although they were on the same campus in Tucson, McDonald was in the Meteorology Department and Kuiper was at the Lunar and Planetary Laboratory he had founded a few years earlier; in mid-1966, McDonald began a correspondence with Kuiper about the possibilities of the extraterrestrial hypothesis. Kuiper agreed that UFO reports had not been analyzed with the necessary scientific competence but rejected that hypothesis on the basis of insufficient evidence. "I would draw a comparison with the law of conservation of energy in classical physics. One would certainly not challenge the validity of this law on the basis of fragmentary observations by unskilled observers made during intervals that are often only matter of seconds," he wrote. While he agreed that the UFO phenomenon had been given "superficial treatment," Kuiper argued that "a conclusion so extraordinary as yours would require a scientific backing many orders of magnitude stronger than available. In my judgment the only defensible position a scientist can take here is that there are unexplained (terrestrial) atmospheric phenomena."[61] McDonald was not convinced and had no illusions about the effect on his career: "That publicly espousing such an hypothesis, even in the pussyfooting language of 'least undesirable hypothesis' is professionally risky is very, very clear to me," he replied to Kuiper. "I take that step because of having looked rather thoroughly at the evidence."[62] In April 1967 Kuiper gave a paper at the Arizona Academy of Science–AAAS meeting in Tucson in which he stated that McDonald was performing a service to science by raising the

overwhelming conviction that the unidentified flying objects are extraterrestrial vehicles. Many things now fit together." University of Arizona Library, Kuiper Archives, Box 6. McDonald relates how he came to the UFO debate in detail in the 1968 congressional hearings (footnote 73), 33ff.

60 James McDonald, "Unidentified Flying Objects: Greatest Scientific Problem of our Times," address to the American Society of Newspaper Editors, April 1967 (Washington, D.C., 1967). On McDonald see the unpublished dissertation by Paul McCarthy, "Politicking and Paradigm Shifting: James E. McDonald and the UFO Case Study," University of Hawaii, c. 1975, copy in Philip Klass Collection, American Philosophical Society Library, Philadelphia, of special interest because McCarthy had access to McDonald's papers.

61 Kuiper to McDonald, October 8, 1966, Kuiper Archives, Box 6.

62 McDonald to Kuiper, October 10, 1966, Kuiper Archives, Box 6.

standards of analysis. Kuiper compared UFOs to the Martian canals and the controversy over organic matter in meteorites, leaving no doubt that the extraterrestrial hypothesis was implausible to him.[63] But in his private correspondence, Kuiper was more harsh toward McDonald.

The new wave of UFOs also brought great interest in the popular press. Science editor John Lear's series of articles in the fall of 1966 in the *Saturday Review* focused attention on the phenomenon. Lear not only brought to readers' attention the ideas of Jacques Vallee on UFOs and those of Carl Sagan on extraterrestrial intelligence, he also managed to obtain an edited version of the Robertson Committee's report (not officially released until several years later), and appraised the public of the Air Force–sponsored study.[64]

Just as Condon was about to begin his study, another figure entered the fray on the side of skepticism. Philip Klass, a senior editor at *Aviation Week and Space Technology,* claimed that at least some UFOs were plasma phenomena. When he queried Kuiper about this explanation 2 years later, Kuiper answered that this was an idea first proposed by his colleague Ferdinand de Wiess. He confided to Klass that while McDonald "is very knowledgeable about cloud physics . . . I have, to my regret, found him increasingly dogmatic and unreasonable about the UFO problem."[65] Klass's plasma theory was only the first shot for him; over the next quarter century, he would take over the skeptical attitude where Donald Menzel left off.

Lear's articles also brought to the public a more sensational and disturbing aspect of the UFO debate during this time – the story of sightings in New Hampshire and the claimed abduction of Betty and Barney Hill by occupants of a spacecraft that had allegedly landed in that state, a story

[63] Gerard Kuiper, Presentation at Arizona Academy of Science–AAAS Meeting, April 29, 1967; copy in Kuiper Archives; also reprinted in *Condon Report* (footnote 77, this chapter) as Appendix C, 839–843. Kuiper sent copies of this to many astronomers and others.

[64] Vallee and Sagan are discussed in John Lear, "What Are the Unidentified Aerial Objects?" *Saturday Review* (August 6, 1966), 41–42, with an excerpt from Sagan; the Robertson Report is excerpted and discussed in John Lear, "The Disputed CIA Document," *Saturday Review* (September 3, 1966), 45–50, and the Condon study in an untitled article, *Saturday Review* (December 3, 1966), 87–89.

[65] Kuiper to Klass, December 17, 1968, Kuiper Archives, Box 6. Kuiper also called the Condon study "a fantastic waste of time" in this letter. John Lear discusses Klass's explanation, and Klass's entry into the UFO field, in "Scientific Explanation for the UFOs?" *Saturday Review* (October 1, 1966), 67–69. Klass's first article on UFOs appeared in the magazine for which he worked as an editor, "Many UFOs Are Identified as Plasmas," *Aviation Week and Space Technology* (August 22, 1966), 48–50; his personal UFO papers, like those of Menzel and Condon, are deposited at the American Philosophical Society Library (approximately 8 linear feet). On Klass's career involvement with UFOs, see Steven J. Dick, interview with Philip J. Klass, December 12, 1992.

documented and popularized by Lear's fellow editor at *Saturday Review*, John Fuller. Although such stories must have made the Condon researchers more incredulous than ever, Harvard historian Oscar Handlin, reviewing Fuller's book *Incident at Exeter*, suggested that "there is . . . nothing inherently implausible about extraterrestrial visitors."[66]

With the beginning of the Condon study, Hynek too began his journey away from skepticism. Feeling it necessary to relate the gist of his work on the subject since 1949 to his scientific colleagues "who could not be expected to keep up with so seemingly bizarre a field," Hynek published a letter in *Science* in which he stated seven common misconceptions about UFOs. He argued that while the Air Force had no evidence supporting the extraterrestrial hypothesis, neither was there evidence against it.[67] In response to Hynek, William Markowitz, an astronomer just retired as director of the Time Service Division of the U.S. Naval Observatory, wrote a widely cited article in which he sided with Menzel and Klass, and left no doubt about the kind of evidence he required to accept the extraterrestrial hypothesis: "I have no quarrel with anyone who wishes to believe that UFOs are under extraterrestrial control. As for me, I shall not believe that we have ever been visited by any extraterrestrial visitor – either from the moon, from a planet of our solar system, or from any other stellar system – until I am shown such a visitor."[68] If there was ever any doubt, the responses to the Markowitz article ensured that even among scientists, there would be no unanimity on UFOs.[69]

With such activity raging around it, all looked toward the Condon study, which became the focus of the UFO debate in the latter half of the 1960s, the center around which most activity in the field gravitated. The study is of interest both for how it was undertaken and for the conclusions reached. On the first point, there is no doubt that the prosecution of

66 John Fuller, *Incident at Exeter: Unidentified Flying Objects over America Now* (New York, 1966), and *The Interrupted Journey: Two Lost Hours Aboard a Flying Saucer* (New York, 1966); Oscar Handlin, "Reader's Choice," *Atlantic Monthly*, 218 (August 1966), 117.

67 Hynek, "UFOs Merit Scientific Study," *Science* (October 21, 1966), 329. Hynek also criticized the Air Force for public consumption in "Are Flying Saucers Real?" *Saturday Evening Post* (December 17, 1966), 17–21. See also Philip J. Klass, "The Conversion of J. Allen Hynek," *The Skeptical Inquirer* (Spring 1979), 49.

68 William Markowitz, "The Physics and Metaphysics of UFOs," *Science*, 157 (1967), 1274–1279: 1279; on Markowitz's career see also Steven J. Dick, oral history interview with William Markowitz, U.S. Naval Observatory Library. On Markowitz's earlier brush with the subject of UFOs see Keyhoe, *Flying Saucers from Outer Space*, 151.

69 For the responses see Richard J. Rosa, "Letters," *Science* (December 8, 1967), 1265; William T. Powers, ibid., 1265, Jacques Vallee, ibid., 1266. Lear reported on Hynek and Markowitz in "UFOs and the Laws of Physics: Concerning Views of J. Allen Hynek and William Markowitz," *Saturday Review* (October 6, 1967), 59.

the study was somewhat tumultuous. Disagreement might have been expected, based on the widely varying opinions on the scientific merit of the subject, but no one could have foreseen quite the tumult that developed. Condon's original proposal seemed straightforward enough. He would spend about half of his time as director of the UFO project. A University of Colorado administrator, Robert Low, would be the scientific coordinator and principal investigator. They would be joined by several colleagues in the university's Psychology Department, including its chairman, Stuart Cook, and David R. Saunders. The study would also have the benefit of two nearby national institutions: the Environmental Science Services Administration (ESSA), with the expertise of astronomer Franklin Roach, and the cooperation of the National Center for Atmospheric Research, headed by astronomer Walter Orr Roberts. The project, which would eventually cost about $500,000 over 2 years, would examine Blue Book files, talk to its staff members, and conduct interviews, field studies, and tests. The contract (between the University of Colorado and the Air Force Office of Scientific Research) indicates that Condon realistically expected no single explanation to emerge. It would be surprising if an unambiguous physical interpretation were found for all cases, he noted, and more reasonable to expect "that the phenomena reflect a variety of perceptual and cognitive processes superimposed on a variety of physical stimuli." The methodology would "lean in the direction of quantification and experimentation and away from the effort simply to enlarge the already substantial body of opinion and impression," although Condon explicitly recognized that the physical scientists might differ in their methodologies from the behavioral scientists.[70]

It all sounded very promising, but during the course of 1967–1968, the study was wracked by internal dissension and outside controversy. The details have been examined elsewhere, but the controversy was fueled at least in part by the question of how to approach the extraterrestrial hypothesis. One faction, led by psychologist Saunders, wanted to concentrate on that hypothesis; the other, led by Low, believed the solution lay in the psychology of the witnesses.[71] When an internal memorandum written by Low was leaked to McDonald by Saunders, and interpreted as biased against a fair study, Condon fired Saunders and another staff member. By the end of the study, only 3 of the 12 original staff members remained. When in the summer of 1968 these problems hit the pages of

[70] John Lear, "Research in America: Dr. Condon's Study Outlined," *Saturday Review* (December 3, 1966), 87–89, reprints the main part of the work statement in the contract. Copies of the full contract are in Condon Papers, Box O7.

[71] See Jacobs, *The UFO Controversy in America*, 225–263, for the details of the controversy.

Science, the journal of the AAAS (an organization that Condon had headed), Condon refused to talk to its reporters and resigned his AAAS membership.[72]

A few days later, the House Committee on Science and Astronautics, aware of the controversy over the continuing Condon study, held hearings on the UFO problem, broadening still more the scientific participation. Indiana Representative J. Edward Roush, who had shown a personal interest in the UFO problem and even corresponded with Kuiper about it, chaired the symposium. "We approach the question of unidentified flying objects as purely a scientific problem, one of unanswered questions. Certainly the rigid and exacting discipline of science should be marshaled to explore the nature of phenomena which reliable citizens continue to report," Roush noted in opening the discussions. The testimony that followed from Hynek and McDonald, as well as from Cornell astronomer Carl Sagan, University of Illinois sociologist Robert L. Hall, civil engineer James Harder, and scientist Robert M. L. Baker, Jr., surely gave Roush and his congressional colleagues a lesson on how nonrigid and inexact science could be.[73] Hynek argued that one should not confuse "psychically unbalanced individuals and pseudoreligious cultist groups" who hoped for salvation from UFOs with those who sought scientific explanations. "Ridicule is not part of the scientific method and the public should not be taught that it is," he concluded.[74] McDonald, criticizing Menzel's explanations as "very, very far removed from what are well-known principles and quantitative aspects of meteorological optics," and characterizing Klass's plasma explanations as completely unreasonable, concluded that "UFO's are entirely real and we do not know what they are, because we have laughed them out of court. The possibility that these are extraterrestrial devices, that we are dealing with surveillance from some advanced technology, is a possibility I take very seriously." Asked on what evidence he based his belief in the extraterrestrial hypothesis, McDonald emphasized that it was not a belief, but rather "The hypothesis I presently regard as most likely."[75]

[72] Philip M. Boffey, "UFO Project: Trouble on the Ground," *Science* (July 26, 1968), 339–342, and comment by Lewis M. Branscomb, letter, *Science*, 161 (September 27, 1968), 1297. Condon describes the dispute, and the circumstances of his resignation from the AAAS, in Condon to Sagan, October 6, 1969. Condon believed Boffey had been sent to Boulder "to maximize the scandal," despite personal appeals to *Science* editor Philip Abelson and AAAS president Roberts.

[73] U.S. Congress, House of Representatives, Committee on Science and Astronautics, *Symposium on Unidentified Flying Objects, Hearings before the Committee on Science and Astronautics, U.S. House of Representatives, 90th Congress, 2d session, 29 July, 1968.*

[74] Ibid., Hynek testimony, 3–17: 6–7.

[75] Ibid., McDonald testimony, 18–32: 26. McDonald's testimony is followed by a lengthy prepared statement, 32–86.

By contrast, Carl Sagan, having just 2 years earlier published with Shklovskii *Intelligent Life in the Universe*, found the extraterrestrial hypothesis for UFOs not at all persuasive. He was much more impressed with the psychological explanations of Jung and others:

> The interest in unidentified flying objects derives, perhaps, not so much from scientific curiosity as from unfulfilled religious needs. Flying saucers serve, for some, to replace the gods that science has deposed. With their distant and exotic worlds and their pseudoscientific overlay, the contact accounts are acceptable to many people who reject the older religious frameworks. But precisely because people desire so intensely that unidentified flying objects be of benign, intelligent, and extraterrestrial origin, honesty requires that, in evaluating the observations, we accept only the most rigorous logic and the most convincing evidence. At the present time, there is no evidence that unambiguously connects the various flying saucer sightings and contact tales with extraterrestrial intelligence.

Sagan therefore supported a moderate investigation of UFOs, but suggested that it would be much better for Congress to support the spacecraft investigations of life and radio searches for intelligent signals.[76]

In the midst of all the internal dissension and external scrutiny, the Condon study ground on, and its final report was a massive multiauthored volume, delivered to the Air Force in November 1968 and released in January, 1969.[77] The bulk of its 967 pages in the paperback edition consisted of a description of the work of the project, case studies, and scientific explanations. Condon himself wrote three sections, which amounted to only about 10 percent of the total report: the 6-page "conclusions and recommendations," a 43-page "summary of the study," and a 50-page history of "UFOs, 1947–1968." For a project director to write such a small portion of the final report is not surprising; what is surprising – and a measure of the gravity with which Condon treated his task – is that Condon himself apparently took the trouble to research and write a history

[76] Ibid., Carl Sagan testimony, 86–92: The quotation is from Sagan's article "Unidentified Flying Objects," in *Bulletin of the Atomic Scientists*, 23 (1967), 43.

[77] Edward U. Condon (project director) and Daniel S. Gillmor (editor), *Final Report of the Scientific Study of Unidentified Flying Objects*, (Boulder, Colo., 1968), 3 vols., 1465 pp., 80 plates. A hardcover edition (967 pp.) was also published by Dutton in association with the University of Colorado Press (New York, 1969). The most accessible version is the Bantam paperback (New York, 1969), 967 pp., which is cited here, also published by Corgi Books (London, 1969). According to Condon, by 1973, after sales of 100,000 in paperback and 15, 000 in hardback, the volume was out of print. Condon Papers, Box 1.

(albeit not entirely accurate) of the UFO phenomenon during the two decades prior to his study. Condon's conclusions, which opened the report, disappointed many a UFO enthusiast. Despite its best efforts, Condon stated, "Our general conclusion is that nothing has come from the study of UFOs in the past 21 years that has added to scientific knowledge. Careful consideration of the record as it is available to us leads us to conclude that further extensive study of UFOs probably cannot be justified in the expectation that science will be advanced thereby."[78] Although this was enough for officials who wished to put an end to the subject, Condon's further elaboration is more revealing. He went on to talk about how science works, and how each scientist must make a personal decision as to what research will be most fruitful. Scientists should not uncritically accept the conclusion of the report, he stated, urging that if new ideas for "clearly defined, specific studies" were forthcoming, they should be supported. Condon made clear that he did not believe any extensive further effort should be made to study UFOs at the present time, but that "this may not be true for all time." He noted that the study focused almost entirely on physical science, first because of priorities and also because "we found rather less than some persons may have expected in the way of psychiatric problems related to belief in the reality of UFOs as craft from remote galactic or intergalactic civilizations. We believe that the rigorous study of the beliefs – unsupported by valid evidence – held by individuals and even by some groups might prove of scientific value to the social and behavioral sciences."[79]

In Condon's second contribution, the "summary of the study," he addressed the extraterrestrial hypothesis. Distinguishing between the extraterrestrial hypothesis (ETH) that UFOs may represent alien spacecraft and the extraterrestrial actuality (ETA) that such spacecraft are an observational fact, Condon agreed that "direct, convincing and unequivocal evidence of the truth of ETA would be the greatest single scientific discovery in the history of mankind. Going beyond its interest for science, it would undoubtedly have consequences of surpassing significance for every phase of human life." Condon and his colleagues, however, found that "no direct evidence whatever of a convincing nature now exists for the claim that any UFOs represent spacecraft visiting Earth from another civilization." Although it would be exciting to believe, "when confronted with a proposition of such great import, responsible scientists adopt a cautiously critical attitude toward whatever evidence is adduced to sup-

[78] Ibid., 1.
[79] Ibid., 4.

port it."[80] This attitude should not be confused with a basic opposition to the idea, Condon explained: "the scientists' caution in such a situation does not represent opposition to the idea. It represents a determination not to accept the proposition as true in the absence of evidence that clearly, unambiguously and with certainty establishes its truth or falsity."

Decisive observations are sometimes hard to come by, Condon emphasized, and therefore progress in science can be painstakingly slow. Like Markowitz 2 years before, he left no doubt of the kind of hard evidence he would like to see: "The question of ETA would be settled in a few minutes if a flying saucer were to land on the lawn of a hotel where a convention of the American Physical Society was in progress, and its occupants were to emerge and present a special paper to the assembled physicists, revealing where they came from, and the technology of how their craft operates. Searching questions from the audience would follow." On theoretical grounds, Condon also argued that "It is regarded by scientists today as essentially certain that ILE [intelligent life elsewhere] exists, but with essentially no possibility of contact between the communities on planets associated with different stars. We therefore conclude that there is no relation between ILE at other solar systems and the UFO phenomenon as observed on Earth."[81] Condon must have realized that it was stretching the argument to assume with certainty that extraterrestrial civilizations billions of years old could not cross interstellar space. Much greater was his insistence that very strong evidence was needed to accept the extraterrestrial hypothesis.

Most of the scientific world shared this view of the canons of evidence. Therefore, when in October 1968 the Air Force requested the National Academy of Sciences to "provide an independent assessment of the scope, the methodology, and the findings of the study," the outcome was not surprising.[82] The president of the National Academy (Frederick Seitz, a student of Condon at Princeton in 1930) appointed a panel chaired by Yale astronomer and former U.S. Naval Observatory Scientific Director Gerald M. Clemence.[83] The panel met on December 2, 1968, and January

80 Ibid., 25.

81 Ibid., 29. In the 1970s and 1980s, the feasibility of interstellar travel would become an important topic in the context of SETI.

82 Alexander H. Flax, Assistant Secretary of Research and Development, to Frederick Seitz, President, National Academy of Sciences, October 29, 1968, in Clemence Papers, U.S. Naval Observatory Archives. The Academy had agreed in October 1966 to review the *Condon Report,* and was officially informed at its October 5, 1968, Council meeting that the Air Force would formally request an Academy review at the end of October.

83 Clemence had been selected earlier and had suggested potential committee members by the end of September; Clemence to Seitz, September 27, 1968, Clemence Papers, USNO Archives. For Seitz on Condon see *Topics in Modern Physics,* eds. W. E. Brittin and H.

6, 1969, the latter with Dr. Condon in attendance. In a letter to Clemence, panel member and astronomer Donald Shane spoke for the predominant scientific view and discussed the potential consequences of relaxing the canons of evidence: "There is no positive evidence that the UFOs represent extra-terrestrial artifacts. The presumptive evidence based on our knowledge of the solar system, as well as on energy and time considerations, argue overwhelmingly against this interpretation, unless we are willing to admit principles of nature quite outside the realm of our present understanding. . . . If we do make such admissions, the field is wide open."[84] In the end, after about 6 weeks of study, the panel agreed with all the findings of the Condon Committee, concluded that "on the basis of present knowledge the least likely explanation of UFOs is the hypothesis of extraterrestrial visitations by intelligent beings," and further agreed that "no high priority in UFO investigations is warranted by data of the past two decades."[85]

The reaction of the broader scientific community, however, would not be unanimous. To be sure, there were those who supported the Condon Report. "Condon Group Finds No Evidence of Visits from Outer Space" headlined *Science,* the official journal of the AAAS .[86] *Science* emphasized the report's "massive documentation," its conclusion that seemingly hard evidence was often not hard at all, and that most cases tended to be explicable when enough evidence was available. As for the residue still unexplained, *Science* favorably cited the National Academy review's opinion that much more convincing evidence would be needed for so extraordinary a hypothesis as extraterrestrials. Elsewhere, MIT physicist Philip Morrison – who in 1959 had proposed the radio search for extraterrestrial intelligence and was not known for his lack of imagination – wrote,

Odabasi (Boulder, Colo., 1971), xxi–xxiii. Seitz was president of the National Academy from 1962 to 1969.

84 C. D. Shane to Clemence, December 4, 1968, Clemence Papers, U.S. Naval Observatory. The members of the panel were Gerald M. Clemence (chairman), H. R. Crane (U. of Michigan), David M. Dennison (U. of Michigan), Wallace Fenn (U. of Rochester), H. Heffer Hartline (Rockefeller U.), E. R. Hilgard (Stanford U.), Mark Kac (Rockefeller U.), Francis W. Reicheldorfer (Washington, D.C.), William W. Rubey (UCLA), C. D. Shane (U. of California, Santa Cruz), and Oswald G. Villard, Jr (Stanford U.). The Air Force reimbursed the National Academy of Sciences about $13,000 for the study. McDonald asked to appear before the committee but was limited to sending his published works on UFOs.

85 National Academy of Sciences Panel, "Review of the University of Colorado Report on Unidentified Flying Objects" (Washington, D.C., 1969), 6 pp. Reprinted in *Icarus,* 11 (November 1969), 440–443, and discussed by *New York Times* journalist Walter Sullivan in the preface to the *Condon Report.* The Academy transmitted its report to the Air Force on January 8, 1969.

86 Philip Boffey, "UFO Study: Condon Group Finds No Evidence of Visits from Outer Space," *Science* (January 17, 1969), 260–262.

"One comes away edified, amused, admiring and well satisfied . . . Science is the stronger for this sincere and expert effort to deal with a public concern."[87] "I want to compliment you and your associates with having produced an outstanding report on a subject which poses unusual problems for a scientist," Kuiper wrote Condon from Arizona.[88]

But the dismissal of such a high-profile and publicly scrutinized phenomenon as UFOs was not going to be that easy. Even Condon realized that his conclusions would not be uncritically accepted. "Scientists are no respecters of authority," Condon wrote in the summary of his report. "Our conclusion that study of UFO reports is not likely to advance science will not be uncritically accepted by them. Nor should it be, nor do we wish it to be." Scientists should be given free reign to decide "the directions in which scientific progress is most likely to be made."[89] In this respect, scientists certainly did not disappoint Condon. Scientific critics were vociferous from the time the Condon Report was released, among them, not surprisingly, David Saunders and James McDonald. During a press conference sponsored by NICAP shortly after the release of the Condon Report, Saunders – the psychologist associated with the Condon project until his dismissal – and McDonald criticized it strongly. Saunders released his book *UFOs? Yes! Where the Condon Committee Went Wrong*, an insider's story with disturbing revelations about the management of the project,[90] while McDonald returned with renewed vigor to excoriate his colleagues for bad science. The Condon Report, he told the University of Arizona's newspaper, was "$500,000 worth of the bum scientific advice the Air Force has been getting for 20 years." McDonald was disturbed at the mismatch between the results of the case studies and Condon's conclusion about the extraterrestrial hypothesis, the first of many to level this criticism. Condon's sole aim seemed to have been to prove or disprove the extraterrestrial hypothesis, McDonald complained. "The extraterrestrial hypothesis remains a scientific *hypothesis*, and the tragedy is that the unprecedented opportunity given to Condon has been wasted, as far as any really substantial illumination of that one of a number of possible UFO hypotheses," McDonald wrote team member William Hartmann.[91]

87 Morrison, "The Condon Report on Unidentified Flying Objects," *Scientific American*, 220, no. 4 (April, 1969), 139–140.

88 Kuiper to Condon, February 10, 1969, Box 6, Kuiper Archives.

89 *Condon Report*, 2.

90 Saunders and Harkins, *UFOs?* as cited in note 58.

91 McDonald to Hartmann, February 10, 1969, Kuiper Archives, Box 6; *Arizona Daily Wildcat* (February 3, 1969), 1, 12, copy in Kuiper Archives. Condon revealed much about the goals of his study when he wrote, "I want to narrow the focus to ETH [the Extraterrestrial Hypothesis] because that is what gives the subject its great popular interest." Condon to Sagan, October 6, 1969. Condon did not believe that atmospheric optics or radar anomalies would be advanced by studying the UFO phenomenon.

The contrast in the scientific opinion of the Condon Report was starkly drawn in *Icarus*, a planetary science journal edited by Carl Sagan. In the same issue in which it reprinted the favorable National Academy review of the report, astronomer H.-Y. Chiu gave an approving review of Condon, while McDonald gave a dissenting view.[92] Ufology, Chiu claimed, should now be regarded as a pseudoscience. And although disproving pseudoscience takes a heavy investment of scientific talent, the UFO question had nevertheless now been given the same scientific study as other areas of inquiry, and its devotees should be satisfied. In a parting shot at the extraterrestrial hypothesis, Chiu argued that a trillion spaceships per year would have to be dispatched to all the stars in our galaxy to make it likely that our Earth has been visited. Some, Chiu concluded, will find this truth a bitter pill; "they will undoubtedly continue their ufological career, perhaps with even greater vigor and bitterness toward scientists." Chiu was right about that, as was evident in the same issue of *Icarus*. In his counter-review, McDonald attacked not only the Condon Report but also the Academy review: "Few scientific subjects receive Academy endorsement on such a superficial basis. It is my considered opinion . . . that the Academy will be hard put to account for having placed its stamp of approval on such a superficial report on a problem that has deeply puzzled the public for over two decades." The Condon Report, he pointed out, analyzed less than 1 percent of the available reports, and even then concentrated too much on trivial cases. "The Colorado project was supposed to explain the tough ones, not the easy ones," he noted, but even so, fully one-third of the cases Condon studied remained unexplained. McDonald characterized the study as 'seriously deficient," the Academy review as a "rubberstamping," and predicted that their conclusions would in the long run seem "almost incredible." One cannot fail to see the sincerity McDonald felt when he wrote, "For the real nature of the UFO phenomena seems so scientifically challenging and cries out for such top-caliber attention that a major study which led only to downgrading it to even lower levels than it had enjoyed in all preceding years seems hard to understand. I cannot understand it."

Although McDonald's reaction to the Condon Report was largely predictable, he was far from alone in his criticism. In particular, Condon must have been surprised and chagrined when an appointed UFO Subcommittee of the American Institute of Aeronautics and Astronautics (AIAA), a professional organization of 25,000 aerospace scientists and

[92] James E. McDonald, review of the *Condon Report, Icarus,* 11 (November 1969), 443–347; H.-Y. Chiu, review of the *Condon Report,* ibid., 447–450. The same issue also carried a review by Canadian astronomer Peter Millman of the 1968 House Hearings, ibid., 439–440.

engineers no less distinguished than their AAAS colleagues, turned its back on Condon's conclusions after its own study. Chaired by Joachim Kuettner of ESSA Research Labs in Boulder – one of the laboratories supposed to lend support to the Condon team – the AIAA subcommittee on UFOs had been formed in 1967 and offered its first statement just after the Condon Report was made public: "The Committee has made a careful examination of the present state of the UFO issue and has concluded that the controversy cannot be resolved without further study in a quantitative scientific manner and that it deserves the attention of the engineering and scientific community." The subcommittee was "greatly perturbed" at the lack of scientific analysis of UFO observations prior to the Condon study and found the Condon Report to be "the most scientifically oriented investigation published on the UFO problem," but concluded, "we find it difficult to ignore the small residue of well-documented but unexplainable cases which form the hard core of the UFO controversy." It noted that while the signal-to-noise ratio of about 5 percent was very low, it could not be ignored: "The issue seems to boil down to the question: Are we justified to extrapolate from 0.99 to 1.00, implying that if 99% of all observations can be explained, the remaining 1% could also be explained; or do we face a severe problem of signal-to-noise ratio (order of magnitude 10^{-2})?" The subcommittee found that the Condon study had made no serious analysis in this direction, and therefore "strongly feels that, from a scientific and engineering standpoint, it is unacceptable to simply ignore substantial numbers of unexplained observations and to close the book about them on the basis of premature conclusions." Thus, while the AIAA's UFO subcommittee found no basis for either rejecting or accepting the extraterrestrial hypothesis, because no basis existed for assessing the probabilities, it also found no basis for Condon's judgment that no further scientific studies of UFOs were needed. To some it seemed a more objective review of Condon's study than that of the National Academy, and it was followed up with what the committee considered some of the best evidence. In a series of articles for its journal *Astronautics and Aeronautics,* James McDonald and Gordon Thayer (the latter who had worked on the Condon study and now worked for ESSA) presented the details of two case studies which they concluded were still completely unresolved. It was to be McDonald's last article; on June 13, 1971, he was found dead from suicide in the desert near Tucson.[93]

[93] "AIAA Committee Looks at UFO Problem," *A&A,* 6 (December 1968), 12; "UFO, An Appraisal of the Problem," *A&A,* 8 (November 1970), 49–51. See also the issues for July 9, 1971, 66, and September 1971, 60 for the articles by McDonald and by Thayer. Klass, who claimed that the committee was loaded with "believers" and without a single

In the wake of congressional hearings, the Condon Report, and the diverse reaction it precipitated, the capstone to scientific interest in UFOs during the late 1960s was a symposium on UFOs held in Boston in December 1969 in conjunction with the annual AAAS meeting. The very fact that the meeting was held at all was a testimony to changed times, but organizing it was not easy. Both Markowitz and Condon were invited to participate; both declined. In their introduction to the proceedings of the meeting, Sagan and Thornton Page (who, we recall, had participated in the 1953 Robertson panel) noted that they had approached the AAAS to hold such a meeting the year before the Condon Report was published, but postponed it in part because of opposition "from some very distinguished scientists" and in part because the Condon Report was not yet available. The organizing committee included Sagan, Page, Morrison, and Walter Orr Roberts, and the editors pointed out that the symposium "could not have been held without the steadfast courage (sometimes in the face of very heated opposition) of the American Association for the Advancement of Science and, in particular, of the Association's then president, Professor Roberts."[94] Some scientists, they held, viewed UFOs as an unscientific subject analogous to astrology, and therefore not to be discussed. Sagan and Page argued that organizations such as the AAAS had the obligation to arrange for discussions on just those subjects that catch the public eye, as a means of demonstrating the scientific method.

Among the 15 invited papers the now familiar positions of Menzel, McDonald, and Hynek were represented, as were the ideas of two of the Condon researchers, Franklin Roach and William K. Hartmann. Carl Sagan argued that the extraterrestrial hypothesis was possible but unlikely, while Frank Drake concluded that "Some aspects of perception work very well, some do well given certain qualifying conditions, and some fail completely."[95] Philip Morrison, with his penchant for zeroing in on the significant issues, argued that the debate came down to the nature of scientific evidence. "Reproducibility" was not enough, for one could not reproduce an aurora or eclipse, nor was "hard evidence" enough (or Darwin would have been in trouble). The prime requirement for responsi-

skeptic, later presented his own explanation for these cases, as did Menzel. For Klass's rebuttal of the AIAA Committee's "best evidence," see his *UFOs Explained*, chapters 19, 20, and 21.

94 Carl Sagan and Thornton Page, eds., *UFOs: A Scientific Debate* (Ithaca, N.Y., 1973), xii. Roberts was head of NCAR, one of the institutions that had pledged its resources to the Condon study. The controversy over holding the symposium is documented in the Menzel Papers, Condon File, American Philosophical Society Library, Box 3, and in the very illuminating letter from Condon to Sagan, October 6, 1969.

95 Carl Sagan, "The Extraterrestrial and Other Hypotheses," in Sagan and Page, *UFOs* 265–275; Frank Drake, "On the Abilities and Limitations of Witnesses of UFO's and Similar Phenomena," in Sagan and Page, *UFOs*, 247–257.

ble evidence, he held, drawing a parallel with the nineteenth-century acceptance of meteorites as extraterrestrial, was "independent and multiple chains of evidence, each capable of satisfying a link-by-link test of meaning." Neither the extraterrestrial hypothesis nor any other explanation of UFOs had multiple chains of evidence or a link-by-link test.[96]

By the end of 1969, the spectrum of scientific opinion on UFOs was well filled. At one end, McDonald, Friedman, a few other scientists, and much of the media and public stood for the extraterrestrial hypothesis. At the other extreme, Donald Menzel, Philip J. Klass, and the official Condon Report believed science would not be further advanced by any study of UFOs – no matter what the hypothesis. In the middle, Hynek and Sagan, the AIAA committee, and many mostly silent astronomers felt that further study might be beneficial, if difficult. As for the Air Force, it ended Project Blue Book on December 17, 1969.[97] Less than a year after the end of his landmark study, Condon gave a brief, light-hearted (some would say flippant) account of his experiences during the study, emphasizing some of the stranger stories he had encountered.[98] It was his last published word on UFOs before his death 4 years later. Though he had charted a distinguished career in quantum physics, the public – and not a few scientists – would remember him primarily for the lasting stamp he put on the UFO debate during 2 years at the height of the controversy in which extraterrestrials had played a dominant – perhaps too dominant – role.

6.3 AFTERMATH: THE NATURE OF EVIDENCE AND THE DECLINE OF THE EXTRATERRESTRIAL HYPOTHESIS IN PHYSICAL SCIENCE

By 1970 the Condon Study was over, the U.S. Air Force was out of the UFO business, and the U.S. Congress was apparently pacified, or at least preoccupied with more immediate concerns such as Vietnam, whose reality – in contrast to that of UFOs – was unquestioned. In retrospect, a quarter century after the beginning of the modern UFO era with the Kenneth Arnold sightings in 1947, we see in the early 1970s the beginnings of a distinct transformation of the UFO debate. UFO sightings were not over – the 1973 national wave kept the controversy going, as did

[96] On the parallel with meteorites see Ron Westrum, "Science and Social Intelligence about Anomalies: The Case of Meteorites," *Social Studies of Science*, 8 (1978), 461–493.

[97] At the request of the participants of the AAAS symposium on UFOs, particularly Menzel, Project Blue Book files were preserved intact in the Air Force Archives at Maxwell Air Force Base, Alabama (Sagan and Page, *UFOs*, 295–301, especially 301). Microfilm copies of about 80,000 pages of material were deposited in the National Archives, Washington, D.C., and made public in 1976.

[98] Condon, "UFOs I Have Loved and Lost," *Proceedings of the American Philosophical Society*, 113 (December 6, 1969), 425–427.

regional waves thereafter[99] – but the hypothesis that UFOs represented alien spacecraft was entering a period of decline, at least from the point of view of physical science and official government attention. Whereas in the 1960s establishment science finally addressed the subject of UFOs with an official study, a National Academy of Sciences review, an AAAS symposium, the persistent prodding of one member of the National Academy (McDonald), and occasional reports and reviews in its mainstream journals like *Science*, such activity rapidly decreased in the 1970s. One may argue whether or not the scientific community had taken the extraterrestrial hypothesis seriously enough in these activities of the 1960s, or whether it should have broadened the scope of its scientific method in addressing the problem, but the historical fact remains that in the aftermath of the Condon Study the hypothesis declined in mainstream science, as measured by scientific attention and discussion in established scientific journals. At the same time, this was decidedly not the case outside science, where the extraterrestrial hypothesis of UFOs remained as alive and well as ever. In this section, we examine the reasons for the decline of the extraterrestrial hypothesis in science, and its continuing viability in American society, notwithstanding the view of science.

At least three circumstances joined in precipitating the decline of the extraterrestrial hypothesis of UFOs in mainstream science. First, even taking into account the failings of the Condon Study, no incontrovertible evidence was produced in favor of the hypothesis, and its champions realized that to maintain the extraterrestrial theory lacking such evidence was an obstacle to further study of the UFO phenomenon. Second, "New Wave" theories of UFOs led away from the extraterrestrial hypothesis and toward ever more ethereal ideas associated with the New Age movement that grew in the last quarter of the twentieth century, confirming to scientists that this was a subject to be avoided. Finally, the rise of claims that, if true, would have proven the extraterrestrial hypothesis – ancient astronauts, spaceship crashes, contactees, and abductees – were based on evidence (such as statements under hypnosis) that most scientists could not accept, thereby bringing the entire extraterrestrial hypothesis into disrepute. Pervading all of these factors was the question of scientific method and the nature of evidence, questions that mean a great deal to working scientists but often all too little to the public. Therein is the

[99] On the 1973 wave, see Jacobs, *The UFO Controversy in America*, chapter 10. The wave did not extend to 1974 (Jacobs, private communication, December 1992). Why no national waves have occurred since 1973 is another relevant piece of data that future historians of the UFO phenomenon must take into account. With the Air Force out of the UFO business, no systematic numbers on UFO sightings are kept, although groups like the Mutual UFO Network (MUFON) still act as a clearinghouse for information.

entering wedge for the dichotomy of opinion between scientists and the public that once again grew wider during the last quarter of the century.

J. Allen Hynek is a prime example of one who was open to the extraterrestrial hypothesis but realized he lacked the proof in any standard sense of scientific method. Hynek's first book-length treatment of the problem, which quickly became a classic in the field, clearly demonstrates why he moved away from the alien as an explanation for the phenomenon he had grappled with for almost a quarter century. *The UFO Experience: A Scientific Inquiry* (1972) was perhaps a book Hynek should have written long before, considering his lengthy association with the Air Force study. In it he introduced his "strangeness-probability" diagrams as a measure of the importance of particular UFO cases. He classified UFO reports into six prototypes, three of which were in the "distant" category (nocturnal lights, daylight discs, and radar and visual reports) and the other three of which were "close encounters." Hynek showed how far he had come since his skeptical days when he analyzed in a serious way not only the distant category but also the close encounters, bringing into the UFO lexicon the phrases "close encounters of the first kind" (close experiences without physical effects), "close encounters of the second kind" (measurable physical effects), and "close encounters of the third kind" (animated creatures reported). In a chapter "Science Is Not Always What Scientists Do," Hynek noted (as had McDonald) that the Condon Report contained many unexplained cases, which Condon chose to ignore in his summary and recommendations; that by choosing to equate UFOs with the extraterrestrial hypothesis, Condon wrongly rejected the entire UFO phenomenon; and that Condon himself was preoccupied with the weirdest cases, to the detriment of the better ones.

Significantly, Hynek pointed out that the extraterrestrial hypothesis could not be proven untrue, and he deduced from this not that it was true (as some would), but that the Condon team had adopted the wrong methodology. While in his view the team should have tested the hypothesis "There exists a phenomenon, described by the content of the UFO reports, which presently is not physically explainable," they instead tackled a "hopeless task" – trying to test the extraterrestrial hypothesis. No crucial observation or experiment could prove the latter false, Hynek argued, while the former could have been proven false by explaining the cases which no one else had been able to explain.[100] It was as if the field

100 J. Allen Hynek, *The UFO Experience: A Scientific Inquiry* (Chicago, 1972). I am using the (New York, 1974) edition, where these quotations are found on pages 227–228. For Hynek's view of the *Condon Report* see "The Condon Report and UFOs," *Bulletin of the Atomic Scientists,* 25 (1969), 39. See also the interview with Hynek by Ian Ridpath in *New Scientist* (May 17, 1973), 422–424.

had been painstakingly narrowed to the good data, the signal separated from the noise, and then the signal thrown away. From his own experience, Hynek concluded that the UFO reports described a phenomenon "worthy of systematic, rigorous study," and (like the AIAA) he suggested that this be undertaken without dragging the speculative and emotional extraterrestrial hypothesis into consideration. Following up on this conviction, in 1973 Hynek founded the Center for UFO Studies as a repository for data and study of the UFO phenomenon.[101]

The idea that the UFO problem might have a better chance of scientific study if stripped of its extraterrestrial connotations had some success. Hynek's plea for further study impressed at least one mainstream scientist, whose review of the book for *Science* stood in sharp contrast to that journal's endorsement of the Condon Report a few years before. Bruce Murray, professor of planetary science at Caltech and a future director of the Jet Propulsion Laboratory, wrote, "On balance, Hynek's defense of UFOs as a valid, if speculative, scientific topic is more credible than Condon's attempt to mock them out of existence." And Murray did not stop there. He viewed Hynek's inability to obtain funds from NASA or the National Science Foundation to study UFOs as "a dismal symptom of the authoritarian structure of establishment science." Moreover, Murray castigated *Science* for treating the Condon Report as a news item rather than publishing a rebuttal. "From this juror's point of view at least," Murray concluded, "Hynek has won a reprieve for UFOs with his many pages of provocative unexplained reports and his articulate challenge to his colleagues to tolerate the study of something they cannot understand."[102] He had not, however, won a reprieve for the extraterrestrial hypothesis, and we see the emphasis moving away from that theory even by those who advocated further study of the UFO phenomenon.

The same avoidance of the extraterrestrial hypothesis while pressing for further study is seen in the critique of the Condon Report by astrophysicist Peter Sturrock of Stanford's Institute for Plasma Research. Sturrock pointed out that despite the efforts of a few individuals, the great bulk of the scientific establishment still ignored the UFO problem, regarding it as unproductive and not respectable. The reason for this

[101] Hynek, *UFO Experience,* chapter 13, "The Case Before Us," 242–257. The Center for UFO Studies publishes the *Journal of UFO Studies*, with "scholarly papers and issues forums on ufological matters."

[102] Bruce Murray, "Reopening the Question," review of Hynek's *The UFO Experience*, in *Science*, 177 (August 25, 1972), 688–689. Regarding establishment science, Hynek noted (207–213) that W. T. Powers, "A Critique of the Condon Report," was refused publication in *Science* in 1969, and Hynek reprinted part of it at the close of his book. Powers was the chief systems engineer at Dearborn Observatory – the observatory Hynek headed.

negative attitude, he believed, was that scientists normally looked for their hard data in scientific journals, which almost always refused to publish UFO articles based on the advice of peer reviewers. Furthermore, he emphasized that Condon's methodology seemed to demand a single convincing case, the normal method pursued by physicists, while the UFO phenomenon might demand a method more often adopted by astronomers, the laborious accumulation of data. No single observation in astronomy, he pointed out, proved Kepler's laws or the shape of the galaxy or stellar evolution. For this reason, Condon had rejected some of the data discussed by his own staff. Thus, Sturrock concluded, "the weaknesses of the Condon Report are an understandable but regrettable consequence of a misapprehension concerning the nature and subtlety of the phenomenon. . . . The substance of the Condon Report represents a persuasive case for the view that there is some phenomenological fire hidden behind the smoke of UFO reports, and the Report therefore *supports* the proposition that further scientific study of UFOs is in order."[103] Moreover, to Sturrock it seemed that the National Academy's review of the Condon Report was based mainly on Condon's summary rather than on the bulk of the report, which differed from that summary.

Others, notably Allan Hendry at Hynek's Center for UFO Studies in the late 1970s, made similar pleas for drastically new methodologies to study the UFO phenomenon.[104] By that time, several related themes were becoming clear to critics of Condon and of previous studies: too much attention had been given to the extraterrestrial hypothesis; the UFO phenomenon deserved further study with a much broader focus; and any further study needed carefully to consider broadening its scientific methodology. The highly emotional issue of the extraterrestrial hypothesis came to be seen as an obfuscation by those supporting further study of the UFO phenomenon. Short of a spaceship landing in plain view, the hypothesis did not seem to be susceptible to normal scientific methods. Three decades of sporadic study had resulted in no incontrovertible proof. If, as Hynek had proclaimed, the extraterrestrial hypothesis was unfalsifiable, then by some definitions it was not a scientific hypothesis at all and attention should turn to other theories that were scientific.

UFO debunkers, however, were not about to let their prey slip away that easily. They used adherence to standard methodology not only to

103 P. A. Sturrock, "Evaluation of the Condon Report on the Colorado UFO Project" (Stanford University Institute for Plasma Research Report No. 599, October 1974). 33 pp. The quotation is from p. 28. Sturrock also showed that most of the cases studied were by junior members of the staff, supported by his useful "Breakdown of Activities among Staff."

104 Allan Hendry, *The UFO Handbook: A Guide to Investigating, Evaluating and Reporting UFO Sightings* (New York, 1979).

deny the extraterrestrial hypothesis, but also to deny that the UFO phenomenon was at all novel, significant, or outside the realm of normal explanation. Critical of the UFO promoters, Klass had announced his intention to follow the methodology of Alfred North Whitehead:

> The progress of science consists in observing interconnections and in showing with a patient ingenuity that the events of this ever-shifting world are but examples of a few general relations, called laws. To see what is general in what is particular, and what is permanent in what is transitory, is the aim of scientific thought.

Klass left no doubt that in his view UFOs were prosaic phenomena, and went on to offer his explanations for some of the most celebrated cases, including the two cases the AIAA considered strongest.[105] To the end, Menzel held to his original credo that a scientist must always choose the simplest hypothesis, and using that method in his final book on UFOs published just before his death, he still believed all sightings could be explained on the basis of known natural phenomena.[106] Robert Schaeffer, another debunker, found only one hypothesis left: "It is the familiar null hypothesis, the cornerstone of statistical analysis; UFOs, as a phenomenon distinct from all others, simply do not exist." No one would be happier than he, Schaeffer insisted, should the extraterrestrial hypothesis be proven.

> But wishing will not make it so. And so long as we wish to adhere to the scientific method (that is, to make factual statements about the real world, as opposed to seeking subjective mystical insight), we are forced to face up to the conclusion that UFOs as real and distinct entities simply do not exist. Those who continue to insist otherwise are openly proclaiming their allegiance to a different world-view, one which, although popular, is incompatible with the world-view of science.[107]

105 *UFOs Explained* (New York, 1974), 5; Klass's other works include *UFOs – Identified* (New York, 1968) and *UFOs: The Public Deceived* (Buffalo, N.Y., 1983). Klass believed the Condon study was not well managed, but he endorsed its conclusions in "The Condon UFO Study: A Trick or a Conspiracy," *The Skeptical Inquirer,* 10 (Summer 1986), 328–341.

106 Donald H. Menzel and Ernest H. Taves, *The UFO Enigma. The Definitive Explanation of the UFO Phenomenon* (New York, 1977). In the Introduction to this volume, astronomer Fred Whipple noted, "I share their [the authors'] firm belief that extraterrestrial life and intelligence are relatively frequent in the universe. But a belief or a hope is a far cry from scientific supporting evidence." Regarding Condon, Menzel and his coauthor concluded "we believe the Colorado staff did a good job. They analyzed many sightings and explained a large number of them as either natural phenomena or chicanery."

107 Robert Sheaffer, *The UFO Verdict: Examining the Evidence* (Buffalo, N.Y., 1981), 211–213. Ironically, just as Schaeffer and others were concluding that extraterrestrials

Table 6.2. *Spectrum of scientific cultures on the UFO question*

Explanation	Advocate	Characteristics
Prosaic natural phenomena (astronomical, balloons, etc.) or fraud	Menzel (1952-1976) Hynek (1950s-1965) Condon (1968) NAS panel (1969) Klass (1966-1990s) *Skeptical Inquirer*	Skepticism Caution "Simplest explanation" Need for Strong evidence for theory
Potentially significant new phenomenon: further study needed	AIAA Committee (1968) Hynek (1966-1986) Murray (1972) Sturrock (1974) *Journal of Scientific Exploration* Vallee (1965 on)	Consider accumulation of data (no single confirmation) Extract signal from noise
Extraterrestrial hypothesis	McDonald (1966-1971) Hynek (1968 and sporadically) Vallee (1965-1980) *Journal of UFO Studies* (on occasion)	Risk takers Willing to speculate Willing to consider alternative methods, especially eyewitness accounts

Hynek, Murray, Sturrock, and others – all scientists themselves – would hardly agree; rather than discarding a phenomenon that did not fit within the normal canons of scientific evidence, they championed a broader view of science that could encompass the study of an unknown phenomenon.

Like the debate over the canals of Mars, the UFO controversy thus reveals the existence of many cultures of science, each with its own characteristics, approaches, and canons of scientific evidence (Table 6.2). Even those with broader views of scientific method, however, agreed that while the UFO phenomenon should be pursued, there was no incontrovertible evidence for the extraterrestrial hypothesis and seemingly no way to obtain such evidence. In short, even as NASA laid plans to search for

were not here, others were concluding that they should be here, given the existence of extraterrestrials and the abundance of time for interstellar travel. Since they apparently were not, this was an argument against the existence of extraterrestrials at all, a point the SETI proponents did not accept. Schaeffer contributed to this tradition a skeptical UFO paper entitled "An Examination of Claims that Extraterrestrial Visitors to Earth Are Being Observed" in *Extraterrestrials: Where Are They?* (New York, 1982), 20–28.

extraterrestrial intelligence by means of radio telescopes, the idea that such intelligence had traveled to Earth was not considered a fruitful scientific hypothesis.[108]

The desire to broaden the scope and method of science was laudable but extremely difficult. Faced with finding alternatives to the extraterrestrial hypothesis, many unfortunately replaced it with even wilder theories. A second factor in the decline of the extraterrestrial hypothesis in mainstream science is the appearance of the New Wave UFO theories – the notion that UFOs "are 'metaphysical,' 'supernatural,' or 'interdimensional.' " While these theories were sometimes combined with the idea of extraterrestrial spacecraft, they began to supersede the "nuts and bolts" extraterrestrial hypothesis of spacecraft that traveled by the conventional laws of physics. Referring to the standard spaceship concept, Schaeffer found that "while the pace of UFO sightings quickened in the mid-to-late 1960s, the tenability of the extraterrestrial hypothesis was gradually deteriorating." By the mid-1970s, even former extraterrestrial hypothesis advocates such as Hynek and Vallee began to cast aside the standard spaceship hypothesis for new ones that involved parallel universes, interdimensional travel, and "astral planes." In "The Case against Spacecraft" in his book *Messengers of Deception* (1980), Vallee argued that the spacecraft hypothesis no longer explained the facts of the UFO phenomenon. In a series of books in the late 1980s and early 1990s, Vallee refined his thesis that "The genuine UFO phenomenon . . . is associated with a form of nonhuman consciousness that manipulates space and time in ways we do not understand." This form of consciousness "does not have to be extraterrestrial. It could come from any place and any time, even from our own environment. . . . The entities could be multidimensional beyond space-time itself. They could even be fractal beings. The earth could be their home port."[109] Hynek also seemed sympathetic to such ideas, and if these once relatively conservative ufologists were now attracted by such

[108] How the two ideas could be separated has been commented on by Henry H. Bauer, *Scientific Literacy and the Myth of Scientific Method* (Urbana and Chicago, 1992), 5–8, where Bauer notes that SETI efforts might be considered in some measure successful even with negative results that allow certain frequencies or lifesites to be eliminated, while there is no such measure of success for UFO research.

[109] Jacques Vallee, *Revelations: Alien Contact and Human Deception* (New York, 1991), 259. This volume was the final volume of Vallee's "Alien Contact" trilogy, including *Dimensions: A Casebook of Alien Contact* (Chicago and New York, 1988) and *Confrontations: A Scientist's Search for Alien Contact* (New York, 1990). Vallee's change of mind (and Hynek's as well) is evident beginning in Hynek and Jacques Vallee, eds., *The Edge of Reality: A Progress Report on Unidentified Flying Objects* (Chicago, 1975) and continuing with *Messengers of Deception: UFO Contacts and Cults* (New York, 1980).

ideas, one can well imagine the hypotheses of those with lesser scientific backgrounds.[110]

Although Vallee's ideas were by no means widely accepted, and although the extraterrestrial hypothesis continued to flourish at levels outside mainstream science,[111] the very existence of such New Wave ideas carried UFOs further from the realm of accepted science and into what most scientists considered pseudoscience, with a predictable reaction.

> A dispassionate observer of the current scene can only be astonished by the rapid growth of bizarre beliefs in recent years among wide sectors of the public. This involves everything from belief in "psychic" forces – clairvoyance, precognition, telepathy, psychokinesis, psychic surgery, psychic healing, astral projection, levitation, plant ESP, life after life, hauntings, and apparitions – the widespread conviction that our earth has been visited in the past by ancient astronauts in "chariots of the gods" and is being visited today by extraterrestrial creatures in space vehicles,

wrote philosopher Paul Kurtz in studying these wider phenomena.[112] The scientific reaction to the broad realm of pseudoscience was inevitable, and as UFOs became more and more closely associated with that realm, the scientific reaction was inevitable. When Harvard astronomer Fred Whipple saw Menzel's last book, for example, as "a solid steppingstone out of the morass of unconscious yearning for supernatural intervention," one could hardly blame him.

That extraterrestrial presence on Earth had become a part of the New Age philosophy was recognized by Vallee:

> To the New Age idealists, the announcement that aliens are here would bring the culmination of many decades of dreams.

[110] On Hynek's shifting ideas see David Jacobs, "J. Allen Hynek and the Problem of UFOs," paper given at the History of Science Society Meeting, Washington, D.C., December 1992, 16.

[111] I wish to thank Michael Swords, editor of the *Journal of UFO Studies*, for bringing this to my attention. In his opinion, of the top 20 UFO investigators in North America in 1992 (more than half of whom hold Ph.D.s), all but two would rest along a spectrum of "sold" to "intrigued" with the extraterrestrial hypothesis. Private communication, December 15, 1992. Swords himself wrote many articles favoring the standard extraterrestrial hypothesis, and the theory is still well represented at enthusiast meetings such as those of the Mutual UFO Network (MUFON), UFO journals and newsletters such as the *International UFO Reporter*, and among the public.

[112] Paul Kurtz, "Believing the Unbelievable: The Scientific Response," in *Science and the Panoramal: Probing the Existence of the Supernatural*, eds. George O. Abell and Barry Singer (New York, 1981), vii. See also Philip H. Abelson, "Pseudoscience," *Science*, 184 (June 21, 1974).

> It would validate all their group meditations on mountaintops, the loving hopes, the prayers for peace. It would give all of us something to worship at a time when the leaders of our traditional religions have made fools of themselves, at a time when the younger generation has very few heroes it can look up to.[113]

The fact that aliens were devoutly desired in the New Age movement did not advance the extraterrestrial hypothesis one bit in the scientific community, nor were scientists impressed with Hynek and Vallee's abandonment of that hypothesis for ones even further removed from empirical confirmation. And if Vallee decried the New Age desire for alien salvation, his own ideas of UFOs surely found sympathy in the New Age philosophy.

Finally, even more weighty than the defection of Hynek, Vallee, and others to New Age explanations of the old UFO phenomena, the spectacular claims of new phenomena ironically contributed mightily to the downfall of UFO respectability among scientists. In particular, the rise of belief in ancient astronauts, crashed spaceships, contactees, and abductees made the extraterrestrial hypothesis disreputable; Vallee himself showed the absurdities of many of these claims in his own books. If the UFO evidence was confusing by scientific standards, evidence of ancient visitations was mostly embarrassing. Although the idea had been broached before and would become a publishing phenomenon later, it was the German writer Erich von Däniken who made ancient astronauts a popular theme worldwide during the 1970s. As the Condon committee was finishing its report, von Däniken was writing his *Chariots of the Gods?* (1970), followed by many more books in the 1970s.[114] The question of ancient astronauts was not invalid; given the time scale of the universe, it is not out of the question that Earth could have been visited in the past. The challenge was to come up with substantial evidence, and here almost all scientists, and to its credit even much of the public, found von Däniken wanting.[115]

[113] Vallee, *Revelations*, 251–252.

[114] Erich von Däniken, *Chariots of the Gods? Unsolved Mysteries of the Past* (New York, 1970), trans. by Michael Heron (first published in Germany under the title *Erinnerungen an die Zukunft* in 1968 and in Great Britain in 1969). This was followed by *Gods from Outer Space: Return to the Stars, or Evidence for the Impossible*, originally published as *Return to the Stars* (New York, 1971); *The Gold of the Gods* (New York, 1973); *In Search of Ancient Gods* (New York, 1973); *Miracles of the Gods* (New York, 1975); and *Von Däniken's Proof*, first published as *According to the Evidence: My Proof of Man's Extraterrestrial Origins* (New York, 1977), trans. by Michael Heron.

[115] For critiques of von Däniken see Ronald Story, *The Space-Gods Revealed: A Close Look at the Theories of Erich von Däniken* (New York, 1976); Story, *Guardians of the*

The controversy over crashed spaceships, alien contactees, and abductees was even worse from the point of view of most reputable scientists. Books such as *The Roswell Incident* (1980) and *UFO Crash at Roswell* (1991), purporting to describe a saucer crash in 1947, simply could not be taken seriously by most scientists.[116] As for contactees and abductees, what in the 1950s and 1960s had been a minor diversion to the UFO theme with the claims of Adamski and the Hills came in the 1970s and 1980s to dominate it. Two fishermen in Mississippi claimed to have been given a physical exam aboard a flying saucer in 1973. Budd Hopkins documented his abduction claims in *Missing Time* (1981) and *Intruders* (1987), the latter published by the prestigious Random House. Whitley Strieber's claims of abduction in *Communion: A True Story* (1987) spawned a best-seller. In *Secret Life: Firsthand Accounts of UFO Abductions* (1992), UFO historian David Jacobs argued the case for alien abductions, based largely on statements of subjects under hypnosis. And in 1994, the work of Hopkins, Jacobs, and others was given a further boost when the formerly skeptical Harvard psychiatrist John Mack concluded in *Abductions: Human Encounters with Aliens* that the phenomenon could not be explained psychiatrically, was not possible within the framework of the modern scientific worldview, and was in all likelihood truly explained by alien abduction. All of these claims were amplified by radio and television talk shows and even news reports. Once again, in a repeat of the 1950s UFO phenomenon, most scientists were only too happy to leave this discussion to the media.[117] The phenomenal popular-

Universe? (New York, 1980); and E. C. Krupp, "The von Däniken Phenomenon," *Griffith Observer* (July 1977), 10–20. On the general UFO question see Story, *UFOs and the Limits of Science* (New York, 1981).

116 Charles Berlitz and William L. Moore, *The Roswell Incident* (New York, 1980). Klass, *The Public Deceived,* 279–289, critiques the book. Berlitz was the author of two books on the Bermuda Triangle. Kevin Randle and Donald R. Schmidt, *UFO Crash at Roswell* (New York, 1991). Because of the continuing publicity, in early 1994 Congressman Steven Schiff asked the the U.S. General Accounting Office to investigate the Roswell crash. This led to an Air Force report which revealed that the debris found at Roswell was wreckage from balloons, sensors, and radar reflectors from the secret "Project Mogul," designed to search the high atmosphere for reverberations from nuclear tests. *New York Times,* September 18, 1994, 1.

117 For two viewpoints on abductions in the early 1990s see Vincente-Juan Ballester Olmos, "Alleged Experiences inside UFOs: An Analysis of Abduction Reports," *Journal of Scientific Explanation*, 8 (Spring 1994), 91–105, and the immediately following article, David Jacobs, "Response to Ballester Olmos," 105–109. For critiques of the abduction claims see Philip J. Klass, *UFO Abductions: A Dangerous Game* (Buffalo, N.Y., 1989); on Strieber, Ernest H. Taves, "Communion with the Imagination," *Skeptical Inquirer*, 12 (February 1987), 90–96. It is perhaps relevant that before publishing "Communion," Strieber was a well-known author of horror novels. In contrast to the 1950s, the scientific skeptics were now organized, and many other articles in *Skeptical Inquirer* (published by the Committee for the Scientific Investigation of Claims of the Para-

ity of the subject demonstrated the continued willingness of the public to adopt, without physical evidence, even the most extreme beliefs of the extraterrestrial hypothesis – close encounters of the third kind – or at least to use them for entertainment value. In short, such beliefs demonstrated allegiance to a different worldview than the scientific, as normally understood.

Lost in a sea of paranormal claims that brought it into disrepute among scientists, the extraterrestrial hypothesis after Condon swung back toward the pre-1965 media excesses and became virtually the exclusive property of authors with mixed motives and unscientific credentials. As such widespread claims proliferated, the UFO phenomenon clearly fell from the province of scientific discourse. The hopes of those who wished to expand the scope of science were largely swamped by the absurd claims of those whose motives were closer to profit than truth. The signal-to-noise ratio problem that had plagued the UFO phenomenon from the beginning was now joined by the signal-to-noise problem of unscrupulous authors.

By the end of the century, the question was not whether extraterrestrials were visiting Earth, but whether there was anything at all to the UFO phenomenon that science could productively illuminate. Neither normal scientific method nor the norms of interaction among scientists held out much hope. Gerard Kuiper believed that the scientist did not avoid UFO research from fear of ridicule but rather because he

> knows how limited and cut-up is the time he can devote to research, between his numerous other duties. He selects his area of investigation not because of pressures but because he sees the possibility of making some significant scientific advance. . . . A scientist would consider the discovery of evidence of life on another planet as perhaps the greatest contribution he could make and one that might earn him the Nobel Prize. But this is no reason for him to chase every will-o'-the-wisp.[118]

There is a practical ring of truth to Kuiper's statement, yet it undoubtedly underestimates the role of scientific peer pressure, which, combined with the elusiveness of the problem and the drift toward pseudoscience, is

normal [CSICOP], founded by philosopher Paul Kurtz) take a skeptical attitude toward such claims. Proponents of a broader scope for science and the scientific method may be found in the *Journal of Scientific Exploration*, a publication of the Society for Scientific Exploration, founded by Peter Sturrock. Sturrock describes the origin of his society as related to the UFO phenomenon in "Guest Column: A Survey and a Society," *Journal of Scientific Exploration*, 8 (Spring 1994), 129–134.

[118] Condon, *Final Report*, 839–890.

likely to render further research slow at best.[119] The continued interest of scientists in the aftermath of the Condon Study is underscored in a poll of astronomers' attitudes toward UFOs undertaken by Sturrock in 1977, the first since Hynek's 25 years before. It showed that 53 percent of responding astronomers believed that the UFO problem deserved further study. Sixty two respondents even claimed they had seen or obtained an instrumental record of an event they believed to be related to UFOs. But only 13 percent of all respondents could see any way to resolve the problem of identifying UFOs.[120] If an objective of science was "a consensus of rational opinion," Sturrock concluded, it had failed in the case of UFOs.

Thus, in addition to career considerations and peer pressure, the relative inaction of scientists by the end of the century seemed an admission that a solution to the problem might have to await further developments in science. While the UFO phenomenon may be beyond the understanding of twentieth-century physics, Hynek liked to say, "there will surely be, we hope, a twenty-first century science and a thirtieth century science, and perhaps they will encompass the UFO phenomenon as twentieth century science has encompassed the aurora borealis, a feat unimaginable to nineteenth century science, which likewise was incapable of explaining how the sun and stars shine."[121] Meanwhile, the history of the debate finds abundant blame on all sides: the unwillingness of many scientists to actively engage the UFO phenomenon is certainly understandable in terms of peer pressure and career advancement, but the desire of some to squash the subject without considering the evidence flies in the face of scientific curiosity that supposedly led them to science to begin with. On the other side, the outrageous claims and hoaxes that presently flood the field are unworthy of scientific attention, and no one should be surprised if scientists fail to engage every will-of-the-wisp report. In the middle of these extremes may yet be a phenomenon that requires study, if only one can find it in the midst of the twin human failings in perception and deception.

In the meantime, lacking any definitive resolution, at the very least the UFO phenomenon holds an important place in the history of the mythic

[119] A fascinating case of peer pressure even among heavyweights in science is Menzel's letter to Harold Urey of July 31, 1969 (Condon papers), in which Menzel chastised Urey for implying in an interview with *Forbes* that he disagreed with the Condon findings.

[120] Peter A. Sturrock, "Report on a Survey of the Membership of the American Astronomical Society concerning the UFO Problem" (1977), reprinted in *Journal of Scientific Exploration*, 8 (Spring, Summer, and Autumn 1994), 1–45, 153–195, 309–346. These results were consistent with a more limited survey conducted by Sturrock of the AIAA membership, *Astronautics and Aeronautics*, 12, no. 5 (1974), 60. For a critique of Sturrock's survey, see Klass, *The Public Deceived*, 57–63.

[121] Hynek, *UFO Experience*, 262.

imagination. Increasingly, toward the end of the century, some scholars and scientists began to view it in just this way. One study called the phenomenon "the first myth to develop in the modern, high-tech, instant global communications world." Another asked, "Do fairies, ghosts, and extraterrestrials exist as living beings – or are they some product of the human mind? The obvious answer to which the evidence overwhelmingly, unemotionally and logically points is a resounding no! They are mental constructs." Even Vallee admitted that the desire for extraterrestrials could be viewed as a modern version of demons and elves.[122]

Whether or not any definitive explanation of the UFO phenomenon is forthcoming, historically the effect of the extraterrestrial hypothesis of UFOs on the extraterrestrial life debate was multifaceted. There is no doubt that it brought scientists out of the closet on a subject they otherwise might never have addressed. It was the public's chief exposure to the subject of extraterrestrial life, and even as more sober and scientific SETI programs were undertaken, UFOs remained prominent in the public mind. For scientists the decline of the extraterrestrial hypothesis of UFOs did not mean the decline of belief in extraterrestrial life; even Menzel had always kept the two distinct. Even so, it is remarkable that proponents of extraterrestrial intelligence would be able to push ahead with their SETI programs even as the UFO phenomenon brought the subject of extraterrestrials on Earth into disrepute.

The attempts of scientists to understand the UFO phenomenon have close parallels with the rest of the extraterrestrial life debate. We recall (Chapter 4) Peter van de Kamp's comment on the extreme difficulty of observing extrasolar planetary systems: "Should we not come to the rescue of a cosmic phenomenon trying to reveal itself in a sea of errors?" he asked. Though van de Kamp was dealing with instrumental measurements, and although the word "cosmic" may prove to be irrelevant to the UFO debate, the phenomenon has surely been plagued by a sea of errors no less than other, more prosaic attempts to find life beyond the Earth. Whether or not a true natural phenomenon is trying to reveal itself only the future will tell. But if and when extraterrestrial intelligence is discovered in the distant reaches of space, it will surely be recalled that for a time in the twentieth century, more than a few Earthlings believed the extraterrestrials had actually frequented the skies of their home planet, crashed or landed, and abducted millions of its citizens.

[122] J. Spencer, *Perspectives* (London, 1989); J. Vallee, *Confrontations* (New York, 1990); Keith Thompson, *Angels and Aliens: UFOs and the Mythic Imagination* (Reading, Mass., 1991); Carl Sagan, "What's Really Going On?" *Parade* (March 7, 1993), 4–7. See also Olmos, "Alleged Experiences."

7

THE ORIGIN AND EVOLUTION OF LIFE IN THE EXTRATERRESTRIAL CONTEXT

Life only avails, not the having lived. Power ceases in the instant of repose; it resides in the moment of transition from a past to a new state, in the shooting of the gulf, in the darting to an aim.

Emerson[1]

There is every reason now to see in the origin of life not a "happy accident" but a completely regular phenomenon, an inherent component of the total evolutionary development of our planet. The search for life beyond Earth is thus only a part of the more general question which confronts science, of the origin of life in the universe.

A. I. Oparin (1975)[2]

A full realization of the near impossibility of an origin of life brings home the point [of] how improbable this event was. This is why so many biologists believe that the origin of life was a unique event. The chances that this improbable phenomenon could have occurred several times is exceedingly small, no matter how many millions of planets in the universe.

Ernst Mayr (1982)[3]

. . . Nowhere in all space or on a thousand worlds will there be men to share our loneliness. There may be wisdom; there may be power; somewhere across space great instruments, handled by strange, manipulative organs, may stare vainly at our floating cloud wrack, their owners yearning as we yearn. Nevertheless, in the nature of life and in the principles of evolution we have had our answer. Of men elsewhere, and beyond, there will be none forever.

Loren Eiseley (1957)[4]

Although the origin of life is today viewed as central to the question of life in the universe, prior to the middle of the twentieth century, theories of the origin and evolution of life played very little role in the extraterrestrial life debate. The reason is found, in part, in the fact that extraterrestrial life had historically been the province of astronomers rather than biologists, despite a few exceptions, including A. R. Wallace. The determination of

[1] Ralph Waldo Emerson, "Self Reliance," in Emerson, *Essays and Lectures* (New York, 1983), 271.

[2] A. I. Oparin, "Theoretical and Experimental Prerequisites of Exobiology," in *Foundations of Space Biology and Medicine*, eds. M. Calvin and O. G. Gazenko (Washington, D.C., 1975), vol. 1, 321–367: 367.

[3] Ernst Mayr, *The Growth of Biological Thought: Diversity, Evolution and Inheritance* (Cambridge, Mass., 1982), 583–584.

[4] Loren Eiseley, *The Immense Journey* (New York, 1957), 162.

conditions on the planets and the search for planets around other stars were physical questions logically prior to any discussion of biological details. And when it came to biology, astronomers for the most part simply assumed that since life had originated on Earth, it would originate on any other planet under proper conditions. Percival Lowell and other astronomers were therefore content to determine planetary conditions rather than to study mechanisms of how life might have originated.

But the lack of biological input into the extraterrestrial life debate in the first half of the century was also a reflection of the state of biology itself. The subject of extraterrestrial life was simply beyond the concern of most biologists during this period, for the very good reason that biology itself was not yet a mature discipline. Modern biology as an autonomous science is generally considered to have begun only with the events precipitated by the publication of Darwin's *Origin of Species* in 1859, and not until a century later were the biological sciences unified via the evolutionary synthesis. As one historian of biology has perceptively written, "By 1955 biology had become not only a unified science, and an empirical science, but a mature science secure of its foundations and well positioned within the positivist ordering of knowledge – intermediate between the physical sciences and the social sciences." Not only was biology still immature during the first half of the century, it was also inferior to physical science in the sense that, unlike Newtonian physics, it lacked universality. Confined to the Earth, no "biological law," however unified, could compete for status with the universality of Newton's laws. Just at the time the unification of the biological sciences was completed, however, the dawn of the Space Age brought the possibility of extending biology beyond the Earth and, by searching for life on Mars (and eventually beyond the solar system), aspiring to the same status of universality held by the physical sciences. The impressive progress in the study of terrestrial biology did not allow one to separate the contingent from the necessary in living systems, including its chemical basis, the role of proteins and nucleic acids, and the probability of life's origin in the first place. "To the extent that we cannot answer these questions," wrote a panel of top biologists commissioned by the U.S. National Academy of Sciences to consider the problem of life on Mars, "we lack a true theoretical biology as against an elaborate natural history of life on this planet. . . . The existence and accessibility of Martian life would mark the beginning of a true general biology, of which the terrestrial is a special case."[5] This search for a "true general biology" was the promise and the

[5] C. S. Pittendrigh, Wolf Vishniac, and J. P. T. Pearman, *Biology and the Exploration of Mars, Report of a Study Held under the Auspices of the Space Science Board, National*

hope of exobiology, and we shall see it embodied in this chapter in the search for the origins and evolution of life.

Given the attention that historians and philosophers of science have lavished on the Darwinian revolution, it is surprising that the history of ideas of the origin of life has received so little analysis. The two substantial studies that have been written make it clear that by the beginning of the twentieth century the problem of the origin of life had reached an impasse.[6] Darwin himself could shed little light on the subject of origins. Having expounded his evolutionary theory in the *Origin of Species,* Darwin concluded only that "There is grandeur in this view of life, with its several powers, having been originally breathed by the Creator into a few forms or into one; and that, whilst this planet has gone cycling on according to the fixed law of gravity, from so simple a beginning endless forms most beautiful and most wonderful have been, and are being evolved."[7] A dozen years later he added only a passing remark about a "warm little pond" in which life might have originated. Most biologists were dissatisfied with such a vague statement; if the most important concept of nineteenth-century biology was the origin of species, one of the many important subjects left for the twentieth century would be the origin and early evolution of life itself.

The impasse at the beginning of the century was largely due to the work of Louis Pasteur. Careful experiments by Pasteur in 1860 had been unable to prove the prevailing view that life originated from nonlife by spontaneous generation. As one historian pointed out, this left three possibilities: one could deny Pasteur's results, a very isolated position by the end of the century; one could view Pasteur's results as valid to a point, but not applicable to extremely simple organisms in the distant past; or one could take Pasteur to be strictly valid and seek an origin of life from beyond the Earth. The lack of experimental proof made the first option

Academy of Sciences, National Research Council, 1964–1965 (Washington, D.C., 1966), 5. On biology as a unified and mature science see V. B. Smocovitis, "Unifying Biology: The Evolutionary Synthesis and Evolutionary Biology," *Journal of the History of Biology,* 25 (1992), 1–65: 55. For the broad outlines of the origins and development of modern biology see William Coleman, *Biology in the Nineteenth Century: Problems of Form, Function and Transformation* (New York, 1971), and Garland Allen, *Life Science in the Twentieth Century* (New York, 1975).

[6] The two chief historical studies on the origin of life are John Farley, *The Spontaneous Generation Controversy from Descartes to Oparin* (Baltimore and London, 1977), and Harmke Kamminga, "Studies in the History of Ideas on the Origin of Life from 1860," Ph.D. thesis, University of London (November 1980). Many of the pioneering papers in the origin of life are reprinted in David W. Deamer and Gail R. Fleischaker, *Origins of Life: The Central Concepts* (Boston, 1994).

[7] Darwin, *Origin of Species* (London, 1859), last sentence of Darwin's Conclusion. Darwin's "warm little pond" remark is in a letter from Darwin to botanist Joseph Hooker in 1871.

increasingly unlikely by the end of the nineteenth century, leaving biologists to focus on a possibly spontaneous origin of simple organisms in the distant past or to postulate an origin of life from beyond the Earth.[8]

In this chapter we do not attempt to present a full history of theories of the origin and evolution of life. Rather, we focus on the delayed, but profoundly significant, effect of these theories on the extraterrestrial life debate and examine, conversely, how the Space Age search for extraterrestrial life influenced the study of the origins of life on Earth – in short, how origins of life and extraterrestrial life studies became integrated in the course of the century. We shall see how, during the first two decades of the twentieth century, the idea that life had reached the Earth from outer space – the "panspermia theory" – lent credence to the idea that life was also spread throughout the universe. We shall see that when after a few decades the panspermia theory was found to be problematic, it was followed by the development of an Earth-bound "chemical theory," with only tenuous and latent links to extraterrestrial life. Only with the rapid development of the life sciences in the second half of the twentieth century, and the beginning of the Space Age, did origins of life considerations become a crucial component of the extraterrestrial life debate, with the latter debate, in turn, having a substantial effect on the study of origins of life. During this period, experiments in the synthesis of organic compounds, analytic techniques applied to meteorites, and the observation of organic molecules in a variety of environments in outer space – all interpreted within the framework of the new chemical theory – combined to achieve notable successes in origins of life studies while leaving the field tantalizingly distant from any final explanation. By the closing decades of the century, the terrestrial and extraterrestrial theories of the origins of life on Earth vied for dominance, both now fully integrated into the discussion of life in the universe. Finally, we shall see how a few evolutionary biologists – fresh from their victory concerning the evolutionary synthesis in the terrestrial context – ventured into outer space, invoking Darwinian principles and historical contingency to conclude that extraterrestrial morphologies – especially intelligence – were in all probability completely unlike those found on Earth.

The stance of the evolutionists emphasizes that, aside from the experimental and observational aspects, as well as crucial broader questions such as the nature of the Earth's primitive atmosphere, philosophical questions inherent in the subject of the origins and evolution of life on

[8] Kamminga, "History of Ideas," 9–10. It is a testimony to the gradual evolution of ideas that even Pasteur himself wrote in 1878 that, having looked for life arising from nonlife for 20 years, "I have not yet found it, although I do not think that it is an impossibility." Jacques Nicolle, *Louis Pasteur: The Story of His Major Discoveries* (New York, 1961), 75–76.

Earth also profoundly affected the prospects for life in the universe. The issues of chance and necessity (also known as "contingency" and "determinism"), of the probability of seemingly rare events occurring given enough time and space, and the nature of life itself were more than mere abstractions; in many ways, they were the root of the problem. As we shall see, any solution to the question of life's origin and development – on Earth or in outer space – therefore depended as much on philosophical attitudes toward such questions as on empirical factors.

The application of theories of the origin and evolution of life on Earth to the universe at large was not a foregone conclusion; biologists might well have limited themselves to the surface of the Earth, as they had for centuries. Yet, the historical record shows that in the second half of the twentieth century biologists for the first time not only significantly expanded their interests beyond the Earth, but also found that expansion crucial to their enterprise of discovering the origin of life on Earth. Debates over the designation of the new discipline as "exobiology" (preferred by some biologists) or "bioastronomy" (preferred by many astronomers) signaled the marriage – or at least the courtship – of biology and astronomy for the first time in history. The development and consummation of that courtship in this century is the subject of this chapter.

7.1 ARRHENIUS AND PANSPERMIA: AN EXTRATERRESTRIAL THEORY OF THE ORIGIN OF LIFE

In the context of the demise of the spontaneous generation theory of the origin of life, already in the first years of the twentieth century the Swedish chemist and physicist Svante Arrhenius suggested that life might have originated from outer space. Arrhenius (1859–1927) delighted in unorthodox ideas; in his doctoral thesis of 1884 he had proposed that electrolysis occurred in a solution when the molecules of electrolytes (such as NaCl) dissociated into positively and negatively charged atoms to carry the electric charge – a concept heretical to the standard Daltonian structureless atom. Arrhenius received the lowest possible passing grade from his professors for this bold idea, but his full elaboration of the dissociation theory over the remaining years of the nineteenth century entitles him to be recognized (with Ostwald and Van't Hoff) as one of the founders of physical chemistry. And the idea that brought discouragement in 1884 earned him the Nobel Prize in chemistry in 1903, a lesson for him that the hurdles of scientific tradition were worth the trouble.[9]

[9] On Arrhenius see *Dictionary of Scientific Biography* (*DSB*), (New York, 1970), 1, 296–302; J. Walker, "Arrhenius Memorial Lecture," delivered May 10, 1928, *Journal of the*

Following his work in electrochemistry, Arrhenius turned his knowledge of chemistry and physics to the explanation of a wide variety of astronomical and meteorological phenomena, where he showed the same originality. The basic idea of life from space was not new, as Arrhenius was well aware. Indeed, the idea of meteorites as carriers of life dated back to Berzelius in 1834. Flammarion popularized the idea in an early edition of his book on the plurality of inhabited worlds (1864), and under the influence of Flammarion the physician H. E. Richter had advanced it in 1865. Both Helmholtz and William Thomson (Lord Kelvin) had independently supported it beginning in 1871, even if only in passing. Many naturalists still clung to the idea of spontaneous generation in 1871, Thomson found, but in his opinion, "science brings a vast mass of inductive evidence against this hypothesis of spontaneous generation. . . . Careful enough scrutiny has, in every case up to the present day, discovered life as antecedent to life. Dead matter cannot become living without coming under the influence of matter previously alive. This seems to me as sure a teaching of science as the law of gravitation."[10]

But how, then, did life originate? Pointing out (in stark contrast to Darwin's implication about the origin of life) that science should always choose a solution consistent with the ordinary course of Nature rather than invoke any supernatural Creative Power, Thomson drew an analogy with the seeds that alight on newly formed lava or on volcanic islands springing up from the sea, bringing vegetation. Every year, he held, millions of meteorites fall on the Earth, probably the fragments of planets once verdant with vegetation.

> Hence and because we all confidently believe that there are at present, and have been from time immemorial, many worlds of life besides our own, we must regard it as probable in the highest degree that there are countless seed-bearing meteoric stones moving about through space. If, at the present instant, no life existed upon this earth, one such stone falling upon it

Chemical Society (London, 1928), 1380–1401. A bibliography of his writings appears in E. H. Riesenfeld, *Svante Arrhenius* (Leipzig, 1931), 93–110.

[10] H. E. Richter, "Zur Darwin'schen Lehre," *Schmidt's Jahrbucher der in- und auslandischen gesammten Medicin,* 126 (1865), 243–249; H. von Helmholtz, *Populäre Wissenschaftliche Vortrage* (Braunschweig, 1876), vol. 3, 101, translated into English as "On the Origin of the Planetary System," *Popular Lectures on Scientific Subjects,* vol. 2 (London, 1893); W. Thomson, "Presidential Address to the British Association for the Advancement of Science," *Nature* (August 3, 1871), 262–270: 269–270, and "The Age of the Earth as an Abode Fitted for Life," *Mathematical and Physical Papers,* 5 (Cambridge, 1911), 205–230. The botanist and bacteriologist Ferdinand Cohn also elaborated Richter's idea in 1872. For more details see Kamminga, "History of Ideas," chapter 3, and Kamminga "Life from Space – A History of Panspermia," *Vistas in Astronomy,* 26 (1982), 67–86.

> might, by what we blindly call natural causes, lead to its becoming covered with vegetation.[11]

Like Thomson's theory, Arrhenius's ideas of life from space – developed between 1903 and 1906 – were "suggested by the failure of repeated attempts made by eminent biologists to discover a single case of spontaneous generation of life."[12] But he significantly changed Thomson's theory to account for its faults and take into account new developments. Arguing that life would be unlikely to withstand the collision of two worlds, and that life associated with meteors would in any case almost certainly be destroyed on the meteor's descent through the atmosphere, Arrhenius found in the concept of "radiation pressure" an answer to previous objections of the panspermia theory. In his *Lehrbuch der kosmischen Physik* (Treatise of Cosmic Physics, 1903), he had made use of the idea of radiation pressure for physical explanations; he now turned to the same concept as a major component in the origin of life.[13]

Although Arrhenius had developed his ideas on panspermia beginning in 1903, they received their fullest expression in his book *Worlds in the Making: The Evolution of the Universe* (1908). As its title implies, this book was Arrhenius's view of the evolution of the universe, from the formation of solar systems to the origin of life. Arrhenius was among the first to conjoin physical and biological evolution, including a view of many solar systems, with life.[14] Expanding the arguments from his popular article of 1903, Arrhenius pointed out that the radiation pressure evident in the deflection of the comet's tail from the sun, or in the radial structure of the solar corona, would also have a propellant effect on small particles in space. Arrhenius argued that there were "living seeds" of this small size and calculated that if such seeds were detached from Earth, they would cross the orbit of Mars in 20 days, that of Jupiter in 80 days, and that of Neptune in 14 months, and reach the nearest star (Alpha

[11] Thomson, "Presidential Address," 270. We shall take up the "meteorite and life" controversy later in this chapter.

[12] Svante Arrhenius, "The Propagation of Life in Space," *Die Umschau*, 7 (1903), 481, reprinted in Donald Goldsmith, *The Quest for Extraterrestrial Life* (Mill Valley, Calif., 1980), 32–33. The idea first appeared to an American audience in Arrhenius, "Panspermy: The Transmission of Life from Star to Star," *SciAm*, 96 (March 2, 1907), 196, from which this quote is taken.

[13] Arrhenius, *Lehrbuch der kosmischen Physik* (Leipzig, 1903). Arrhenius ("Propagation of Life") cites Maxwell's 1873 theoretical prediction of radiation pressure and its recent experimental confirmation (1901) by the Russian scientist Lebedev. The American physicists Ernest Fox Nichols and Gordon Ferrie Hull proved this in the same year.

[14] Arrhenius, "The Celestial Bodies, in Particular the Earth, as Abodes of Organisms" and "The Spreading of Life through the Universe," chapters 2 and 8 of *Worlds in the Making* (New York, 1908), originally published as *Varldarnas utveckling* (Stockholm, 1906).

Centauri) after 9000 years. He argued that the times of interplanetary transit were short enough for the seeds to preserve their germinating power; while the interstellar times were much more problematic, the very low temperature of interstellar space (−220 °C), he believed, might actually "suspend the extinction of the germinating power, as it arrests all chemical reactions." Arrhenius believed the survival of such spores was supported by the experimental work of Marshall Ward, Roux, Duclaux, and others. "In this way life would be transferred from one point of a planetary system, on which it had taken root, to other locations in the same planetary system, which favor the development of life."[15] In short, Arrhenius divorced the germs of life from meteorites, added radiation pressure, and in doing so infused new life into the panspermia hypothesis.

Arrhenius's concept of panspermia seems to have been more widely cited than accepted. His idea inspired considerable experimental activity to determine the viability of spores and bacteria. The French plant physiologist Paul Becquerel (1879–1955), working in the cryogenic laboratory of Kamerlingh Onnes in Leiden, showed that bacteria retained their germinating power for as long as 2 years under extreme cold in a vacuum. He also showed, however, that ultraviolet radiation from the Sun had a deadly effect.[16] While some argued that the effect of UV rays was not instantaneous, and might be delayed long enough for some germs to survive as they receded from the Sun,[17] the objections were serious enough to cast doubt on the panspermia theory. Moreover, with the increasingly widespread acceptance in the 1920s of James Jeans's idea that planetary systems were extremely rare, the sources of life also disappeared.

Arrhenius himself, however, held the idea of panspermia to the end of his life; his last article marshaled evidence in its favor.[18] It is clear that for him it had become more than an isolated theory; it was also part of an inte-

[15] Arrhenius, *Worlds in the Making,* 221–228. Arrhenius (220) cites Schwarzschild on the size of bodies that would be most strongly affected by radiation pressure.

[16] Kamminga, "History of Ideas," 114; P. Becquerel, "L'action abiotique de l'ultraviolet ef l'hypothèse de l'origine cosmique de la vie," *Comptes rendus de l'Academie des sciences, Paris,* 151 (1910), 86–88, also reported in "Panspermy and Life Germs," *SciAm,* 103 (October 15, 1910), 297; Becquerel, "La vie terrestre provient-elle d'un autre monde?," *Bulletin de la Societé astronomique de France* (Paris, 1924), 393–417. Becquerel later wrote "La suspension de la vie des spores des bacterie . . .," *Compte Rendus,* 231 (1950), 1392–1394, and in the same year became involved in the controversy over life on Mars: "Nouvelles possibiliters experimentales de la vie sur la planete Mars," *Bulletin de la society astronomique de France* (Paris, 1950), 351–355. On Becquerel see *DSB,* 1, 561.

[17] A. Berget, "Appearance of Life on Worlds and the Hypothesis of Arrhenius," *Smithsonian Institution Annual Report,* 1912–1913, 543–551; and "The Transmission of Life to a Dead World Despite the UV Ray," *Current Opinion,* 56 (April 1914), 286.

[18] Arrhenius, "Die thermophilen Bakterien und der Strahlungsdruck der Sonne, *Zeitschrift für phys. Chemie,* 130 (1927), 516–519.

grated worldview. From the beginning, the implications of panspermia for life on other planets was clear to Arrhenius: "What holds true for the Earth also holds true for the other planets which are accessible to life." And again, "This idea involves another, which appeals to me very strongly, namely, that all organisms in the universe are related and the process of evolution is everywhere the same."[19] Although Arrhenius by no means accepted that life existed everywhere – his opposition to Lowell's theory of Martian canals about 1910 is striking – in *The Destinies of the Stars* (1918) he argued for an abundance of life among the stars:

> it was undoubtedly a great truth that Giordano Bruno gave his life for, because it is highly probable, nay almost certain, that around the countless suns which dot the firmament spin dark bodies, although unfortunately our most powerful lenses do not reveal them. A number of these unseen stellar bodies shelter living beings, which even might have climbed to a higher point on the ladder of evolution than have the inhabitants of the Earth.[20]

Although the idea of panspermia never was widely accepted and died a quiet death during the 1920s, it had served not only as an original theory of the origin of life, but also as one that clearly implied that life was abundant throughout the universe. Moreover, the idea that all organisms were related, and that the processes of evolution were everywhere the same, were concepts that would be later revived. When panspermia returned in the 1960s, neither its roots nor its implications for extraterrestrial life would be forgotten. In the meantime, enormous progress would be made in the understanding of life on Earth, a prerequisite to unraveling its origins.

7.2 INTERLUDE: THE RISE OF THE CHEMICAL THEORY AND ITS LATENT EXTRATERRESTRIAL IMPLICATIONS, 1924–1957

Between 1924 and the dawn of the Space Age, the connections between origins of life studies and extraterrestrial life reached a low point. Not

[19] Arrhenius, *Die Umschau*, in Goldsmith, *The Quest*, 32; Arrhenius, "Panspermy," 196.
[20] Arrhenius, *The Destinies of the Stars* (New York, 1918), 255–256, translated from the Swedish (1915) by his student J. E. Fries. It is some measure of the popularity of Arrhenius's writings that the Swedish edition, published in November 1915, went through three editions in 2 months. The English translation, delayed somewhat by the war, included additional subject matter. Crowe's remarks on Arrhenius and Lowell are in Crowe, *The Extraterrestrial Life Debate* (Cambridge, 1986), 538, 544–545. See also Arrhenius, *The Evolution of Worlds* (Leipzig, 1921).

only did panspermia – with its strong extraterrestrial connections – seem unviable after all, life in the universe also began to appear problematic for other reasons. The Martian canal furor had abated beginning about 1910, and although observations of Mars at its close approach in 1924 left some hope for Martian vegetation, life in the solar system seemed increasingly unlikely, or at least inaccessible enough that biologists should not spend much time worrying about it. Moreover, life outside the solar system also began to appear improbable as Jeans's hypothesis of the rarity of planetary systems, stemming from 1916, was beginning to take root. For these reasons alone, as well as for reasons internal to the discipline, biologists felt no compulsion to discuss extraterrestrial life. Nor was any new theory of the origins of life likely to focus on life beyond the Earth.

There were other reasons as well for continued lack of biological interest even as the biological sciences were questioning their foundations and searching for unification. Biologists knew little about astronomy and undoubtedly did not feel comfortable treading on astronomical territory; A. R. Wallace's example, with its quickly outdated conclusions, was not encouraging. For those biologists who did venture beyond their home planet, the biological point of view was very different from the astronomical. W. D. Matthew, a paleontologist at the American Museum of Natural History, drew attention to this difference as early as 1921. Responding to comments on life beyond the Earth in the journal *Science,* Matthew found it "noticeable that, as usual, the astronomers take the affirmative and the biologists the negative side of the argument." Astronomers, he felt, have a more open mind toward hypotheses that cannot be definitely disproved, because they know that by mathematical and deductive methods these hypotheses will be eventually confirmed or disproved. The biologist, on the other hand, deals with evidence that is inductive and experimental. His subjects are too complex for mathematical analysis:

> Always he is compelled to adopt toward the illimitable numbers of possible explanations, a decidedly exclusive attitude, and to leave out of consideration all factors that have not something in the way of positive evidence for their existence. If he fails to do so, he soon finds himself struggling hopelessly in a bog of unprofitable speculations. A critical rather than a receptive frame of mind is the fundamental condition of progress in his work.

Moreover, Matthew continued, the astronomer sees the vast extent of the universe, probably with many planets, and under proper conditions "sees life as possible, and by the incidence of the laws of chance probable or almost certain, if they be duplicated often enough." The biologist, on the other hand, has at the forefront of his mind the evolution of life on Earth. He sees that although favorable conditions for life existed for hundreds of millions of years on Earth, life had originated only once, or at most half a dozen times. The origin of life appeared to him to require not only favorable conditions, but also "some immensely complex concatenation of circumstances so rare that even on earth it has occurred probably but once during the aeons of geologic time." And if the immensely complex circumstances for the fundamental substance of life had occurred on Earth only once since Cambrian times, it appears to the biologist infinitely less probable that they have occurred outside the Earth.[21] With the caveat that Matthew's conclusion must have been based on a small sample of biologists who broached the subject, this cautious attitude emphasizing the complexity of life goes a long way toward explaining the reticence of biologists to enter the debate about life beyond the Earth.

Thus for three decades proponents of the developing chemical theory of the origins of life would make few claims for extraterrestrial life, preferring to explain what they knew needed to be explained on Earth rather than trying to explain something that may not have existed at all. But once biologists and chemists did make such claims in the mid-1950s – perhaps signaling not only a shift in their belief but also the first steps in the search for a generalized and universally applicable biology – the stage was set for the origin of life to become a significant component of the extraterrestrial life debate. And when it did, it brought with it a history of concepts and controversy unique to biology. It is therefore important that we examine at least in outline how this new theory came about, how the extraterrestrial connection originated, and the nature of the biological debate that came with it.

Beginning with only one aspect of the last issue, if one were going to talk about the origins of life, it was important to answer the question "What is life?" In addition, we need to see the question of origins in the larger context of that debate over the nature of life itself. In the broadest sense, scientists at two poles of thought had historically joined in combat: those who believed life was reducible to physics and chemistry and those who believed its essence was a "vital force." Early in the twentieth century, the two leaders of "neovitalism" were Henri Bergson in France and Hans Driesch in Germany; the influence of their work, however, spread far beyond national boundaries. As one American biochemist put it dis-

21 W. D. Matthew, "Life in Other Worlds," *Science* (September 16, 1921), 239–241. Matthew (b. Feb. 19, 1871) was at the American Museum from 1895 to 1926, when he became professor of paleontology and head of the Paleontology Department at the University of California, Berkeley.

dainfully in the 1920s when referring to them, vitalism "asserts that the phenomena of life are not determined by law-abiding forces, but by a form of activity the effects of which are unpredictable, and which, consequently, must be regarded from a formal point of view as chaotic and beyond the range of science." The theme of vitalism versus mechanism, in a bewildering variety of forms, echoed down the corridors of biological history and was still very much alive at the beginning of the twentieth century, influencing not only the origins of life issue, but also many other aspects of biology.[22]

For mechanists inquiring into the nature of life, an answer to the question "What is the material basis for life?" was therefore not only essential to progress in their research program, it was also the foundation of their philosophy. Darwin's staunchest supporter, T. H. Huxley, gave his influential answer in a lecture delivered in Edinburgh in 1868 entitled "On the Physical Basis of Life," occupying an intellectual territory where Darwin had feared to tread and inspiring ever more refined discussion even into the next century. The significance of Huxley's lecture was not his identification of the substance known as "protoplasm" as the basis for all plant and animal life on Earth (this had been conjectured for a decade or more), but his assertion that this protoplasm assumed its living properties without any intervening vitalist principle. In searching further for the physical basis of life, Huxley emphasized that all protoplasm was composed of carbon, hydrogen, oxygen, and nitrogen, which combined into water, carbonic acid (carbon dioxide), and nitrogen compounds, all three of which were necessary to life. "These new compounds, like the elementary bodies of which they are composed, are lifeless," he emphasized. "But when they are brought together, under certain conditions, they give rise to the still more complex body, protoplasm, and this protoplasm exhibits the phenomena of life. I see no break in this series of steps in molecular complication." Physical scientists seemed especially pleased that biology might be reducible to physics; in another famous address 5 years later, the physicist John Tyndall agreed that nonliving matter offered the "promise and potency of every form of terrestrial life."[23]

[22] For the history of the vitalism–mechanism controversy in the context of nineteenth-century biology see Coleman, *Biology in the Nineteenth Century.* The work of Bergson reached English audiences in his *Creative Evolution* (New York, 1911), and that of Hans Driesch in *The Science and Philosophy of the Organism,* 2 vols. (London, 1908). The quotation is from Leonard Troland, "The Chemical Origin and Regulation of Life," *The Monist,* 24 (1914), 92–133.

[23] T. H. Huxley, "On the Physical Basis of Life," reprinted in *Collected Essays,* (New York, 1968), vol. 1, 130–165: 150–154. The address was delivered on November 8, 1868, was first published in Britain in the *Fortnightly Review* (1869), 5, 129–145, and reached the United States in the New York newspaper *The World* (February 18, 1869) under the title "New Theory of Life." In the same year that he gave this address, Huxley claimed to

This concept of protoplasm as the unit of life held sway until almost the turn of the century, when the rise of colloidal biochemistry and the discovery of viruses turned the focus to smaller entities. "Beginning about 1897 the protoplasm theory was replaced by a radically new theory of life processes, which ascribed each chemical change occurring in the living cell not to the whole protoplasm but to a specific intracellular enzyme," wrote one historian of biology. Enzymes and their catalyzing action had been known for a long time under the name "fermentation"; indeed, Pasteur's work on spontaneous generation in 1860 had grown out of his research on lactic and alcoholic fermentation involving yeast. But in 1897 Eduard Buchner extracted an enzyme from yeast cells and showed that it could catalyze a complex physiological reaction outside the cell – in the absence of protoplasm. The triumph of isolating this "zymase" enzyme, as he called it, gave hope that all vital processes could be linked to specific enzymes, which he believed were specialized cases of the substance long known as "protein." It also served to catalyze a new discipline; between 1901 and 1905 the term "biochemistry" came into common usage, biochemical journals were founded, and the enzyme theory became the central dogma of the new biochemistry. Buchner's breakthrough with zymase "can be said to have initiated the development of biochemistry as a discipline separate from general physiology," wrote another historian.[24]

With these developments, the possibility was raised that protein enzymes could be the units of life rather than protoplasm as a whole, which some now began to view as only a passive, if highly structured, substance in which enzymatic reactions occurred. This was explicitly proposed by

have discovered a primitive organism, named by him *Bathybius haeckelli,* that he believed was undifferentiated protoplasm. In 1875 chemists found that it was an inorganic chemical. On this episode see P. F. Rehbock, "Huxley, Haeckel, and the Oceanographers: The Case of *Bathybius haeckelii,*" *Isis,* 66 (December 1975), 504–533. On the history of the idea of protoplasm see Gerald Geison, "The Protoplasmic Theory of Life and the Vitalist–Mechanist Debate," *Isis,* 60 (Fall 1969), 273–292. Tyndall's 1874 address is *Address delivered before the British Association assembled at Belfast, 1874* (London, 1874).

24 Robert E. Kohler, Jr., "The Enzyme Theory and the Origin of Biochemistry," *Isis,* 64 (June 1973), 181–196: 185, 191–192; Garland Allen, *Life Science,* 157; Farley, *Spontaneous Generation Controversy,* 158. On the rise of biochemistry see especially Henry M. Leicester, *Development of Biochemical Concepts from Ancient to Modern Times* (Cambridge, Mass., 1974), and J. S. Fruton, *Molecules and Life – Historical Essays in the Interplay of Chemistry and Biology* (New York, 1972). The nature of proteins was not well understood in 1900, although it had been recognized since 1820 that they could be broken down into amino acids. See Allen, *Life Science in the Twentieth Century,* 158–159, 162ff., and Fruton's essays on "The Nature of Proteins, Molecules and Life," 87–179. The essays of Fruton, professor of biochemistry at Yale, are particularly illuminating and are supplemented in Fruton, *A Skeptical Biochemist* (Cambridge, Mass., 1992), which has as one of its themes the tension between the outlooks of the historian and the experimental scientist.

Harvard biochemist Leonard Troland (1889–1932), who argued that enzymes, making use of their property of "autocatalysis" (catalyzing their own production) and surrounded by an oily liquid produced in the early ocean, might be "primitive living bodies," even if "they cannot be regarded as living in the full sense in which this term is applied to the complex substance of animal and plant cells." Troland went on to propose many other roles for enzymes, including a genetic one, all of which served to dismiss the vitalist philosophy.[25]

Still others regarded the newly discovered viruses, part of the germ theory first proposed by Pasteur, as the units of living matter. In this camp was biochemist Archibald Macallum, who argued that protoplasm consists of colloidal suspensions, each particle of which "is in a definite sense, alive," and which occur individually as viruses. To Macallum the origin of life hinged on an organism "which consists of a few molecules only and of such a size that it is beyond the limits of vision with the highest powers of the microscope." Amino acids and proteins had formed countless times in the past, he argued, until "one giving the right composition resulted in ultramicroscopic particles endowed with the chemical properties of ultramicroscopic organisms." Such discussions reached a peak by the 1920s with the work of Felix d'Herelle on bacterial viruses, known as "bacteriophages," which he had independently discovered in 1917. By 1929 the American geneticist H. J. Muller (who had been influenced by Troland's work) added a genetic approach when he suggested that the first living matter "consisted of little else than the gene, or genes," where the gene was the unit of heredity.[26]

Others, however, still believed the cell was the unit of life, among them the distinguished American biologist Edmund B. Wilson. More than 50 years after Huxley's address, he lectured on the same subject with the same title, now giving structure to Huxley's formless protoplasm (Figure 7.1). Wilson (1856–1939), who made fundamental contributions to both embryology and the study of the cell, recalled how, as students at Yale, both he and William Sedgwick "fell under the spell of Huxley's Edin-

[25] L. T. Troland, "The Chemical Origin and Regulation of Life," *The Monist*, 24 (1914), 92–133: 103–105. See also Troland, "Biological Enigmas and the Theory of Enzyme Action," *The American Naturalist*, 51 (1917), 321–350. Troland's work is discussed in Kamminga, "History of Ideas," 285ff.

[26] A. B. Macallum, "On the Origin of Life on the Globe," *Transactions of the Canadian Institute*, 8 (1908), 435–436, and H. J. Muller, "The Gene as the Basis of Life," *Proceedings of the International Congress of Plant Science* (1929), 1, 897–921, reprinted in *Studies in Genetics: The Selected Papers of H. J. Muller* (Bloomington, Ind., 1962), both discussed in Farley, *Spontaneous Generation Controversy*, 159–162. Sixteen of the 20 amino acids were known by 1900. On the changing conception of the gene see Bruce Wallace, *The Search for the Gene* (Ithaca, N.Y., 1992), and Fruton, "The Chemical Nature of the Gene," in *Molecules of Life*, 225–261.

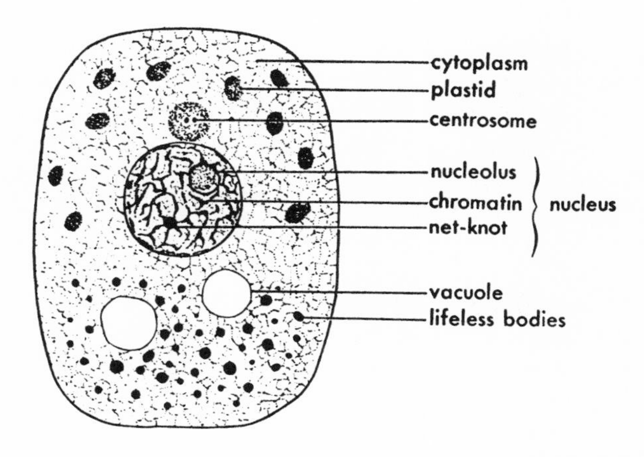

Figure 7.1. The generalized cell, as interpreted by biologist E. B. Wilson in 1922. Wilson was among those who believed the cell was the unit of life.

burgh address," which "aroused immediate and widespread public attention" because of its materialistic implications, even though Huxley claimed he was no materialist. By 1922 Wilson characterized Huxley's heresy as "an orthodox platitude," but he went on to elaborate the concept of protoplasm in light of new knowledge. We now knew, Wilson emphasized, that protoplasm was not chemically a single homogeneous substance, but "the seat of varied and incessant chemical transformations." But did protoplasm, or the parts of the cell, or the entire cell, constitute life? In the view of both the biochemist and the cytologist, Wilson answered, life was a property of the cell as a whole. It was this view of the cell as a colloidal system composed of chemicals that Wilson believed offered great promise of future progress in biology:

> Shall we then join hands with the neo-vitalists in referring the unifying and regulatory principle to the operation of an unknown power, a directive force, an archaeus, an entelechy or a soul? Yes if we are ready to abandon the problem and have done with it once for all. No, a thousand times, if we hope really to advance our understanding of the living organism. . . . To maintain that observation and experiment will not bring [us] nearer to a solution of the puzzle would be to lapse into the dark ages.[27]

[27] E. B. Wilson, "The Physical Basis of Life," *Science*, 57 (March 9, 1923), 277–286. This address was the first Sedgwick Memorial Lecture, delivered in Boston December 29, 1922, to the AAAS. The study of the cell was well advanced by the beginning of the

In this manner, the question "What is life?" oscillated between mechanism and vitalism, between the cell and its constituents, between hope and frustration about understanding the origin of life. By 1937 the British biologist Norman W. Pirie, in an article still widely cited in the second half of the century, reviewed the definitions of life and argued that the terms "life" and "living" were meaningless, and that the transition from nonliving to living was comparable to the gradual transition from green to yellow in the spectrum or from the acid to the alkaline condition in chemistry, except that life could not be defined in terms of only one variable. "Until a valid definition has been framed," he concluded, "it seems prudent to avoid the use of the word 'life' in any discussion about border-line systems and to refrain from saying that certain observations on a system have proved that it is or is not 'alive.' "[28] Despite the likely wisdom of this conclusion, the importance of the question to the origin of life generated ever more sophisticated definitions for the remainder of the century.

The changes in knowledge of the cell and its components, and the accompanying diversity of opinion about the nature of life, naturally affected opinion on the origin of life. Although Huxley clearly stated that protoplasm was the unit of life, he did not broach the question of how it had originated. Historian John Farley found that in the last decades of the century, there was universal agreement that life could be manifested only at the level of the cell, and with the discovery of the increasing complexity of cell structure, the idea of the spontaneous generation of life seemed more remote than ever. "Between 1880 and 1905," he wrote, "the prevailing view, that even the simplest living organism was extremely complex, opened up a hiatus between life and nonlife that seemed impossible to bridge by any fortuitous meeting of molecules. During the following twenty-five years, however, the gap was narrowed to such an extent that there appeared to be no discontinuity." The spontaneous generation of a protein enzyme, virus, or gene seemed easier than the creation of the entire cell, leading Farley to compare the 1920s with the 1860s in its acceptance of the sudden generation of life from nonlife, although now at a much lower level of complexity.

How these changes fed more specifically into the origin of life debate is clear in the work of those who explicitly took up the subject. During the

twentieth century. Wilson had written *The Cell in Development and Heredity* (New York, 1896), a volume which Allen, *Life Science in the Twentieth Century,* note. that "more than almost any other, epitomized the importance that biologists were beginning to attach to the cell as the fundamental unit of physiology, heredity, development, and even evolution," 123–124.

[28] N. W. Pirie, "The Meaninglessness of the Terms 'Life' and 'Living,' " in *Perspectives in Biochemistry,* eds. J. Needham and D. R. Green (Cambridge, 1937), 11–22.

first decades of the twentieth century, while it was realized that biochemistry might hold the key to understanding the origin of life on Earth, progress was extremely slow. As one British discussion in 1912 pointed out, the shift from the natural history approach of Tyndall and Huxley to the first biochemical understandings of the composition of the cell had shed some light on how life might have originated on Earth. But these were still little more than "vague speculations in the present state of our knowledge concerning the chemistry of the protein-compounds on the one hand, and the metabolism and modes of life of the simplest living things on the other. Whether life originated on the earth itself, as biologists have generally supposed, or was brought in some way to the earth from infinite space, as some physicists have suggested, its first origin involves a synthesis of protein-substances in Nature by some process as yet totally unknown." While one participant saw progress in the realization that chemistry was the root of the problem rather than the metaphysical vitalism advanced by Bergson and others, he was not impressed with the biologists' grasp of the subject: "It is clear that, so long as biologists are satisfied with the modicum of chemistry which is now held to serve their purpose, they will never be able to escape from the region of vague surmise."[29] This criticism proved prophetic throughout the first half of the century: the origin of life was an interdisciplinary problem made all the more difficult by the youth of biology and its tenuous connections to other disciplines. Only those such as Troland in the developing field of biochemistry might have the knowledge to shed light on the subject of origins of life, and they were often busy with the more tractable problems of understanding life as it already existed.

By 1929 the widely read biology text of H. G. Wells, Julian Huxley, and G. P. Wells concluded that "the great majority of biologists agree in thinking that probably all the life upon the earth had its origin from the matter of the earth at a definite time in the earth's history." They did not agree on the nature of the first form, although D'Herelle's recently discovered bacteriophage seemed to the authors a likely missing link between the living and the dead. Whatever its nature, the authors speculated that

[29] The quotations are from a discussion on the origin of life at the British Association for the Advancement of Science meeting at Dundee, September 10, 1912. The comment on Tyndall and Huxley is from the Presidential Address at this meeting: E. A. Schaefer, "Life: Its Nature, Origin, and Maintenance," reprinted in *Annual Report of the Board of Regents of the Smithsonian Institution* (Washington, D.C., 1913), 493–525. The "vague speculations" quote is from E. A. Minchin, "Introductory Remarks to Joint Discussion with Section K on the Origin of Life," in *Report of the Eighty Second Meeting of the British Association for the Advancement of Science, Dundee, 1912* (London, 1913), 510–511. The remark on biologists' knowledge of chemistry is from H. E. Armstrong, "The Origin of Life: A Chemist's Fantasy," in *Annual Report,* 527–541.

it had arisen only once, in the surface waters of the warm early Earth, under the action of sunlight.[30]

The beginning of real progress – in the form of a substantive and testable chemical theory – came only in the 1930s and from an unexpected quarter. The author of the new theory was the Russian biochemist Aleksandr Ivanovich Oparin (1894–1980). A graduate of Moscow University in 1917, Oparin was from the beginning of his career thrust into the Russian Revolution, which was to set the stage for so much of the politics of the twentieth century. After 4 years of postgraduate research and several years of lecturing in the Department of Plant Physiology at Moscow University, he became a professor there in 1929. Many of his papers during this period were on plant enzymes and their role in metabolism. One of the organizers of the Institute of Biochemistry of the USSR Academy of Sciences, Oparin worked as assistant director of the Institute under his mentor, Aleksei N. Bakh, from 1935 and became its director in 1946, a post he retained for the rest of his life. By virtue of his activities before and during this time, he became one of the pioneers of biochemistry in the Soviet Union.[31]

Why a new theory of the origin of life should have arisen at this time and place has been the subject of considerable speculation. Little is known of specific influences other than those cited in his work and the general political milieu in which he worked. The effect of the Communist philosophy of dialectical materialism on science, and in particular on biology, has been the subject of much discussion among historians. Although Oparin was clearly influenced by this philosophy at various stages in his life, it is difficult to say just how influential it was in the origin of his theory.[32] Whatever the motivations, Oparin's first paper on the origins of life was written in 1924, and J. D. Bernal has called it "the first and principal modern appreciation of the problem." Well aware of the history of theories of spontaneous generation, Oparin believed Pasteur's experiments had demonstrated "beyond doubt" that all living organisms

[30] H. G. Wells, Julian Huxley, and G. P. Wells, "The Origin of Life," in *The Science of Life* (New York, 1929), 649–653. The authors noted the recent experiments of Baly at Liverpool, demonstrating that under the action of sunlight, sugars and other organic substances could be produced from a mixture of water, CO_2, and ammonia. This was a precursor to the experiments in prebiotic synthesis that became popular beginning in the 1950s.

[31] Biographical details on Oparin are scant, but see Mark B. Adams, *DSB* (1990), 18, 695–700, and its references; John Turkevich, *Soviet Men of Science* (New York, 1963), 275; E. Broda, Oparin obituary in *Transactions in Biological Sciences* (November 1980), iv–v; and Kamminga, "History of Ideas," 229ff.

[32] On both the general context in the Soviet Union and its relation to Oparin, see especially Loren R. Graham, *Science and Philosophy in the Soviet Union* (New York, 1972), chapter 7, "Origin of Life."

developed from other living things. But this left the question of how the first living things arose. In addressing that question, Oparin reviewed panspermia theories, including the ideas of Richter, Helmholtz, and Thompson (although he was apparently not yet familiar with those of Arrhenius). He did not reject panspermia, but agreed with others that such a theory only pushed the question of origins to some other planet. Oparin was thus aware of the broader problem of life in the universe, but was more interested in the basic question of whether there is a fundamental difference between the living and the dead. Oparin thought not and noted that chemists "can now prepare artificially almost all the substances which are encountered in organisms."[33] He went on to discuss the characteristics of life: structure, metabolism, reproduction, and response to stimuli, concluding that none of them was inherent only in living things. For example, the more metabolism was studied, the more Oparin was convinced that it included "nothing that cannot be explained in terms of the general laws of physics and chemistry."[34] The same was true of the other properties of life. He was particularly impressed with the colloidal properties of protoplasm and was confident of the role of colloidal biochemistry in explaining protoplasm. To discover the conditions under which the properties of life were conjoined, Oparin felt, would be to explain the origin of life.

The remainder of Oparin's essay discussed the formation of the Earth and its consequent conditions. Stimulated by the recent identification of methane (CH_4) in the atmospheres of the giant planets, Oparin believed the early Earth had a reducing (hydrogen-dominated) atmosphere; thus, the problem was the synthesis of organic compounds, the formation of polymers, and the origin of the first cells under such conditions.[35] Although he did not explicitly mention extraterrestrial life, Oparin's discussion was couched in a cosmic context, and it was clear from the discussion that exactly the same thing that had happened on Earth would happen any-

33 A. I. Oparin, *Proiskhozhdenie zhinzy* (Moscow, 1924), translated by Ann Synge as "The Origin of Life," in J. D. Bernal, *The Origin of Life* (Cleveland and New York, 1957), Appendix 1, 199–234, and reprinted in Deamer and Fleischaker, *Origins of Life*, 31–71. It is followed in the Bernal reprint by Bernal's commentary, 235–241, in which Bernal calls the 1924 work the first modern appreciation of the problem. See also V. C. Vaughn, "A Chemical Concept of the Origin and Development of Life," *Chem. Rev.*, 4 (1927), 167–187.

34 Oparin, "Origin of Life," 214–216.

35 The idea of organic synthesis in the context of the formation of the Earth had been discussed before Oparin. See, for example, T. C. Chamberlin and R. T. Chamberlin, "Early Terrestrial Conditions That May Have Favored Organic Synthesis," *Science*, 28 (December 25, 1908), 897–911, reprinted in Deamer and Fleischaker, *Origins of Life*, 15–29. Here Chamberlin (whose theory of Earth formation we discussed in Chapter 4) states that the prevalent view was that the gap between organic compounds and life could not be bridged. Unlike Oparin's atmosphere, the Chamberlins' was not reducing.

where else in the universe under the same conditions. This implication, however, remained latent in Oparin's 1924 work; intent on the Earth's history, he was not ready to generalize his theory to the entire universe, which in any case (unknown to Oparin at this early stage) British astronomer James Jeans was proclaiming devoid of habitable planets.

Oparin's ideas remained unknown in the Western world until 1938, when an English translation appeared of his 1936 book, a greatly enlarged version of the 1924 pamphlet.[36] Here Oparin greatly elaborated his theory, now drawing on a variety of fields, from astronomy and geology to biochemistry, to support his earlier work. Noting that earlier students of the origin of life had "tacitly accepted as axiomatic" that CO_2 was the primary carbon compound on the early Earth, Oparin found this conclusion "entirely unfounded." In countering this belief, he cited the 1932 results of the German physicist Rupert Wildt, confirmed by Theodore Dunham, showing that Jupiter's atmosphere contained methane and ammonia. The same was true of Saturn. These and other facts from meteorite analysis supported his assumption that "carbon, at least in part, first appeared on the Earth's surface in the reduced form, particularly in the form of hydrocarbons."[37] In attempting to explain protoplasm and the origin of the cell, he turned especially to the work of Bungenburg de Jong on "coacervates" as models for protocells.[38] And he drew on a variety of experimental work, including his own, in deciding the general pathway that chemical evolution might have taken toward life.

Much as Oparin elaborated his theory in 1936, he still made no attempt to generalize his conclusions to other planets, even if he had used physical observations of the planets to reach his crucial conclusion of a reducing atmosphere of methane and ammonia. He had even less reason now for extending the origin of life theory than he had in 1924, for he was now familiar with Jeans's theory of the rarity of planetary systems, which he adopted as the theory of the origin of the Earth and mentioned in connec-

[36] A. I. Oparin, *Vozhiknovenie zhizny na aemle* (Moscow, 1936), translated as *The Origin of Life* (London, 1938), republished (New York, 1952) by Dover. The 1952 edition is a reprint of the 1938 edition with a new Introduction; all references will be to this edition.

[37] Oparin, *Origin of Life*, 64–82: 82. Oparin cites astronomer Henry Norris Russell's presidential address before the AAAS in December 1934, printed in *Nature*, 3406 (1935), 9, II, as well as A. Adel and V. Slipher, *Physical Review*, 46 (1934), 902.

[38] Oparin, *Origin of Life*, 150ff. Oparin cites articles of Budenberg de Jong [*sic*] in 1931–1932 issues of *Protoplasma*. H. G. Bungenburg de Jong had first described coacervates in 1932 as a mixture of protein and gum arabic. Bernal, *Origin of Life*, 126ff., later gave some of the history of the idea of coacervates as "the nearest we can come to cells without introducing any biological – or at any rate any living biological – substance." But he rejected both coacervates and "proteinoids" as probably irrelevant to the origin of life.

tion with the faults of the panspermia theory (which by now he associated with Arrhenius). Oparin noted that the rarity of planetary systems offered no support for the panspermia hypothesis, since any germs in the panspermia hypothesis would require hundreds of thousands or millions of years to be transported from one planetary system to another.[39] Nor did Oparin place any faith in the recently reported evidence for traces of living organisms in meteorites, believing them to be terrestrial contaminants.[40] But he did take the opportunity to express his first opinion about life in the universe. The lack of support for panspermia, he concluded, "does not mean that life exists only on our Earth. We still have too little information to deny completely the possibility of existence of organisms on some other planets, whirling around stars similar to our Sun. But there can be no doubt that these worlds inhabited by living organisms are much farther removed from our solar system than are the nearest stars."[41]

This was only a passing remark; extraterrestrial life was not Oparin's research program. Rather, he expressed his more limited goal as follows:

> To establish the possibility for generation of life in the dim past of the Earth's history, it is necessary first of all to prove the possibility of a primary formation of organic substance on our planet and, secondly, to trace the further evolution of this substance. Contemporary science enables us to furnish a more or less definite answer to both of these problems.[42]

By contrast, while Oparin could and did place the Earth's history in astronomical context in order to determine its primitive conditions for life, contemporary science did not yet permit a definite answer to the question of life on other planets, and for this reason he did not address it in any detail.

By the time Oparin's work became known in the Western world, the British biochemist J. B. S. Haldane (1892–1964) had provided a brief

39 Oparin, *Origin of Life,* 42–43, 82–90. Oparin cited Jeans, "Contemporary Development of Cosmic Physics," *Smithsonian Bulletin* (1927), *The Stars in Their Courses* (1931), and "Origin of the Solar System," translated into Russian in *Mirovednje,* 19 (1930). Jeans's theory was relevant because it meant that the Earth was formed from gases tidally pulled from the Sun. On the panspermia hypothesis he now cited Arrhenius, *Lehrbuch der kosmischen Physik* (1903) and *Worlds in the Making* (1908).

40 Oparin, *Origin of Life,* 38. Oparin cites in particular the work of Charles B. Lipman, a bacteriologist at the University of California, Berkeley, who claimed in the early 1930s that specimens from eight different meteorites, when placed in cultures, produced rods and coccoid cells. See John G. Burke, *Cosmic Debris: Meteorites in History* (Berkeley, Calif., 1986), 312–313, for details of this controversy, which would arise again in the 1960s.

41 Oparin, *Origin of Life,* 42.

42 Oparin, *Origin of Life,* 63.

independent account of the origin of life similar to Oparin's. Appointed to a position in biochemistry at Cambridge University under Frederick Hopkins in 1921, Haldane remained there until 1933, during which time he composed an eight-page essay on the origin of life. Stimulated by the ongoing question of whether bacterial viruses (bacteriophages) were living entities (he thought they might be an intermediate form of life), Haldane proceeded to speculate on the origin of life entirely independently of Oparin. Much has been made of the fact that Haldane too was a Marxist; for us, the interest in his essay lies in the fact that his theory was similar to Oparin's and that he too limited his discussion to the Earth. Unlike Oparin he believed that the early atmosphere contained no oxygen, a fact that allowed the UV rays of the sun to reach the Earth's surface without being stopped by ozone. And unlike Oparin, he believed that the atmosphere contained carbon in the oxidized form of CO_2 rather than the reduced form of methane (CH_4).[43] But Oparin and Haldane shared the basic precept that the overall early atmosphere was reducing, that organic substances would be synthesized in such an atmosphere, and that these organics accumulated until the primitive ocean (in Haldane's words) "reached the consistency of hot dilute soup." Moreover, they agreed independently that these organics would form larger and more complex molecules, resulting in "the first living or half living things" such as viruses, and eventually in unicellular organisms such as bacteria. In short, they agreed that life could arise from nonlife, and despite their differences, their theory became known as the "Oparin–Haldane theory" of the origins of life.[44]

By 1940, when the British Astronomer Royal Sir Harold Spencer Jones wrote *Life on Other Worlds,* he remarked that "It seems reasonable to suppose that whenever in the Universe the proper conditions arise, life must inevitably come into existence. This is the view that is generally

[43] J. B. S. Haldane, "The Origin of Life," *Rationalist Annual* (1929), reprinted in Bernal, *Origin of Life,* 243–249, followed by Bernal's commentary, and in Deamer and Fleishaker, *Origins of Life,* 73–81. Haldane gives an interesting commentary on the work of Herelle and other contemporaries, concluding (246) that while no one supposes an enzyme is alive, "the bacteriophage is a step beyond the enzyme on the road to life, but it is perhaps an exaggeration to call it fully alive." See also "The Origin of Life" in *The Inequality of Man and Other Essays* (London, 1932), reprinted in Haldane, *On Being the Right Size* (Oxford, 1985), 100–112. On Haldane see Ronald W. Clark, *JBS: The Life and Work of J. B. S. Haldane* (New York, 1968), and *DSB.* On Haldane's Marxism see Graham, *Science and Philosophy,* 264, and on the differences between Oparin and Haldane see Kamminga, "History of Ideas," 246–250.

[44] As proof that organic substances are synthesized from a mixture of water, CO_2, and ammonia, Haldane cited (as did Wells, Huxley, and Wells in their textbook of the same year) the experiments of E. C. C. Baly and his colleagues at Liverpool, published 2 years earlier.

accepted by biologists."[45] This is a notable difference from Matthew's assessment of biological opinion on the subject 20 years earlier, and while the generality of the statement must be very limited, that it was made at all is likely a reflection of the Oparin–Haldane hypothesis developed in the meantime. In any case, as we shall see later, the inevitability of the development of life, as well as the nature of life, was another of the many philosophical questions that would pervade the debate.

Oparin's book received mixed reviews, but it was not until the mid-1940s that the Western world took any note of the Oparin–Haldane thesis and not until the early 1950s that it really took off. In the United States, Norman Horowitz (later one of the primary investigators for the Viking biology experiments) used it as the basis for his theory of biochemical synthesis.[46] And in Britain, among the first to embrace the theory was J. D. Bernal (1901–1971), professor of physics (1938–1963) and of crystallography (1963–1968) at the University of London. Bernal was best known for his work on X-ray crystallography and the atomic structure of compounds. His interest in the connections between physical science and life went back to his days at Cambridge, and his interest in the origin of life stemmed from Haldane. Bernal provides some insight into why, in at least at one leading scientific center in Britain, the new theory was not taken up earlier; he was put off from studying the origin of life at Cambridge, he recalled, because "the atmosphere at Cambridge at that time was, and possibly still is, opposed to any idea of speculation – it was not sound science. [Ernest] Rutherford's dictum was: 'Don't let me catch anyone talking about the Universe in my laboratory,' " a reference to the empirical philosophy of the pioneering nuclear physicist, which may have served physics well but precluded the study of topics such as the origin of life.[47] It was not until 1947 that Bernal delivered a lecture titled "The Physical Basis of Life" in the tradition of T. H. Huxley and E. B. Wilson. But unlike his predecessors, Bernal concentrated on origins, rather than structure and mechanism, adding another modicum of support to the struggling theory

45 Harold Spencer Jones, *Life on Other Worlds* (New York, 1940), 57.

46 N. H. Horowitz, "On the Evolution of Biochemical Syntheses," *PNAS,* 31 (1945), 153–157, reprinted in E. A. Shneour and E. A. Otteson, *Extraterrestrial Life* (Washington, D.C., 1966), 82–86, and in Deamer and Fleischaker, *Origins of Life,* 341–345. See also Horowitz, "The Origin of Life," in *Frontiers of Science,* ed. E. Hutchings, Jr. (New York, 1958). Reviews of Oparin's book appeared in N. W. Pirie, "Organisms and the Earth," *Nature,* 142 (September 3, 1938), 412–413, and in B. Harrow, "The Origin of Life," *Science,* 88 (July 15, 1938), 58. Pirie was very skeptical of Oparin's claims, arguing that a suspension of judgment was better than such speculation.

47 J. D. Bernal, *The Origin of Life* (Cleveland and New York, 1967), Preface, ix–x. The reference is to the physicist Ernest Rutherford. On Bernal see C. P. Snow in *DSB,* 15 (1978), supplement 1, 16–20.

of organic synthesis in a reducing atmosphere. Modestly claiming that most of what he said was not new, but based on Haldane (1929), Oparin (1938), the French microbiologist Andre Lwoff (1943), and his colleague A. Dauvillier (1947), Bernal, too, limited his discussion to the Earth, even though Jeans's theory was just beginning to topple and Weizsäcker was beginning to revive the nebular hypothesis.[48]

The most important influence of the Oparin–Haldane hypothesis, however, was the experimental work it inspired, although nearly three decades after Oparin's original 1924 paper.[49] That hypothesis suggested that it was now possible to investigate a kind of "spontaneous generation" at a much more refined level than Pasteur's experiments of almost a century before – namely, the generation of complex organic molecules by chemical synthesis.[50] Although organic molecules had been synthesized in 1828, Melvin Calvin and his associates in 1950 began the first experiments specifically designed to synthesize organic compounds under conditions suitable for investigating the origins of life on Earth. Calvin later recalled that "a curious concatenation of circumstances" brought about the experiment, including his own interest in the origin of life, the availability of labeled carbon (C-14) to trace the carbon atom through the reaction, and the availability of the nearby cyclotron as an energy source. It was not the Oparin–Haldane theory, however, that guided Calvin's team, although they cited Oparin's 1938 volume for the general idea that life had originated in an organic milieu. Rather, the team irradiated a solution of CO_2 and water vapor with ionizing radiation from the Berkeley, California, cyclotron. Although a small amount of hydrogen made the conditions mildly reducing, it was just the kind of atmosphere that

[48] J. D. Bernal, *The Physical Basis of Life* (London, 1951). The book, based on the Guthrie Lecture delivered to the Physical Society in 1947, was first published in *Proceedings of the Physical Society* (1949). Bernal's original lecture had been attacked in a letter of comment by N. W. Pirie, and Bernal answered Pirie by notes in the 1951 printed work. For Pirie's further attack on Bernal's work, see footnote 105.

[49] As Oparin, *Origin of Life*, shows, a great deal of experimental work applicable to (if not inspired by) the origin of life problem had taken place prior to the publication of his book in 1938, even with claims to have produced life. See especially chapters 1 and 3.

[50] It is essential to realize that organic molecules are not equivalent to life, despite their name. The term "organic molecules" was originally applied to substances isolated from plants and animals to distinguish them from other substances of mineral origin. Organic molecules contained carbon and hydrogen, and perhaps oxygen and nitrogen, and had properties very different from those of inorganic molecules. In modern usage, "organic chemistry" applies to the reactions of molecules containing carbon, while "biochemistry" refers to the chemistry of living things. Thus meteorites and interstellar molecules (which are not believed to be biogenic) are the province of organic chemistry but not of biochemistry. Stated in these terms, origin of life studies attempt to understand how Nature's organic chemistry led to biochemistry.

Oparin had argued against. Simple carbon compounds such as formic acid and formaldehyde were formed under these conditions.[51]

It was left for Stanley Miller, a graduate student of Harold Urey at the University of Chicago, to undertake the first experiments on prebiotic synthesis using the reducing atmosphere suggested by Oparin. Urey had proposed the idea of a reducing atmosphere composed of methane, ammonia, water, and hydrogen based on the formation of the Earth from the solar nebula, a theme he enunciated in his landmark book *The Planets*, published in 1952.[52] Miller – who had planned a thesis under physicist Edward Teller until Teller announced that he was leaving to set up Lawrence Livermore Laboratory in California – came to the idea of prebiotic synthesis after hearing Urey lecture on the primitive reducing atmosphere. After some persuasion, Miller convinced Urey (who was skeptical about the success of the experiment for a thesis topic) to allow him to undertake the experiment on a trial basis. For preparation he read Urey's article, published only a few months before. To the surprise of both Urey and Miller, the experiment was successful, synthesizing amino acids in large quantities.[53] This landmark experiment is generally considered the begin-

[51] The results of the 1951 experiments are reported in W. M. Garrison, D. C. Morrison, J. G. Hamilton, A. A. Benson, and M. Calvin, "The Reduction of Carbon Dioxide in Aqueous Solutions by Ionizing Radiation," *Science*, 114 (October 19, 1951), 416, reprinted in Shneour and Otteson, *Extraterrestrial Life*, 65–69. The experiment should also be seen in the context of Calvin's research on the chemical mechanism of photosynthesis, in particular the attempt to determine whether photosynthesis (conversion of CO_2 to O_2) or respiration (conversion of O_2 to CO_2) occurred first. Calvin also reminisced about the experiment 20 years later in his volume *Chemical Evolution* (New York and Oxford, 1969), 124, where he viewed this work as "the first experiments in the modern sequence" of origin of life studies, and regretted that the experiment had contained no nitrogen and an atmosphere "not fully reducing but only slightly reducing," thus missing the major step forward taken by Miller 3 years later.

[52] H. C. Urey, *The Planets* (New Haven, Conn., 1952). Urey had published the idea a few months earlier in "On the Early Chemical History of the Earth and the Origin of Life," *PNAS*, 38 (April 15, 1952), 351–363, reprinted in Deamer and Fleischaker, *Origins of Life*, 83–95.

[53] Miller describes the circumstances surrounding his interaction with Urey and the experiment in "The First Laboratory Synthesis of Organic Compounds under Primitive Earth Conditions," in Jerzy Neyman ed., *The Heritage of Copernicus: Theories "Pleasing to the Mind"* (Cambridge, Mass., 1974), 228–242. His results were first reported in Stanley Miller, "A Production of Amino Acids under Possible Primitive Earth Conditions," *Science*, 117, no. 3046 (May 15, 1953), 528, reprinted in Deamer and Fleischaker, *Origins of Life*, 147–148, and elaborated in "Production of Some Organic Compounds under Possible Primitive Earth Conditions," *Journal of the American Chemical Society*, 77 (May 12, 1955), 2351; and "The Formation of Organic Compounds on the Primitive Earth," *ANYAS*, 69 (1957), 260–275, reprinted in Shenour and Otteson, *Extraterrestrial Life*, 166–181. It should be noted that this was by no means the first synthesis of amino acids, only the first under plausible primitive Earth conditions. The first amino acid synthesis (glycine) was accomplished by Lob in 1913.

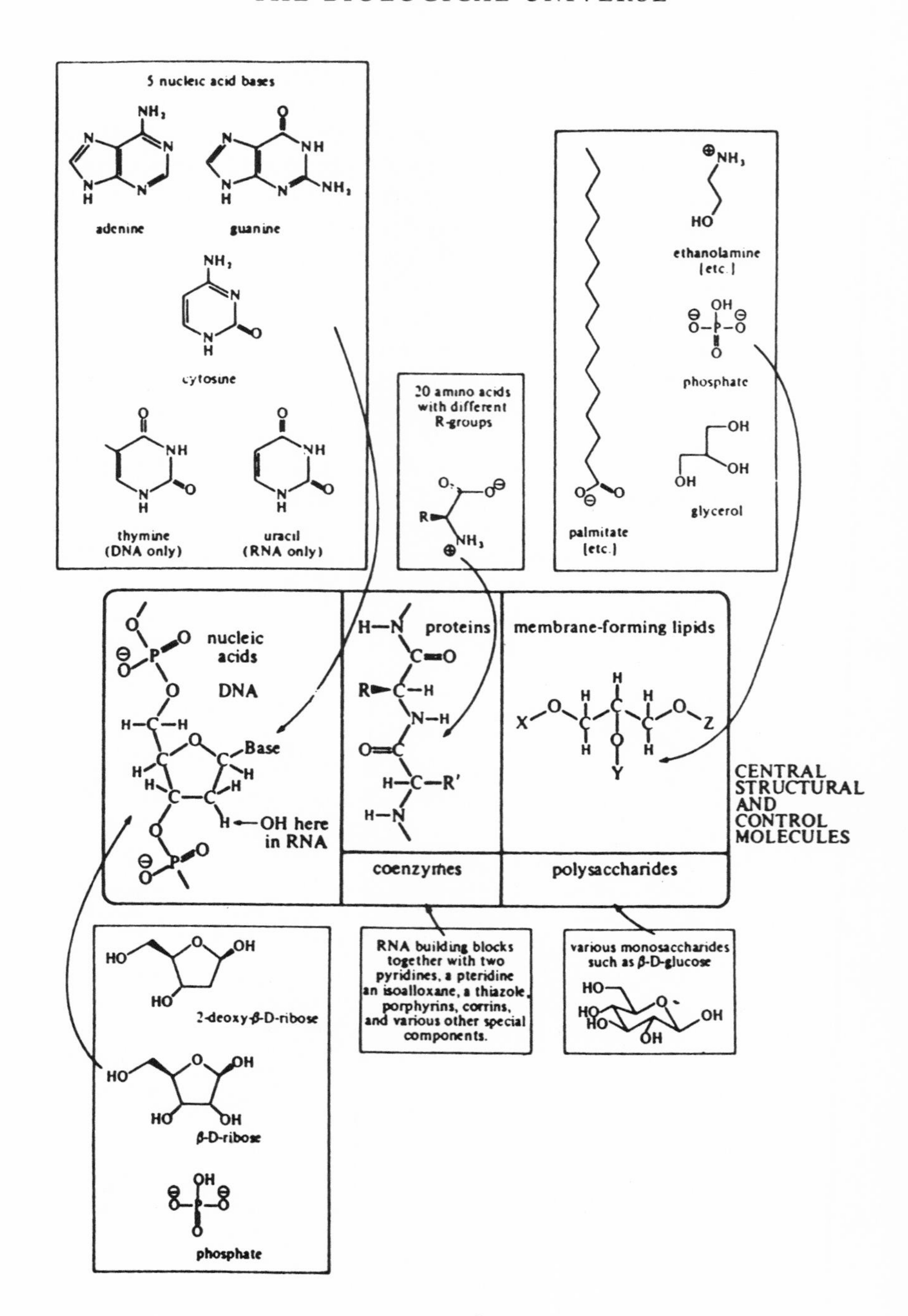
5 nucleic acid bases
adenine
guanine
cytosine
thymine
(DNA only)
uracil
(RNA only)
20 amino acids
with different
R-groups
palmitate
[etc.]
ethanolamine
[etc.]
phosphate
glycerol
nucleic
acids
DNA
Base
OH here
in RNA
proteins
membrane-forming lipids
coenzymes
polysaccharides
CENTRAL
STRUCTURAL
AND
CONTROL
MOLECULES
2-deoxy-β-D-ribose
β-D-ribose
phosphate
RNA building blocks
together with two
pyridines, a pteridine
an isoalloxane, a thiazole,
porphyrins, corrins,
and various other special
components.
various monosaccharides
such as β-D-glucose

ning of the modern experimental era for the origins of life, in which many laboratories participated during the last half of the twentieth century.

It is not surprising that Miller and Urey did not mention extraterrestrial life in their early papers; they were experimental accounts, not speculation about implications. Nor did other laboratory workers speculate on extraterrestrial life in the early 1950s. Plenty of excitement, after all, was being generated in the study of biology on Earth, particularly with regard to nucleic acids. Nucleic acids, the last of the four basic substances in living organisms to be isolated, had been discovered in 1869 during studies of the cell nucleus. By the beginning of the twentieth century, it was known that they consisted of a sugar (ribose), phosphate, and a base, and by the 1920s two kinds of nucleic acids were recognized, eventually named "ribonucleic acid (RNA)" and "deoxyribonucleic acid (DNA)." While the chemistry of the nucleic acids was well known, their physical structure was not; it was assumed to be a simple construction that played only a supportive role in life, while the all-important hereditary genes were composed of protein. Nucleic acids began to assume a central role only after 1944, when Oswald Avery proved that the gene was composed of DNA rather than protein and identified its chemical structure. And in 1953 – the year of the Miller–Urey experiment – James Watson and Francis Crick broke the genetic code, demonstrating the physical structure of the nucleic acids, the famous double helix.[54] Thus, by 1953 the general structures were known of the building blocks of life – proteins, lipids (fats), polysaccharides, and nucleic acids – as well as the structures of their constituents (Figure 7.2). This breakthrough in the molecular biology of the gene was potentially of great importance to understanding the origins of life; although the effect would be delayed, nucleic acids, like proteins, were poised to move to center stage in that arena.

Though Miller, Urey, and their experimental colleagues did not yet extend their conclusion beyond the Earth, it is surprising how quick others were to do so. Already in 1954, Harvard biologist George Wald,

[54] The history of these momentous events has by now been recounted many times. A good starting point is Horace Freeland Judson, *The Eighth Day of Creation: The Makers of the Revolution in Biology* (New York, 1979), especially 27–31, and Watson's own classic, *The Double Helix* (New York, 1968). A sugar–base unit is known as a "nucleoside"; a sugar–base–phosphate unit is called a "nucleotide." Of the five bases, guanine and adenine are called "purines," while thymine, cytosine and uracil are called "pyrimidines."

Figure 7.2 (*opposite*). The building blocks of life (central horizontal panel): nucleic acids, proteins, lipids and polysaccharides, and some of their constituents (upper and lower panels). These were known by the 1950s. From Cairns-Smith (1982), by permission of Cambridge University Press.

reviewing the status of the chemical theory of the origin of life on Earth, noted that the story really did not end with the Earth. By bringing the origin of life within the realm of natural phenomena, the chemical theory implied a further conclusion. It is the essence of natural phenomena, Wald wrote, "to be repetitive, and hence, given time, to be inevitable. This is by far our most significant conclusion – that life, as an orderly natural event on such a planet as ours, was inevitable. The same can be said of the whole of organic evolution. All of it lies within the order of nature, and apart from details all of it was inevitable." And, Wald pointed out, not only on Earth, but on any planet similar to ours in size and temperature. Probably influenced by his Harvard colleague Shapley, Wald pointed out that 100,000 Earth-like planets were now believed to exist in our galaxy alone. With 100 million galaxies, the number of Earth-like planets in the universe would be at least 10 million million. "What it means to bring the origin of life within the realm of natural phenomena is to imply that in all these places life probably exists – life as we know it. . . . Wherever life is possible, given time, it should arise."[55] This widely read article, with its view of the inevitability of the formation of life, was thus among the first to proclaim the new gospel of abundant life throughout the universe from the point of view of the biologist.

In the same year, 1954, Haldane himself proposed that the coming Space Age would be able to answer such questions as the existence of "astroplankton" in the dust of the Moon, and he saw the Space Age as a means of testing various theories of the origin of life. If astroplankton were found on the Moon, this would support the panspermia hypothesis. If no evidence of life was found on Mars or Venus despite their favorable conditions, this would argue against the chemical theory. If life were found there similar to Earth's, this would count against panspermia if and only if Martian organisms were built of looking-glass sugars and amino acids. One might also separate hypotheses by synthesizing organisms, working out the probability that such synthesis could have taken place in a given time. If such synthesis were found to be extremely improbable, the Earth with its life might be unique. And if life were found on Mars, this empirical fact would overrule any theoretical calculations.[56]

By early 1953 Oparin too had first broached the subject of life on other planets. Interestingly enough, it was in response to materialists who criticized him, among other things, for believing that life arose only once on Earth, attributing to it a uniqueness seen as opposed to materialist philosophy. Responding to this criticism in the journal *Problems of Philoso-*

[55] George Wald, "The Origin of Life," *SciAm*, 191 (August 1954), 44.

[56] Haldane, "The Origins of Life," *New Biology*, 16 (April 1954), 12–27, reprinted in Shneour and Otteson, *Extraterrestrial Life*, 70–80.

phy, Oparin held that life was now arising not on Earth, but on other planets.[57] It was no accident that subsequent editions of his book *The Origin of Life* were entitled *The Origin of Life on the Earth.*

In 1956 Oparin collaborated with the Soviet astronomer V. Fesenkov in writing the book *Life in the Universe.* Here Oparin and Fesenkov acknowledged that Jeans was wrong in the sense that life may not be exceptional after all.[58] Independently of Wald, he came to the same conclusion: "In its constant development matter pursues various courses and may acquire different forms of motion. Life, as one of these forms, results each time the requisite conditions for it are on hand anywhere in the Universe."[59]

By 1957 the origin of life community was clearly gathering strength, as evidenced by the fact that, spurred by the International Union of Biochemistry, the First International Conference on the Origin of Life on the Earth was held in Moscow in August 1957. In the Foreword to the *Proceedings* of the meeting, Oparin, the chief organizer, saw the meeting as a sign "that the previous negative attitude of scientists to the problem has now finally ceased to exist and the question of the origin of life has become a field of intensive experimental work."[60] The symposium laid out for discussion an entire research program with five steps: (1) formation of primitive organic compounds; (2) transformation of primary organic compounds; (3) origin of proteins, nucleoproteins, and enzymes; (4) origin and structure of metabolism; and (5) evolution of metabolism. But the *Proceedings* show that 2 months before the launch of Sputnik, extraterrestrial life was still a very minor theme in the origin of life community, despite the book of Oparin and Fesenkov the previous year. Though much was made of the conditions after the Earth was formed, including a paper by Fesenkov, only Urey's paper on planetary atmospheres and origin of life (read by Stanley Miller in Urey's absence) extended the conclusions to other planets.[61]

57 Graham, *Science and Philosophy,* 277, citing Oparin, "K voprosu o vozniknovenii zhizni," *Voprosy filosofii,* no. 1 (1953), 138–142.

58 A. I. Oparin and V. G. Fesenkov, *Zhizn' vo Fselennoi* (Moscow, 1956), published in English as *Life in the Universe* (New York, 1961), 121, 149; see Graham, *Science and Philosophy,* 277–278.

59 Oparin and Fesenkov, *Life in The Universe,* 239.

60 *Proceedings of the First International Symposium on the Origin of Life on Earth,* held at Moscow, August 19–24, 1957; the English–French–German edition was edited for the International Union of Biochemistry by F. Clark and R. L. M. Synge, and for the Academy of Sciences of the USSR by A. I. Oparin, A. E. Braunshtein, A. G. Pasynskii, and T. E. Pavlovskaya (New York and London, 1959), x.

61 H. C. Urey, "Primitive Planetary Atmospheres and the Origin of Life," in Oparin et al., *Proceedings,* 16–22. In addition, A. I. Lebedinskii, in a brief discussion, pointed out that "the successful solution of the problem of the origin of life on the Earth may be made far

It had taken three decades from the origin of the chemical theory with Oparin in 1924 until the chemical theory was even weakly conjoined with the idea of extraterrestrial life in the work of Haldane, Wald, Oparin, and Fesenkov. Even then, it was largely in passing, and the Moscow conference shows that the origin of life community did not yet embrace the study of extraterrestrial life as part of its research program; problems much more basic to the theory were at stake. But the logic of Wald and others emphasizing the inevitability of life under proper conditions – wherever they existed – was destined to carry the day for much of the rest of the century. The heroic era of origins of life studies, from Oparin's essay of 1924 to the Moscow conference of 1957, not only brought the subject of the beginnings of life on Earth out of lethargy and scientific disrepute, it also laid the foundation for the strong connections to extraterrestrial life that would soon follow with the dawn of the Space Age. During that time, biochemistry and molecular biology had unraveled the secrets of life on Earth, including the genetic code in DNA and RNA. The new knowledge from an increasingly unified, if not yet universal, discipline of biology was incorporated into a new theory of the origin of life, along with astronomy and geochemistry. The crucial concepts of the reducing atmosphere, the hot dilute soup of accumulating organic chemicals, and, above all, the faith in the synthesis of ever more complex organic molecules culminating with biochemistry and life passed into the conceptual repertoire of the generation about to travel to the Moon.

7.3 THE INTEGRATION OF ORIGIN OF LIFE AND EXTRATERRESTRIAL LIFE STUDIES IN THE SPACE AGE

Without doubt, the Space Age was the single most important driving force behind the substantial link forged beginning in the late 1950s, between the chemical theory of the origin of life and the existence of extraterrestrial life. Had there been no Space Age, it is likely that the chemical theory would have remained Earthbound, as it had for three decades since its development by Oparin. Even the few tentative steps in the years prior to 1957, including Haldane's 1954 essay (which referred specifically to astronautics and its implication for origins of life) and the volume of Oparin and Fesenkov, were influenced by knowledge of the coming Space Age. Not only did the years immediately following 1957 see the rapid application of origin of life studies to problems raised by the

easier by a study of the conditions under which life exists on Mars," 113. It is significant that the title of the conference confined the discussions primarily to the Earth, so it is not surprising that the connection to extraterrestrials was still limited.

Space Age, and the application of Space Age data to origin of life studies, modified panspermia theories also gradually came back into vogue at this time, strengthening even more the link between the origin of life and extraterrestrial life. In short, the Space Age saw the integration of two subjects that had been almost entirely separate since the heyday of the Arrhenius hypothesis at the beginning of the century. Exactly how this integration came about, both institutionally and substantially, is the subject of this section.

Within the context of the Space Age, one may distinguish at least four interrelated ways in which this integration of origin of life studies and extraterrestrial life studies occurred. First, biological scientists (including primarily biochemists and geneticists) were naturally drawn into pressing Space Age issues such as planetary contamination and life detection, issues that suddenly became national concerns. Intellectually, they saw that these issues held the potential to contribute to long-standing biological problems, and the considerable NASA funding suddenly available for biological research was for many a novel and difficult-to-refuse proposition. For the first time, biologists and space scientists worked together toward a common goal. In the process, NASA transported origins of life studies – and their accompanying biologists – into outer space. Second, beyond those specific Space Age goals, theories and experiments in the origin of life were increasingly seen in an extraterrestrial context, whether or not they were funded by NASA or driven by specific Space Age goals. Origins of life was one of the few areas in which biological science could aspire to universality. That aspiration did not begin or end with the Viking landings but had an internal dynamic of its own, felt in varying degrees by origins of life practitioners. Continuing a process initiated by the Space Age, conferences on the origin of life, previously confined to terrestrial concerns, expanded to the realm of the extraterrestrial, while conferences on extraterrestrial life, traditionally largely astronomical, included an increasingly biological component. Third, even had the aspiration for a universal biology not been present, the discovery of amino acids in carbonaceous meteorites, and of complex organic molecules (though not at the level of amino acids) in interstellar molecular clouds, comets, and interplanetary dust, forced biological interest in the extraterrestrial realm, and sustained it even after the issues of planetary contamination and life detection were resolved (to the satisfaction of most) by the Viking landers. Finally, philosophical issues such as chance, necessity, and the nature of life, endemic to the origin of terrestrial life, were not only shared, but were even more crucial in the extraterrestrial realm. We shall review in turn the roles of each of these four factors.

Space Age Issues

As we have seen in Chapter 3 in the context of solar system studies, the Space Age, with its related concern for the prevention of contamination of the extraterrestrial environment and its hope for the detection of life beyond the Earth, raised immediate substantive issues of interest to those studying the origins of life. The contamination issue originally emanated from the biological side of the National Academy of Sciences rather than from NASA, and from the beginning it naturally drew on experts from the biological sciences. We recall that it was the National Academy that sponsored two committees, one headed by geneticist Joshua Lederberg and the other by biochemist Melvin Calvin, to study the problem of extraterrestrial life, with special emphasis on contamination. It was NASA, of course, that led the way for programs to detect life on Mars, eventually, by necessity, becoming deeply involved with the contamination issue as well. Nor were these interests exclusively American. If the extraterrestrial environment was going to be contaminated, it could just as well happen via Russian microbes as American ones – a terrible irony considering Oparin's work. Similarly, Russian spacecraft could just as well be the first to discover life on Mars; even though those spacecraft missed Mars and, in any case, carried out no sophisticated life detection experiments, the issues were discussed in the Soviet Union as well as in the United States. These extraterrestrial components of the origin of life thus had international ramifications, even if only in a bipolar sense in terms of the two space powers.

We cannot describe here in full how the biological community became involved with the issue of extraterrestrial life via NASA; that is an important issue of discipline emergence, about which more will be said in Chapter 9. It will suffice to trace the entry into the field of Melvin Calvin (Figure 7.3), the pioneer in photosynthesis and origins of life research who, according to the first head of NASA's exobiology office, "did more than anyone else to establish the posture of the nation in bracing itself for the search for extraterrestrial life."[62] As we have seen, Calvin (1911–), inspired by George Gaylord Simpson's *The Meaning of Evolution* (1949), had been working on the problem of chemical evolution since 1950 at the University of California, Berkeley, and had done experiments on prebiotic evolution using hydrogen, CO_2, and water. Unlike Urey and Miller, who 3 years later performed their famous

[62] Shirley Thomas, *Men of Space: Profiles of Scientists Who Probe for Life in Space* (Philadelphia and New York, 1963), vol. 6, 60, quoting Freeman Quimby.

Figure 7.3. Organic chemist Melvin Calvin, pioneer in photosynthesis and origin of life research, Nobel Prize winner, and exobiology enthusiast. Courtesy Graphic Arts Department, University of California, Lawrence Berkeley Laboratory.

experiment using methane and ammonia in a reducing atmosphere, Calvin did not produce amino acids, leaving his experiments less well known (at least to the public) than those that produced complex organic components. His fame had to await his work in another related area – photosynthesis – for which he received the Nobel Prize in 1961. But, Calvin recalled many years later, it was while performing the experiments on prebiotic evolution in about 1950 that he came to the private conclusion that "there must have been evolutionary processes of the same sort elsewhere in the solar system, and perhaps elsewhere in the universe." The reason for this belief was that "there was no question in my mind about the Darwinian behavior of molecules. And if that were true, then that meant it must be true in the entire universe."[63]

Although the extraterrestrial component does not appear in Calvin's lectures on the origins of life in 1956, by late 1958 Calvin was chairing the Panel on Extraterrestrial Life to consider problems of contamination by spacecraft. And by early 1959, he was participating in the Lunar and Planetary Exploration Colloquia being held in California. It was here that he gave his first detailed statement on the relation of origins of life research to extraterrestrial environments. After discussing in detail the latest research on the origins of life, Calvin came to the point of most immediate concern to those interested in extraterrestrial life:

> What are the probabilities that cellular life as we know it may exist at other sites in the universe than the surface of the earth? In view of the chemistry of carbon (and a few of its near neighbors) and the consequences this has given rise to in the environment to be found on the surface of the earth, all that is required to provide some kind of answer to this question is an estimate of the number of other sites, or planets, in the universe which may have the environmental conditions to support such life.[64]

These estimates Calvin found in Hoyle (1955) and Shapley (1958), adopting the latter's estimate of 100 million suitable planets in the universe. Like Wald and others before him, Calvin assumed that when conditions were present, life would evolve, a premise he carried even to the case of advanced intelligence.[65]

Calvin is only one example of how a biochemist entered the realm of the extraterrestrial: experiments in photosynthesis and chemical evolution led to an interest in origins of life; the belief that chemical evolution was occurring beyond the Earth led to the conviction that life also must

[63] Interview with Calvin (1981), in Swift, *SETI Pioneers* (Tucson, Ariz., 1990), 120–122, 129–135. For background, see also the interview in Thomas, *Men of Space,* 28–61.

[64] Calvin, "Origin of Life on Earth and Elsewhere," *Proceedings of Lunar and Planetary Exploration* Colloquium, 1, no. 6 (April 1959), 8; see also *Annals of Internal Medicine,* 54, no. 5 (1961), 954–976. The 1956 lecture is found in Calvin, "Chemical Evolution and the Origin of Life," *American Scientist,* 44 (July 1956), 248.

[65] Calvin integrated all these points in his important book *Chemical Evolution: Molecular Evolution towards the Origin of Living Systems on the Earth and Elsewhere* (New York and Oxford, 1969). As we shall see in the next chapter, Calvin was also a member of the Green Bank conference on extraterrestrial intelligence in 1961, about which he wrote in "Communication: From Molecules to Mars," *AIBS Bulletin* (October 1962), 29–44, and "Talking to Life on Other Worlds," *Science Digest,* 53 (January, 1963), 14–19, 88–89. Calvin was also involved in the problem of extraterrestrial life from the point of view of meteorites (see the later discussion). Among Calvin's students prominent in exobiology were Cyril Ponnamperuma and Elie Shneour.

be originating there; and the origins of life on planets such as Mars became a prime goal of the Space Age, allowing Calvin not only to explore his beliefs, but also to bring along many others as well. Each biologist would travel his own unique career path to outer space. Whether they were forward-looking geneticists such as Lederberg or biologically inclined astronomers such as Sagan, the Space Age allowed each to bring his interests to bear on the problem of extraterrestrial life and a select few to participate on the Viking science teams.[66] Key biologists perceived the mutual benefits of their involvement in matters extraterrestrial, and not only inserted themselves in appropriate committees but also began to bring along some of their biological colleagues.

NASA further encouraged the involvement of biologists by establishing not only ad hoc panels of consultants, but an entire Life Sciences program. When in 1960 the space agency began that program, it became an important center of activity for origin of life studies, particularly as applicable to the space program. We recall that already in July 1959 Administrator Glennan had formed a Biosciences Advisory Committee, chaired by Dr. Seymour Kety, to examine NASA's long-term interests in the life sciences. The January 1960 report endorsed activities far beyond the immediate operational interests of the space program, including "basic biologic effects of extraterrestrial environments . . . and identification of complex organic or other molecules in planetary atmospheres which might be precursors or evidence of extraterrestrial life."[67] It was during the early formation of NASA's exobiology program that Calvin was drawn into NASA's activities as chairman of the Space Biology Committee.[68]

Among the recommendations of the Kety committee was the establishment of a NASA life sciences research facility, which was eventually won by Ames Research Center near San Francisco. Here NASA set up its exobiology program, first under Harold P. "Chuck" Klein (later of Viking fame) and then under Richard S. Young when Klein became head of life sciences

[66] One of Sagan's earliest papers, for example, "On the Origin and Planetary Distribution of Life," *Radiation Research*, 15, no. 2 (August 1961), 174–192, reprinted in Shenour and Ottesen, *Extraterrestrial Life*, acknowledges support by the "Panel on Extraterrestrial Life, Armed Forces – National Research Council Committee on Bio-Astronautics, National Academy of Sciences." For more on the activities of these committees in the context of the growth of the discipline of exobiology, see Chapter 9.

[67] "Report of the NASA Bioscience Advisory Committee," January 25, 1960. On the development of life sciences at NASA from several different perspectives, see (in addition to Chapter 3) Ezell and Ezell, *On Mars* (Washington, D.C., 1984); John Pitts, *The Human Factor: Biomedicine in the Manned Space Program* (Washington, D.C., 1985); and Elizabeth A. Muenger, *Searching the Horizon: A History of Ames Research Center, 1940–76* (Washington, D.C., 1985), NASA SP-4304, Chapter 5, 91–114.

[68] Biological issues were also discussed in other forums; see, for example, the "Astrobiology Session" chaired by Norman Horowitz in *Proceedings of the Lunar and Planetary Exploration Colloquium*, 2, no. 1 (September 1959), 2.

at Ames. Cyril Ponnamperuma (a student of Calvin) came in as a postdoctoral associate and ultimately headed a branch in chemical evolution; experiments on the origin of life were an important part of the research. By the mid-1960s, origin of life studies were not only strongly integrated into the space program, they were essential to one of its most highly visible programs – the search for life on Mars. The work of Ponnamperuma and others at NASA's life sciences laboratory was not pure research but was related to spaceflight goals.[69] When Miller and Horowitz reviewed experimental progress in origin of life studies for the Space Studies Board of the National Academy of Sciences, it was no mere abstract exercise, but an essential task with the end in view that the ideas "were important for the design of experiments to detect life on Mars." The same pattern was repeated again and again during the 1960s, and the culmination with the Viking program is well known.[70] Because of its decision to search for life on Mars, NASA nourished exobiology, and a preeminent part of exobiology was the study of the origins of life.

At the same time, it is clear that biologists had their own agenda. Aside from urgent issues such as planetary contamination and life detection, the biological community was quick to see the potential benefits of the space program to its own work, including specifically origins of life issues, which raised the possibility of a generalized biology. Already in 1959, Stanley Miller and Harold Urey had given a justification for both exobiology and origins of life studies when they concluded, in another of their papers on organic synthesis:

> Surely one of the most marvelous feats of the 20th century would be the firm proof that life exists on another planet. All the projected space flights and the high costs of such developments would be fully justified if they were able to establish the existence of life on either Mars or Venus. In that case, the thesis that life develops spontaneously when the conditions are favorable would be far more firmly established, and our whole view of the problem of the origin of life would be confirmed.[71]

[69] Steven J. Dick, interview with H. P. Klein, September 15, 1992, 3–4; Klein to Dick, private communication, October 10, 1994; C. Ponnamperuma, "Chemical Evolution and the Origin of Life," *Nature,* 201 (January 25, 1964), 337–340; and "Life in the Universe – Intimations and Implications for Space Science," *Astronautics and Aeronautics,* 3 (October 1965), 66–69.

[70] Stanley L. Miller and N. H. Horowitz, "The Origin of Life," in Pittendrigh et al., *Biology,* 64; J. J. Oro, "Investigation of Organo-Chemical Evolution," in *Current Aspects of Exobiology* (Oxford, 1965), 13–76; and Miller, "Extraterrestrial Life," in *Lectures in Aerospace Medicine,* Brooks Air Force Base, 1962.

[71] Miller and Urey, "Organic Compound Synthesis on the Primitive Earth," *Science,* 130 (July 31, 1959), 245–251: 251, reprinted in Deamer and Fleischaker, *Origins of Life,* 149–155.

Thus, barely 6 months after the formation of NASA, at least some forward-looking biologists realized they were literally entering a new world. Biochemists such as Calvin and Horowitz were not only essential to NASA, they were also extremely interested in how the search for extraterrestrial life could inform their own research.

Aspirations for Universality

Even as the space program catalyzed the integration of origin of life and extraterrestrial life studies, theories and experiments previously viewed in the context of life's beginning on Earth were increasingly seen in an extraterrestrial context, raising the two fields of study to a more permanent level of integration. In the Soviet Union, Oparin, himself not involved in the Soviet or American search for life on Mars, wrote at length on the origin of life in space, and a few years later called exobiology and the origin of life "the inseparable connection." Noting that the origin of life on Earth was no longer seen as a happy accident but rather as a normal evolutionary process, Oparin held that "Study of the conception of life on Earth amounts to an investigation of only one example of an event which must have occurred countless times in the world. Therefore, an explanation of how life appeared on Earth should strongly support the theory of existence of life on other bodies in the universe."[72]

After the mid-1960s, treatises on the origin of life almost always encompassed extraterrestrial life, and treatises on exobiology had the origins of life as one of their principal foundations.[73] And by 1970 it was no surprise at all that a National Academy committee surveying the life sciences should include a substantial discussion of extraterrestrial life in its chapter on the origin of life; indeed, its absence would have been considered a serious omission.[74]

While research on the synthesis of organic compounds and theories of the origins of life could proceed without regard to the question of extraterrestrial life, at a deeper conceptual level the two were becoming increas-

72 Oparin, "The Origin of Life in Space," *Space Science Reviews,* 3 (1964), 5–26; "Theoretical and Experimental Prerequisites of Exobiology," in *Foundations of Space Biology and Medicine,* vol. 1, 321–367. The latter article, including 654 references, is a major work.

73 For example, Stanley Miller and Leslie Orgel, *The Origins of Life on the Earth* (Englewood Cliffs, N.J., 1974), 191–218; C. Ponnamperuma, *Exobiology* (Amsterdam, 1972); and Ponnamperuma, *The Origins of Life* (London, 1972).

74 "The Origins of Life," chapter 5 in *Biology and the Future of Man,* ed. Philip Handler (London, 1970), 163–201. The chapter concludes with a section on "Results of a Survey of Radio Astronomers on Activities and Attitudes toward the Search for Extraterrestrial Intelligent Life." The panel members, who included Horowitz, Sagan, and Drake, are identified on 929ff.

ingly intertwined; biology was pursuing unification and universality at the same time. Nowhere is the nature of the integration more evident than in the book *Exobiology* (1972) by Cyril Ponnamperuma, a student of Calvin involved with NASA biology from early in his career. *Exobiology* was one of a series of volumes in which the general editors (including Lederberg's mentor, E. L. Tatum) proclaimed that "the sharp boundaries between the various classical biological disciplines are rapidly disappearing" and that the biological specialist needed to keep abreast of a broad number of fields; it is significant that by 1972 exobiology was viewed as one of the "frontiers of biology" that needed to be thus absorbed. In his Preface, Ponnamperuma wrote that laboratory experimentation on the origins of life was one of three approaches (including spacecraft and SETI) to the search for extraterrestrial life. The implications of such experiments were not confined to the Earth, he emphasized; to the contrary, "our primary objective becomes the understanding of the origin of life in the universe. This is the scientifically broader question before us. If we can understand how life began on the earth, we can argue that the sequence of events which lead to the appearance of terrestrial life may be repeated in the staggering number of planetary systems in our universe."[75] The contents of the volume bear out his statement: In addition to an opening chapter by Oparin on life in the universe and closing chapters on various aspects of life beyond the Earth, the major portion of the volume reported on ideas and experiments undertaken as the basis for understanding the origin of both terrestrial and extraterrestrial life. Such an integration became common practice, and it was only strengthened by the observation of organic molecules in meteorites and other outer-space environments. It is therefore important that we at least summarize developments in the laboratory and in theories of the origin of life since the time of the Miller–Urey experiment, to show not only what was being integrated, but also how the results affected the chances of life beyond the Earth.

Despite the great promise of prebiotic synthesis beginning with the experiments of the 1950s, by the end of the century most of the steps to life had not been duplicated in the laboratory under conditions believed to exist on the primitive Earth. Just how far laboratory syntheses remained from actually producing life (media hype notwithstanding) is clear from Figure 7.4, which shows in general outline major concepts in the evolution of chemical and biological complexity, including the uncertainties about the relative roles of proteins and nucleic acids at the interface between chemical and biological evolution. Following the Miller–

[75] Ponnamperuma, *Exobiology*.

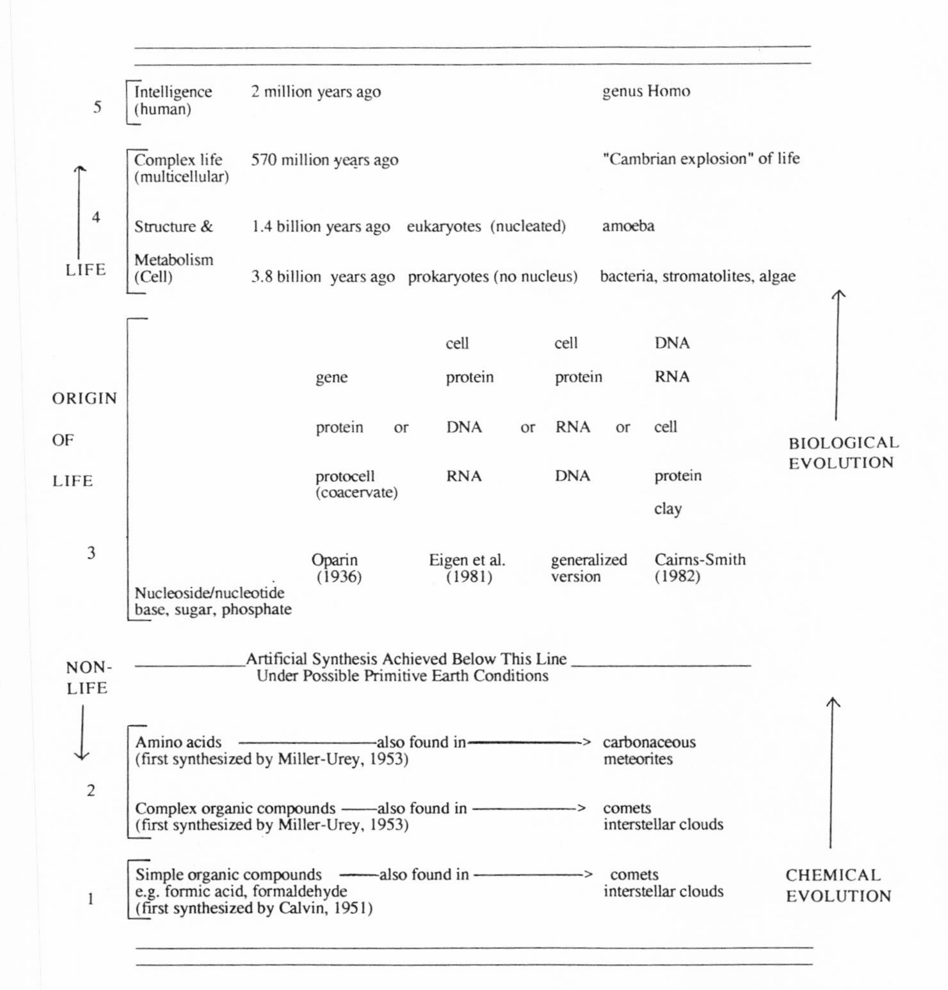

Figure 7.4. Evolution of chemical and biological complexity. At levels 1 and 2, artificial synthesis is easily achieved under possible primitive Earth conditions. At level 3, it is uncertain whether RNA, DNA, proteins, or the cell framework came first, though Oparin's proposed sequence dominated for 40 years. The "gene" of his theory was known to be composed of RNA and DNA after 1944. Life has been described as beginning at level 3 with the proposed RNA "random replicator" of Eigen's theory, or the clay crystal directing the synthesis of protein, or at level 4 with the cell. At level 5 intelligence evolved gradually and is not easily defined. The human brain has 16 orders of magnitude more atoms than the cell.

Urey experiment, it became clear that amino acids were not difficult to synthesize; by early 1970, 17 of 20 amino acids occurring in proteins had been synthesized under possible primitive Earth conditions. Yet, 25 years later, no Miller–Urey experiment had produced the much more complex proteins, nucleic acids, polysaccharides, or lipids that make up a cell, although some had been synthesized under non-Earth-like conditions. In fact, none of the nucleotides or nucleosides that compose the nucleic acids had been synthesized under such conditions. When one considers the chemical complexity of these molecular building blocks, as shown in Figure 7.2, coupled with the knowledge that while amino acids contain fewer than 100 atoms, nucleic acids have thousands and the simple bacterium has some 10^{11}, this lack of progress is not surprising and the prospects for progress are daunting.

Given these circumstances, alternative pathways of organic synthesis were suggested within the Oparin–Haldane paradigm. One of those most vigorously advocated, by chemist Cliff Matthews, proposed that hydrogen cyanide (HCN) polymerized nucleosides and nucleotides directly, rather than via amino acids, a proposal that received support from astronomical observation of abundant HCN in various space environments.[76] But despite the proposal of this and other alternative pathways to life, Miller and Orgel's statement in 1974 that "nucleoside synthesis under plausibly prebiotic conditions has proved to be unexpectedly difficult, so much so that no really satisfactory method has been reported," remained true in the 1990s.[77] By whatever method one chose, no Miller–Urey experiment came close to duplicating the "simple cell" of a bacterium.

More fundamentally, as biochemist Robert Shapiro pointed out in a critical analysis of origin of life theories in 1987, three out of four of the parts of the mature Oparin–Haldane hypothesis, including the reducing atmosphere, the accumulated hot dilute soup of organic chemicals, and the development of life out of this soup, lacked confirmation. "The evidence and the opinions of those scientists most concerned with the subject have moved in the opposite direction" from the Oparin–Haldane hypothesis, he wrote, detecting "an erosion of the paradigm" and a "re-

[76] Clifford N. Matthews, "Hydrogen Cyanide Polymerization: A Preferred Cosmochemical Pathway," in J. Heidmann and M. Klein, eds., *Bioastronomy: The Search for Extraterrestrial Life – The Exploration Broadens* (Berlin, 1991), 85–87, including some of the objections to Matthews's proposal. Matthews had advocated this role for HCN at least since 1967; see C. N. Matthews and R. E. Moser, "Peptide Synthesis from Hydrogen Cyanide and Water," *Nature*, 215 (September 16, 1967), 1230–1234.

[77] Miller and Orgel, *Origins of Life*, 112. On alternative pathways see the papers in section 4, "Alternative Biologies? Alternative Pathways?" in G. Marx, ed., *Bioastronomy, the Next Steps* (Dordrecht, 1987), 161–235.

treat from the hypothesis." And as for progress in prebiotic synthesis with the aim of reaching the biochemical level, he emphasized that "a mixture of simple chemicals, even one enriched in a few amino acids, no more resembles a bacterium than a small pile of real and nonsense words, each written on an individual scrap of paper, resembles the complete works of Shakespeare."[78]

The major assumption that the Earth's primitive atmosphere had been strongly reducing grew weaker with increasing geochemical evidence that it may have contained carbon dioxide, water vapor, free nitrogen, and some hydrogen rather than being rich in ammonia and methane, leaving it only mildly reducing.[79] As early as the 1960s, Miller and Horowitz indicated only a weak commitment to a reducing atmosphere when they commented that "It would probably not be necessary to accept the hypothesis of the reducing atmosphere if organic compounds could be synthesized under oxidizing conditions." But experiments by that time had already shown no significant synthesis of organic molecules under such oxidizing conditions. The concept of the strongly reducing atmosphere was therefore retained until the evidence that volcanic outgassing had produced an atmosphere rich in CO_2 seemed to outweigh the evidence for the CH_4 atmosphere presumed to have been left over from the solar nebula. By 1980 the former was viewed as a "serious alternative" to the reducing atmosphere that had been assumed in 30 years of prebiotic experiments; yet, new experiments by Miller and others using mildly reducing atmospheres showed that amino acids were much more difficult to produce than in a strongly reduced atmosphere,

[78] Shapiro, *Origins: A Skeptic's Guide to the Creation of Life on Earth* (Toronto and New York, 1987), 110–113. See also Antoine Danchin, "From Stars to Mineral to Life: Is the Paradigm Changing?" in J. and K. Tran Thanh Van et al., eds., *Frontiers of Life* (Gif-sur-Yvette Cedex, France, 1992), 399–414. On the lack of prebiotic synthesis beyond amino acids, at least under putative primitive conditions, see Shapiro, *Origins,* 104, 172, 281. This despite the fact that Calvin (*Chemical Evolution,* 105) estimated that by 1967, 15 to 20 laboratories around the world were working on the problems of prebiotic synthesis compared to 1 (his own) in 1950. For a review of prebiotic synthesis experiments as of the mid-1970s, see Norman W. Gabel and C. Ponnamperuma, "Primordial Organic Chemistry," in C. Ponnamperuma, *Exobiology,* 95–135, and Miller and Orgel, *Origins of Life.*

[79] R. A. Kerr, "Origin of Life: New Ingredients Suggested," *Science,* 210 (1980), 42–43. Kerr noted that "No geological or geochemical evidence collected in the last 30 years favors a strongly reducing primitive atmosphere," but that until the 1970s, neither geochemists nor evolutionists had attempted to overturn the idea proposed by Urey in 1952. Miller and Orgel, *Origins of Life,* asserted that it was not yet possible to choose between the one extreme view that all of the carbon in the Earth's primitive atmosphere was in the form of methane (highly reduced) and the other extreme view that it was in the form of CO_2 (highly oxidized). At that time, they believed organic synthesis would occur even in a more weakly reduced atmosphere containing CO_2, CO, and H_2, or CO_2 and H_2.

making clear that an understanding of the origin of life even on Earth was far from complete.[80]

Although not everyone was ready to abandon essential elements of the Oparin–Haldane hypothesis, and certainly not the idea of chemical evolution, there was an increasing search for alternatives outside of the paradigm. One of them, proposed by Nobelist Manfred Eigen and his colleagues in Germany, began with a different prebiotic soup, including small proteins, lipids, and nucleotides. RNA rather than DNA was proposed as the first replicative molecule (a supposition first suggested in the late 1960s by Carl Woese, Leslie Orgel, and Francis Crick and supported by research in molecular biology), and through a series of uncertain steps, proteins and finally cells were supposed to be produced. One advantage of the theory was that the RNA replicator, termed the "random replicator" by Shapiro, might be considered a living unit; though still an intricate molecule, it might be more easily synthesized than proteins or DNA. Another advantage was that it was shown that RNA not only could replicate but also could catalyze specific reactions, thus carrying out the function previously thought to be the preserve of protein enzymes. Quite apart from other details of Eigen's work, the RNA-first concept was gaining wide acceptance by the end of the century.[81]

Nor was it at all clear whether nucleic acids had originated before or after the self-containing cell in which they were eventually situated. Biochemist Sidney Fox demonstrated experimentally in the 1960s that when amino acids were heated, bubblelike protein structures he termed "proteinoids" or "microspheres" were formed, exhibiting lifelike properties of the cell. There seemed to be no reason why such empty cells could not have formed as a framework before the nucleic acids, eventu-

[80] Oxidizing experiments are discussed in Miller and Horowitz, "The Origin of Life," in Pittendrigh et al., *Biology*, 51. See also S. L. Miller and G. Schlesinger, "Carbon and Energy Yields in Prebiotic Syntheses Using Atmospheres Containing CH_4, CO and CO_2," *Origins of Life*, 14 (1984), 83–89.

[81] Manfred Eigen, W. Gardiner, P. Schuster, and R. Winkler-Oswatitsch, "The Origin of Genetic Information," *SciAm* (April 1981), 88–118; T. R. Cech, "A Model for the RNA-Catalyzed Replication of RNA," *PNAS*, 83 (1986), 4360–4363, reprinted in Deamer and Fleischaker, *Origins of Life*, 371–374; and Cech, "RNA as an Enzyme," *SciAm*, 255, 5 (November 1986), 64–75. See also Eigen, *Steps towards Life: A Perspective on Evolution* (Oxford, 1992). Shapiro, *Origins*, 163–66, critically discusses the work of Eigen and his colleagues, and Deamer and Fleischaker, *Origins of Life*, 338–339, cite the work of Woese, Orgel, and Crick. The RNA-first idea is now widely disseminated in textbooks; see, for example, G. Zubay, *Biochemistry* (Reading, Mass., 1983), 1227, where it is argued that "circumstantial evidence suggests that RNA-like polymers probably preceded DNA." But Freeman Dyson, *Origins of Life* (Cambridge, 1985), argues for an updated Oparin theory.

ally capturing the latter during new cell formation; indeed, Oparin had claimed that such a self-enclosed structure (he preferred coacervates) was necessary to concentrate the molecules. In reviewing progress on origin of life studies by 1988, Fox revealed both his starting point and his conclusion when he stated that, in his view, "The tide of general thinking was beginning to turn, back to the days when T. H. Morgan had said to a young graduate student, 'Fox, all the important problems of biology are problems of protein.' "[82]

Another, even more radical alternative, proposed by the Scottish chemist A. Graham Cairns-Smith, required neither the reducing atmosphere nor the dilute soup and made use of the crystalline properties of clay minerals to argue for a mineral origin of life. Clays had been given a role in the origin of life before, notably with Bernal's suggestion in 1947 that clays had helped concentrate organic molecules by sweeping them from the oceans. But Cairns-Smith – explicitly noting "little enthusiasm for the idea now, either among geologists, geochemists or planetary astronomers," of a reducing atmosphere – saw clay as part of the living being itself, a kind of mineral life rather than carbon life. He went on to propose how such mineral life could replicate and evolve, culminating in the "genetic takeover" of the clay minerals by organic molecules and the beginning of biological evolution as we know it. This was an innovative but radical answer to the weaknesses of the standard theory; though many scientists were willing to rethink the reducing atmosphere of the Oparin–Haldane theory, most were not willing to go so far as to abandon carbon life. Although exotic biochemistries were a practical problem discussed during the search for life on Mars, in the end the Viking experiments tested for life of the terrestrial variety, and at least one modern review of the alternatives in the context of extraterrestrial life concluded that, apart from the proposal of Cairns-Smith, no plausible alternatives existed outside of carbon-based life.[83]

[82] Sidney Fox, *The Emergence of Life: Darwinian Evolution from the Inside* (New York, 1988), 142. Fox's results were first reported in S. W. Fox and S. Yuyama, "Effects of the Gram Stain on Microspheres from Thermal Polyamino Acids," *Journal of Bacteriology*, 85 (1963), 279–283, and developed over the next three decades.

[83] G. C. Pimentel et al., "Exotic Biochemistry in Exobiology," in Pittendrigh et al., *Biology*, 243–251; Ronald D. Brown, "Exotic Chemical Life," in Marx, *Bioastronomy*, 179–185. The unique properties of water and carbon for life were emphasized in Harvard biochemist L. J. Henderson's *The Fitness of the Environment: An Inquiry into the Biological Significance of the Properties of Matter* (New York, 1913), reprinted (Boston, 1958, and Gloucester, Mass., 1970), with an Introduction by George Wald. Cairns-Smith's ideas on clay mineral life were proposed in 1982 in *Genetic Takeover and the Mineral Origins of Life* (New York, 1982) and eloquently elaborated in Cairns-Smith, *Seven Clues to the Origin of Life* (Cambridge, 1985). See also Andre Brack, "Origins of Life: State of the Art," in *Frontiers of Life* (note 78 above), 416–420.

While not all origin of life practitioners saw their work in an extraterrestrial context, the relevance of these theories and experiments to extraterrestrial life is clear from many of the participants, quite aside from the general claims of connection stated by Oparin, Ponnamperuma, and others. The role of astronomical observations, the continued faith in the power of chemical evolution, and the implications for life beyond the Earth are all evident, for example, in Matthews's statement that

> laboratory and extraterrestrial studies increasingly suggest that hydrogen cyanide polymerization is a truly universal process that accounts not only for the past synthesis of protein ancestors on Earth, but also for chemistry proceeding elsewhere today within our solar system, on satellites around other stars, and in the dusty molecular clouds of the Milky Way and other spiral galaxies. The existence of this preferred pathway adds greatly to the probability that life is widespread in the universe.[84]

Similar sentiments were expressed in other ways by many participants. Moreover, conferences on the origin of life (Table 7.1) and on extraterrestrial life (Table 9.1 in Chapter 9) increasingly sought to establish the mutual relevance of the two domains, precipitating not only a merging of concepts but also a mingling of researchers in the two fields. We have already seen that extraterrestrial life played only a very minor role in the first conference on origin of life in Moscow in 1957. As we would expect from the activity of the Space Age, however, by the time of the second international meeting in 1963 there was considerably more impetus from the problem of extraterrestrial life; indeed, NASA was one of the sponsors of the meeting, which was opened by the head of its Exobiology Branch, Freeman H. Quimby. The meeting, which for the first time brought together J. B. S. Haldane and Oparin, also included NASA exobiologists Cyril Ponnamperuma and Richard S. Young, as well as others who now undertook their experiments with NASA funding. Sidney W. Fox, whose Institute for Space Biosciences at Florida State University hosted the event, noted that the conference was also remarkable for the active participation of eminent biochemists, in view of the fact that "prebiological chemistry has not always been regarded as fully respectable by all conventional biochemists."[85] In

[84] Matthews, "Hydrogen Cyanide Polymerization."
[85] Sidney W. Fox, ed., *The Origins of Prebiological Systems and of Their Molecular Matrices* (New York, 1965), xv–xvi.

Table 7.1. *Selected conferences on the origin of life*

Date	Sponsor	Location	Proceedings
August 1957	International Conference on the Origin of Life on the Earth International Union of Biochemistry	Moscow	Oparin et al., *Proceedings of the First International Symposium on the Origin of Life on the Earth* (1959)
October 1963	Florida State U., U. of Miami, NASA Second International Conference on the Origin of Life	Wakulla Springs, Fla.	Fox, *The Origins of Prebiological Systems and of Their Molecular Matrices* (1965)
May 5-8 1968	NASA, Smithsonian Institution, NewYork Academy of Sciences	Princeton. N.J.	Margulis, *Proceedings of the Second Conference on Origins of Life: Cosmic Evolution, Abundance, and Distribution of Biologically Important Elements* (1971)
April 1970	Third International Conference on the Origin of Life	Pont-a-Moussan, France	Bivet and Ponnamperuma, *Molecular Evolution I*
June 25-28 1973	Fourth International Conference on the Origin of Life	Barcelona, Spain	Oro, Miller, Ponnamperuma, Young, *Cosmochemical Evolution and the Origin of Life*, 2 vols. (1974) (First ISSOL meeting)[a]
1977	Fifth International Conference on the Origin of Life	Kyoto, Japan	
October 18-20 1978	Fourth College Park Colloquium on Chemical Evolution	College Park, Md.	Ponnamperuma and Margulis, *Limits of Life* (1980)
1980	Sixth International Conference on the Origin of Life	Jerusalem	
October 29-31 1980	Fifth College Park Colloquium on Chemical Evolution	College Park, Md.	Ponnamperuma, *Comets and the Origin of Life* (1981)

Table 7.1 *(cont.)*

June 1-12 1981	NATO Advanced Study Institute	Maratea, Italy	Ponnamperuma, *Cosmochemistry and the Origins of Life* (1983)
July 10-15 1983	Seventh International Conference on the Origins of Life	Mainz Germany	Dose, Schwartz, and Thiemann *Proceedings of the 7th International Conference on the Origins of Life* (1984)
July 21-25 1986	Fifth ISSOL Meeting and Eighth International Conference on the Origin of Life	University of California Berkeley	OLEB[b]
	NASA/UC Berkeley		
July 3-8 1989	Sixth ISSOL Meeting	Prague, Czechoslovakia	OLEB[b]
July 23-27 1990	NASA: 4th Symposium on Chemical Evolution and the Origin and Evolution of Life	NASA Ames	
August 13-17 1990	Gordon Conference on the Origin of Life	Plymouth, New Hampshire	Held every 3-4 years
October 14-19 1991	Third Blois Recontre, "Frontiers of Life"	Chateau de Blois, France	Tran Thanh Van *Frontiers of Life* (1992)
	CNRS,[c] NASA, others		
1993	Seventh ISSOL Meeting	Barcelona, Spain	OLEB[b]
April 25-29 1994	Fifth Symposium on Chemical Evolution and the Origin and Evolution of Life	NASA Ames	
August 22-26 1994	Gordon Conference on the Origin of Life	Newport, R. I.	

[a] ISSOL is the International Society for the Study of the Origin of Life, founded in 1969 and incorporated in 1972.

[b] During the 1980s and 1990s key papers and abstracts of the triennial ISSOL meetings appear in special issues of the journal *Origins of Life and Evolution of the Biosphere* [OLEB], the official journal of ISSOL.

[c] Centre National de la Recherche Scientifique, France

the same year, Bernal infiltrated the International Congress on Biochemistry with a paper on the cosmic aspects of the origin of life.[86] By 1968 NASA was sponsoring its own conference on the subject, and at subsequent international meetings (usually held every 3 or 4 years), extraterrestrial considerations became a significant aspect of the discussions.

Similarly, conferences on extraterrestrial life drew on origins of life experts from the start. As we shall see in the next chapter, Melvin Calvin played a significant role at the first conference on communication with extraterrestrial intelligence, held at Green Bank, West Virginia, in 1961. At the Jet Propulsion Labatory meeting on "Current Research in Exobiology" in 1963, the school of chemical synthesis of organic molecules was represented by Juan Oró; the meteorite controversy by Staplin; and simulated extraterrestrial environments, contamination, and life detection experiments each by its own experts. Since then, virtually every conference on the general problems of extraterrestrial life has included origin of life as a major component. Both in the transfer of concepts and in the mingling of two scientific communities, such conferences were an essential glue in the integration of origin of life and extraterrestrial life studies. Via origins of life, extraterrestrial life was the biological sciences' window on its aspirations for universality.

Revival of Panspermia

Yet another pathway to integration was the revival of panspermia in many forms. Just as Lord Kelvin and Arrhenius had turned to panspermic theories of life from outer space after spontaneous generation was disproven, so the increasing uncertainties of Earthbound prebiotic synthesis led some to turn once again to outer space. Although Carl Sagan emphasized the difficulties of the classical panspermia idea in light of modern astronomy, the trend was toward a neopanspermia radically different from Arrhenius's version. The central focus of the new panspermia in the last four decades of the twentieth century was not life itself drifting through space, but prebiotic chemicals. Observational evidence of organic synthesis in meteorites, comets, dust, the atmospheres of the Jovian planets or their satellites, and interstellar molecular clouds showed Nature to be proficient at building the chemistry of life.[87] Such observations raised the possibility that organic synthe-

[86] J. D. Bernal, "Cosmic Aspects of the Origin of Life," 5th International Congress of Biochemistry, Moscow (1963), 3, 3–11.

[87] For a review see J. Oró, *Space Life Sciences*, 3 (1972), 507–550. Sagan's review of classical panspermia was reported in "Life's Spread through the Universe," *S&T*, 23 (April 1962), 183, 197; see also I. Shklovskii and C. Sagan, *Intelligent Life in the*

sis, which Oparin and Haldane postulated had occurred in the Earth's primitive atmosphere, might have taken place even before the formation of the Earth. Already in the early 1960s, as Oró and others gave birth to the field of organic cosmochemistry, Bernal had made just this suggestion.[88] The discovery that life on Earth may have begun more than 3.5 billion years ago, just at the time of heavy bombardment of cometary and meteoritic material, raised interest to new heights.

Once again, however, the interpretations of these observations were ambiguous. Whether, and if so just how, organic synthesis in space was related to the origin of life on Earth were open questions. So was its lesson for the existence of life on other planets. There was no doubt that the detection of organic molecules in interstellar space and a variety of interstellar and interplanetary environments were first-rate scientific discoveries. What is particularly interesting in the present context, however, is the use made of these discoveries in the extraterrestrial life debate. In short, we shall find that while the existence of organics in space was firmly established, their lesson for life would become a subject of considerable debate. Exobiologists lost no opportunity to point out that the ease with which organics were apparently formed boded well for extraterrestrial life; critics were quick to emphasize the considerable distance between even complex organic molecules and life.

The first claims of organic matter from space came from the study of meteorites, a connection made in the 1870s and 1880s, with another brief spurt of activity in the 1930s. That certain stony meteorites contained carbon was known from the work of the Swedish chemist J. J. Berzelius in 1834, but it would turn out that these "carbonaceous chondrites" were the rarest form of meteorite.[89] Only in the 1960s did analytic techniques allow the nature of the more complex compounds to be determined. Once again, Melvin Calvin and his colleagues were ahead of their time. Having obtained from the Smithsonian Institution a piece of the carbonaceous

Universe (San Francisco, 1966), 207–212.

[88] J. Oró, "Comets and the Formation of Biochemical Compounds on the Primitive Earth," *Nature,* 190 (1961), 389–390. Oró, "Studies in Experimental Organic Cosmochemistry," *ANYAS,* 108 (1963), 464–481; Oró, "Synthesis of Organic Compounds by High-Energy Electrons," *Nature,* 197 (March 9, 1963), 971–974. Bernal in *The Origins of Prebiological Systems and Their Molecular Matrices,* ed. S. W. Fox (New York, 1965), 65–88.

[89] See Burke, *Cosmic Debris: Meteorites in History* (Berkeley, Calif., 1986), 166–173, 312ff; Crowe, *The Extraterrestrial Life Debate,* 400–406; Walter Sullivan, *We Are Not Alone* (New York, 1964), 124–148; and, in connection with Oparin's knowledge of the 1930s claims, Oparin, *Origins of Life,* and Burke, *Cosmic Debris.* Of 700 meteorites collected by 1963 after an observed fall, only 20 were carbonaceous. For a listing of their properties and dates of fall see Brian Mason, "Organic Matter from Space," *SciAm,* 208 (March 1963), 43–49. The term "chondrite" refers to the small, round bodies of magnesium silicates found in these meteorites.

chondrite that had fallen near Murray, Kentucky, in 1950, in the opening days of the 1960s they reported at the First International Space Science Symposium in France "the presence in meteorites of complex organic materials, some of them apparently uniquely pertinent to life processes."[90] Among the molecules reported was a chemical similar to cytosine, one of the four bases in the DNA molecule. Notably absent were any traces of amino acids. The lesson Calvin drew from the analysis, which demonstrated the power of the techniques of infrared and ultraviolet spectroscopy, chromatography, and mass spectroscopy, was that the question "as to whether or not there were possibly prebiotic forms out on astral bodies other than the earth, seems to be answered, at least tentatively, in the affirmative."

Following this claim, at a meeting of the New York Academy of Sciences in March 1961, the team of Bartholomew Nagy, Warren Meinschein, and Douglass Hennessy reported that material from the Orgueil carbonaceous meteorite, which fell near Orgeuil, France, in 1864, included "paraffinoid hydrocarbons" characteristic of living organisms. Nagy and Hennessy were colleagues in the Chemistry Department at Fordham University, and Meinschein, a petroleum chemist at the Esso Research and Engineering Company in nearby Linden, New Jersey, was a key member of the team because the technique used for the meteorite analysis was similar to that used for petroleum research. Based on the similarity of the detected distribution of hydrocarbons to that found in animal products such as butter, the authors concluded that these hydrocarbons represented biogenic activity and further inferred that "biogenic processes occur and that living forms exist in regions of the universe beyond the earth." Bernal was quick to point out the consequences of this claim. Writing in *Nature* 3 weeks later, he urged that scientists not ignore these analyses just because of widespread media publicity; "whatever the interpretation put on them they are of cardinal importance to science." The evidence, Bernal noted, allowed only "that the meteorite material may be of organismal origin and not that it must be so." It was possible that the carbonaceous material was formed by some form of life or by wholly inorganic processes, he pointed out. If the former, the material must have been produced by life on the planetary body from which the meteorite arose. Furthermore, he suggested that meteorites could be the source of material for the first synthesis of life on Earth.[91]

[90] Melvin Calvin and Susan Vaughn, "Extraterrestrial Life: Some Organic Constituents of Meteorites and Their Significance for Possible Extraterrestrial Biological Evolution," in *Space Research* (New York, 1960), 1171–1191.

[91] B. Nagy, D. J. Hennessy, and W. G. Meinschein, "Mass Spectroscopic Analysis of the Orgueil Meteorite: Evidence for Biogenic Hydrocarbons," *ANYAS*, 93 (1961), 25–35; J.

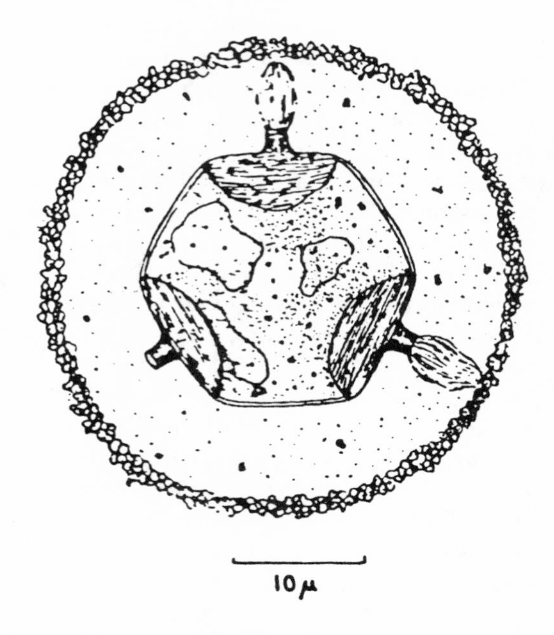

Figure 7.5. Sketch of an "organized element" found in carbonaceous meteorites, believed by Nagy and Claus to be possible remnants of living organisms. Reprinted with permission from Claus and Nagy, *Nature* (1961). Copyright 1961 Macmillan Magazines Limited.

Even as this controversy was heating up, Nagy and George Claus, a microbiologist at New York University Medical Center, made a further startling claim of quite a different kind. They asserted that two of the meteorite samples they examined (the Orgueil and Ivuna) contained five types of "organized elements," structures they identified as possible remnants of organisms "resembling fossil algae" (Figure 7.5). This was an interpretation disputed by most. Philip Morrison, the Cornell physicist who 3 years earlier had suggested a means for communication with extraterrestrial intelligence, concluded that this was an example of the snow-

D. Bernal, "Significance of Carbonaceous Meteorites in Theories on the Origin of Life," *Nature,* 190 (April 8, 1961), 129–131. U.S. Geological Survey microbiologist Frederick Sisler reported bacteria in the Murray meteorite but realized that this could have been caused by contamination, beginning another controversy; see Sullivan, *We Are Not Alone,* 127ff., and *Proceedings of the Lunar and Planetary Exploration Colloquium,* 2, no. 4 (November 15, 1961,), 67–73. Nagy and his colleagues were opposed by University of Chicago chemist Edward Anders, who reviewed the subject in "Meteoritic Hydrocarbons and Extraterrestrial Life," *ANYAS,* 93 (August 29,. 1962), 651–664. For a contemporary review of these events see Mason, "Organic Matter."

flake phenomenon, intricately patterned but hardly alive. And Edward Anders and Frank Fitch at the University of Chicago showed that at least some of the results were likely due to contaminants.[92]

A raft of papers in the March 1962 issue of *Nature* showed just how difficult were the interpretations of these photographs and how important the results to the origin of life. Bernal supported the microfossil theory, and although Urey was more skeptical, he supported further work on the subject and even arranged for a research post for Nagy at the University of California. Another chemist reiterated that even if the organic molecules in meteorites are not evidence of life, still they may hold the key to the origin of life on Earth – as evidence of the composition of the primitive solar nebula.[93] Although by 1975 Nagy himself had come to the conclusion that a biological interpretation of the so-called organized elements in meteorites was only a "remote possibility," the intervening variety of conclusions about the same phenomenon demonstrated that for the origin of life, no less than for other issues we have examined in previous chapters, interpretation was problematic even when the object was at hand and subject to experiment. More than contamination was involved in the interpretations. Urey, for example, while admitting the uncertainties, saw both organics and organized elements as relating to theories of the origin of the Moon. He believed the meteorites might have come from the Moon, which "became contaminated temporarily with water and life-forms from Earth early in its history," an idea that might favor the Moon's having been torn from the Earth rather than captured by it. "Very considerable modifications of conventional hypotheses regarding the origin of meteorites and the composition of the lunar surface are involved," he wrote 3 years before the first Apollo landing.[94] Like

[92] G. Claus and B. Nagy, "A Microbiological Examination of Some Carbonaceous Chondrites," *Nature,* 192 (November 18, 1961), 594–596; Philip Morrison, "Carbonaceous 'Snowflakes' and the Origins of Life," *Science,* 135 (February 23, 1962), 663–664; E. Anders and F. W. Fitch, "Search for Organized Elements in Carbonaceous Chondrites," *Science,* 138 (December 28, 1962), 1392–1399.

[93] The March 24, 1962, issue of *Nature* (vol. 193, 1119ff.) includes papers by Urey and many of the principals involved. Anders and Fitch published a review in "Search for Organized Elements in Carbonaceous Chondrites" and still another in "Observations on the Nature of the 'Organized Elements' in Carbonaceous Chondrites," *ANYAS,* 108 (1963), 495–513, reprinted in Shneour and Otteson, 29–47. See also Anders, "On the Origin of Carbonaceous Chondrites," *ANYAS,* 108 (1963), 514–533, 611–612.

[94] H. C. Urey, "Biological Material in Meteorites: A Review," *Science,* 151 (January 14, 1966), 157–166: 166; Urey, "The Origin of Organic Molecules," in *The Nature of Biological Diversity,* ed. J. M. Allen (New York and London, 1963). In 1975, Nagy wrote regarding a biological interpretation of organized elements that "The chemical and morphological data may be construed as suggestive but certainly not conclusive in this respect." Although there was no consensus, Nagy continued, "There seems to be a growing tendency to consider them indigenous, non-biological organic particles which

Urey, others had a vested interest in a resolution to the meteorite problem, which ensured its contentious, drawn-out nature.

Despite the apparent lack of organized elements in carbonaceous meteorites, the early 1970s advanced spectacularly beyond the claims of simple organic content. The fall of a carbonaceous chondrite near Murchison, Australia, in 1969 provided the opportunity to analyze the meteorite with a minimum of contamination concerns, one of the primary ambiguities in previous experiments. A sample analyzed at NASA's Ames Research Center – still NASA's chief life sciences laboratory – gave unambiguous results: 74 amino acids, the primary building blocks of life, 55 of which did not occur on Earth, and including "right-handed" forms not found in terrestrial living systems. Subsequent confirmation in other carbonaceous meteorites, including many relatively uncontaminated in the highlands of Antarctic, led many to believe that this was their standard composition, and that others in which it was not found had been altered with the passage of time. Although most believed the amino acids in carbonaceous meteorites to be cosmochemical (originating in the solar nebula, for example) and not biogenic, the fact that they had been generated at all beyond the Earth was seen by many as further evidence of the possibilities of life originating in extraterrestrial environments.[95]

Even as the controversy raged over meteorites, comets had long been known from optical observations to possess simple carbon-bearing molecules such as C_2, C_3, and CH. By 1974 observations of the comet Kohoutek had brought the first definitive detection of the organic molecules hydrogen cyanide (HCN) and methyl cyanide (CH_3CN).[96] While

may be associated with mineral matter." Bartholomew Nagy, *Carbonaceous Meteorites* (Amsterdam, 1975), 680–684.

[95] K. A. Kvenolden et al. "Evidence for Extraterrestrial Amino–Acids and Hydrocarbons in the Murchison Meteorite," *Nature,* 228 (December 5, 1970), 923–926, reprinted in Deamer and Fleischaker, *Origins of Life,* 209–212; K. A. Kvenvolden, J. G. Lawless and C. Ponnamperuma, "Non-Protein Amino Acids in the Murchison Meteorite," *PNAS,* 68 (1971), 486–490; J. G. Lawless, C. E. Folsome, and K. A. Kvenvolden, "Organic Matter in Meteorites," *SciAm,* 226 (June 1972), 38–46; C. Ponnamperuma, *Cosmochemistry and the Origin of Life* (Dordrecht, 1983), chapter 1. Although claims for amino acids in meteorites had been made in the early 1960s, contamination and other concerns made the results uncertain (see Nagy, *Carbonaceous Meteroites,* 427). The early Antarctic results were reported in 1979 in R. K. Kotra, A. Shimoyama, C. Ponnamperuma, and P. E. Hare, *Journal of Molecular Evolution,* 13 (1979), 179–183. For still mysterious reasons, one of the properties of terrestrial life is that its amino acid molecules are all "left handed," with no mirror images. Outside the living world, molecules are "racemic," having equal numbers of right- and left-handed molecules. An early paper discussing this asymmetry in the context of exobiology is Lubert Stryer, "Optical Asymmetry," in Pittendrigh et al., *Biology,* 141–146; this "chirality" property is a theme often referred to in the extraterrestrial life debate.

[96] L. E. Snyder, "A Review of Radio Observations of Comets," *Icarus,* 51 (July 1982), 1–24; Fred L. Whipple, "The Nature of Comets," in C. Ponnamperuma, ed., *Comets and*

Oró and others continually stressed the known existence of organics in comets, and the possible existence of biochemical compounds such as amino acids and purines, they also emphasized that it was unlikely that chemical evolution had progressed beyond that seen in carbonaceous chondritic meteorites. Just how cometary organic materials affected the origin of life on Earth remained problematic. Aside from a direct impact such as some believe occurred in Siberia in 1908, some suggested that they might enter the Earth in the form of interplanetary dust particles that originated with comets. It has been estimated that over 10,000 tons of such dust fall on the Earth each year. As one astronomer summarized it:

> In the context of the origin of life, cometary dust is of interest because it is a vehicle in which organic materials released from comets can survive for extended periods of time in space, and it is also a form in which cometary materials can non-destructively enter planetary atmospheres. If comets actually contain life then comet dust could be a vehicle for panspermia.

While he called this dust "universal ambrosia," he also pointed out that the survival of any organic material hitting the Earth at high velocities was in question. Others were less cautious, including astronomer Fred Hoyle and his colleague Chandra Wickramasinghe, who claimed that the organic material on comets might actually be in the form of bacteria.[97]

By the late 1960s the first organic molecules, including ammonia (NH_3), formaldehyde (H_2CO), and hydrogen cyanide (HCN), had also been found in molecular clouds in outer space. Three decades later, some 65 of the more than 100 molecular species discovered in space had been identified as organic, as evidenced mainly by their characteristic radio signature. The discoverers wrestled with the question of the relation of these organic molecules to life: "The predominance of organic species

the Origin of Life (Dordrect, 1981), 1–20: 6. For the general background, see Donald K. Yeomans, *Comets: A Chronological History of Observation, Science, Myth and Folklore* (New York, 1991), especially 219–231. In his inventory of carbon in the solar system, Oparin (*Origin of Life,* 71) was already aware of the existence of C_2, cyanogen, and hydrocarbons in cometary tails, citing C. Rafferty, *Philosophical Magazine,* 32 (1916), 546.

97 A. Lazcano-Araujo and J. Oró, "Cometary Material and the Origins of Life on Earth," in Ponnamperuma, *Comets,* 191–225; D. E. Brownlee, "Interplanetary Dust – Its Physical Nature and Entry into the Atmosphere of Terrestrial Planets," in Ponnamperuma, *Comets,* 63–70: 70; Hoyle and Wickramasinghe, "Comets – A Vehicle for Panspermia," in Ponnamperuma, *Comets,* 227–239; Ellen J. Zeman, "Complex Organic Molecules Found in Interplanetary Dust Particles," *Physics Today* (March 1994), 17–19. For a science fiction treatment of even more advanced life on a comet, see David Brin and Gregory Benford, *Heart of the Comet* (Toronto and New York, 1986).

and their similarity to the products obtained in the synthesis of amino-acids in the study of the origin of life suggests a very close parallel between interstellar clouds and prebiotic chemistry," astronomer David Buhl noted in reviewing the early discoveries in 1971. However, he was quick to add that "It is difficult to establish a direct connexion, but the similarity of the organic molecules produced in laboratory experiments to the molecular composition of interstellar clouds suggests a dominant direction in the chemical synthesis which proceeds despite the very different environment of the cool, low density interstellar clouds." Carl Sagan came to the same conclusion, arguing that interstellar organic molecules were unlikely to have survived the formation of the Earth, and that although some fell to Earth in the first billion years of its history, most came from atmospheric chemistry.

> Accordingly, the contribution of interstellar organic chemistry to problems in biology is not substantive but analogical. The interstellar medium reveals the operation of chemical processes which, on Earth, and perhaps on vast numbers of planets throughout the universe, led to the origin of life, but the actual molecules of the interstellar medium are unlikely to play any significant biological role.[98]

Meanwhile, in our own solar system, evidence had been accumulating of organics in some of the giant gaseous planets and their satellites. As early as in his classic book on the planets in 1952, Urey had suggested that organic molecules might cause Jupiter's multicolored appearance. By 1971 Sagan concluded in a review article (mostly based on chemical synthesis experiments) not only that "it seems rather clear that very large quantities of organic molecules must exist on Jupiter today, that quite complex organic reaction chains are operating, and that organic molecules are an important presumptive source of the Jovian coloration," but also that "there appear to be some interesting exobiological opportunities on Io, Europa, Ganymede, Callisto, Titan, Triton and Pluto." Even after the launching of the Voyager spacecraft, however, the complexity of Jupiter's organics and the issue of whether organic or inorganic chemistry caused its coloration were still in question. Attention increasingly focused on Saturn's satellite Titan, which the Voyager's infrared spectrometer

[98] David Buhl, "Chemical Constituents of Interstellar Clouds," *Nature,* 234 (December 1971), 332–334; Carl Sagan, "Interstellar Organic Chemistry," *Nature,* 238 (July 14, 1972), 77–80; Gerrit L. Verschuur, "Interstellar Molecules," *S&T,* 83 (April 1992), 379–384, including a list of interstellar molecules discovered with the dates of discovery. For the early discoveries see A. H. Barrett, "The Beginnings of Molecular Radio Astronomy," in K. Kellermann and B. Sheets, eds., *Serendipitous Discoveries in Radio Astronomy* (Green Bank, Md., 1983), 280–290.

showed to have a rich assortment of at least nine simple organic molecules ranging from hydrocarbons to HCN. As in the case of Jupiter, the level of complexity of organics remained unknown, but Sagan and his colleagues led the way in laboratory experiments of simulated Titan atmospheres showing that complex organic solids known as "tholins" were the likely candidate for causing the reddish color of Titan's atmosphere.[99] Ironically, it turned out that more organics existed beyond Mars than on Mars, the planet on which $1 billion had been spent to search for life.

By the mid-1970s, then, evidence for amino acids in meteorites, and for lesser organics in comets and interstellar clouds, had been well established, and the existence of at least simple organics in the atmospheres of some of the Jovian planets and satellites was soon to be proven. Although the discovery of organic molecules in space was fascinating in showing nature at work on prebiotic chemistry, and although by the early 1990s molecules with up to 13 atoms ($HC_{11}N$) had been found, with the exception of carbonaceous chondrites, the existence of organics even at the level of amino acids had not been proven in outer space. In this sense, most interstellar chemistry, at least as observed by humans, had not reached the level of the Miller–Urey experiment in 1953. With the exception of the organized elements controversy in the early 1960s, and Hoyle and Wickramasinghe's theories (discussed further later), all of the discussion of organics in outer space environments was strictly prebiotic. Still, even the formerly skeptical Anders was among those claiming organics from interplanetary dust particles and meteorites, and by the end of the century there was some consensus that the source of organic material for the beginning of life on Earth could have been outer space. As plausible relationships between extraterrestrial organic molecules and their delivery to Earth were proposed (Figure 7.6), a refined version of Arrhenius's panspermia gained increasing support.[100]

A much bolder theory of actual life, rather than prebiotic chemicals, coming to Earth was put forth by Fred Hoyle (whom we last saw in Chapter 4 arguing for the abundance of planetary systems beginning in

[99] Carl Sagan, "The Solar System beyond Mars: An Exobiological Survey," *Space Science Reviews*, 11 (1971), 827–866: 848–849 and 861; C. Sagan, W. R. Thompson, and B. N. Khare, "Titan's Organic Chemistry," in M. D. Papagiannis, ed., *The Search for Extraterrestrial Life: Recent Developments* (Dordrecht, 1985), 107–121.

[100] For a discussion of the relative importance of various sources of prebiotic organic molecules, see Christopher Chyba and Carl Sagan, "Endogenous Production, Exogenous Delivery and Impact-Shock Synthesis of Organic Molecules: An Inventory for the Origins of Life, *Nature*, 355 (January 9, 1992), 125–132, reprinted in Deamer and Fleischaker, *Origins of Life*, 123–130; J. Mayo Greenberg, "Chemical Evolution of Interstellar Dust – A Source of Prebiotic Material," in Ponnamperuma, *Comets*, 111–127; Edward Anders, "Pre-biotic Organic Matter from Comets and Asteroids," *Nature*, 342 (November 16, 1989), 255–257.

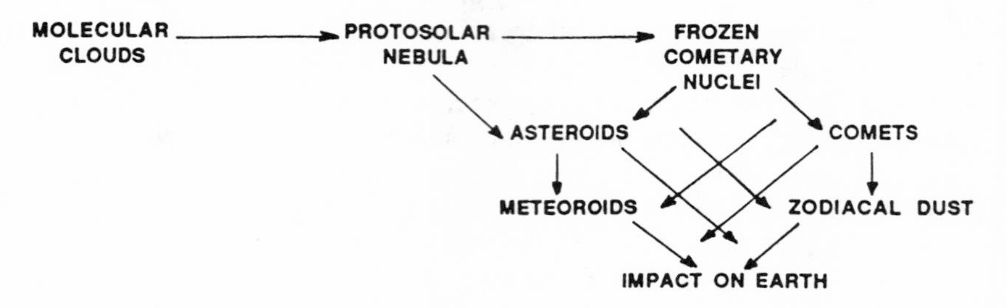

Figure 7.6. Possible relationships between sources of organic molecules and their delivery to Earth. From A. Chantal Levasseur-Regourd, "Cometary Studies," in J. Heidmann and M. Klein, eds., *Bioastronomy* (1991), 110. Copyright 1991 Springer Verlag GmbH & Co. KG.

1955, and in Chapter 5 in regard to his science fiction) and his colleague Wickramasinghe, who developed several different theories from 1977 to 1981. Rejecting the experimental synthesis of organics since the work of Miller and Urey in 1953 as irrelevant to the conditions present during the origin of life on Earth, in their book *Lifecloud* (1978) the authors argued for the astronomical basis for the origin of life, not just for prebiotic molecules, a proposition closer to Arrhenius's original theory but by no means identical. They described how their work on interstellar dust led them to believe that its spectroscopic signature could only be explained by its being composed of cellulose, a polysaccharide biochemical (the substance of wood) that was the most abundant terrestrial organic substance. More specifically, they claimed not only that organic molecules exist, but also that dust clouds form nucleic acids and proteins. These biomolecules were assembled into still more complex forms on planetesimals like comets, which showered the Earth during its first billion years – giving the Earth not only its volatile materials in the atmosphere, but also life in the form of living cells. Hoyle and Wickramasinghe also argued in favor of the meteoritic organized elements of Claus and Nagy, based on their theory of comets. In particular, they argued that Claus and Nagy's observations were just what one would expect when "primitive living organisms evolve in the mixture of organic molecules, ices and silicate smoke which make up a comet's head."[101]

By 1979 Hoyle and Wickramasinghe had carried their theory another step, interpreting interstellar absorption features in the UV to be evidence

[101] Hoyle and Wickramasinghe, *Lifecloud: The Origin of Life in the Universe* (New York, 1978), 91ff, 110–111.

of frozen bacteria, viral particles, and algae in outer space, microorganisms that occasionally caused epidemics on Earth. Critics were quick to point out that "It is extremely difficult . . . to visualize any possible mechanisms by which such hypothetical organisms could originate, survive and evolve under the harsh environmental conditions of the interstellar medium."[102] By 1981 they were arguing in *Evolution from Space* that the sudden bursts of new life forms on Earth are caused by the arrival of genes from outer space. Thus, according to Hoyle and his colleague, not only the origin of life, but also its evolution, was greatly influenced by events in outer space. It was not quite the *Black Cloud,* but neither was it presented as science fiction. Biologist Lynn Margulis called the first book "wanton, amusing promiscuous fiction." While this demonstrates that the quest of the biological sciences for universality had its limits, a NASA study entitled *The Evolution of Complex and Higher Organisms,* while not endorsing Hoyle, did support the more general concept that events in outer space might well have affected the origin and development of life on Earth.[103]

Nor was Hoyle's the wildest panspermic theory of the origin of life. Already in 1973 no less a scientist than Nobelist Francis Crick, of DNA fame, proposed with biologist Leslie Orgel the idea of "directed panspermia," whereby life was sent by spaceship from other planets to intercept Earth, among others. Unlike Hoyle and Wickramasinghe, however, Crick and Orgel did not claim that their hypothesis had amassed enough evidence to be taken as true or even likely. It was only a logical possibility, perhaps the ultimate elaboration of panspermic theory that had begun a century earlier.[104]

102 F. Hoyle and C. Wickramasinghe, *Diseases from Space* (London, 1979). Hoyle, A. H. Olavesen, and N. C. Wickramasinghe, "Identification of Interstellar Polysaccharides and Related Hydrocarbons," *Nature,* 271 (January 19, 1978), 229–231. Hoyle and Wickramasinghe, "Biological Chromophroes and the Interstellar Extinction at Ultraviolet Wavelengths," *Astrophysics and Space Science,* 65 (1979), 241–244, and "On the Nature of Interstellar Grains," ibid., 66 (1979), 77–90, and ibid., 68 (1980), 499–503. Antonio Lazcano-Araujo R. and J. Oró, "Cometary Material and the Origins of Life on Earth," in Ponnamperuma, *Comets,* 191–225. Shapiro critiques the Hoyle and Wickramasinghe work in *Origins,* 233–247.

103 Hoyle and Wickramasinghe, *Evolution from Space* (New York, 1981). On Margulis, see Shapiro, *Origins,* 236. On panspermia see also Hoyle and Wickramasinghe, *Space Travellers: The Bringers of Life* (Cardiff, 1981) and Hoyle, *The Intelligent Universe* (New York, 1984). The NASA study is *The Evolution of Complex and Higher Organisms* (Washington, D.C., 1985).

104 F. Crick and L. E. Orgel, "Directed Panspermia," *Icarus,* 19 (July 1973), 341–346, reprinted in Goldsmith, *The Quest,* 34–37; F. Crick and L. E. Orgel, *Life Itself* (London, 1982). For the general idea of life on Earth having arisen from extraterrestrial activity, Crick and Orgel cite Thomas Gold's "Cosmic Garbage" in *Air Force and Space Digest* (May 1960), 65, an idea also elaborated in the Strugatsky brothers' science fiction novel *Roadside Picnic* (1972; English translation 1977).

Philosophical Issues

Finally, aside from the three areas of detection of life on Mars, exobiologists' aspirations for universality in their theories of the origin of life, and the discovery of organic molecules in a variety of space environments, studies of the origin of life on Earth and of extraterrestrial life were united by philosophical questions underlying all theories of origin – questions that repeatedly thrust themselves to the forefront and loomed even larger in the context of the origin of life beyond the Earth. The old question "What is life?," a property that Huxley had invested in protoplasm and others in the enzyme, the virus, the gene, or the cell, never ceased to be a concern, either in the terrestrial or in the extraterrestrial context. N. W. Pirie, the British biologist who had previously argued that the term "life" was meaningless, reiterated while attacking Bernal's *Physical Basis of Life* in 1953 that no more was known then about the environment in which life originated than in the days of Huxley and Tyndall, despite the speculations of Oparin and Haldane. Not only was the nature of the atmosphere unknown, the raw materials to be synthesized and the role of proteins were equally mysterious, and Pirie concluded that "pure Irish whimsy" must have led Bernal to write so much "blarney" about the origin of life.[105] Pointing out that "physics is as fickle a science as any other," Pirie undoubtedly felt the same about physicist Erwin Schrödinger's influential book *What Is Life?* The role of physicists aside, Pirie's skeptical view of origins of life studies was based largely on his belief in scientific ignorance about the nature of life, a point of view that he never ceased to stress and that still has its adherents today.[106]

The question of the nature of life was obviously even more critical when applied to extraterrestrial life. Bernal remarked in his 1947 lecture that, unlike chemistry and physics,

> terrestrial limitations obviously beg the question of whether there is any more generalized activity that we can call life. . . . Whether there are some general characteristics which would apply not only to life on this planet with its very special set of

[105] N. W. Pirie, "Vital Blarney," *New Biology*, 12 (April 1952), 106–112. The Pirie–Bernal argument is described further in "The Evolution of Life in the Universe," *JBIS*, 12 (July 1953), 180–182.

[106] See also Pirie, "The Nature and Development of Life and of Our Ideas about It," *The Modern Quarterly*, 3 (1948), 82–93; Pirie, "On Making and Recognizing Life," *The New Biology*, no. 16 (April 1954), 41–53; "The Origins of Life," *Nature*, 180 (November 2, 1957), 886–888; Pirie, "Some Assumptions Underlying Discussions on the Origins of Life," *ANYAS*, 69 (1957), 369–376.

> physical conditions, but to life of any kind, is an interesting, but so far purely theoretical question.

Bernal recalled that when he discussed the problem with Einstein, the latter concluded that any generalized description of life would include things we call life only in a poetical manner: "Any self-subsisting and dynamically stable entity transforming energy from any source, or as Haldane put it, 'any self-perpetuating pattern of chemical reactions,' might be called 'alive' in this sense."[107] In a lecture to the British Interplanetary Society in 1952, Bernal extended his speculations about the origin of life to the universe, arguing that "the biology of the future would not be confined to our own planet, but would take on the character of cosmobiology."[108] The differences between life forms on Earth and Mars, he predicted, would be revealing; and indeed, 10 years later, the nature of life was an important question in designing experiments for the search for life on Mars.

The philosophical question that dominated the problem of the origin of life beyond the Earth, however, was not so much the nature of life as the role of chance and necessity in its formation. On the latter question, after all, hinged the whole enterprise of exobiology, for if life was a chance occurrence with very low odds, it need never happen again, no matter how big the universe and how expansive the time available. One might never come to the question of the nature of life in the universe if it could not exist in the first place. In entering that debate, biologists were tackling one of the oldest questions in philosophy; the ancient Greek atomist Democritus had written that "Everything existing in the Universe is the fruit of chance and necessity." From physics to ethics, and occasionally even in early debates on the spontaneous generation of life, this philosophical question loomed but seldom brought consensus.[109]

It is one of the hallmarks of the extraterrestrial life debate that exobiologists, perhaps driven by their desire for a universal biology, placed their faith in the necessity, or at least the high probability, of the origin of life under proper conditions. Eschewing, in the tradition of Huxley, any idea that life was separated from nonlife by a vitalist property, the principle of the high probability of the chance formation of life became

[107] Bernal, *Physical Basis,* 14–15.

[108] Bernal's lecture to the British Interplanetary Society on November 8, 1952, was described in detail by A. E. Slater, "The Evolution of Life in the Universe," *JBIS,* 12 (May 1953), 114–118. The BIS was chaired at this time by Arthur C. Clarke.

[109] On chance and necessity (which also bear on the issue of determinism), see Steven M. Cahn, "Chance," and Richard Taylor, "Determinism," in Paul Edwards, ed., *The Encyclopedia of Philosophy* (New York, 1967), 73–75, 359–373. See also Ernest Nagel, *The Structure of Science* (New York, 1961), chapter 10.

one of their major assumptions, and one they sought to justify as widely held. One sees it already in the first modern treatment of extraterrestrial life by British Astronomer Royal Sir Harold Spencer Jones, who wrote in 1940 that although the nature of the steps from complex organic substances to the first living cell were unknown, "Nevertheless, it seems reasonable to suppose that whenever in the Universe the proper conditions arise, life must inevitably come into existence. This is the view that is generally accepted by biologists." We recall that this was the view of George Wald, who in his influential 1954 article on the origins of life called the inevitability of life on Earth and in the universe "by far our most significant conclusion." The same assumption was held by organic chemist Melvin Calvin, who wrote that all that was needed for an estimate of the probability of cellular life as we know it was an estimate of the number of planets with conditions similar to Earth's. Necessity, or predestination, was the view of many of the exobiological experimentalists working on prebiotic synthesis, including Ponnamperuma. And Shklovskii and Sagan, while admitting that laboratory investigations provided only a "likely story" of the origin of life, and one that needed to be checked by investigating living systems beyond the Earth, nevertheless showed where their sympathies lay by championing even the search for intelligent life.[110]

The assumption of necessity adopted by most exobiologists may be traced at least to the early part of this century in the context of life's origins. In arguing for his enzyme theory of life's origin, for example, Harvard biochemist Leonard Troland wrote in 1914:

> we are forced to say that the production of the original life enzyme was a chance event. . . . The striking fact that the enzyme theory of life's origin, as we have outlined it, necessitates the production of only a single molecule of the original catalyst, renders the objection of improbability almost absurd . . . and when one of these enzymes first appeared, bare of all body, in the aboriginal seas it followed as a consequence of its characteristic regulative nature that the phenomenon of life came too.[111]

Given enough time or space, or a simple enough entity, or the need for only a single first molecule, exobiologists could argue that an event governed by chance was transformed into necessity when the laboratory was

[110] Spencer-Jones, *Life on Other Worlds,* 57; Wald, "Origin of Life"; Calvin, "Origin of Life"; Shklovskii and Sagan, *Intelligent Life,* 244. On Ponnamperuma and other predestinists see Shapiro, *Origins,* 107–109, 187–188.

[111] Troland, "Chemical Origin."

the immense, and immensely old, universe. "Wherever life is possible, given time, it should arise," Wald wrote.

But as we have seen, the roots of the opposite view also ran deep. Troland's contemporary, W. D. Matthew, had already departed from the view of necessity by 1921 and argued, in the context of extraterrestrial life, that the chances of life's origins were extremely small. *This* was the position, he held, that was accepted by most biologists. It was the view championed by the French scientist Lecomte du Noüy in his widely read and admired *Human Destiny* (1947). There du Noüy calculated that a single simple protein molecule would take 10^{243} billion years to form by the workings of chance, 243 orders of magnitude more time than was available based on the age of the Earth. That single molecule, he emphasized, was very far from life itself and even more remote from humanity. And in any case, he argued, the laws of chance used in the physical world do not apply to biological phenomena; a materialistic theory of the universe was unsatisfactory. Du Noüy, like others, had his own necessity that denied any role for chance: in examining science and its consequences, he concluded that "these consequences lead inevitably to the idea of God." Placed in the broader context of "human destiny," du Noüy's arguments were undoubtedly more influential than all other technical scientific articles on the subject combined.[112]

The idea of the low probability of the chance formation of life, not necessarily accompanied by any religious conclusions, was fortified in the second half of the century as knowledge of the complexity of life increased. Some biologists – even some exobiologists – became increasingly concerned about the odds of life having arisen by chance. In the same paper in which he wrote that "comparing the chemical structure of extraterrestrial life with that of life on Earth" was the only way to answer the question of life's origins, biochemist Norman Horowitz also wrote that "It is assumed by some biologists, and in my experience, by most astronomers who consider the matter, that the probability of the origin of life given favorable conditions – i.e., conditions resembling those of the primitive Earth – is practically unity. I think that this optimistic estimate may be far from the mark." Because there was as yet no satisfactory theory for the origin of the nucleic acid–protein system found on Earth, Horowitz wrote, "an objective estimate, based on known chemistry and known biology, would lead to a probability for the origin of life of close to zero." Only the discovery of extraterrestrial life might change that estimate, and

112 Lecomte du Noüy, *Human Destiny* (New York, 1947), xvi, 33–37. Two decades later, Shklovskii and Sagan (*Intelligent Life*, 28) cited this conclusion, but went on to argue "other possibilities" and to cite the synthesis of amino acids and other organic constituents.

for Horowitz the Viking results confirmed his skeptical assessment, remarkably similar to that of Matthew 50 years earlier. Those results, however, did not seem to shake the faith of exobiologists. Although a few, like Horowitz, were convinced after the Viking spacecraft failed to find organic molecules on Mars that the conditions for the origin of life were extremely special, most exobiologists put their faith in observation beyond the solar system, preferring not to extrapolate from Mars to the the rest of the universe.[113]

The same theme of the workings of chance, and of its low odds in the origins of life, pervaded the influential book *Chance and Necessity* (1971) by the French biologist Jacques Monod. Monod, eventually director of the Pasteur Institute in Paris, and awarded the Nobel Prize with Andre Lwoff and Francois Jacob in 1965 for explaining the genetic regulation of protein synthesis in cells, drew on the latest results in molecular biology to illustrate how chance predominated in all aspects of life's development. Sanger's first description in 1952 of the protein known as "insulin," Monod noted, was "both a revelation and a disappointment," since the sequences composing the protein showed no general law of assembly, only the blind workings of chance. And although these blind workings were then faithfully replicated except for mutations, and although natural selection operated on those mutations under demanding conditions in the process of evolution, life was "the product of an enormous lottery presided over by natural selection, blindly picking the rare winners from among numbers drawn at utter random." Although life was explicable by physical principles (Monod was no vitalist), it was not predictable, either in the fact of its origin or in the fact or direction of its evolution. In Monod's view, the biosphere of the Earth was a unique occurrence that could not be deduced from first principles, no matter how successful a universal theory might be in other domains. If life appeared on the Earth only once, which the unity of biochemistry argued, then the chance of its occurring before the event, its a priori probability, "was virtually zero." Although humanity recoiled from this idea because of a feeling of destiny, Monod warned, that was a feeling against which "we must be constantly on guard." "The universe was not pregnant with life, nor the biosphere with man. Our number came up in the Monte Carlo game," Monod

[113] N. H. Horowitz, "The Biological Significance of the Search for Extraterrestrial Life," in *The Search for Extraterrestrial Life*, vol. 22 of *Advances in the Astronautical Sciences*, ed J. S. Hanrahan (New York, 1967), 3–13. Among the biochemists who changed their minds in the wake of Viking was Sidney Fox (of protenoid fame), who commented that "earlier optimism about extraterrestrial life is far less justified than it has appeared to be, mainly on the basis of data that no one seems to have paid enough attention to." Sidney Fox, "Humanoids and Proteinoids," *Science*, 144 (1964), 954, reprinted with additional comments in 1980 in Goldsmith, *The Quest*, 222–224.

concluded, while not drawing any de Noüyian theological conclusions. Although Monod did not address the subject of life beyond the Earth, if chance reigned on Earth, one could hardly reach any conclusions about life beyond the Earth when any theory of the origins of life could not lead to predictable results.[114]

By the 1980s, even some astronomers began to be impressed with the improbability of the chance formation of life. One was the irrepressible Hoyle, who argued that life could not have been produced by a random sequence of events and then drew the de Noüyian conclusion that it is derived from a cosmic intelligence. Beginning with their book *Evolution from Space,* Hoyle and Wickramasinghe showed themselves unalterably opposed to the Earth-bound chemical theory on the grounds of probability. The chance of obtaining an enzyme in functioning form from the random ordering of the 20 amino acids they calculated as 10^{20}. But there were about 2000 enzymes, "and the chance of obtaining them all in a random trial is only one part in $(10^{20})^{2000} = 10^{40,000}$, an outrageously small probability that could not be faced even if the whole universe consisted of organic soup." The authors were frank about the implications: "If one is not prejudiced either by social beliefs or by a scientific training into the conviction that life originated on the Earth, this simple calculation wipes the idea entirely out of court. But if one is so prejudiced it is possible, in the fashion of a grand master with a lost game of chess, to wriggle ingeniously for a while," making a series of postulates for which there is no evidence. In his brief autobiography, Hoyle again referred to the intransigence of those who did not face up to these odds:

> I estimated (on a very conservative basis) the chance of a random shuffling of amino acids producing a workable set of enzymes to be less than $10^{-40,000}$. Since the minuteness of this probability wipes out any thought of life having originated on the Earth, many whose thoughts are irreversibly programmed to believe in a terrestrial origin of life argue that the enzyme estimate is wrong. It is – it is too conservative.

In a concluding chapter to their *Evolution from Space* entitled "Convergence to God," Hoyle and Wickramasinghe argued that the idea that life was assembled by a cosmic intelligence had a vastly greater probability.[115]

114 J. Monod, *Chance and Necessity* (New York, 1971), 42–43, 138, 144–146. Monod made much of the idea that proteins were "the essential molecular agents of teleonomic [purposeful] performance in living beings" in the sense of oriented, coherent, and constructive activity. This he traced, in turn, to stereospecific and noncovalent complexes.

115 Hoyle and Wickramasinghe, *Evolution from Space* (New York, 1981), 24, 129ff.; Fred Hoyle, "The Universe: Past and Present Reflections," *Annual Reviews of Astronomy and Astrophysics,* 20 (Palo Alto, Calif., 1982), 4–5.

In his book advocating panspermia, Francis Crick, the physicist turned molecular biologist, also questioned the chances of life originating on Earth. Taking a more moderate position than Hoyle, Crick concluded that "it is impossible for us to decide whether the origin of life here [on Earth] was a very rare event or one almost certain to have occurred." Without direct experimental support, those who championed the certainty of life, according to Crick, had only a hollow argument. In a chapter on "A Statistical Fallacy," he concluded that life would most probably not have originated on Earth by the chemical theory given a second chance.

> If the earth started all over again (with only small variations so that events would not repeat themselves exactly), would we expect to see life beginning for a second time? More to the point, if a planet rather similar to the earth exists elsewhere, what are the chances that life could get going there? Even in these cases there is a strong psychological urge to believe that such events must be highly likely because of the example of life on earth. Unfortunately, this argument is false.[116]

The plain fact is, Crick argued,

> that the time available was too long, the many microenvironments on the earth's surface too diverse, the various chemical possibilities too numerous and our own knowledge and imagination too feeble to allow us to be able to unravel exactly how it might or might not have happened such a long time ago, especially as we have no experimental evidence from that era to check our ideas against.

Similarly, the skeptical biochemist Robert Shapiro, in a chapter on "The Odds" in his *Origins,* calculated the chances of the spontaneous generation of the simplest known organism, the bacterium. Assuming that a billion years was available for the origin of life on Earth and that an ocean 10 km deep covered the planet, he found that 10^{51} trials were available to form the bacterium. Calculating then how many trials would have been required, he concluded that the estimate of Hoyle and Wickramasinghe of $10^{40,000}$ was too low. Rather, he agreed with Yale University physicist Harold Morowitz that the odds are really 1 in $10^{100,000,000,000}$, or 1 in 10 to the hundred billionth.[117] At these odds, Shapiro remarked, all the time and space in the universe would make no differ-

[116] *Life Itself,* 87–90.

[117] Shapiro, *Origins,* 125–131. The conclusions of Morowitz are given in his *Energy Flow in Biology* (New York, 1968).

ence; "If we were to wait, we would truly be waiting for a miracle." The only hope was that the origin of life was at a level considerably below that of the bacterium, and this is why the RNA random replicator was proposed. However, to create RNA with 20 nucleotides, or 600 atoms rather than the millions required for the bacterium, Shapiro calculated the odds at 1 chance in 10^{992}, still very poor odds indeed. If one accepted these calculations, then the problem of the origin of life on Earth having occurred by the chance assembly of atoms even at the level of the random replicator remained intransigent and the odds of life beyond the Earth extremely low. To those like Hoyle, Crick, and Shapiro, it seemed that Lecomte du Noüy had *underestimated* just how low the probability was. Still, each could draw his own conclusion. For Shapiro, at least, his calculations did not necessarily mean that the origin of life itself was improbable, but simply that the random replicator mechanism was a poor one; he preferred a more gradual route that did not rely on particular magic molecules.

As some of these same authors had pointed out, if chance and necessity loomed large in the origin of life, its effect was enormously magnified in evolution, which encompassed many steps after life's origin. In his *Time's Arrow and Evolution* (1951), biologist Harold F. Blum showed how physical nature – in the form of the second law of thermodynamics – placed constraints on the options available for biological evolution. The second law dictated that the universe tends toward disorder, and Blum considered living organisms as thermodynamic systems. The unique place of the Earth in the solar system should make one cautious about life in the universe, he argued. But "if he considers the complexity of living systems and the combination of physical limitations and apparent accident that have characterized the course of organic evolution, he may be still more critical of the idea that life exists elsewhere in the universe." The same argument applied not only to the fact of evolution, but to morphology of its products. Even if life did exist elsewhere, "it probably has taken a quite different form. And so life such as we know may be a very unique thing after all, perhaps a species of some inclusive genus, but nevertheless a quite distinct species." Four years later, Blum reaffirmed that belief when he concluded that "For close parallelism of biological evolution among the planets Time's Arrow would have had to play a much more directly deterministic role than now seems likely."[118]

In choosing chance and contingency over necessity and determinism,

[118] Harold F. Blum, *Time's Arrow and Evolution* (Princeton, N.J., 1951), 211–212, and 212A in the second edition (1955). Blum, "Perspectives in Evolution," *American Scientist*, 43 (October 1955), 595–610. We recall that Blum's book greatly impressed future Viking project scientist Gerald Soffen, who went to Princeton to study under Blum.

Blum anticipated the opinion of the majority of evolutionists who would enter the debate in the Space Age. Like Blum, many would extend their thinking backward from biological to chemical evolution. Harvard evolutionist Ernst Mayr, for example, wrote in 1982 that "A full realization of the near impossibility of an origin of life brings home the point how improbable this event was. This is why so many biologists believe that the origin of life was a unique event. The chances that this improbable phenomenon could have occurred several times is exceedingly small, no matter how many millions of planets in the universe."[119] As we shall see in the next section, the issue of chance and necessity would dominate the thinking of Mayr and other evolutionists who commented on extraterrestrial life.

There were, of course, ways around what exobiologists considered such pessimistic arguments, and they were championed by more than just exobiologists. In his defense of Darwinism, Oxford zoologist Richard Dawkins argued that natural selection was like a "blind watchmaker" who, through a cumulative effect, could build complex entities with no purpose in view. And while the origin of life via cumulative selection was not a probable event, and while one could not use its presence on Earth to argue for its probability in the universe, nevertheless, in Dawkins's view, "The origin of life on a planet can be a very improbable event indeed by our everyday standards, or indeed by the standards of the chemistry laboratory, and still be sufficiently probable to have occurred, not just once but many times, all over the universe." He pointed out that our view of probabilities was conditioned by our relatively short life span, and that the judgment of probabilities by an alien with a life span of a million centuries would be quite different. Although this did not address the odds calculated by people like Hoyle and Shapiro, it did at least make a point about the subjective nature of our idea of probability. Biologist Peter Mora offered a more radical way out when he argued in "The Folly of Probability," one of the papers delivered at the 1963 Florida conference on "Prebiological Systems," that while probability was a proven concept in physics and mathematics, it should not be applied to the question of the origin of life. The latter, Mora argued, might require a teleological explanation taking into account the purpose of living systems. Mora fully realized the vitalist implications of his argument, and even if he did not agree with Mora, Bernal commented that this paper, and its subsequent discussion joined by Haldane and Oparin, among others, "posed the most fundamental questions of the theory of the origins of life that have

[119] Ernst Mayr, *The Growth of Biological Thought: Diversity, Evolution and Inheritance* (Cambridge, Mass., 1982), 583–584.

been raised at this conference, or, as far as I know, elsewhere." By the 1990s, work on the concepts of self-organization and complexity offered further hope to origin of life optimists.[120] Still, in the view of most exobiologists, the only way to resolve the dilemma of chance and necessity was to search for life beyond the Earth and, after the Viking results, to continue the search.

Clearly, in their search for a universal biology, exobiologists placed their faith in necessity and determinism, even with all of the advances in biochemistry and molecular biology and even in the wake of the Viking results. Although the experimental synthesis of the amino acids and the discovery of complex organic molecules in space environments were clearly not the same as life itself, these syntheses instilled in them a faith that the process would take place. Synthesizing life itself, some emphasized, was not necessarily the goal of biochemists; the head of NASA's early effort in exobiology pointed out in 1963 that just as astronomers could seek an explanation of the origin of the solar system without making one, the primary objective for biochemists was an adequate explanation of the origin of life without necessarily creating life.[121] But given the lack of progress in synthesizing even proteins under possible primitive Earth conditions, and in the face of better understood odds of chance formation, even the latter objective remained elusive at the end of the century.

As we contemplate the evolution of biological complexity as depicted in Figure 7.4, we see how little hard evidence was deduced for the likelihood of extraterrestrial life from the point of view of origin of life studies in the twentieth century. Meteors, comets, and interstellar molecules contained only organic compounds at steps 1 and 2. The hope of the early 1960s that meteorites might contain fossil remnants of life at step 4 is now discredited. Artificial synthesis has been achieved for organic molecules at steps 1 and 2 under possible primitive Earth conditions and at the lower levels of step 3 under other conditions. But the assumption of a reducing atmosphere is now in question, and thereby the relevance of those experiments to the origin of life. The transition from step 2 to 3 – the origin of nucleic acids and proteins – is not understood, nor are the

[120] P. T. Mora, "The Folly of Probability," in *The Origins of Prebiological Systems*, ed. S. W. Fox, (New York, 1965), 39–64; Richard Dawkins, *The Blind Watchmaker* (New York, 1987), 139–166. Dawkins (143) remarked that it was his guess that life was common in the universe. Dawkins also believed that once natural selection began, only a "relatively small amount of luck" was required for the evolution of life and intelligence. For work on self-organization and complexity in the context of origin of life see Stuart Kauffman, *At Home in the Universe: The Search for the Laws of Self-Organization and Complexity* (New York and Oxford, 1995).

[121] Freeman H. Quimby, "Introductory Remarks," in Fox, ed., *Origins*, 1–3.

transitions from 3 to 4 (origin of the cell) or from 4 to 5 (origin of intelligence).

Yet, despite the scientific problems and the philosophical uncertainties, by the 1990s the four factors that we have documented in this section – the search for life on Mars, the aspiration of exobiologists for a universal biology, the discovery of organic molecules in space, and shared philosophical problems – had consummated the marriage of origin of life and extraterrestrial life studies; their integration was no longer a goal but a fact. Nowhere is this more evident than in a study on "The Search for Life's Origins" by the National Academy of Sciences, written more than three decades after the same body had expressed concern about planetary contamination – a concern that sparked biologists' interest in life in outer space as a prelude to the search via spacecraft. The "Planetary Biology and Chemical Evolution" aspects emphasized in the report's subtitle were now inseparable from the study of the origins of life. The origins of life issue was now dominated by the cosmic history of the biogenic elements and compounds, the implications of early planetary environments for chemical evolution and the origins of life, and even the search for life outside the solar system. Recommendations for further study included spacecraft missions to Mars (still the highest priority despite the Viking results), to comets and asteroids, to Titan and the outer gas planets, and to Earth-orbiting facilities for studying the interstellar medium and collecting dust particles, as well as a raft of related ground-based studies. Experiments in prebiotic synthesis would also continue; the report fully recognized that because of evidence indicating that the primitive Earth's atmosphere may have been composed of carbon dioxide, nitrogen, and water vapor, rather than the highly reducing atmosphere assumed by Urey and Miller, "the question of the synthesis of organic compounds on the prebiotic Earth is far from settled and must be reexamined."[122]

These goals would carry the search for life's origins well into the next century. The progress in obtaining data relevant to the origin of life during the 30 years since the dawn of the Space Age, from the study of comets and interstellar matter to the Viking biological experiments, was balanced by the great uncertainties still remaining. By the end of the century the quest for the origins of life on Earth was poised to continue, now fully integrated into the goals of exobiology, with a small cadre of biologists still enthusiastically pursuing their quest for universality and

[122] National Research Council, *The Search for Life's Origins: Progress and Future Directions in Planetary Biology and Chemical Evolution* (Washington, D.C., 1990), 80. The study was carried out by the Academy's Committee on Planetary Biology and Chemical Evolution, part of the Space Science Board, and was chaired by Viking Biology team head H. P. Klein. A list of committee members is given on page iii.

equal status with the physical sciences that a universal biology would bring. In this pursuit, however, they would have to deal with the evolutionary biologists, who were largely pessimistic about the morphology, if not the very existence, of extraterrestrials, placing a damper on the aspirations of those who sought to extend the hard-won principles of terrestrial biology to the universe at large or to incorporate those principles into a more generalized biology.

7.4 EVOLUTION AND EXTRATERRESTRIALS: CHANCE AND NECESSITY REVISITED

If the leap from the synthesis of amino acids to the origin of life was large, the leap from first life to intelligence was, in the eyes of some (though not all), even more monumental. But it is a matter of terrestrial history that chemical evolution not only begat biological evolution, biological evolution also begat intelligence. Beyond the origin of life discussed in the preceding sections, the issue of chance and necessity was at the foundation of three related questions in the evolution of extraterrestrial life: Would intelligence identical to humanity evolve on other planets? If not, would intelligences different in form evolve? And if the latter, would these intelligences have the ability to communicate with others so different in form? In light of the SETI endeavor begun in 1960, these were not merely academic questions, but issues relevant to science policy and the odds of a return on taxpayers' money. Busy with problems on Earth, evolutionists gave only limited attention to the question of evolution in the extraterrestrial context. Nevertheless, among those who did enter the debate – including some of the pioneers of the evolutionary synthesis – patterns emerge suggesting that the quest for universality in biology had its limits.

Alfred Russel Wallace, cofounder with Darwin of what later became known as the "theory of natural selection," had directly addressed the relevance of evolution to extraterrestrials in his book on plurality of worlds at the beginning of the century. In an appendix for the 1904 edition of his book entitled "An Additional Argument Dependent on the Theory of Evolution," Wallace pointed out that almost all biologists believed that no species had ever originated more than once on Earth. Because the environment, which affects natural selection, is not the same in any two places, "an identical specific evolution cannot take place a second time." Even less probable, then, was a whole series of identical developments from the dawn of life to the development of the human organism. If any one change had taken place in the line of descent to humans, humans would never have come into existence. The development of humanity depended on millions of distinct modifications; the chances against all of these modifications occurring independently even in two distinct parts of our planet are "almost infinite." Therefore, on another planet, where environmental conditions are even more diverse, Wallace concluded that the evolution of a species identical to humanity would be "infinitely improbable."

But how about an intelligence equal to that of humanity, but in very different form? To Wallace this seemed "quite inadmissible," again from observation that of all the life on Earth, none had developed intelligence or showed the proclivity to do so.

> The mere assertion, therefore, that a being possessing man's intellectual and moral nature combined with a very different animal form, might have been developed, is wholly valueless. We have no evidence for it, while the fact that no other animal than man has developed his special faculties even to a lower degree, is strong evidence against it.[123]

If the physical improbabilities set forth in his book were 1 million to one, Wallace concluded, then the evolutionary improbabilities of the development of intelligence were less than 100 million to one, "and the total chances against the evolution of man, or an equivalent moral and intellectual being, in any other planet, through the known laws of evolution, will be represented by a hundred millions of millions to one." To those who agreed that this was true for a universe of natural law, but perhaps not true given the action of a Creator, Wallace argued that we had no knowledge of the Creator's purpose, and he might well be satisfied with the millions of souls already produced on Earth and the millions more that would be in the future.

Wallace felt sure that this argument would "appeal to all biological students of evolution," and in this he seems to have been correct – to a point. Extending his 1921 argument about the unlikelihood of the origins of life on another planet, paleontologist W. D. Matthew at the American Museum of Natural History in New York City agreed with Wallace that the evolution of humanlike creatures on another planet was remote – and constructions like cities or canals even more remote. In the unlikely event that any intelligence did evolve, natural selection guaranteed that "it probably – almost surely – would be so remote in its fundamental character and its external manifestations from our own, that we could not interpret or comprehend the external indications of its existence, nor even

[123] A. R. Wallace, *Man's Place in the Universe,* 4th ed. (London, 1904), Appendix, 326–336: 334.

probably observe or recognize them." Indeed, beginning in the late 1920s, the many editions of *The Science of Life* by H. G. Wells, G. P. Wells, and Julian Huxley pressed the same point even for nonintelligent life while showing that the search for a universal biology, one conceivable in terms of life on Earth, was not yet underway. Life on Mars would be so different from life on Earth, they argued, that one would have to term it something like "Beta Life," "an analogous thing and not the same thing. It may not be individualized; it may not consist of reproductive individuals. It may simply be mobile and metabolic. It is stretching a point to bring these two processes under one identical expression." Moreover, they argued, although one could conceive of silicon instead of carbon or sulfur instead of oxygen acting under different temperatures and pressures to produce consciousness, these would not merely be different forms of life but would strain the very meaning of the word "life." Nor does this seem to have been a matter of semantics, but rather a statement that the knowledge of life on Earth could tell us little about life in the universe.[124]

Thirty years later, the unification of biology and the evolutionary synthesis complete, the position of Wallace and Matthew was adopted practically unchanged in terms of the morphology of extraterrestrial life and intelligence but not in terms of the likelihood of its existence. Anthropologist Loren Eiseley was among the first to reecho the diverse morphology concept, writing in 1957, "Of men elsewhere, and beyond, there will be none forever." Eiseley's conclusion was reached in the context of Darwin, who, he noted, saw clearly that the development of life was not a pattern imposed from without, but one that had been modified by natural selection "along roads which would never be retraced." Expanding to the context of extraterrestrial life, he wrote that "Every creature alive is the product of a unique history. The statistical probability of its precise reduplication on another planet is so small as to be meaningless."[125]

But Eiseley and others from the biological sciences were not so unanimous about two other questions: the chances of intelligence occurring in some form, and whether one could reach a mutual understanding with such intelligence. The American geneticist H. J. Muller believed that higher life forms would develop on other planets and agreed that they "may be expected to have followed radically different courses in regard to many of the features" of their evolution, as evidenced by

[124] W. D. Matthew, "Life in Other Worlds," *Science*, 44 (September 16, 1921), 239–241: 241; H. G. Wells, Julian Huxley, and G. P. Wells, "Is There Extra-terrestrial Life?" in *The Science of Life* (New York, 1929), 11–13. This latter text greatly influenced Johsua Lederberg, among others of his generation.

[125] Eiseley, *Immense Journey*, 160–161.

differences among advanced life forms on Earth. "How much greater, then, might such differences be between the forms of Earth and those of another planet. These differences, affecting their whole internal economy, including the biochemistry within their cells, would also be expressed in their gross anatomy and in their outer form." Still, Muller believed "it would certainly be capable of achieving much mutual understanding with our own, since both had been evolved to deal usefully with a world in which the same physico-chemical and general biological principles operate."[126]

The most detailed and influential treatment of the problem came from the evolutionist George Gaylord Simpson, whose article "The Nonprevalence of Humanoids" was written as the United States was preparing to spend massive amounts of money to search for life on Mars.[127] Aware of Matthew's 1921 article, Simpson argued that evolutionists needed to be heard by those who espoused exobiology – a " 'science' that has yet to demonstrate that its subject matter exists!" Simpson took to task both physical scientists and biochemists engaged in exobiology who assumed, "usually without even raising the question," that once life arose on another planet, its course would be similar to that on Earth. He pointed out that biologist Harold Blum had distinguished two possibilities: the deterministic one, in which evolution must be similar to Earth's, and the opportunistic one, in which life had many possible courses.[128] Exobiologists, Simpson noted, seemed to fall into the deterministic camp. But he found no support for this in evolutionary biology on Earth, the only empirical evidence available. To the contrary, he argued that the fossil record showed no necessity that evolution proceed from protozoa to humans; rather, most early life forms became extinct, indicating that nature experiments with its life forms. Moreover, the processes of evolution by mutation and recombination were not directed. "If the causal chain had been different, *Homo sapiens* would not exist," Simpson concluded, mimicking Wallace's argument 60 years earlier. "No species or any larger group has ever evolved, or can ever evolve, twice."[129] Extend-

[126] H. J. Muller, "Life Forms to Be Expected Elsewhere Than on Earth," *The American Biology Teacher*, 23 (October 1961), 74–85: 80, 84. On Muller see Elof Carlson, *Genes, Radiation and Society: The Life and Work of H. J. Muller* (Ithaca, N.Y.), 1981.

[127] George Gaylord Simpson, "The Non-Prevalence of Humanoids," *Science*, 143 (February 21, 1964), 769–775, reprinted in Simpson, *This View of Life* (New York, 1964), and Goldsmith, *The Quest*, 214–221. Simpson took the term "humanoid" from science fiction, to which he admitted he was "addicted."

[128] Simpson cites Blum's recently published review of two books on the origin of life, "Negentropy and Living Systems," *Science*, 139 (February 1, 1963), 398, but undoubtedly knew of Blum's *Time's Arrow*.

[129] Here Simpson cites his *The Meaning of Evolution* (New Haven, Conn., 1949) and "The History of Life," in *The Evolution of Life*, ed. Sol Tax (Chicago, 1960). See also Simpson,

ing the morphology argument to the more general question of the existence of extraterrestrials, Simpson was pessimistic to the point that he ended with a practical suggestion. Pointing to the expenditure for space exobiology, he said, "Let us face the fact that this is a gamble at the most adverse odds in history. Then if we want to go on gambling, we will at least recognize that what we are doing resembles a wild spree more than a sober scientific program."

One of the architects of the unification of biology, the evolutionist Theodosius Dobzhansky, while pointing out that chance and necessity were not mutually exclusive in evolution, agreed with Simpson. He stated that physical scientists such as Laplace usually adopted the determinist view but argued that "Laplacean determinism sheds no light on evolutionary history." Humanity was "invented," not predestined, according to Dobzhansky, and although natural selection would invent many other forms on other planets, the reinvention of humanity was unlikely. A biologist living in the Eocene period, he argued, could not have predicted the emergence of humanity. Of humanity's 100,000 genes, perhaps half had changed since that time, and the chance of the same 50,000 genes being changed in the same ways and in the same sequence was zero.

> Natural scientists have been loath, for at least a century, to assume that there is anything radically unique or special about the planet Earth or about the human species. This is an understandable reaction against the traditional view that Earth, and indeed the whole universe, was created specifically for man. The reaction may have gone too far. It is possible that there is, after all, something unique about man and the planet he inhabits.[130]

The views of Simpson and Dobzhansky were carried even further by another pioneer of the evolutionary synthesis, Harvard biologist Ernst Mayr. Mayr argued not only the unlikelihood of the human form emerging on another planet, but also "the incredible improbability of genuine intelligence emerging." Even though quantum mechanics had placed de-

"Some Cosmic Aspects of Organic Evolution," in *Evolution und Hominisation,* ed. G. Kurgh (Stuttgart, 1962), 6–20.

[130] Theodosius Dobzhansky, "Darwinian Evolution and the Problem of Extraterrestrial Life," *Perspectives in Biology and Medicine,* 15 (Winter 1972), 157–175: 173–175. In his 1972 article, Dobzhansky remarked (158) that "Evolutionary biologists have mostly steered clear of these issues [of extraterrestrial life], probably finding them too speculative." On Dobzhansky, who (like H. J. Muller and Sidney Fox) worked in the laboratory of T. H. Morgan, see Mark B. Adams, ed., *The Evolution of Theodosius Dobzhansky* (Princeton, N.J., 1994), and Francisco J. Ayala, "Dobzhansky," in *DSB,* Supplement, 233–242 and references therein. On Dobzhansky's role in the unification of biology see Smocovitis, "Unifying Biology," 20ff.

terminism in doubt, Mayr was impressed by how physical scientists were "still think along deterministic lines," while evolutionists were "impressed by the incredible improbability of intelligent life ever to have evolved, even on earth." Mayr pointed out that for 3 billion years after the first prokaryotes formed, life did not evolve. Then when eukaryotes (nucleated cells) originated in the Cambrian period, four kingdoms of life developed in quick succession: the protists, fungi, plants, and animals. But intelligence did not even begin to develop in any of the kingdoms except that of the animals. Even then, there were hundreds of branching points that led to humanity. These chance events implied not only that humanity was improbable, but also intelligence itself: "There were probably more than a billion species of animals on earth, belonging to many millions of separate phyletic lines, all living on this planet earth which is hospitable to intelligence, and yet only a single one of them succeeded in producing intelligence." Nor, he argued, could the "convergent evolution" of an organ such as the eye be applied to intelligence itself, because while the history of life on Earth confirms the former, it denies the latter. Finally, Mayr found the idea that extraterrestrials would have the "technology and mode of thinking of late twentieth century man" to be "unbelievably naive." In a remarkable parallel to Simpson's arguments against funding for NASA's life search on Mars, Mayr used these evolutionary arguments 30 years later to argue against NASA's SETI programs.[131]

Harvard paleontologist Stephen Jay Gould agreed with Wallace and Simpson and other evolutionists about the nonprevalence of humanoids. One of the main points of his book *Wonderful Life* (1989) is that evolution is a

> staggeringly improbable series of events, sensible enough in retrospect and subject to rigorous explanation, but utterly unpredictable and quite unrepeatable. . . . Wind back the tape of life to the early days of the Burgess Shale; let it play again from an identical starting point, and the chance becomes vanishingly small that anything like human intelligence would grace the replay.[132]

[131] Ernst Mayr, "The Probability of Extraterrestrial Intelligent Life," in Edward Regis, Jr., ed., *Extraterrestrials: Science and Alien Intelligence* (Cambridge, 1985), 23–30. For Mayr's anti-SETI campaign, see his letter "The Search for Intelligence," *Science*, 259 (March 12, 1993), 1522. Speaking of the NASA project, he wrote, "I find it astounding that the go-ahead for this project was given without more broad-based consultation and that such a highly dubious endeavor is supported by NASA in this time of appalling federal debt." Six months later, all funding for the NASA SETI project was terminated by Congress.

[132] Gould, *Wonderful Life: The Burgess Shale and the Nature of History* (New York, 1989), 14, 301. This book has the contingency of life as its major theme, focusing on the

The primary insight won from the Burgess Shale, the richest fossil field of the Cambrian explosion of life about 570 million years ago, is that evolution was not a result of continual proliferation and progress of life, but of the decimation of species "probably accomplished with a strong, perhaps controlling, component of lottery." This view of the contingency of life was bolstered by the increasing acceptance toward the end of the century of the "mass extinction" hypothesis, whereby species were decimated by random occurrences, including the collision of large meteors or asteroids with the Earth.[133]

But at the same time, Gould made the point specifically in regard to extraterrestrial intelligence that lack of detailed repeatability did not necessarily imply lack of extraterrestrials, a distinction some had failed to make.

> When we use "evolutionary theory" to deny categorically the possibility of extraterrestrial intelligence, we commit the classic fallacy of substituting specifics (*individual* repeatability of humanoids) for classes (the probability that evolution elsewhere might produce a creature in the *general* class of intelligent beings). I can present a good argument from "evolutionary theory" against the repetition of anything like a human body elsewhere; I cannot extend it to the general proposition that intelligence in some form might pervade the universe.

Though some evolutionists (such as Simpson and Mayr) did make that extension, Gould pointed out that four evolutionists (including himself) had signed a pro-SETI petition in the belief that some finite chance existed of finding such intelligence. He noted that the phenomenon of "convergence" in terrestrial biology, whereby nature had independently invented the general phenomena of the eye, flight, and other adaptive forms many times (though different in detail), might indicate that nature had invented intelligence on many planets. Though some had argued that its appearance only once on Earth indicated the rarity of intelligence as one

Cambrian explosion. Acceptance of contingency in extraterrestrial evolution raises the problem of the recognition of alien life; on this aspect see "A Discussion on the Recognition of Alien Life," organized by N. W. Pirie, *Proceedings of the Royal Society of London, Series B,* vol. 189 (1975), 137–274, and Pirie's "Introductory Remarks in a Discussion on the Recognition of Alien Life," 139–141. This discussion included papers by Sagan, geneticist C. H. Waddington, J. E. Lovelock, and A. G. Cairns Smith, among others.

133 For a review up to 1990 see David Raup, *Extinction: Bad Genes or Bad Luck?* (New York, 1991).

of Nature's general solutions, Gould did not believe this conclusion could be extended from one planet to the universe.[134]

Thus, among evolutionists who cared about the issue (which was by no means most evolutionists), there was a diversity of opinion as to whether intelligence existed beyond the Earth, along with virtual unanimity that if it did exist, the forces of natural selection would produce morphologies vastly different from the humanoid form. There was also uncertainty about the capability of such different intelligent morphologies for mutual recognition and communication. Evolutionists therefore lent credibility to the quest for a universal biology only to the extent that they believed that if life originated, natural selection would operate throughout the universe under whatever conditions life found itself. They did not share the optimism of exobiologists that life would originate, nor did they take for granted that mutual understanding would be easy or even possible. "We are not alone in the universe, and do not bear alone the whole burden of life and what comes of it," Wald wrote in his 1954 essay. "Life is a cosmic event – so far as we know the most complex state of organization that matter has achieved in our cosmos. It has come many times, in many places – places closed off from us by impenetrable distances, probably never to be crossed even with a signal."[135] In his article on the nonprevalence of humanoids, Simpson made an even more fundamental argument about communication: "I therefore think it extremely unlikely that anything enough like us for real communication of thought exists anywhere in our accessible universe."

On the other hand, exobiologists of the SETI variety, while largely accepting the argument of morphological diversity of the evolutionists, continued their programs despite the uncertainties regarding the existence and communicability of extraterrestrials. Beginning at their earliest

[134] Stephen Jay Gould, "SETI and the Wisdom of Casey Stengel," in *The Flamingo's Smile: Reflections in Natural History* (New York, 1985), 403–413. The essay originally appeared in *Discover Magazine* (March 1983). Aside from himself, Gould cited Tom Eisner (Cornell), David Raup (University of Chicago), and Edward O. Wilson (Harvard) as having signed Sagan's pro-SETI petition. Wilson (of sociobiology fame) has remarked that "The laws of chance alone dictate that there is life around some of the stars, and because evolution is such an enormously creative force there are probably also advanced civilizations." Charles J. Lumsden and E. O. Wilson, *Promethean Fire: Reflections on the Origins of Mind* (Cambridge, Mass., 1983), 53. Noting that SETI proponents' assumption of the similarity in terrestrial and extraterrestrial behavior and intelligence was "unpopular with paleontologists and biologists because the Phanerozoic record shows no evidence of such predictability or consistency," Raup wrote that "Whereas I agree with this objection, I remain a strong supporter of the SETI effort. Only by discovering biological systems elsewhere in space will we really have the means to know whether our own biological system has predictable patterns not yet recognized" (*Extinction*, 40–41).

[135] Wald, "Origin of Life," 53.

conferences, they included discussions in the evolutionary arena on issues like intelligence in dolphins, and went on to broader issues such as interspecies communication and factors affecting the rate of evolution of intelligence. Despite the occasional warning that cognition (as opposed to lower intelligence) could be "exceedingly rare" in the universe, they forged ahead, confident that empiricism was better than speculation. In his Pulitzer Prize-winning book *The Dragons of Eden* (1977), Sagan reflected the view of most exobiologists:

> Once life has started in a relatively benign environment and billions of years of evolutionary time are available, the expectation of many of us is that intelligent beings would develop. The evolutionary path would, of course, be different from that taken on Earth. The precise sequence of events that have taken place here . . . have probably not occurred in precisely the same way anywhere else in the entire universe. But there should be many functionally equivalent pathways to a similar end result.

It was, Sagan suggested, a matter of natural selection that "smart organisms by and large survive better and leave more offspring than stupid ones," although he admitted that the capacity for self-destruction by technological civilizations left matters uncertain. But the longevity of civilizations aside, exobiologists for the most part accepted contingency in evolution, while still opting for necessity in believing that some form of intelligence would develop.[136]

In their consensus on the contingent nature of the evolution of life, evolutionists and exobiologists alike were in agreement with biochemists who had probed the molecular depths of life, only to find blind chance at work during its origin. Whether in its origin or its evolution, contingency seemed to be the theme of life. While this left wide open the possibilities for extraterrestrial intelligence, unless one accepted the theological con-

[136] Carl Sagan, *The Dragons of Eden: Speculations on the Evolution of Human Intelligence* (New York, 1977), 230. Reports on dolphin intelligence began with John Lilly at the Green Bank conference on extraterrestrial intelligence in 1961 and continued as an occasional feature of these conferences through the end of the century. The rate of evolution of intelligence was discussed at a 1979 conference on extraterrestrial life by Dale A. Russell, "Speculations on the Evolution of Intelligence in Multicellular Organisms," in John Billingham, ed., *Life in the Universe* (Cambridge, Mass., 1981); an entire session of the 1987 Bioastronomy conference was devoted to the question "Is Intelligence an Inevitable Evolutionary Trait?" in Marx, *Bioastronomy*, 235–302; see especially W. H. Calvin, "Fast Tracks to Intelligence," 237–245. The rarity of advanced intelligence termed "cognition" was stressed at a 1979 conference by C. Owen Lovejoy, "Evolution of Man and Its Implications for General Principles of the Evolution of Intelligent Life," in Billingham, ed., *Life in the Universe*, 317–329.

clusion of du Noüy, Hoyle, and Wickramasinghe, the broader implications for terrestrial intelligence were staggering, since the vast majority of terrestrials agreed with du Noüy's general idea of "human destiny." In arguing that life on Earth was contingent, Monod had written, "We would like to think ourselves as necessary, inevitable, ordained from all eternity. All religions, nearly all philosophies, and even a part of science testify to the unwearying, heroic effort of mankind desperately denying its own contingency." Yet this was the one point on which biochemists, evolutionists, and many exobiologists agreed.

Considering the immense difficulties in the study of the origin and evolution of life, and the predilection of evolutionists for natural selection presiding over the blind workings of chance rather than predestination, it may seem remarkable that anyone should seriously consider the possibility of communication with advanced intelligence. Yet the final scientific story remaining to be told in the saga of the extraterrestrial life debate is just this: that the search for extraterrestrial intelligence became a serious scientific project in the last half of the twentieth century. There is no greater indication of the strength of the empirical tradition among those exobiologists who supported SETI, of their deeply held basic assumptions about the nature of the universe, and of their continued aspirations for a universal biology, of which terrestrial life was presumed to be only a particular case.

8

SETI: THE SEARCH FOR EXTRATERRESTRIAL INTELLIGENCE

. . . The presence of interstellar signals is entirely consistent with all we now know, and if signals are present the means of detecting them is now at hand. Few will deny the profound importance, practical and philosophical, which the detection of interstellar communications would have. We therefore feel that a discriminating search for signals deserves a considerable effort. The probability of success is difficult to estimate; but if we never search, the chance of success is zero.

Cocconi and Morrison (1959)[1]

This, then is the paradox: all our logic, all our anti-isocentrism, assures us that we are not unique – that they must be there. And yet we do not see them.

David Viewing (1975)[2]

Sing all ye citizens of heaven above!

Adeste Fidelis,
Traditional Christmas hymn

Two great ideas took hold in the last half of the twentieth century with regard to extraterrestrial intelligence – interstellar communication and interstellar colonization – both imaginative and bold concepts that nevertheless were amenable not only to serious discussion but also to observational confirmation. In contrast to the Viking project, the setting now moved from one planet in the solar system to the entire Galaxy, whose bounds were measured in tens of thousands of light years rather than the light hours that defined our own provincial planetary system. This was a realm clearly beyond the bounds of terrestrial spacecraft but nevertheless under the scrutiny of astronomers at ever-increasing levels of sophistication. Although the idea of interstellar communication began before Viking was conceived, in the post-Viking era it was the last best hope for testing the hypothesis of extraterrestrial life for the remainder of the twentieth century. That hope was seriously challenged by those who believed that if extraterrestrial intelligence existed, it

[1] G. Cocconi and P. Morrison, "Searching for Interstellar Communications," *Nature*, 184 (September 1959), 844, reprinted in D. Goldsmith, *The Quest for Extraterrestrial Life* (Mill Valley, Calif., 1980), 102–104.

[2] David Viewing, "Directly Interacting Extra-Terrestrial Technological Communities," *JBIS*, 28 (1975), 735–744.

would have colonized the Galaxy and arrived on Earth long ago. In either case, by focusing on intelligence rather than microorganisms, the uncertainties – and the implications for humanity – were now immensely magnified.

Of the two ideas, the search aspect was more amenable to observation, thanks to the discovery of radio waves and the development of radio astronomy. All the problems inherent in the search for planetary systems, in research on the origin of life, and in theories of the evolution of intelligence and technology – in short, all the problems associated with the possibilities of extraterrestrial intelligence – could be leapfrogged if only a method were found for direct communication with such intelligence. It is little wonder, then, that a practical method for doing just that became a kind of Holy Grail, the focus of a considerable research effort and the source of contentious debate after the dawn of the Space Age. Exactly how this came about is a story full of surprise and drama, an endeavor that eventually opened truly new areas of inquiry in the age-old quest for humanity's place in the universe. That it was challenged by the forces of interstellar colonization, who emphasized the startling observation of the absence of extraterrestrials on Earth to argue that extraterrestrials in fact did not exist, made for an interstellar debate of gargantuan interest for those interested in humanity's status in the cosmos.

In this chapter we first consider the prelude to interstellar communication, a now forgotten series of sporadic events we shall characterize as the "era of interplanetary communication" in the first half of the century. Second, we detail the serious effort to open the electromagnetic spectrum for interstellar communication, a bold effort propelled by only a bare minimum of rationale and resources. Third, we document the subsequent and much more substantial rationales developed for interstellar communication as scientists from many disciplines attempted to answer the basic question of whether or not the effort should be made at all, as well as the crisis in this rationale provoked by the idea of interstellar colonization in the mid-1970s. Finally, we examine the historical development of strategies and programs actually undertaken despite the crisis, with particular focus on the rise and fall of the most comprehensive interstellar communication program of all, developed in the United States by NASA, only to be dismantled by ridiculing political forces. Building on knowledge of our own solar system, as well as on the more speculative ideas about other planetary systems and the origin and evolution of life, the search for extraterrestrial intelligence was in many ways a convergence of all previous elements of the extraterrestrial life debate, with the added spectacular dimension of communication. It was perhaps foreordained that the search for such intelligence would be a controversial endeavor, and history does not disappoint in this respect. The ultimate in the search for humanity's place in the universe, as it oscillated between imagination and restraint, the search for intelligence encompassed all the ambiguities of the extraterrestrial life debate.

8.1 PRELUDE: THE ERA OF INTERPLANETARY COMMUNICATION

Although the idea of visual signals to the planets[3] dates to the early nineteenth century,[4] in 1887 Heinrich Hertz demonstrated the existence of a method which, it was soon realized, could be developed into a form of communication of much greater power: radio waves. The idea of radio communication on Earth had barely been developed when the Serbian-born American physicist and engineer Nikola Tesla (1856–1943) not only extended the idea to the planets, but also claimed that he might actually have intercepted such communications. While conducting experiments on the wireless transmission of power in Colorado Springs, Colorado, in 1899, Tesla observed unusual electrical disturbances that, he later recalled, "positively terrified me, as there was present in them something mysterious, not to say supernatural." After he had ruled out well-known electrical disturbances from the Sun and the Earth's atmosphere, a bold idea sprang to Tesla's fertile mind: "It was some time afterward when the thought flashed upon my mind that the disturbances I had observed might be due to an intelligent control. Although I could not decipher their meaning, it was impossible for me to think of them as having been entirely accidental. The feeling is constantly growing on me that I had been the first to hear the greeting of one planet to another."[5]

[3] Substantial parts of this section first appeared in Steven J. Dick, "Back to the Future: SETI before the Space Age," 44th Congress of the International Astronautical Federation, October 16–22, 1993, Graz, Austria, paper IAA.9.2–93–790.

[4] On the early ideas of Gauss (1822), Johann Joseph von Littrow (1842), and Charles Cros (1869), see Michael J. Crowe, *The Extraterrestrial Life Debate 1750–1900: The Idea of a Plurality of Worlds from Kant to Lowell* (Cambridge, 1986), 205–207, 393–400, and F. Drake and D. Sobel, *Is Anyone Out There? The Scientific Search for Extraterrestrial Intelligence* (New York, 1992), 170–172. Crowe also discusses early-twentieth-century ideas of visual signals, with further references.

[5] Nikola Tesla, "Talking with the Planets," *Collier's Weekly*, 26, no. 19, 4 (1901), reprinted in *Current Literature* (March 1901), 359–360. As early as 1896, the *New York Sun* and *New York Herald* had reported on Tesla's interest in interplanetary communication. The interviews are reprinted in *The Electrical World* (April 4, 1896), 369. On Tesla see Margaret Cheney, *Tesla: Man Out of Time* (New York, 1981); I. Hunt and W. W. Draper, *Lightning in His Hand: The Life Story of Nikola Tesla* (Denver, 1964); J. J. O'Neill, *Prodigal Genius: The Life of Nikola Tesla* (New York, 1944); and K. M. Swezey, "Nikola Tesla," *Science*, 127 (May 16, 1958), 1147–1159.

Tesla's announcement was greeted with skepticism,[6] and it was two decades before the revival of the idea of interplanetary communication came from another radio pioneer, Guglielmo Marconi (1874–1937). On January 20, 1919, The *New York Times* ran a front-page headline "Radio to Stars, Marconi's Hope," in which Marconi expressed the belief that the eternal property of radio waves "makes me hope for a very big thing in the future . . . communication with intelligences on other stars . . . It may some day be possible, and as many of the planets are much older than ours the beings who live there ought to have information for us of enormous value." Suggesting that mathematics might be used as a language of communication, Marconi even claimed he had received unexplained signals that might have originated from the stars. This claim prompted the *Times,* in a lengthy editorial, to advise that humanity should "Let the Stars Alone," lest we receive knowledge "for which we are unprepared precipitated on us by superior intelligences." It also prompted Tesla to reiterate his earlier claim, and further to claim that life on innumerable planets was a "mathematical certainty."[7]

A year later, The *New York Times* (quoting the *London Daily Mail*) reported that Marconi had on occasion detected with his radio equipment, both day and night, "very queer sounds and indications, which might come from somewhere outside the Earth." Morse letters reportedly occurred often, but no message was decipherable. Because the signals occurred simultaneously in the London and New York receiving stations and because they were of equal intensity, Marconi inferred that they originated at very great distances. "We have not yet the slightest proof of their origin," he noted, while speculating that it could be the Sun. But when the press raised the contentious question about other planets, Marconi replied, "I would not rule out the possibility of this, but there is no proof. We must investigate the matter much more thoroughly before we venture upon a definite explanation."[8]

For the next few weeks the *New York Times* followed up on the story almost daily, sometimes on the front page. British Astronomer Royal Frank Dyson agreed that it was quite possible to receive radio waves from other planets. Marconi repeated that "They are sounds. They may be signals. We do not know." He ruled out atmospherics and noted that they

[6] L. I. Anderson, "Extra-Terrestrial Radio Transmissions," *Nature,* 190 (April 22, 1961), 374; R. S. Ball, "Signalling to Mars," *Living Age,* 229 (1901), 277–284; "Nicola Tesla and His Talk with Other Worlds," *Colorado Springs Gazette,* March 9, 1901; E. S. Holden, "What We Know about Mars," *McClure's,* 16 (1901), 439–444.

[7] *NYT,* "Radio to Stars, Marconi's Hope," January 20, 1919, p. 1; *NYT,* "Let the Stars Alone," January 21, 1919, p. 8; Nikola Tesla, "That Prospective Communication with Another Planet," *Current Opinion,* 66 (March 1919), 70–71.

[8] *NYT,* "Marconi Still at Sea on Mysterious Sounds," January 27, 1920, p. 7.

occurred only at very long wavelengths of 100 km, three to four times those of normal commercial use. Radio engineers were quoted as being highly skeptical that the signals emanated from another planet, especially from an intelligent source, and argued that they were atmospheric disturbances induced by the Sun. The director of the Paris Observatory pointed out that if New York and London received these messages, so should the French, yet nothing abnormal had been heard from the Eiffel Tower's radio station. The U.S. Navy Department, with its advanced radio communications system, was reported to be keeping an open mind.[9]

The well-known inventor Charles Steinmetz, described as a leading authority on radio waves, denied that the signals really came from Mars but held that "if the United States . . . should go into the effort to send messages to Mars with the same degree of intensity and thoroughness with which we went into [World War I,] it is not at all improbable that the plan would succeed." The effort, he noted, would require all the electric power of the country be concentrated into one sending station, with 1000-foot towers, at a cost of perhaps $1 billion. C. G. Abbot, director of the Smithsonian Astrophysical Observatory, argued that if Marconi's signals were from another planet, Venus was a much more likely source than Mars, since Mars was too cold and lacked water. And Elmer Sperry, head of the Sperry Gyroscope Company, boasted that his company could send a message to Mars using 150 or 200 Sperry searchlights of a billion candlepower.[10] Meanwhile, in Europe, the French Academy of Sciences agreed to act as a judge for a 100,000-franc prize "to be given for the best means of making a sign to a heavenly body and the receipt of a reply." And no less a scientific icon than Albert Einstein was quoted from an interview as believing that Mars and other planets might be inhabited, but that Marconi's signals were due either to atmospheric disturbances or to experiments with other wireless systems.[11]

Two weeks later, *Scientific American* argued that while Marconi's conjectures should not be dismissed, there was absolutely no proof of Martians; that it was unlikely that they would develop a Morse code as on Earth; that they could not transmit over the 50-million-mile distance separating Earth and Mars; and that the Eiffel Tower, the U.S. Navy, and

[9] *NYT,* "Astronomer Thinks Mars Could Signal," January 28, 1920, p. 5; "Marconi Testing His Mars Signals," January 29, 1920, p. 1.

[10] *NYT,* "First Mars Message Would Cost Billion," January 30, 1920, p. 18; "Suggests Venus Is Source of Signals," January 30, 1920, p. 18; "Opposing Views on Mars Signals," January 31, 1920, p. 24.

[11] *NYT,* "Might Talk to Mars on Waves of Light," February 1, 1920, p. 1; "Offers a $20,000 Prize for Sign to a Planet – French Academy of Sciences to Decide Winner – Einstein Would Use Rays of Light," February 2, 1920, p. 24; "Italian Scientist Tells of Odd Signals," February 6, 1920, p. 11.

other radio stations had not heard the Marconi signals, "although they have searched for them." The writer believed that atmospheric disturbances or sunspots were the most likely sources of the signals. Suspecting even the Japanese or the "the Russian Bolsheviki, who have turned to radio as a convenient means of propagating their cause at home and abroad," he concluded that "this matter deserves careful study when a scientist of Mr. Marconi's standing takes it so seriously." Marconi's conjecture also inspired an interesting attempt in *Scientific American* to determine what might actually be communicated to Martians by dots and dashes in the absence of a common language (Figure 8.1).[12]

Less than a year later, the London manager of the Marconi Wireless Telegraph Company reported that Marconi was convinced he had intercepted new messages from Mars aboard his yacht *Electra* in the Mediterranean, and that it would be only a matter of time before two-way communication was established. Marconi, the manager noted, was convinced of this because he received regular signals with a wavelength of 150 km, some 10 times larger than the wavelength of radio waves transmitted by Earth stations. Marconi, listening at wavelengths never before heard, found that "their regularity disproved any belief that they were caused by electrical disturbances. The only resemblance to the code used on this planet is in the letter 'V' of the international code. These 'V' splashes were continued time after time, much after the manner of station calls or test signals sent out from radio stations." Tesla, displaying admirable restraint for a change, was quoted as noting that wireless stations also sent out undertones, which might interfere with normal wavelengths and create very long waves, a phenomenon he had noticed increasingly since 1907 as more powerful radio stations were established.[13]

Marconi's interest in interplanetary communication peaked during a trip from Southhampton, England, to New York City aboard the *Electra* from May 23 to June 16, 1922. *The New York Times* noted that Marconi "spent the time crossing the Atlantic performing many electrical experiments, principally by listening for signals from Mars." Marconi admitted that "they might have come from any region in the universe where electrons are in vibration." It is notable, however, that when Marconi addressed the Institute of Radio Engineers and the Institute of Electrical

[12] *SciAm*, "Those Martian Radio Signals," 122 (February 14, 1920), 156; H. W. Nieman and C. W. Nieman, "What Shall We Say to Mars?" *SciAm* (March 20, 1920), 122, 298. See also C. F. Talman, "Are Martians People?" *SciAm* (March 6, 1920), 301. While the *Scientific American* criticisms were logical for their time, we know today that at least one of the arguments – the inability to communicate over interplanetary distances – was what science fiction writer Arthur C. Clarke would call "a failure of imagination."

[13] *NYT*, "Marconi Sure Mars Flashes Messages," September 2, 1921, p. 1; "Messages from Mars Scouted by Experts," September 3, 1921, p. 4.

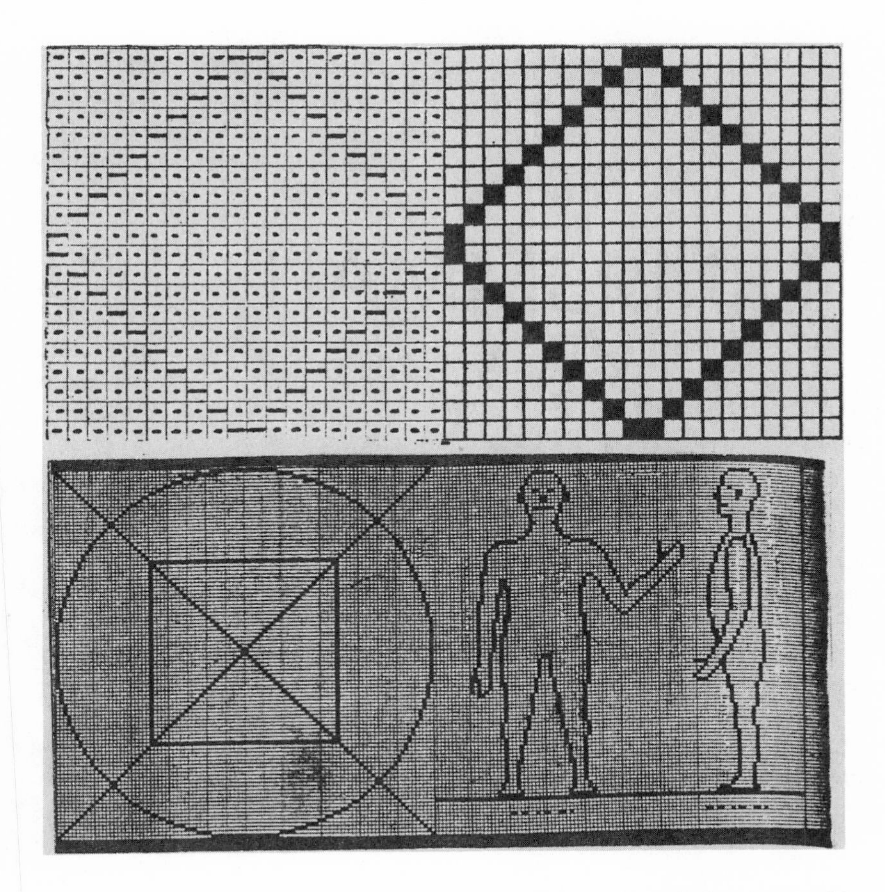

Figure 8.1. A 1920 proposal for communication by dots and dashes arranged in blocks. Top: using strips of telegraph tape and plotted on graph paper. Bottom: A larger number of blocks allows more complicated messages. Schemes for interstellar communication 40 years later were strikingly similar, using the concept of prime numbers to determine the grid. From "What Shall We Say to Mars?" by H. W. Nieman and C. W. Nieman, March 20, 1920. Copyright 1920 by *Scientific American,* Inc. All rights reserved.

Engineers on the subject of "Radio Telegraphy" on June 20, he discussed long-range radio communication but had nothing to say on the subject of interplanetary radio communication.[14] No more was heard from Mar-

[14] *NYT,* "No Mars Message Yet, Marconi Radios; Ends Yacht Trip 'Listening In' on Planet Today," June 16, 1922, p. 19, reprinted in Goldsmith, *The Quest,* p. 80. Guglielmo Marconi, "Radio Telegraphy," *Proceedings of the Institute of Radio Engineers,* 10

coni on the subject, as he went on to a busy life with more certain prospects of success than contacting extraterrestrials.

In the midst of these Marconi episodes, another thread of the story was developing: wireless interplanetary communication from a balloon. The prime mover in this daring enterprise, which combined an imaginative idea with a bold technology, was the well-known American astronomer David P. Todd. Todd (1855–1939) had been professor of astronomy and navigation and director of the Observatory at Amherst College from 1881 to 1920, and was by now almost 70 years old, with a well-deserved reputation for innovative (some said wild) ideas. A veteran of 12 solar eclipse expeditions from 1878 to 1914, Todd had invented automatic apparatus for photographing solar eclipses; proposed a search for a trans-Neptunian planet; and was for some years closely associated with Percival Lowell, in 1907 leading the Lowell expedition to Chile to photograph the "canals" of Mars. His interest in aeronautics sparked while working with S. P. Langley in 1880 on the design of experimental aircraft, Todd had made several balloon flights, championed astronomical observations from heights, and pioneered the development of early aeronautical organizations. In 1925, in conjunction with the U.S. Army Air Service, he made the first photographs of the solar corona from an airplane.[15]

While Todd undoubtedly followed with great interest Marconi's statements on interplanetary communication, his own interest in the subject predated that of Marconi. As early as 1909, Todd had suggested that Martians might communicate with the Earth using Hertzian waves. *Scientific American* reported that "It is his plan to take the most sensitive wireless telegraph receivers he can find up in a balloon, in order to diminish any obstructive influence that the atmosphere may exert, and listen for signals in space." The justifiably skeptical editors of the magazine wondered "how Prof. Todd can tell whether his signals come from Mars, or whether the receivers have not simply responded to electrical waves sent out from the sun." Three weeks later, their skepticism had not diminished. Anticipating in rudimentary form the problem of radio frequency interference (RFI), they pointed out that about 2000 wireless stations were scattered over the Earth and that the Earth's atmosphere might also be a source of electrical signals. Quite aside from errant radio signals, they also criticized the proposed life support system and won-

(1922), 215–238. Among the biographies of Marconi, see W. P. Jolly, *Marconi* (New York, 1972). Another radio pioneer, Sir Oliver Lodge, had gone on record as preferring visual signals to the planets rather than radio signals. *NYT*, "Lodge's Signal to Mars," February 4, 1920, p. 13; O. Lodge, "Our Place in the Universe," *The Commonwealth*, 8 (April 1903), 110–111.

15 On Todd's involvement with Lowell see W. G. Hoyt, *Lowell and Mars* (Tucson, Ariz., 1976), 189–191, 196–198.

dered whether instruments could be made light enough to be carried aloft along with the people.[16]

Although Todd apparently failed to carry out his 1909 plan, by 1920 he had not only revived the plan but elaborated it after considerable thought. In articles datelined Omaha, Nebraska (where the Army Signal Corps undertook its balloon experiments), The *New York Times* for April 15 and 18 of that year reported that Todd, "after more than five years of preparation, during which time he has studied the proposition from every conceivable angle," had set the date of April 23 for a balloon ascent to try to communicate with Mars. The article continued:

> A specially constructed balloon piloted by one of the greatest balloon experts in the world, all the facilities of the War Department's chief balloon school, experts from the Rockefeller Institute, apparatus from Johns Hopkins University, the very latest inventions in wireless telegraphy and wireless impulses and specially built instruments of different kinds will assist professor Todd in putting to a test the growing belief that the Martians are trying to communicate with the people of Earth.

Moreover, Todd intended to send signals to Mars at the same time, using the balloon as a relay station. In subsequent days, however, the *Times* reported only on a separate ground-based attempt (also from Omaha) at radio communication with Mars. The fate of the balloon ascent was left hanging in a cryptic statement that while construction of the balloon was progressing, "their experiment will be held in abeyance . . . until sanctioned by the U.S. Government."[17]

Though unsuccessful with his balloon experiments,[18] Todd by 1924 had more successfully pressed yet another bold project related to inter-

[16] *SciAm,* "More About Signalling to Mars" (May 15, 1909), 371. *Sci Am,* "Prof. David Todd's Plan of Receiving Martian Messages," 100 (June 5, 1909), 423. As evidence of the slim margin between acclaim and ridicule in science, 3 years later the Austrian physicist V. F. Hess made a daring series of 10 balloon flights to about 15,000 feet, discovering cosmic rays using ionization chambers. For this feat he received the Nobel Prize in 1936.

[17] *NYT,* "Effort to Signal Mars," April 15, 1920, p. 10; *NYT,* "To Try This Week to Talk to Mars," April 18, 1920, sect. 2, p. 1. *NYT,* "Radio Expert Hopes to Get Mars Signal," April 21, 1920, p. 17; *NYT,* "Listens for Mars Signal," April 22, 1920, p. 2; *NYT,* "No Sounds from Mars Greets Experimenters," April 23, 1920, p. 17. For the context of balloon flight see T. D. Crouch, *The Eagle Aloft: Two Centuries of the Balloon in America* (Washington, D.C., 1983), and D. DeVorkin, *Race to the Stratosphere: Manned Scientific Ballooning in America.* (New York, 1989).

[18] Further research, however, might uncover more of the story. See, for example, Todd's entry in the *National Cyclopedia of American Biography; Literary Digest,* "Radio Messages from Mars?" 82, no. 10 (September 6, 1924), 28–29; and *NYT,* "Professor David Todd, Astronomer, Dies," June 2, 1939, p. .

planetary communication. The *New York Times* reported that Todd had obtained informal assurances from the U.S. Army and U.S. Navy that they would observe a period of radio silence on August 22 and 23, when Mars was at its closest approach. In an effort to obtain worldwide cooperation, Todd also discussed radio silence with the State Department and several embassies. In addition, the Chief Signal Officer of the Army, Major General Charles M. Saltzman, said Army operators would "listen in" for signals from Mars, and the Chief of the Code Section stood by, "ready to translate any peculiar messages that might come by radio from Mars" (Figure 8.2). On August 21, 1924, the Chief of Naval Operations of the U.S. Navy "sent a dispatch to the twenty most powerful stations under his command, from Cavite in the Philippines to Alaska, the Canal Zone and Puerto Rico, telling them to avoid unnecessary transmissions and to listen for unusual signals. A similar order was sent to the Army stations." Department of Commerce officials stood ready to cooperate if asked, showing surprising open-mindedness. Some experts were more skeptical; the chief of the radio laboratory of the National Bureau of Standards declared that the Earth's atmosphere would prevent any signals from reaching the ground.[19]

On the same day that the Chief of Naval Operations issued his order, the Point Grey Wireless Station in Vancouver, British Columbia, reported unusual signals, consisting of four distinct groups of four dashes each. A close watch was to be kept and "eminent British scientists" notified in the event of anything unusual. The following day the signals were heard again in Vancouver, and were now reported to have been heard at the same time of day for more than 4 weeks. Moreover, the Associated Press reported that British wireless experts, using "a twenty four tube set erected on a hill at Dulwich," had heard at a wavelength of 30 km sounds "likened to harsh dots, but they could not be interpreted as any known code. The noises continued on and off for three minutes in groups of four and five dots." The *New York Times* reported that representatives of the Marconi Company and London universities were present. But the Marconi Company, apparently having had a change of heart (or of personnel) from 2 years earlier, reportedly now regarded attempts to signal Mars as "a fantastic absurdity."[20]

It is difficult to assess the effect of these episodes on scientists other

[19] *NYT*, "Asks Air Silence When Mars Is Near: Prof. Todd Obtains Official Aid in Washington Despite Doubts of Its Efficacy," August 21, 1924, p. 11; *NYT*, "Listening for Mars: Heard Anything?" August 22, 1924, p. 12; *NYT*, "Radio Hears Things as Mars Nears Us," August 23, 1924, p. 1. See also Walter Sullivan, *We Are Not Alone* (New York, 1964), 178–179. A revised edition of this book was published in 1966 and 1993, but all page references will be to the first edition unless otherwise specified.

[20] *NYT*, "Radio Hears Things as Mars Nears Us," August 23, 1924, p. 1.

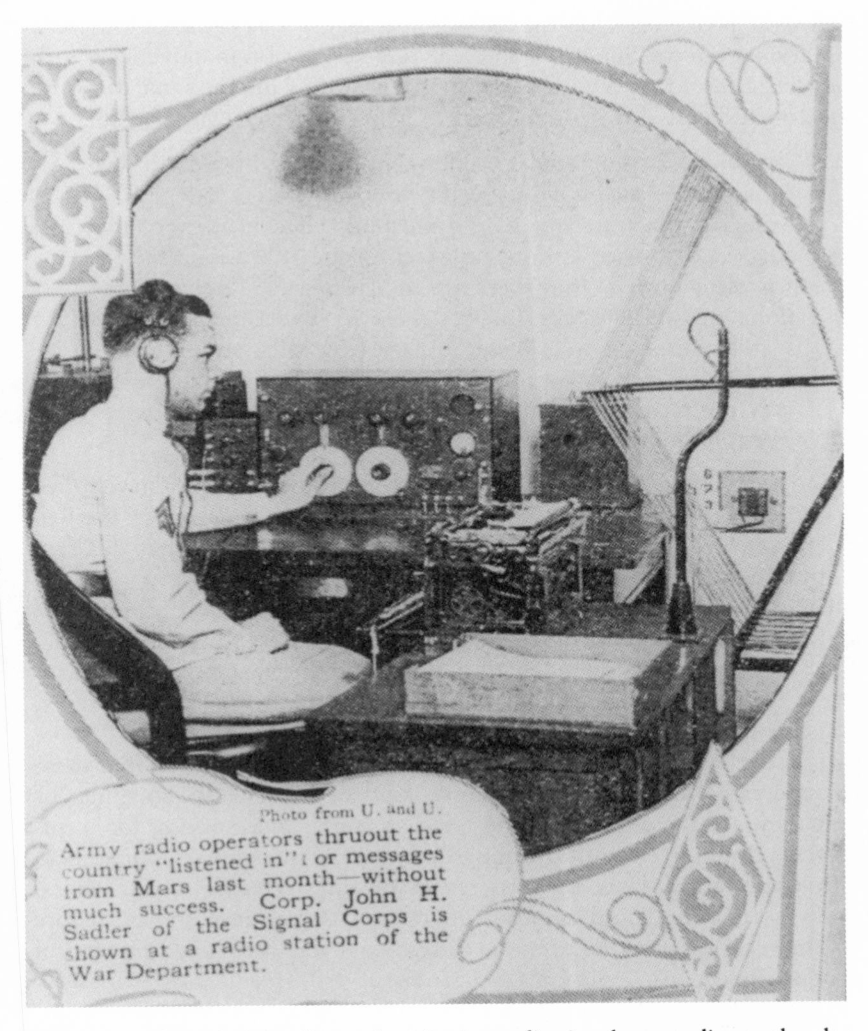

Photo from U. and U.
Army radio operators thruout the country "listened in" for messages from Mars last month—without much success. Corp. John H. Sadler of the Signal Corps is shown at a radio station of the War Department.

Figure 8.2. The U.S. Army listens for Martian radio signals, according to the plan of David P. Todd, as pictured in *Radio Age* for October 1924.

than science fiction writers. One can well imagine that most thought Todd crazy. But it is of interest that in 1929 physicist E. O. Hulburt at the U.S. Naval Research Laboratory, with the problem of interplanetary communication in mind, investigated the types of radio waves Martians might use based on the expected Martian atmosphere. The previous year,

Hulburt had published a paper on the ionization in the Earth's Kennelly–Heaviside layer based on the assumption that it was caused by the UV radiation from the Sun and had come to a conclusion in agreement with the facts inferred from wireless telegraphy. He now did the same for Mars and concluded:

> For wireless communication between Mars and the earth one should perhaps use waves well below 100 meters in length in order to penetrate our own atmosphere. But in view of the suggested poor conditions on Mars for the utilization of these waves, it may be that there are no short-wave receiving stations on Mars, except possibly those for experimental or research purposes. . . . From the present calculations, quite apart from other considerations it is concluded that only a very optimistic experimenter would look for successful wireless communication between the earth and Mars.[21]

Ten years after Marconi had reported on radio telegraphy in the *Proceedings of the Institute of Radio Engineers,* Karl G. Jansky of the Bell Telephone Laboratories reported in the same journal that he had detected a strange radio static that he could not attribute to any known source. This he interpreted the following year as coming from beyond the solar system, a claim that was greeted with skepticism by most astronomers. That interplanetary communication had not been forgotten with the announcement of this discovery is evident in the *New York Times*'s (May 5, 1933) front-page article reporting it, which concluded by saying, "There is no indication of any kind, Mr. Jansky replied to a question, that these galactic radio waves constitute some kind of interstellar signalling, or that they are the result of some form of intelligence striving for intragalactic communication." Although the seed of interstellar communication was planted from the first report of extraterrestrial radio waves, radio astronomy would require a quarter century of development before communication between even the closest stars was considered a practical proposition.[22]

[21] E. O. Hulburt, "Ionization in the Atmosphere of Mars," *Proceedings of the Institute of Radio Engineers,* 17 (September 1929), 1523–1527.

[22] K. G. Jansky, "Directional Studies of Atmospherics at High Frequencies," *Proceedings of the Institute of Radio Engineers,* 20 (1932), 1920–1932, reprinted in W. T. Sullivan III, *Classics in Radio Astronomy* (Dordrecht, 1982), 10–22, and K. G. Jansky, "Electrical Disturbances Apparently of Extraterrestrial Origin," *Proceedings of the Institute of Radio Engineers,* 21 (1933), 1387–1398, reprinted in Sullivan, *Classics,* 23–35. On the background to Jansky see W. T. Sullivan III, "Karl Jansky and the Discovery of Extraterrestrial Radio Waves," in Sullivan, *The Early Years of Radio Astronomy: Reflections Fifty Years after Jansky's Discovery,* (Cambridge, 1984), 4–42. *NYT,* "New Radio Waves Traced to Centre of the Milky Way," May 5, 1933, p. 1.

Although most astronomers ignored Jansky's result, and although Harvard astronomy instructor Fred Whipple and graduate student Jesse Greenstein failed in an attempt in 1937 to explain Jansky's results in terms of radiation from dust at the galactic center, neither they nor anyone else considered Jansky's discovery as interstellar signaling, since no regularity was detected in the signal. Gone were the days of Marconi when any mysterious transmission could be so interpreted, perhaps because the likelihood of intelligence on Mars had declined and the planet in any case was not at one of its favorable oppositions. But the allure of discussing interplanetary communication at least in theoretical terms was not gone, and it is fascinating to find that in the same year that Whipple and Greenstein published their article, their Harvard colleague, Donald Menzel, published an article on interplanetary communication by shortwave. Menzel, whom we last saw in Chapter 6 in connection with his skeptical attitude regarding UFOs, had come to Harvard as an astronomer in 1933, but perhaps more to the point, by that year he had received his amateur radio license – the result of an interest beginning at the age of 9, when he built a wireless transmitter and receiver.[23] Menzel's interest in interplanetary communication thus appears not to have come from Jansky, but from his amateur radio background, where it was a subject of considerable interest. "The question of radio-communication with distant planets still holds supreme charm for all red-blooded radio experimenters," the editors of *Short Wave and Television* wrote in their December 1937 issue, the issue in which appeared Menzel's article "Can We Signal Mars by Shortwave?" In one of those curious closures in history, Menzel's immediate spur may well have been an article regarding Tesla's invention of a method for signaling interplanetary inhabitants.[24]

In his article, Menzel made no case for intelligence on Mars and reported that "the general consensus of opinion is that no very high degree of intelligent life exists in our solar system." The thrust of his article was therefore not the transmission or receipt of actual signals, but the question of whether one might be able in theory to communicate via radio signals sent from beyond the Earth. If we received a radio message from Mars, Menzel argued, radio technology implies knowledge of mathematics and physical science, and mathematics forms a natural starting point. Using dots and dashes, one could begin by transmitting problems in addition and

[23] On Menzel see L. Goldberg, *S&T,* (April 1977), 244–251.

[24] D. Menzel, "Can We Signal Mars?" *Short Wave & Television* (December 1937), 406–407, 450–451. The founder and editor of the magazine was none other than Hugo Gernsback, the "father of science fiction." For the Tesla article see *The Sky,* "Signalling the Inhabitants of Other Planets," (October 1937), 27; see also John J. O'Neill, *Prodigal Genius: The Life of Nikola Tesla* (New York, 1944), 242–243.

answers, to which the Martian would reply with their own. One could then advance to abstract numbers like pi and the relative distances of the planets from the Sun. The alphabet could be transmitted by transmitting a series of paired numbers, coordinates on a graph. Since one could proceed in this way to more and more complex information, Menzel saw "no obvious limit to the information that could be exchanged."

The same issue of *Short Wave and Television* included the cautious opinion of radio pioneer Lee de Forest, the more imaginative interpretations of Tesla, and a remarkable analysis by American Telephone and Telegraph (AT&T) staffer Joseph L. Richey of the optimal wavelengths and the power required to send a signal to Mars. Intended as a complementary article to Menzel's, it concentrated not on linguistic considerations, but on the "quantitative and engineering" aspect. Although Richey himself, as a public information employee, was unable to undertake such a study himself, he drew on the technical expertise of future radio astronomy pioneer G. C. Southworth. Southworth (1890–1972), who would work on radar during World War II and in 1942 detect the first thermal radio waves from the Sun, was then at Bell Telephone Labs. Based on Southworth's input, Richey set forth four major factors to take into account in interplanetary communication: (1) electromagnetic waves of some sort would be required; (2) large amounts of energy would be needed because of interplanetary distances; (3) the approach and recession of Mars from the Earth need to be taken into account; and (4) absorption of the signal by the atmospheres of both Mars and the Earth needs to be addressed. Because the ionosphere is difficult to penetrate at wavelengths longer than 10 meters, and because water vapor and carbon dioxide absorb parts of the infrared, whereas UV and shorter wavelengths are either absorbed by the atmosphere or ionize it, Richey concluded that the "ultra-short or hyper-short" radio wavelengths below 10 meters and above the infrared would be best, as well as the visible spectrum, 50 percent of which he estimated penetrated the Earth's atmosphere. Considering the radio wavelengths for communication, Richey and Southworth calculated that a simple "half-wave radiator " antenna would require 780,000 kilowatts of power for communication when Mars is at its closest to Earth. Using an array of transmitting antennas under certain conditions, 1300 kilowatts of power would be required to communicate at that distance and 60,000 kilowatts at its furthest distance. And this was only radio telegraphy; radio telephony would require about four times greater power. "Communication with Mars is too difficult for us to consider with radio instrumentalities, and economically and technically not feasible with present-day equipment," Ritchey concluded. Such communication would be impossible "until something new develops, that will make radiation and power requirements a much easier proposition to realize."[25]

Although by the 1930s these discussions of interplanetary communication were only theoretical exercises, during this decade changing views of the universe were laying the conceptual, if not the technical, groundwork for interstellar communication. One such view was expressed by amateur radio pioneer Hiram Percy Maxim, founder of the American Radio Relay League. In 1932 he wrote an article in *Scientific American* in which his interest in astronomy combined with his amateur radio background to give interstellar communication a prominent role. Unconcerned about the technical power requirements, although he undoubtedly knew them, Maxim simply assumed that attempts at interstellar communication were inevitable in the enlarged universe discovered by Hubble and others. The presence of intelligence in the universe, he believed, was hidden only by our own backward state. Such intelligence "could, indeed, have been sending out signals for centuries, yet we would have been too backward to detect them. The only means of interstellar communication of which we have any knowledge is radio, and as that has only just appeared on earth it is in a very primitive state." Yet Maxim was sure it would happen:

> Radio waves represent our first tool with which it may prove possible to carry a signal across the great reaches of astronomical space. If among the billions of stars there are some which have a family of satellites such as our own sun has, then it is reasonable to expect that some of these satellites would be in a zone of temperatures where life could exist. If life does exist somewhere else, and it is reasonable to expect that it does, then some day someone is likely to encounter, by means of radio, an extra-terrestrial intelligence. What a sublimely dramatic moment it will be for those concerned when this first interstellar contact is made![26]

Across the Atlantic, perhaps recalling the days of interplanetary communication attempts inspired by Marconi's speculations, even a theologian expressed a similar view. In a symposium on the evolution of the universe published in *Nature* in 1931, the Bishop of Birmingham, E. W. Barnes, held that James Jeans's theory of the rarity of planetary systems was too pessimistic, that evolution would have produced advanced be-

25 J. L. Richey, "Communicating with Mars – A Few Technical Considerations," *Short Wave and Television* (December 1937), 452–454.

26 H. P. Maxim, "What Is It All About?" *SciAm* (April 1932), 199–203. Maxim, *Life's Place in the Cosmos* (New York and London, 1933), 148–160.

ings on the planets of many stars, and that wireless communication with them might be possible. Failing the invention of a super-telescope, he wrote:

> there remains the possibility of wireless communication. . . . I have no doubt that there are many other inhabited worlds, and that on some of them beings exist who are immeasurably beyond our mental level. We should be rash to deny that they can use radiation so penetrating as to convey messages to the earth. Probably such messages now come. When they are first made intelligible a new era in the history of humanity will begin.[27]

There is no doubt that Maxim and Barnes were ahead of their time. Although there were undoubtedly more discussions of interplanetary communication in the amateur radio literature, and perhaps a few more scattered references to interstellar communication, its time had not yet come. The era of interplanetary communication drew to a close, and as World War II closed in on the world, hard-nosed radio engineers like Grote Reber realized that there was much more to radio astronomy than interplanetary signals, while scientists like Bernard Lovell and J. S. Hey, working around the world with radar, had deeper concerns with the terrestrial rather than the extraterrestrial. But less than 25 years after the pessimistic assessment of interplanetary communication presented in *Shortwave and Television*, the first attempt at *interstellar* communication would be undertaken. While the interplanetary communication era passed quietly into history, the new discipline of radio astronomy led unwittingly but inevitably to the modern era of interstellar communication.

8.2 CORNELL, OZMA, AND GREEN BANK: THE OPENING OF THE ELECTROMAGNETIC SPECTRUM FOR SETI

The idea of interstellar communication – or the "search for extraterrestrial intelligence (SETI)," as it became widely known after 1976 – had its origins neither in the era of interplanetary communication nor in the space program that had given birth to the Viking search for life on Mars. Rather, interstellar communication found its origin in the "new astronomy," which recognized that celestial objects radiated energy in many regions of the electromagnetic spectrum other than the optical portion

[27] E. W. Barnes, *Supplement to Nature*, 128 (1931), 719–722.

familiar to our everyday experience. Space science and the new astronomy would eventually become intimately linked, but it was not this linkage that gave birth to the modern SETI.

It was two academic physicists, Giuseppe Cocconi and Philip Morrison at Cornell University, who stumbled on the subject of interstellar communication almost as an aside to their primary research. Cocconi, a graduate of the University of Milan in 1937, had spent the following year in Rome working with Enrico Fermi building a cloud chamber, one of the early detectors for high-energy physics. Fermi, who received the Nobel Prize in that year and escaped to the United States before the outbreak of war, inspired Cocconi to continue working in cosmic ray research and particle physics. This he did at Milan and Catania until 1947, when Hans Bethe invited him to join the physics faculty at Cornell.[28] Morrison, on the other hand, was the intellectual son of another theoretical physicist – J. R. Oppenheimer – under whom he received his Ph.D. in theoretical physics from the University of California, Berkeley, in 1940. During the war, Morrison worked at Los Alamos on the testing and design of the first atomic bomb, an effort headed by Oppenheimer and substantially aided by Fermi. He then joined the physics department at Cornell some 2 years before the arrival of Cocconi.[29] Cornell at this time was a hotbed for physics and astronomy, "always a place where imaginative people felt at home," one physicist present during the period recalled.[30] For more than 10 years, Cocconi and Morrison each followed his own research interests until events drew them together in 1959.

In 1958, during a chamber music performance at Cornell, Morrison had first pondered the promise of gamma ray astronomy, publishing his results late that year.[31] His colleague Cocconi, who had been studying gamma rays emitted by the Cornell synchrotron, one day in spring of 1959 broached the subject of gamma rays as signals in a conversation with his wife, also a physicist at Cornell. "We were thinking that a narrow burst of gamma radiation could be a signal that can travel far and straight in galactic space and be peculiar enough to be recognized," he recalled many years later.[32] The next day, Cocconi went to see Morrison. Morrison recalled:

[28] Interview with Cocconi, February 1983, in David W. Swift, *SETI Pioneers: Scientists Talk About Their Search for Extraterrestrial Intelligence* (Tucson, Ariz., 1990), 49–53.

[29] Interview with Morrison, June 1981, in Swift, *SETI Pioneers,* 19–48.

[30] Interview with Freeman Dyson, January 1984, in Swift, *SETI Pioneers,* 312–326: 325. In the conclusion to his volume, Swift comments on the importance of Cornell for many SETI pioneers, 393–394.

[31] Philip Morrison, "On Gamma-Ray Astronomy," *Nuovo Cimento,* 10, 7 (March 16, 1958), 858–865.

[32] Interview with Cocconi in Swift, *SETI Pioneers,* 51.

> We were thinking about gamma rays of natural origin when we realized that we knew how to make them, too. We were making lots of them downstairs at the Cornell synchrotron. So Cocconi asked whether they could be used for communication between the stars. It was plain that they would work, but they weren't very easy to use. My reply was enthusiastic yet cautious. Shouldn't we look through the whole electromagnetic spectrum to find the best wavelength for any such communication? That was the germ of the idea.[33]

After some thought, Morrison suggested that the longer wavelengths in the radio region would be better because there would be less interference. Within a few days, Cocconi had calculated that the 250-foot Jodrell Bank Telescope in England – the largest in the world – was just about at the point of being able to signal, or receive Earth-type signals from, the nearest star.

To have an idea is one thing, but to bring it before one's peers in print and suggest that it be acted on is quite another, especially when it is a speculative idea outside one's own field of expertise. Nevertheless, Cocconi and Morrison plunged ahead:

> After we talked about it a bit, we realized that it was rather a good idea. We felt that it should be called to the attention of the community of scientists who could criticize it, and also to the radio astronomers who would have to carry it out in some way. We were not that confident that we knew all about it. We were very amateurish in radio astronomy. So we thought it should be published and criticized and elaborated on by experienced radio astronomers.

They began writing in Ithaca, New York, and when Cocconi went on sabbatical at the Centre Européen Recherche Nucléaire (CERN) near Geneva, Morrison joined him in the summer of 1959 to finish their landmark paper, "Searching for Interstellar Communications." In late June, Cocconi even wrote Sir Bernard Lovell at Jodrell Bank, suggesting that he might wish to undertake a search with his radio telescope, but received a discouraging reply. Still they decided to publish in *Nature,* which had a reputation for printing speculative papers; even so, Morrison took the precaution of sending the paper via P. M. S. Blackett, a

[33] Interview with Morrison, June 1981, in Swift, *SETI Pioneers,* 22–23. Morrison has also recounted the origins of the SETI idea in *Cosmic Search* (January 1979), 7–9, and in "Twenty-Five Years of the Search for Extraterrestrial Communications," in M. D. Papagiannis, ed., *The Search for Extraterrestrial Life: Recent Developments* (Dordrecht, 1985), 13–19.

professor of physics at Imperial College with a reputation for imagination, who enthusiastically forwarded their article to *Nature,* where it was published in September 1959.[34]

In their paper, Cocconi and Morrison admitted that present theories did not yield reliable estimates of the probabilities of planet formation, origin of life, or evolution of advanced societies. But they noted that around one star, the Sun, an advanced society had developed on one planet, and that on another (Mars) some type of life might have developed. Despite lack of knowledge of the lifetime of civilizations, the authors found it "unwarranted to deny" that such societies might have very long lifetimes, perhaps comparable to geological time. If so, our Sun would appear as a likely life site, and they might be beaming a signal and patiently awaiting an answer. If one accepted this, the problem was to determine on what channel an extraterrestrial civilization might be signaling. Only electromagnetic waves would not be dispersed in the galactic plasma, they argued. But the spectrum of such waves was extremely large. Because of absorption in planetary atmospheres, Cocconi and Morrison quickly narrowed the possibilities to the visible, radio, and gamma ray regions. But because the visible and gamma ray regions would have required too much power or techniques too complicated to be practical, they concluded that "the wide radio band from, say, 1 Mc to 10^4 Mc/sec [Mc = million cycles], remains as the rational choice."[35] This was still a broad band to search, but then the authors came to an idea that would become pivotal in the twentieth-century SETI debate:

> Just in the most favored radio region there lies a unique, objective standard of frequency, which must be known to every observer in the universe: the outstanding radio emission line at 1420 Mc/s (= 21 cm) of neutral hydrogen. It is reasonable to expect that sensitive receivers for this frequency will be made at an early stage of the development of radio astronomy. That would be the expectation of the operators of the as-

34 G. Cocconi and P. Morrison, "Searching for Interstellar Communications," *Nature,* 184 (September 1959), 844, reprinted in Goldsmith, *The Quest,* 102–104. Cocconi to Sir Bernard Lovell, June 29, 1959, reprinted in Sir Bernard Lovell, *The Exploration of Outer Space* (New York, 1962), Appendix, 82–84. See also Lovell, "Search for Voices from Other Worlds," *NYT,* sect. 6 (December 24, 1961), pp. 10, 29–31. By 1962 Lovell had a more favorable attitude toward extraterrestrial life, as stated in chapter 5 of *Exploration* and in testimony to the U.S. Congress, but the Jodrell Bank telescopes were never used for the search Cocconi requested.

35 Cocconi and Morrison, "Interstellar Communications," 102. One Mc refers to 1 Megacycle, or one million cycles. In later terminology, 1 Mc/sec is equivalent to 1 Megahertz (MHz), or 1 million cycles per second. By comparison, 1 Gigahertz (Ghz) is 1 billion cycles per second, so that 10,000 Mhz = 10 Ghz.

> sumed source, and the present state of terrestrial instruments indeed justifies the expectation. Therefore we think it most promising to search in the neighborhood of 1420 Mc/s.

One problem was that any artificial signal would have to compete with the natural 21-cm radiation. Cocconi and Morrison calculated that to generate and receive a signal that would rise above this background would not be possible with the 250-foot Jodrell Bank Telescope, but it would with the 600-foot telescope then planned by the Naval Research Lab at Sugar Grove, West Virginia. They also noted a second problem: the orbital motion of unseen planets would Doppler shift a signal up or down by about 300 kc/sec (100 km/sec). Aside from drawing attention to the general problem of interstellar communication, Cocconi and Morrison therefore made two important suggestions: the idea of what would later be called communication "beacons" aimed directly at the Earth, and the idea of a "magic frequency" on which the communication beacon might occur. The first idea could certainly be questioned and the latter subject to second-guessing, but without doubt they stimulated a dialogue among scientists that would last throughout the century.

Cocconi and Morrison suggested that the closest Sun-like stars be examined first – some seven within 15 light years of the Sun. And they closed with a challenge:

> The reader may seek to consign these speculations wholly to the domain of science fiction. We submit, rather, that the foregoing line of argument demonstrates that the presence of interstellar signals is entirely consistent with all we now know, and that if signals are present the means of detecting them is now at hand. Few will deny the profound importance, practical and philosophical, which the detection of interstellar communications would have. We therefore feel that a discriminating search for signals deserves a considerable effort. The probability of success is difficult to estimate; but if we never search, the chance of success is zero.

With that tantalizing parting shot, the two physicists went their separate ways – Cocconi back to Cornell and then permanently to CERN in 1962 (where he would play very little role in the future of SETI), and Morrison back to Cornell and in 1967 to MIT, where he would become an active participant in future SETI discussions.

That the time was ripe for the subject in a broader sense is evident in the fact that Frank Drake, a young astronomer at the National Radio Astronomy Observatory (NRAO) in Green Bank, West Virginia, had independently begun efforts to actually observe signals such as those Cocconi and Morrison described. Here too events conspired to support Drake in what would surely have been seen as a wild scheme at any long-established conservative institution. This the NRAO decidedly was not. In the struggle over whether large government laboratories were better than smaller institutions where individuals might have more control over their research, the idea of large scale won out, and in August 1956 the National Science Foundation contracted with Associated Universities Incorporated (which already ran the Brookhaven National Laboratory) to establish NRAO. Groundbreaking took place in October 1957, and in 1958 a new 140-foot telescope was begun. Into these surroundings Drake came in 1958. He recalled that when he arrived

> the institution had just been formed. There were three scientists there: myself, David Heeschen and John Findlay. No instruments existed. The entire observatory was housed in several farm houses on a four-square-mile area that had been procured for the observatory. It was a very happy and good time, because we had great expectations, we had the go-ahead to build the world's greatest telescopes, and there also seemed to be an infinite supply of money available.[36]

A graduate of Cornell and Harvard, Drake had been interested in extraterrestrial life from an early age, and had been influenced during his Cornell years by a lecture on planetary systems given by Otto Struve. Drake was also impressed by an episode from his graduate student days at Harvard in which he detected a strong narrowband signal while observing the Pleiades. Although the signal turned out to originate on Earth, after that Drake always had in the back of his mind the possibility of an artificial signal.[37]

After Drake (Figure 8.3) came to NRAO, it was therefore natural that he should follow up on this subject. The 140-foot telescope, an immense project going through more than the usual problems of construction, would not be finished for years, so NRAO decided to purchase an 85-foot telescope similar to one already on the drawing boards for the University of Michigan. It was this telescope with which Drake became associated

36 Steven J. Dick, interview with Frank Drake, May 29–30, 1992. Allan A. Needel, "Berkner, Tuve and the Federal Role in Radio Astronomy," *Osiris*, 3 (1987), 261–88, and Needell, "The Carnegie Institution of Washington and Radio Astronomy: Prelude to an American National Observatory," *JHA*, 22 (1991), 55–67.

37 Interview with Drake, June, 1981, in Swift, *SETI Pioneers*, 59; see also F. Drake and D. Sobel, *Is Anyone Out There? The Scientific Search for Extraterrestrial Intelligence* (New York, 1992), 21–42. For an early interview with Drake see Shirley Thomas, *Men of Space: Profiles of Scientists Who Search for Life in Space* (Philadelphia, 1963), 62–89.

Figure 8.3. The 85-foot Howard E. Tatel radio telescope used by Frank Drake at the National Radio Astronomy Observatory in Green Bank, West Virginia, for Project Ozma in April 1960. Part of the Ozma team reassembled for the 25th anniversary of the project, including Drake, standing second from the right.

and which achieved observing capability in early 1959. Drake and his colleague David Heeschen (who had arrived earlier from Harvard) had a list of programs for the telescope, including observations of Jupiter, the galactic center, and many other phenomena that had not yet received much attention in those pioneering days of radio astronomy.[38] In March 1959 – at about the time Cocconi was going to Morrison's office at Cornell – Drake calculated that an 85-foot radio telescope like the one soon to be completed at NRAO allowed for detection of Earth-like radio signals to about 10 light years. It must have been with some trepidation that Drake broached the subject to his colleagues, but this he did one day at the local lunch spot 5 miles from the Green Bank operation. "I suggested that we put together some simple equipment – it would require a narrowband radio receiver, something we didn't usually use in radio astronomy – and search some nearby stars for signals." At that lunch was Lloyd Berkner, president of Associated Universities, Inc., since 1950, and the acting director until a permanent first director could be appointed for NRAO. Berkner supported Drake's project from the beginning: "He was all for it. His whole history had been to push exotic experiments," Drake recalled. "He was sort of an entrepreneur and gambler of science. He liked to do that kind of thing." Indeed, Berkner would be an early strong supporter of the search for life in the solar system.[39]

By April 1959 the 85-foot Tatel telescope was completed, but given his other tasks, it was summer before Drake and his colleagues began to develop equipment for their interstellar communication project. At this point another piece of luck occurred: on July 1, Otto Struve became the first director of NRAO. As we have seen in Chapter 4, Struve was one of the few astronomers of the time who believed intelligent life might be abundant in the universe, due largely to his work on stellar rotation in relation to planetary systems. Struve had first learned at an Associated Universities trustees' meeting that radio transmitters on Earth could send signals that current radio telescopes could detect at a distance of 10 light years. When Drake reversed the problem, and determined that the 85-foot Tatel telescope could detect from 10 light years an extraterrestrial

[38] For the early research goals, equipment, and organization at Green Bank see Associated Universities, Inc., *National Radio Astronomy Observatory Annual Report, July 1, 1959*, and *The National Radio Astronomy Observatory* (New York, October 1959). For further context at Green Bank see Steven J. Dick, interview with Frank Drake, May 29–30, 1992.

[39] Drake interview, Swift, *SETI Pioneers*, 61, 68. In addition to the Swift interview, Drake recounts Project Ozma in "A Reminiscence of Project Ozma," *Cosmic Search* (January 1979), 11–15; "Project Ozma," in K. I. Kellerman and G. A. Seielstad, eds., *The Search for Extraterrestrial Intelligence*, Proceedings of an NRAO Workshop held at the National Radio Astronomy Observatory, Green Bank, West Virginia, May 20, 21, 22, 1985, 17–26; and, most recently, in F. Drake and D. Sobel, *Is Anyone Out There?*, 21–42. One of the best near-contemporary accounts is that of Walter Sullivan, *We Are Not Alone*, chapter 14, "Project Ozma." On Berkner's support for the search for life in the solar system, see Berkner, "Space Flight and Science," *Astronautics* (October 1961), 46–47, 140–150.

signal similar in strength to that produced by the Millstone Hill radar system on Earth, Struve was receptive.[40] Both Berkner and Struve

> were positive and both very supportive. In fact, if we had proposed a more grandiose project, they probably would have supported it. I think it was me that was the conservative in all this. They were older. Both Struve and Berkner were near retirement age and they were willing to risk. They were venturesome; they weren't worried about their career or anybody saying bad things about them. That was all right at their stage.[41]

Two ground rules were laid down by mutual agreement. First, recognizing the sensational nature of the subject, they decided to keep the project as secret as possible because if "this young observatory was trying to establish the concept of national observatories as a good thing, then we had to be very careful not to do things which bring criticism." Second, the equipment built would be useful for more conventional astronomy, in particular for a study of a 21-cm line Zeeman effect, which would indicate the presence of magnetic fields in space. For this

> we would need two channels, good frequency stability, narrow bandwidths, all very similar to the SETI requirement; and, in order that the system would be suitable for [the] 21 centimeter Zeeman effect, we would build it and do the search at the 21 cm line. That is why the frequency was picked, not for any profound reason like magic frequencies or waterhole or anything else. It was a way to prevent criticism of the observatory, and in a way kill two birds with one stone.[42]

The fact that the 21-cm line occurred at an optimum place in the electromagnetic spectrum for detection was, of course, important. Cocconi and

[40] Struve, "Astronomers in Turmoil," *Physics Today* (September 1960), 18–23: 22–23. Bracewell (interview in Swift, *SETI Pioneers*, 143–144) has pointed out that Struve, as a Russian emigré, may also have been influenced by Tsiolkovsky's ideas on the abundance of extraterrestrial intelligence. On Tsiolkovsky's beliefs on the subject, see Alexis Tsvetikov, " 'Next Question' and K. E. Tsiolkovsky," *Science*, 131 (March 18, 1960), 872–874.

[41] Interview with Drake in Swift, *SETI Pioneers*, 68.

[42] Kellermann and Seielstad, *The Search*, 19. On the idea of the cosmic waterhole, see footnote 86. The independent choice of 21 cm by Cocconi, Morrison, and Drake at this time was no surprise. The existence of an emission line in the radio spectrum at 21 cm had been predicted by van de Hulst in 1945 and was discovered by Ewen and Purcell at Harvard in 1951. The 21-cm line was the only line in the radio spectrum known until 1963, when hydroxyl lines were discovered, followed by hundreds of additional lines in later years. For the original papers on the prediction and discovery of the 21-cm line see Sullivan, *Classics*, 297ff.

Morrison had already pointed this out in their 1959 article, and early the following year, Drake emphasized the same point with a diagram that later became known as the "microwave window" or "cosmic window." It was a rudimentary form of a diagram that would be seen again and again in this context (Figure 8.4).[43]

The design of the system – which Drake dubbed "Ozma" because it was searching for distant beings at least as exotic as those in L. Frank Baum's popular *Wizard of Oz* – was relatively simple by present standards. Most important, of course, was the size of the dish – 85 feet – increasing the collecting area more than 10 times over the 25- and 50-foot dishes previously available. In those early days, progress was measured by leaps and bounds; Drake compared the improvement to the difference between Galileo and the 200-inch telescope at Mt. Palomar – and, of course, the 140-foot telescope, when completed, would be much better yet. But size alone would not have allowed the program Drake had in mind. His calculation of 10 light years for the detection range depended on the new, sensitive radio receivers just coming on the scene in 1958–1959. In particular the maser, invented by Charles H. Townes and developed as a solid-state device by Bloembergen at Harvard, and the parametric amplifier, were crucial to any scheme to detect signals at that distance.[44] The maser gave a factor of 100 improvement over previous receivers, and combined with the dish size, the 85-foot telescope offered a factor of more than 1000 improvement. The problem with such sensitive receivers was noise – of the receivers themselves and, in this case, also of the continuum radiation of astronomical objects. For this reason, there was a signal band and a comparison band; by taking the difference between the two, the system was designed to reject all signals except for a possible narrowband signal believed to be characteristic of an interstellar signal of intelligent origin. In technical terms, it was a "DC comparison radiometer"; as Drake stated, the Ozma equipment was a reflection of the Harvard background of the NRAO people.[45] The project was not only low in profile but also low in budget; only $2000 was to be spent on

43 Drake (1960), footnote 49. The development of knowledge of this microwave window is itself of historical interest. See, for example, Drake in Kellermann and Seielstad, *The Search*, 21. John Kraus, *Radio Astronomy* (Powell, Ohio, 2nd ed., 1986), chap. 8, 8, gives a similar diagram with sources for the data in various parts of the spectrum.

44 Drake's fascination with the possibilities of these receivers is evident in his articles at this time: "Radio Astronomy Receivers – I [and II]," *S&T* (November 1959), 26–28, and (December 1959), 87–89.

45 For details on the system see Drake, "Project Ozma," in Kellerman and Seielstad, *The Search*. See also Thomas, *Men of Space*, 64–65, on Drake's assessment of instrumental improvement.

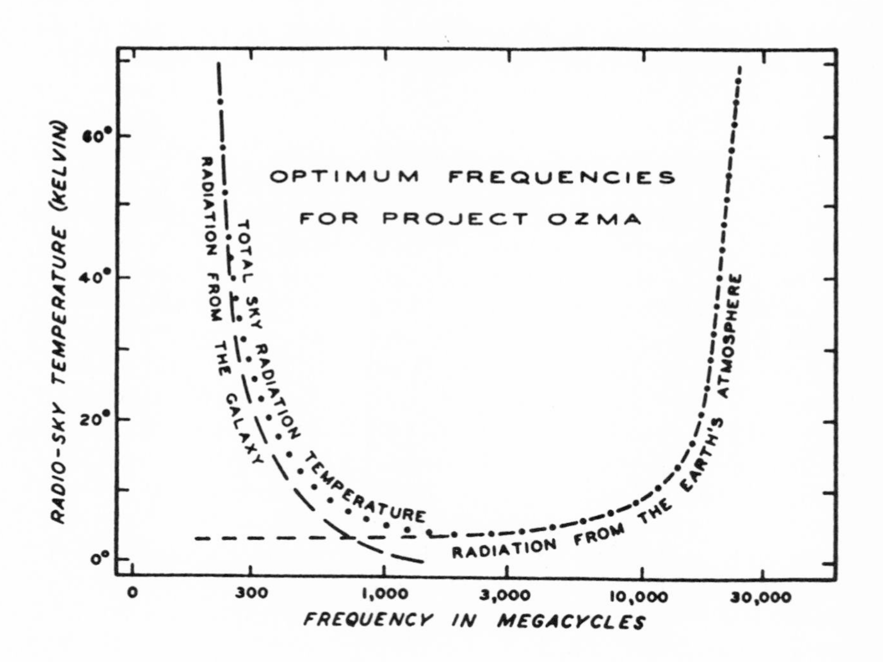
OPTIMUM FREQUENCIES
FOR PROJECT OZMA
RADIO-SKY TEMPERATURE (KELVIN)
60°
40°
20°
0°
RADIATION FROM THE GALAXY
TOTAL SKY RADIATION TEMPERATURE
RADIATION FROM THE EARTH'S ATMOSPHERE
0
300
1,000
3,000
10,000
30,000
FREQUENCY IN MEGACYCLES

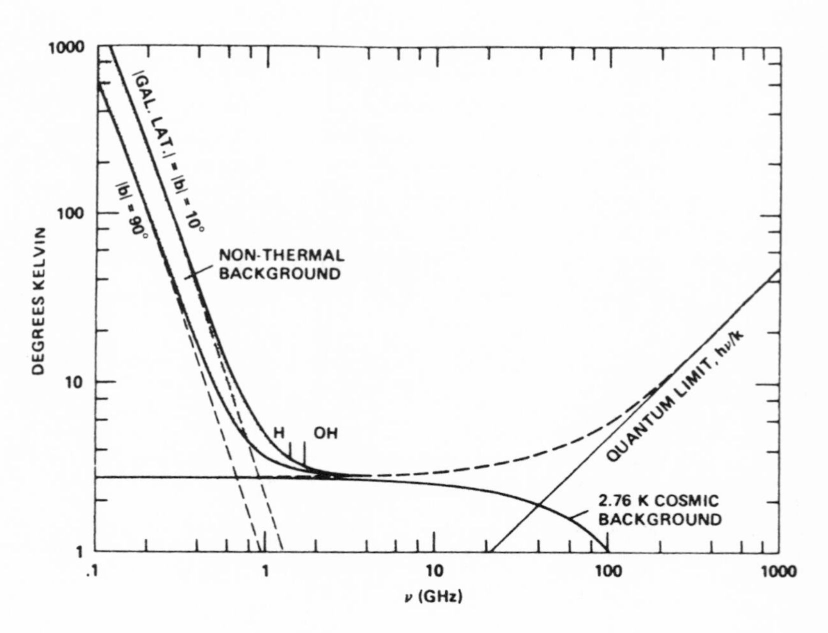
1000
100
10
1
DEGREES KELVIN
|GAL. LAT.| = |b| = 10°
|b| = 90°
NON-THERMAL
BACKGROUND
H
OH
QUANTUM LIMIT, hν/k
2.76 K COSMIC
BACKGROUND
.1
1
10
100
1000
ν (GHz)

parts unique to Ozma, and this was mostly for four narrowband vacuum tube "Kronhite" filters. Most of the other parts of the system were part of conventional radio astronomy, including the chart recorder that recorded the signals.

Drake and the NRAO staff had been working at a relatively relaxed pace on the equipment for the project when Cocconi and Morrison's paper was published in September. Drake was encouraged because the two physicists had come to the same conclusion he had regarding the strategy of the 21-cm line, on theoretical rather than practical grounds. But Struve was distinctly worried that Green Bank would lose the credit for what he thought was an important idea.[46] Convinced of the need to publicize Ozma, Struve first did so during his Karl T. Compton lectures at MIT in November 1959 – less than 2 months after Cocconi and Morrison's paper.[47] This immediately brought press attention; a measure of the popular interest was the cover of the first issue for 1960 of *Saturday Review,* which featured "The Search for Intelligent Life on Other Planets," a story written by John Lear. As it turned out, Lear would be the only reporter to actually observe parts of Project Ozma firsthand.[48]

But there was also scientific fallout now that the cat was out of the bag. For one thing, Drake himself could now talk about the project in the open. At the end of his second paper on radio receivers in December 1959, Drake stated for the first time in print that one of the most exciting possibilities of the new receivers was a search for intelligent interstellar signals. By January 1960, 4 months before Ozma observations began, Drake had published a popular article detailing the meth-

[46] Thomas, *Men of Space,* 62.

[47] As reported in "Anybody Out There?" *Time* (November 23, 1959), 84–85; also William L. Lawrence, "Radio Astronomers Listen for Signs of Life in Distant Solar Systems," *NYT,* sect. 4 (November 22, 1959), p. 11. Curiously, the printed version of the lectures discusses extraterrestrial life but does not mention Ozma. The lectures were printed in Struve, *The Universe* (Cambridge, Mass, 1962). But see Struve's comments 2 weeks later in Struve, "Detecting Radio Signals from Outside Solar System," correspondence in *NYT,* sect. 4 (December 6, 1959), p. 10.

[48] John Lear, "The Search for Intelligent Life on Other Planets," *Saturday Review,* 43 (January 2, 1960), 39–43. Among many others, see Ray Bradbury, "A Serious Search for Weird Worlds," *Life,* 49, no. 17 (October 24, 1960), 116–130, which brought a response from Struve et al. in *Life,* 49, no. 20 (November 14, 1960), 22, and G. DuShane's editorial, "Next Question," *Science,* 130 (December 25, 1959), 30.

Figure 8.4 (*opposite*). Drake's diagram (top) showing optimum frequencies for interstellar radio communication, bounded on one side by radiation from the Galaxy and on the other side by radiation from the Earth's atmosphere. From *Sky and Telescope* (January 1960), p. 141, by permission of Sky Publishing Corporation. A later version (bottom) often seen in SETI literature shows more detail but the same "microwave window."

ods and strategies of Ozma.[49] The new openness also brought more tangible benefits; on reading about the project, Dana Atchley, president of Microwave Associates, offered the use of his company's first copy of a parametric amplifier, which it claimed was the best in existence. In addition, the publicity brought two supportive visitors: the theologian Theodore Hesburgh and the engineer Bernard Oliver, the latter to become another SETI pioneer.[50] Another future SETI pioneer, Sebastian von Hoerner, also was a visiting scientist at NRAO during the period of Project Ozma.

On April 8 Ozma observations began, a routine in which four people participated as observers and several others as telescope operators and technicians. The targets were Tau Ceti and Epsilon Eridani, the two nearest solar-type stars known. On the first day, Tau Ceti was observed until it set at about noon; then the telescope was turned to Epsilon Eridani. "A few minutes went by," Drake recalled. "Then it happened. Wham! Suddenly the chart recorder started banging off scale." The search was not to be that easy; the signal turned out to be due to terrestrial rather than celestial intelligence, and a total of about 200 more hours of observing produced nothing.[51] Although 7200 channels were covered, or 360 kilohertz per star centered on 21 cm in case of Doppler shifts due to the supposed planet's orbital velocity, the remainder of the observations were routine. The project failed to find any intelligent signals of extraterrestrial origin.

Despite its failure, the project fired the imagination of the public, and was widely reported in the press and even in scientific journals.[52] As for the scientific reaction, Struve himself stated in late 1960 that Project Ozma

> has aroused more vitriolic criticisms and more laudatory comments than any other recent astronomical venture, and it has divided the astronomers into two camps: those who are all for it and those who regard it as the worst evil of our generation. There are those who pity us for the publicity we have received

[49] Frank Drake, "How Can We Detect Radio Transmissions from Distant Planetary Systems?," *S&T*, 19 (January 1960), 140–143, reprinted in A. G. W. Cameron, ed., *Interstellar Communication: A Collection of Reprints and Original Contributions* (New York, 1963), 165–175, and Goldsmith, *The Quest*, 114–117.

[50] Oliver dropped in on Drake after reading about Ozma in *Time* magazine; Swift interview with Oliver, *SETI Pioneers*, 94.

[51] Another view of the program is given by telescope operator Fred Crews, "Project Ozma – How It Really Was," in Kellerman and Seielstad, *The Search*, 27–30. Crews states that the first observing run was April 11–19.

[52] *Science*, 131, 1303 (April 29, 1960), For the press reaction in addition to those items cited in footnote 48, see entries in E. F. Mallove et al., *A Bibliography on the Search for Extraterrestrial Intelligence*, NASA RP 1021 (Washington, D.C.), 1978, 7.

> and those who accuse us of having invented the project for the sake of publicity.[53]

Project Ozma had not achieved its hoped-for goal of discovering exotic beings in distant places, but it had stimulated imaginative minds, and an imaginative mind was a difficult thing to stop.

That Struve, nearing the end of his distinguished career, had the courage of his convictions despite the controversy over Project Ozma is evident in his support for an "informal conference" on interstellar communication, held in late 1961 at NRAO.[54] Once again Lloyd Berkner played an important role, for the meeting was sponsored by the Space Science Board of the National Academy of Sciences, a board that Berkner now headed. Moreover, the board included among its members Harold Urey, whose work on origin of life and carbonaceous meteorites had drawn him into the extraterrestrial life debate, and the board's Panel on Exobiology was chaired by Joshua Lederberg. The Green Bank conference may thus be seen as an extension of the Space Science Board's study of the validity of the exploration of life on Mars.[55] Struve served as chairman of the meeting, and in addition to the pioneers – Cocconi, Morrison and Drake – the meeting was attended by astronomers Su-Shu Huang and Carl Sagan (the latter another young member of the Panel on Exobiology), biochemist Melvin Calvin, dolphin researcher John C. Lilly, electrical engineer Bernard M. Oliver, Dana W. Atchley (who had supplied Project Ozma with its parametric amplifier), and J. P. T. Pearman of the National Academy staff. The purpose of the meeting was to examine "the prospects for the existence of other societies in the galaxy with whom communications might be possible; to attempt an estimate of their number; to consider some of the technical problems involved in the establishment of communication; and to examine ways in which our understanding of the problem might be improved."[56] Given the subject matter, it is not surprising that

53 This passage, and Struve's other recollections of Ozma, are contained in his article "Astronomers in Turmoil," *Physics Today* (September 1960), 18–23: 22–23. The turmoil refers to the general turmoil of radical new ideas and opportunities in astronomy, of which SETI was only one.

54 The conference, held November 1–2 at Green Bank, is described in J. P. T. Pearman, "Extraterrestrial Intelligent Life and Interstellar Communication: An Informal Discussion," in Cameron, *Interstellar Communication*, 287–293. Pearman was a biologist on the staff of the National Academy.

55 The Space Science Board was founded in 1958 as an advisory group. For the work of its Panel on Exobiology see Chapter 3, and for Urey's work on origin of life and meteorites, see Chapter 7.

56 Pearman, "Extraterrestrial Intelligent Life," 288. This is the only contemporary official account of the meeting by one of the participants. *NYT* journalist Walter Sullivan gives a good popular account in *We Are Not Alone*, chapter 17.

this was an invitation-only meeting that included almost all of the people interested in the subject at the time but excluded the media.

While Pearman handled much of the logistics of the meeting, the organization of the scientific content fell largely to Frank Drake. Thinking in the days before the meeting about how to proceed, he decided to arrange the discussions of extraterrestrial intelligence around an equation that concisely represented the relevant factors. Thus appeared for the first time an equation (eventually known as the "Drake Equation") that would be used repeatedly in the following decades in attempts to determine the likelihood of communicating civilizations in our galaxy: $N = R^* f_p n_e f_l f_i f_c L$, where each symbol on the right side of the equation represented a factor on the way to the number of communicating civilizations in the Galaxy (N). The first three factors were astronomical, estimating, respectively, the rate of star formation, the fraction of stars with planets, and the number of planets per star with environments suitable for life. The fourth and fifth factors were biological: the fraction of suitable planets on which life developed and the fraction of those life-bearing planets on which intelligence evolved. The last two factors were social: the fraction of cultures that were communicative over interstellar distances and the lifetime (L) of communicative civilizations. The uncertainties, already great enough for the astronomical parameters, nevertheless increased as one progressed from the astronomical to the biological to the social. The equation, Drake himself later recalled,

> was a way of organizing the meeting. I thought we should organize the meeting and categorize topics and establish themes for various sessions at the meeting, and that caused me simply to think about what we needed to discuss and how these things were related. And it was easy to see that they were interrelated in the way that is described by the equation.[57]

Although Drake was the first to put these factors in simple equation form, assessments of the probabilities of extraterrestrial life and intelligence had been sporadically undertaken in the course of twentieth-century discussions of the subject. Particularly important was a discussion by former Harvard Observatory Director Harlow Shapley. On the eve of the events of 1959–1961, in his book *Of Stars and Men,* he noted that if there were 10^{20} stars in the universe, and if only one star in a thousand had a planetary system, one in a thousand of those had planets at the right distance for water and warmth, one in a thousand of those had planets large enough to hold an atmosphere, and one in a thousand

57 Interview with Drake in Swift, *SETI Pioneers,* 73.

of those had the proper chemical composition of air and water, then the combination of all four "one in a thousand" chances left one chance in 10^{12}. Not very much, but with 10^{20} planets, Shapley pointed out, fully 100 million (10^8) planetary systems might still be suitable for organic life in the universe.[58] Shapley felt this to be a very conservative estimate, underestimated by a factor of 10^3 or 10^6, so that 10^{14} planetary systems with life might exist in the universe. Drake was familiar with Shapley's estimate and his method for calculating it; indeed, he had cited it in his 1960 article on interstellar communication.[59] There were differences in details between Shapley and Drake; with a view toward the interstellar communication now under discussion, Drake concentrated on the Galaxy rather than the universe, on intelligence rather than just life, and furthermore on communicative intelligence. But the concept of calculating probabilities was the same.

Probabilities had also been used by radio astronomer Ronald Bracewell in another early discussion of the number of advanced communities in the Galaxy. Bracewell, however, had couched his discussion in graphical rather than equation form.[60] Closer to the Drake Equation was Sebastian von Hoerner's paper on the nature of intelligent extraterrestrial signals and the distances from which they might emanate, written while he was at the NRAO and discussed with Drake. In discussing the distances between civilizations, von Hoerner spoke of "the fraction of all stars which have planets where life can develop, . . . the time needed to develop a technical civilization, . . . the longevity of the technical civilization, . . . the age of the oldest stars, and . . . the fraction of all stars which at present have a technical civilization."[61] Von Hoerner concluded that 1 in 3 million stars might have a technical civilization but that the longevity of a technical civilization (a concept he credited to Bracewell) might be very limited. It was not so much his conclusion that was important as his

58 Harlow Shapley, *Of Stars and Men* (Boston, 1958), chapter 4, "An Inquiry Concerning Other Worlds," 53–75: 73–74.

59 Drake, "Radio Transmissions," 168. Cameron himself, in his Introduction to this volume of reprints and new papers, stated, "Harlow Shapley set the stage for our discussions in his book *Of Stars and Men*" by suggesting that "we take an overly cautious view of the probabilities of life in the universe and do some simple calculations" regarding planetary systems and planets suitable for life. Cameron, *Interstellar Communication*, 1–2. Shapley did not, however, couch his discussion in terms of an equation. Nor did he broach the subject of radio communication, and thus he did not consider the factor of communicative civilizations.

60 R. N. Bracewell, "Communications from Superior Galactic Communities," *Nature*, 186 (1960), 670, reprinted in Cameron, *Interstellar Communication*, 243–248. Bracewell was invited to the meeting but was in Australia at the time.

61 Sebastian von Hoerner, "The Search for Signals from Other Civilizations," *Science*, 134 (December 8, 1961), 1839–1843, reprinted in Cameron, *Interstellar Communication*, 272.

method; with Shapley, Bracewell, and von Hoerner in the background, it was natural that Drake should organize the meeting in terms of probabilities of factors.

When Drake began the Green Bank meeting by writing his equation on the board, he could not have known that he was establishing the paradigm for SETI discussions for at least the last four decades of the twentieth century. But by considering in turn astrophysical, biological, and social factors, he did just that, and the Green Bank meeting was only the first of many occasions on which experts would discuss the factors that Drake proposed. In that regard, at the meeting Struve remained enthusiastic about the number of planetary systems based on his work on stellar rotation and was supported by his former student Su-Shu Huang, who had concluded from his own research on habitable zones around stars that the number of planets in the Galaxy suitable for life was very large. Calvin, the expert on chemical evolution, argued that the origin of life was a common and even inevitable step in planetary evolution, and his already formidable credentials were given another boost when, during the meeting, he was awarded the Nobel Prize for his work on the chemical pathways of photosynthesis. Lilly, who had just come out with his controversial book *Man and Dolphin,* argued that dolphins were an intelligent species with a complex language and that we might even be able to communicate with them. Not only was this relevant to interspecies communication such as would be required with extraterrestrials, if one accepted his claim that dolphins were in some sense intelligent, it meant that intelligence had developed at least twice on Earth – evidence of its fairly common occurrence. On the other hand, it was obvious to all that dolphins were not a technological species, and had they evolved on another planet, they would not be attempting communication by radio.[62] A notable lacuna at the conference, perhaps reflecting the "two cultures" problem, was the absence of any social scientists to discuss the evolution of civilizations and their longevity, two crucial factors in the Drake Equation. Summarizing the results of their discussions, the members of the conference concluded that depending on the average lifetime of a civilization, the number of communicative civilizations in the Galaxy might range from fewer than 1000 to 1 billion. Opting for optimism, most of the members felt that the higher number was likely to be closer to the truth.[63]

[62] John C. Lilly, *Man and Dolphin* (Garden City, N.Y., 1961). For an early interview with Lilly see Thomas, *Men of Space,* 112–144.

[63] Aside from the account in Walter Sullivan and Pearman's account in Cameron, *Interstellar Communication,* documentation on the meeting is sparse, although the papers in the Cameron volume (covering the years 1959–1962) are said to "reflect" the discussion at

By virtue of its conclusions, the Green Bank conference gave the blessing of a group of considerable experts to the theory of Cocconi and Morrison and the observational approach of Drake. Even more than any technical recommendations, it largely removed the stigma associated with the idea of intelligence beyond the Earth. "It was wonderful," Sagan recalled, "these good scientists all saying that it wasn't nonsense to think about the subject . . . the fact that they came showed that they didn't think it was beyond the pale. . . . There was such a heady sense in the air that finally we've penetrated the ridicule barrier. . . . It was like a 180 degree flip of this dark secret, this embarrassment. It suddenly became respectable."[64]

The concept of abundant extraterrestrial intelligence – the rationale that underlay the concepts elaborated and actions undertaken in the previous 2 years – had barely opened the electromagnetic spectrum to SETI, but it clearly opened the door to enormously important questions. As Struve said at the end of the MIT lectures at which he had first announced Project Ozma, "There can be little doubt today that the free will of intelligent beings is not something that exists only on the earth. We must adjust our thinking to this recognition."[65] That was a bold call to action, one of Struve's last before his death in 1963. It was not to go unheeded, even if progress was slow because more sophisticated discussions were needed. Future possibilities seemed endless, but three seminal events – Cocconi and Morrison's paper, Drake's first observations, and the Green Bank conference – laid the foundation for further thought and action.

8.3 A RATIONALE FOR SETI: OPTIMISTS, PESSIMISTS, AND THE DRAKE EQUATION

In the wake of the three seminal events of 1959–1961, two immediate questions faced those who wished to pursue the problem of interstellar communication: (1) what is the likelihood that extraterrestrial intelligence exists? and (2) is interstellar communication via the electromagnetic spectrum really the best method of contact? The Green Bank conference, and specifically the Drake Equation, had given some early, tentative answers to the first question and leaned heavily toward the conclusion that the radio search was the optimal method for interstellar communica-

the meeting. One exception is Calvin's "Talking to Life on Other Worlds," *Science Digest*, 53 (January 1963), 14–19, 88, 89. No photographs exist of the meeting. One further source of information may be the archives of the Space Science Board, located at the National Academy of Sciences in Washington, D.C.

64 Steven J. Dick interview with Carl Sagan, January 6, 1993, 21–22.

65 Otto Struve, *The Universe* (Cambridge, Mass., 1962), 159.

tion, at least if civilizations were so numerous that they could be reached with radio telescope technology. The trajectory of argument within the radio search paradigm arced steadily upward until about 1975, when serious doubts were raised about the existence of extraterrestrial intelligence, and thus about the propriety of attempts at radio communication. Exactly how the Cocconi–Morrison interstellar communication paradigm became ascendant to 1975, was attacked by the proponents of interstellar colonization using some of its own most cherished assumptions, and then remained intact enough to stimulate an increasing number of observational programs is the subject of this section.

The Cocconi–Morrison paper almost immediately stimulated others to elaborate their own ideas. Late in 1959, the physicist Freeman Dyson had proposed an alternative electromagnetic spectrum scenario. On the assumption that other civilizations would be millions of years older than ours, he suggested that they might have redistributed the mass of one of their planets in a spherical shell around their sun for maximum exploitation of their resources. In this case, the resulting object would radiate in the far infrared, a phenomenon amenable to search with existing optical telescopes.[66] This scenario had the advantage of not assuming that advanced civilizations were actively seeking each other, but rather would passively reveal themselves by their cosmic engineering. Another electromagnetic alternative discussed independently of the Green Bank conference but in the same year was the use of the maser (microwave amplification by stimulated emission of radiation) for communication. Writing in *Nature,* R. N. Schwartz and maser inventor Charles Townes (the latter to receive the Nobel Prize for his work in 1964) concluded that the maser was most advantageously used for such a purpose above the Earth's atmosphere. The following year, however, Bernard Oliver – one of the Green Bank conference participants – argued that masers were not competitive with radio systems for interstellar communication.[67]

A more radical proposal came from R. N. Bracewell of the Radio Astronomy Institute at Stanford. Bracewell's interest had been stimulated by the work on planetary systems of Otto Struve and Su Shu Huang at

[66] Freeman J. Dyson, "Search for Artificial Stellar Sources of Infrared Radiation," *Science,* 131, (1959), 1667, reprinted in Cameron, *Interstellar Communication,* 111–114, and in Goldsmith, *The Quest,* 108–109. Dyson gave credit for this idea of an artificial biosphere to science fiction writer Olaf Stapledon, whose *Star Maker* Dyson had come across in London in 1945; the infrared consequences of this, however, were Dyson's idea. See Dyson, *Disturbing the Universe* (New York, 1979), 211.

[67] "Interstellar and Interplanetary Communication by Optical Masers," *Nature,* 190 (1961), 205, reprinted in Cameron, *Interstellar Communication,* 223 and Goldsmith, *The Quest,* 110; B. M. Oliver, "Some Potentialities of Optical Masers," *Proceedings of the Institute of Radio Engineers,* 50, (1962), 135.

Berkeley, where he had come to lecture on radio astronomy at the invitation of Struve in 1954–1955. Although the seed of interest had been planted, Bracewell's thoughts on extraterrestrial intelligence, he recalled, lay dormant until he read the paper of Cocconi and Morrison in 1959.[68] Unless advanced civilizations were extremely abundant, Bracewell considered it unlikely that we could find any within 100 light years using contemporary technology. Rather, he suggested, they might send probes to the stars. With sophisticated probes, they might explore intensely their neighboring planetary systems and spray the nearest 1000 stars with more modest probes, each probe going into orbit about a star and equipped with a radio transmitter to attract the attention of any indigent civilization. In this case, Bracewell suggested, we should pay attention only to those civilizations capable of reaching us and search for signs of probes sent to the solar system by these advanced civilizations.[69] He even suggested that such signals might have already been detected in the form of the mysterious "long-delay echoes," a theme picked up 10 years later by the Scottish researcher Duncan Lunan.[70] Finally, Bracewell suggested that once the probe had found our awakening civilization, it might then be inducted into a "galactic club" of communicating civilizations.

The most radical alternative of all was "manned" interstellar travel, analyzed in 1962 by Sebastian von Hoerner, a radio astronomer who had been visiting at NRAO during the Ozma Project. He concluded that the requirements were so great that "space travel, even in the most distant

[68] Interview with Bracewell in Swift, *SETI Pioneers,* 141–144.

[69] R. N. Bracewell, "Communications from Superior Galactic Communities," *Nature,* 186, (May 28, 1960), 670–671, reprinted in Cameron, *Interstellar Communication,* 243–248. Bracewell elaborated on these ideas in "Life in the Galaxy," in S. T. Butler and H. Messel, eds., *A Journey through Space and the Atom* (Sydney, 1962), reprinted in Cameron, *Interstellar Communication,* 232–242, and much later in "The Opening Message from an Extraterrestrial Probe," *Astronautics and Aeronautics,* 11 (May 1973), 58–60; "Interstellar Probes," in C. Ponnamperuma and A. G. W. Cameron, eds., *Interstellar Communication: Scientific Perspectives* (Boston, 1974), 102–116, and finally in his book *The Galactic Club: Intelligent Life in Outer Space* (San Francisco, 1974). On the later development of Bracewell's ideas about probes, see Swift, *SETI Pioneers,* 149–153. The probe idea was effectively portrayed in Michael McCollum's science fiction novel *Life Probe* (New York, 1983).

[70] Long-delay echoes were reported by C. Stormer, "Short Wave Echoes and the Aurora Borealis," *Nature,* 122 (November 3, 1928), 681, and B. Van der Pol, "Short Wave Echoes and the Aurora Borealis," *Nature,* 122 (December 8, 1928), 878, and discussed in K. G. Budden and G. G. Yates, "A Search for Radio Echoes of Long Delay," *Journal of Atmospheric and Terrestrial Physics,* 2 (1952), 272–281. In Lunan's *Man and the Stars: Contact and Communication with other Intelligence* (London, 1974), Lunan made a case that the LDEs were from an orbiting Bracewell probe (223–262). This was soon debunked, for example, by Stoneley and Lawton, *CETI: Communication with Extraterrestrial Intelligence* (New York, 1976), 42–48.

future, will be confined completely to our own planetary system, and a similar conclusion will hold for any other civilization, no matter how advanced it may be. The only means of communication between different civilizations thus seems to be electro-magnetic signals."[71] Edward Purcell, the Nobel Prize winner who, with Ewen, had discovered the 21-cm line, arrived at the same conclusion and ended his article with the memorable words "All this stuff about traveling around the universe in space suits – except for local exploration, which I have not discussed – belongs back where it came from, on the cereal box."[72]

In the Soviet Union a similar pattern emerged favoring the Cocconi–Morrison approach, but with more attention paid to a Dyson-like alternative relying on supercivilizations. In that country – where the space age had first dawned, the cold war was in full swing and the political differences with the United States seemed irreconcilable – the work of Cocconi, Morrison, and Drake had first fallen on the receptive mind of radio astronomer Iosif S. Shklovskii. Working at the Sternberg Astronomical Institute of Moscow University, in 1960 he had expressed in a lengthy article his belief that interstellar communication by radio "is legitimate and timely." This conclusion he based on the collapse of Jeans's hypothesis of the rarity of planetary systems, on Struve's argument that the stellar rotational slowdown at the F5 spectral type was due to the formation of planetary systems, and on the belief that life would develop on planets with suitable conditions. In arriving at his conclusion that "there are in the galaxy at least a billion planets . . . on which a highly organized and possibly intelligent life may take place," Shklovskii did not yet make use of the probabilistic arguments that would be characteristic of the Drake Equation.[73] But on the basis of the same qualitative arguments given by Cocconi and Morrison, he did not hesitate to support searches for interstellar communication such as that undertaken by Drake.

Shklovskii's interest soon spread to his colleagues, perhaps in part because the subject of life in space was seen as supported by the Communist philosophy. The Soviet astronomer Nicholas Bobrovnikoff wrote:

> The Soviets are emphatic that their materialistic philosophy is in complete agreement with the idea of extraterrestrial civiliza-

[71] Sebastian von Hoerner, "The General Limits of Space Travel," *Science*, 137, 18 (1962), reprinted in Cameron, *Interstellar Communication*, 144. Morrison (footnote 79, 262) had reached the same conclusion in the same year.

[72] Edward Purcell, "Radioastronomy and Communication through Space," USAEC Report, BNL-658, reprinted in Cameron, *Interstellar Communication*, and Goldsmith, *The Quest*, 188–196. See also John W. Finney, "Scientists and Congress Ponder if Life Exists in Other Worlds," *NYT* (March 23, 1962), 1, 16.

[73] Iosif S. Shklovskiy, "Is Communication Possible with Intelligent Beings on Other Planets?," *Priorda*, no. 7, 21 (1960), reprinted in Cameron, *Interstellar Communication*, 5–16.

> tions. According to this philosophy life is a normal and inevitable consequence of the development of matter, and intelligence is a normal consequence of the existence of life. Even the best informed scientists in the USSR, like Oparin and Shklovskiy, must necessarily subscribe to this crude philosophy promulgated more than 100 years ago by Marx and Engels.[74]

But more than Communist ideology seems to have propelled Shklovskii's interest in the subject; for him it was a subject of genuine astronomical interest. At his own initiative, and with the approval of U.S.S.R. Academy of Sciences President Mstislaw Keldysh, in mid-1961 he began work on a book on extraterrestrial life and intelligence for the fifth anniversary of the launching of Sputnik. Published in 1962, *Vselennaia, Zhizn, Razum* (Universe, Life, Mind) received wide circulation in the West after Carl Sagan revised and expanded it for an English edition.[75] An entire section, devoted to extraterrestrial intelligence, gave prominent attention to the radio search, the first book by an astronomer to do so.

Shklovskii's volume stimulated his colleague at Moscow University (and former student), Nicolai Kardashev, to think about the possible levels of advanced societies and, surpassing even the vision of Dyson (with whose work he was familiar), to suggest a classification of civilizations into three types.[76] Type I civilizations would have a technological level similar to ours at present, as measured by total energy consumption. Type II civilizations

[74] N. T. Bobrovnikov, "Soviet Attitudes concerning the Existence of Life in Space," chapter 19 in *Handbook of Soviet Space-Science Research,* ed. George E. Wukelic (New York, 1968), 456. The Russian emigré Alexis Tsvetikov (footnote 40) commented in 1960, however, that Tsiolkovsky's ideas about extraterrestrial civilizations "have not been praised highly by the Soviet Government, for they are considered groundless and anti-materialistic."

[75] Shklovskii and Sagan, *Intelligent Life in the Universe* (San Francisco, 1966). Shklovskii delightfully describes his participation in SETI events in the last chapter of his autobiography, *Five Billion Vodka Bottles to the Moon: Tales of a Soviet Scientist* (New York and London, 1991), translated and adapted by Mary Fleming Zirin and Harold Zirin, with an Introduction by astronomer Herbert Friedman. For further information on Shklovskii, see also Leo Goldberg, "Josif Shklovsky: A Personal Reflection," *S&T,* (August 1985), 109; P. V. Shcheglov, "Iosif Samuilovich Shklovskii," *QJRAS,* 27 (1986), 700–702; Harold Weaver, "The Award of the Bruce Gold Medal for 1972 to Professor J. S. Shklovsky," *Mercury,* 1 (July–August 1972), 6–7; and "Josif Samuilovich Shklovskii: In Memorian," in A Special Issue of *Soviet Astronomy,* 30 (September–October 1986), 495–97.

[76] Kardashev, "Transmission of Information by Extraterrestrial Civilizations," *Soviet Astronomy,* 8 (1964), 217, reprinted in Goldsmith, *The Quest,* 136. On Kardashev, see interview in Swift, *SETI Pioneers,* 178–197, where we find that Kardashev's interest in extraterrestrial life began not with Cocconi and Morrison, but with the "Russian Lowell" Tikhov (182). His interest in astronomy had begun at age 5 with a Moscow Planetarium show involving Giordano Bruno. Steven J. Dick interview with Kardashev, August 9, 1988, 1, 6.

would be capable of harnessing the energy of their own star – for example, the construction of a Dyson sphere. And Type III civilizations would be able to utilize energy on the scale of their own galaxies. Kardashev believed that there was an "extremely low probability" of detecting the Type I civilizations searched for by Project Ozma and suggested that Type II or III civilizations would be better targets. If Shklovskii was the Soviet equivalent to Cocconi–Morrison, then Kardashev was its Drake, but when Kardashev launched his first search in 1963, it was not for Type I civilizations. Rather, as we shall see in the next section, Kardashev searched for traces of the more advanced Type II or III civilization. Although widely influential, Kardashev's classification scheme found no takers in the United States until 1973, when Carl Sagan suggested that it might be better to search for Type II or III civilizations among the nearer galaxies rather than Type I or younger civilizations among the nearer stars.[77]

The appeal to the Soviets of the idea of extraterrestrial intelligence is evident in the Conference on Extraterrestrial Civilizations held in 1964 at the Byurakan Observatory in Soviet Armenia. This meeting, which grew out of discussions by Shklovskii's group at Sternberg, including Kardashev, L. M. Gindilis, and V. I. Slysh, was organized by Kardashev and hosted by the observatory's director, astrophysicist V. A. Ambartsumian. The Soviet conference differed from the Green Bank conference in that it was attended entirely by radio astronomers; its purpose, too, differed in that the aim was "to obtain rational technical and linguistic solutions for the problem of communication with extra-terrestrial civilizations which are much more advanced than the Earth civilization." Accordingly, attention was focused on Kardashev's three types of civilizations, concentrating on the more advanced types while not completely excluding nearby intelligence. Among others who would be important to the Soviet SETI effort, V. I. Slysh elaborated on Kardashev's criteria for artificiality, and V. S. Troitskii from the Gorkii State University even discussed possible detection of signals from other galaxies. During the 1960s, the Soviets kept up a vigorous debate on SETI. Although they did not discuss alternatives like interstellar travel and probes, as the Americans had, and did not use a quantitative Drake Equation approach, they had much the same discussion in qualitative terms, paid more attention to the likelihood that extraterrestrial civilizations would be much more advanced than ours, emphasized communication rather than search, and gave more attention to the problems of message decoding and problems related to the development of civilizations.[78]

[77] Carl Sagan, "On the Detectivity of Advanced Galactic Civilizations," *Icarus,* 19 (1973), 350, reprinted in Goldsmith, *The Quest,* 140–141.

[78] V. A. Ambartsumian, Introduction to G. M. Tovmasyan, ed., *Extraterrestrial Civilizations: Proceedings of the First All-Union Conference on Extraterrestrial Civilizations*

In the United States, by contrast, the discussion after the Green Bank conference centered on the likelihood of communicative extraterrestrial civilizations utilizing radio technology. In this task, the compelling nature of an equation – even one whose parameters were not well known – was not to be denied, as can be seen in the meteoritic career of the Drake Equation following its debut at Green Bank. Only a month after the Green Bank meeting in November 1961, Philip Morrison used a similar equation in a NASA lecture, and he discussed probabilities in more detail (but without the equation) the following year in a meeting of the Philosophical Society of Washington. The equation first saw print not in an article by Drake, but in Pearman's account of the Green Bank conference, published in 1963; in the same volume, Cameron used a similar equation.[79] Sagan was also among the first to publish the equation, and Drake himself used it in a paper presented at a JPL symposium on exobiology in February 1963, published in 1965.[80] Although it was not known at first as the Drake Equation, everyone knew (and most stated) where it had first been used. Perhaps the decisive events in the spread of the Drake Equation were Walter Sullivan's account of it in *We Are Not Alone* (1964) and Sagan's incorporation of it into his translation and expansion of Shklovskii's book (1966). These books ensured the rapid diffusion of the Drake Equation to the public and scientists alike.[81]

Although not immediately used in the Soviet Union, the Drake Equation, with its emphasis on radio communication, focused attention on the electromagnetic radio search paradigm. Already by 1966 this concept,

and Interstellar Communication, May 20–23, 1964 (Jerusalem, 1967), translated from the Russian *Vnezemnye tsivilizatsii* (Erevan, 1965). Soviet ideas about SETI during the 1960s were reviewed in S. A. Kaplan, *Vnezemyne tsivilizatsii: Problemy mezhzvezdnoi svzyazi* (Moscow, 1969), translated in English as *Extraterrestrial Civilizations: Problems of Interstellar Communication* (Jerusalem, 1971).

79 Pearman, "Extraterrestrial Intelligent Life," 287–293. The equation appears on page 288. Cameron uses a similar equation in "Future Research on Interstellar Communication," in Cameron, *Interstellar Communication,* 309–315, and another version in "Communicating with Intelligent Life on Other Worlds," *S&T* (November 1963), 258–261. It is in Cameron's first article (310) that he refers to Morrison's use of the equation during a lecture at the Institute for Space Studies of NASA's Goddard Space Flight Center on December 14, 1961, later published as "Interstellar Communication," *Bulletin of the Philosophical Society of Washington,* 16 (1962), 58, reprinted in Cameron, *Interstellar Communication,* 249–271, and Goldsmith, *The Quest,* 122–131: 127. Cameron was an astronomer at Goddard interested in the origin of the solar system, and Morrison's lecture probably began Cameron's interest in interstellar communication.

80 Sagan, "Direct Contact among Galactic Civilizations by Relativistic Interstellar Spaceflight," *Planetary and Space Science,* 11 (1963), 485, reprinted in Goldsmith, *The Quest,* 205–213. Drake, "The Radio Search for Intelligent Extraterrestrial Life," in *Current Aspects of Exobiology,* ed., G. Mamikunian and M. H. Briggs (Oxford, 1965), 323–345.

81 Sullivan, *We Are Not Alone,* and Shklovskii and Sagan, *Intelligent Life.*

and all of the assumptions that went with it, were sufficiently entrenched for Dyson to label it the "orthodox view" of interstellar communication, which he described as follows:

> Life is common in the universe. There are many habitable planets, each sheltering its brood of living creatures. Many of the inhabited worlds develop intelligence, and an interest in communicating with other intelligent creatures. It makes sense then to listen for radio messages from out there, and to transmit messages in return. It makes no sense to think of visiting alien societies beyond the solar system, nor to think of being visited by them. The maximum contact between alien societies is a slow and benign exchange of messages, a contact carrying only information and wisdom around the galaxy, not conflict and turmoil.[82]

This was certainly the main point of view espoused by what became the bible of the SETI movement, Shklovskii and Sagan's *Intelligent Life in the Universe*.

As anyone who read science fiction knew, this was not the only possible view of the universe – and not even the dominant one in science fiction. In his 1966 article, Dyson elaborated his idea that we should search for extremely advanced technologies, and Carl Sagan even suggested that the idea be followed up.[83] But as Dyson could see even in the mid-1960s, the orthodox view of radio communication was rapidly taking hold. Radio communication seemed relatively tame compared to Dyson civilizations and galactic probes, interstellar travel, and even masers. More to the point, it was a practical method, a logical extension of the new field of radio astronomy, and one that at least some of its practitioners were keen to carry out. For these reasons, the discussion of rationale and strategy within the radio search paradigm continued to broaden.

Nowhere is this more evident than in the first international SETI meeting, at which Sagan also played an important organizing role. Held 10 years after the Green Bank conference, its organizing principle was the Drake Equation, which for all its quantitative look showed how difficult still were concrete results. Held at the Byurakan Astrophysical Observatory in Yerevan, the Soviet Union (in sight of Mount Ararat), the meeting was sponsored by the Academies of Sciences of both the United States and the Soviet Union. The organizers included not only Sagan, Drake, and

[82] Dyson, "The Search for Extraterrestrial Technology," *Perspectives in Modern Physics*, ed. R. E. Marshak, (New York, 1966), 641–655.

[83] Carl Sagan and R. G. Walker, "The Infrared Detectability of Dyson Civilizations," *ApJ*, 144 (June 1966), 1216–1218.

Morrison of the United States, but also Ambartsumian, Kardashev, Shklovskii, and Troitskii of the Soviet Union. Instead of the 11 participants at Green Bank in 1961, 28 Soviets, 15 Americans, and 4 scientists from other nations participated – and from a wide variety of disciplines. The conclusion of the group – very tentatively expressed – was that perhaps a million technical civilizations existed in the Galaxy.[84] Although the participants also discussed astroengineering activity, the further discussions of techniques of contact, message content, the consequences of contact, and the recommendations focused on the radio search method. The participants were obviously self-conscious about their conclusions. One of the more important ideas of the Byurakan conference was the "subjective probability" nature of the Drake Equation. As Shklovskii would later say, the problem was that the public, and some scientists, treated it as if it were an objective probability; this was a goal rather than the present status.[85]

Even while the Byurakan meeting was in progress, in the United States NASA's interest in extraterrestrial intelligence was slowly beginning to stir – with what effect we shall see in the next section. The fruits of their first study, headed by Green Bank veteran Bernard Oliver and NASA scientist John Billingham, became known as "Project Cyclops," whose rationale also relied heavily on the Drake Equation. Oliver, employing what he frankly called some "very approximate and probably optimistic values" in the equation, came to the conclusion that the number of communicative civilizations was approximately equal to the average lifetime of galactic civilizations – and the latter was, of course, wide open to conjecture. As to the question of where to search, the Cyclops report,

84 Carl Sagan, ed., *Communication with Extraterrestrial Intelligence (CETI)* (Cambridge, Mass., 1973), 166. The meeting was later described by two participants: Freeman Dyson, *The New Yorker* (November 6, 1971), and University of Chicago historian William H. McNeill, *Chicago Magazine* (May–June 1972). The inconclusive nature of the conference is ridiculed in Alfred Adler, "Behold the Stars," *Atlantic Monthly,* 234 (October, 1974), 109, reprinted in Goldsmith, *The Quest,* 224–227. The conference is also described in "First Soviet–American Conference on Communication with Extraterrestrial Intelligence," *Icarus,* 16 (April 1972), 412–414. See also "CETI Questionnaire, Responses from Soviet Specialists," *Spaceflight,* 15, no. 4 (1973), 137–138; translated from *Zemlya i Vselennaya,* 1972.

85 Terrence Fine, "Nature of Probability Statements in Discussions of the Prevalence of Extraterrestrial Intelligent Life," in Sagan, *Communication,* 357–361. Another sidelight of the 1971 meeting was a questionnaire on SETI to sound out the opinions of the scientific community on the subject. Among the respondents was the Soviet physicist and political dissident Andrei Sakharov. Sakharov's reply is described in detail in L. M. Gindilis, "Andrei Dmitriyevich Sakharov and the Search for Extraterrestrial Intelligence," in Shostak (footnote 129), 27–33. Sakharov supported international cooperation for "a persistent search for beacon signals" at the level of 5% of the total spending on Earth- and space-based astronomy. He showed an interest in the implications of contact and even outlined briefly a method for communication.

while admitting that "the arguments for using the hydrogen line no longer seem quite so compelling" now that many other radio frequency lines had been discovered, nevertheless confirmed the concept of a "naturally identified frequency" as a method for narrowing the search. It argued that the region of the spectrum from 1 to 3 Gigahertz (1 to 3 billion cycles per second) was the location of "likely beacon frequencies," in particular the portion from 1.420 GHz (the 21-cm hydrogen line) to 1.662 GHz (the OH line). While leaving open other possibilities, Oliver argued that "surely the band lying between the resonances of the disassociation products of water is ideally situated and an uncannily poetic place for water-based life to seek its kind. Where shall we meet? At the water hole, of course!"[86] Combined with the strategy of "magic frequencies," and especially the "water hole," optimistic values in the Drake Equation became a common rationale of most discussions of the subject during the first decade of its career (Table 8.1). As we shall see, and as the last entry in Table 8.1 indicates, there was another side to the story that would be amply represented a decade later.

By 1971, then, the Drake Equation had become a common feature of the SETI movement, and had succeeded in persuading not only the public but also many astronomers of the likelihood of extraterrestrial intelligence. Although there were counterproposals to the radio method,[87] one measure of the seriousness with which some in the scientific community took the existence of extraterrestrials by the mid-1970s is evident in the reaction of Nobel Prize–winning astronomer Sir Martin Ryle. When in 1974 Drake and other astronomers at Arecibo sent a message to some 300,000 stars that comprise the Great Cluster of Hercules (also known as M 13), Ryle objected strenuously. Although the targets were some 25,000 light years distant, Ryle was acting on deeply felt principles when he agitated for the International Astronomical Union to urge that no attempts be made to communicate with other civilizations because of possible hostile consequences.[88]

[86] B. M. Oliver and J. Billingham, *Project Cyclops: A Design Study of a System for Detecting Extraterrestrial Intelligent Life* (Moffett Field, Calif., 1971), 63–64. This may have been the first expression of the "waterhole" idea, which received considerable attention among SETI advocates.

[87] See, for example, D. R. Bates, "Difficulty of Interstellar Radio Communication," *Nature,* 248 (March 22, 1974), 317–318 and "CETI: Put Not Your Trust in Beacons," *Nature,* 252 (December 1974), 432–433.

[88] As reported by Walter Sullivan in "Astronomer Fears Hostile Attack; Would Keep Life on Earth a Secret," *NYT* (November 4, 1976), p. 46, reprinted in Goldsmith, *The Quest,* 267–268. The *Times* responded with an editorial disagreeing with Ryle: "Should Mankind Hide?" *NYT* (November 22, 1976), p. A24. The Arecibo message was described by the staff of the National Astronomy and Ionosphere Center (NAIC), "The Arecibo Message of November, 1974," *Icarus,* 26 (December 1975), 462, reprinted in Goldsmith, *The*

Table 8.1. *Estimates of factors in the Drake Equation for communicative civilizations*

Author	Date	R^*	f_p	n_e	f_l	f_i	f_c	L	N
Green Bank[a]	1963	1-10	.5	1-5	1	1	.1	10^3-10^8	$<10^3$-10^9
Cameron[b]	1963		1	.3	1	1	.5	10^6	2×10^6
Sagan[c]	1963	10	1	1	1	.1	.1	10^7	10^6
Shklovskii and Sagan[d]	1966	10	1	1	1	.1	.1	10^7	10^6
Byurakan[e]	1971	10	1	1	--------	.01	---------	10^7	10^6
Oliver[f]	1971	20	.5	1	.2	1	.5	?	$=L$
Rood and Trefil[g]	1981	.05	.1	.05	.01	.5	.5	10^4	.003

R^* = rate of star formation, f_p = fraction of stars forming planets, n_e = number of planets per star with environments suitable for life, f_l = fraction of suitable planets on which life develops, f_i = fraction of life bearing planets on which intelligence evolves, f_c = fraction of intelligent cultures communicative over interstellar distances, L = lifetime of a communicative civilization, N = number of communicative civilizations in the Galaxy at a given time.

[a] J. P. T. Pearman, footnote 54 in chapter references.

[b] A. G. W. Cameron, "Future Research on Interstellar Communication," in Cameron (1963), 309-315. Cameron used a slightly different form of the Drake Equation, substituting the number of suitable stars in our galaxy (N_s) for the rate of star formation (R^*), and multiplying this and the other standard factors by L_c/L_p, where L_c is the lifetime of the civilization and L_p is the lifetime of the planet during which this civilization can exist, yielding 2×10^6 for the number of communicative civilizations in the Galaxy. There were many such variations of the equation.

[c] Footnote 80.

[d] Footnote 75.

[e] Footnote 84.

[f] Footnote 86.

[g] Footnote 118. Rood and Trefil (p. 123) interpreted their result as meaning that there was only one chance in 300 that a SETI program would be successful. They also had an additional factor f_d (not shown here), the desire for communication, to which they gave a value of .5.

Although suggestions were made for the modification of the Drake Equation, it continued to be used in substantially unchanged form.[89] Users realized its weaknesses, and critics were quick to point them out. The equation, Bracewell noted, boiled down to two factors – the rate of formation of intelligent communities and their longevity, about neither one of which anything was known. Even Oliver referred to it as "a way of compressing a large amount of ignorance into small space."[90]

However shaky its status as a scientific tool, as Shklovskii later said, it was not devoid of meaning. The goal was to turn its subjective probabilities into mathematical probabilities, and one could just see this beginning to happen in the case of at least one of the factors, the number of planetary systems. But most of the other parameters were admittedly beyond the realm of twentieth-century science. Faced with this dilemma, rather than withdraw from the debate, many preferred to fall back on the probabilistic arguments so prevalent in the origins of life debate. Green Bank Director Otto Struve, for example, held that it was admittedly impossible to give a definitive answer to the question of intelligent life in the universe. But if, he argued, there are 50 billion stars with planets in our own Galaxy alone, then

> an intrinsically improbable single event may become highly probable if the number of events is very great. If the probability of finding intelligent life on one planet at a given time is substantially greater than 10^{10}, then it is probable that a good many of the billions of planets in the Milky Way support intelligent forms of life. To me this conclusion is of great philosophical interest. I believe that science has reached the point where it is necessary to take into account the action of intelligent beings, in addition to the action of the classical laws of physics.[91]

Quest, 293. Drake was director of NAIC at the time. Ryle's opposition a decade earlier, in reaction to Soviet suggestions for international cooperation sent to the IAU (where Ryle was chair of the radio astronomy commission) had been based on the efficient use of telescope time.

[89] J. G. Kreifeldt, "A Formulation for the Number of Communicative Civilizations in the Galaxy," *Icarus,* 14 (June 1971), 419–430, discusses two types of Drake Equation and proposes one of his own. Kreifeldt was a member of the Interstellar Communication Group formed for the 1970 summer study at NASA Ames, as described in Section 4 of this chapter. Also R. N. Bracewell, "An Extended Drake's Equation," in J. Billingham and R. Pesek, *Communication with Extraterrestrial Intelligence* (Oxford, 1979), 67–70. See also footnote 117.

[90] June 1983 interview with Bracewell, in Swift, *SETI Pioneers,* 160–161. Oliver and Billingham, *Project Cyclops,* 26.

[91] Struve, "Astronomers in Turmoil," *Physics Today* (September 1960), 23. On Probabilities and the Drake Equation see also Thomas L. Wilson, "Bayes' Theorem and the Real SETI Equation," *QJRAS,* 25 (1984), 435–448.

Just this problem – taking into account the action of intelligent beings – was to become the central focus of the next decade of debate, with unexpected results for SETI.

Although the persistence of optimists and pessimists from the beginning of the debate is evident in the varying values and opinions about the Drake Equation, after 1975 SETI went through what one of its participants called "a major crisis of identity and purpose."[92] This crisis undermined the very foundations of the SETI endeavor, bringing into question the logic of the radio search paradigm by claiming that all searches of the electromagnetic spectrum might well be fruitless. Needless to say, these claims rallied the SETI forces, causing them to react in ways that further clarify their own "SETI culture."

The origin of the crisis was renewed attention to a question casually raised by the pioneering nuclear physicist Enrico Fermi almost 10 years before the modern era of SETI. During a luncheon with colleagues at Los Alamos in 1950, Fermi had simply asked, "If there are extraterrestrials, where are they?" a question that now became known as The "Fermi Paradox."[93] UFO believers would have answered without hesitation that the extraterrestrials *were* here – they had known it all along – and perhaps it was in part the reputation of this group that kept scientists from pursuing the Fermi question for 25 years. That in 1975 the issue was raised in forceful form in two independent articles in the United States and Britain, without at least one of the authors knowing about Fermi's casual question, is some indication of the force of its logic. The fact that there are no intelligent beings from outer space on Earth now, argued Michael Hart and David Viewing in their respective articles, is an observational fact that argues strongly that extraterrestrials do not exist.[94] The basis for this

[92] Papagiannis, "Evolution of Our Thoughts on the Best Strategy for SETI," in Kellermann and Seielstad, *The Search*, 31–36: 33.

[93] Thanks to the detective work of Eric M. Jones, we know that this conversation took place in the summer of 1950 at a lunch at the Fuller Lodge at Los Alamos. Present were Emil Konopinski, Edward Teller, and Herbert York, who 35 years later gave their recollections of the conversation in Eric Jones, "Where Is Everybody?," *Physics Today* (August 1985), 11–13. Two of the participants remembered the remarks being very casual, while York recalled some discussion about probabilities 10 years before the Drake Equation appeared.

[94] Michael Hart, "An Explanation for the Absence of Extraterrestrial on Earth," *QJRAS*, 16 (June 1975), 128–135, reprinted in Goldsmith, *The Quest*, 228–231; and David Viewing, "Directly Interacting Extra-Terrestrial Technological Communities," *JBIS*, 28 (1975), 735–744. Viewing quotes Fermi on p. 741. On Hart as the initiator of this paradox, see Papagiannis, *The Search*, 437. The Fermi remark was quoted as early as 1963 by Carl Sagan, "Direct Contact," 495, but to raise the possibility that contact might have taken place in historical times. In the opening paragraph of the Introduction

conclusion was the assertion that interstellar travel was possible after all, coupled with attention to the time scales involved. The pessimistic views of those like Purcell, Hart claimed, were based on relativistic spaceflight; the use of nuclear propulsion at say, 1/10th the speed of light would have much more reasonable energy requirements.[95] Given the age of the universe and the time needed for intelligence to develop, Hart and Viewing stated, extraterrestrials should have populated the galaxy. At a velocity of 1/10th the speed of light, Hart argued, this would have occurred in a mere 1 million years. Having addressed the physical argument against interstellar flight, Hart went on to argue against sociological and temporal considerations and to reject the view that extraterrestrials were here now. All sociological arguments – that advanced civilizations engage in spiritual contemplation rather than space exploration, that they destroy themselves, or that they set aside planets like the Earth as wildlife preserves – Hart felt were answered because in order to be effective they would have to apply to every race in the galaxy at all times, an unreasonable assumption. If only one race survived, it could have colonized the galaxy given the time scales involved. In Viewing's words, "This, then, is the paradox: all our logic, all our anti-isocentrism, assures us that we are not unique – that they *must* be there. And yet we do not see them."[96] Thus, Hart concluded, the existence of thousands of civilizations in the Galaxy is quite implausible: "though it is possible that one or two civilizations have evolved and have destroyed themselves in a nuclear war, it is implausible that every one of 10,000 alien civilizations had done so." And there might be a few advanced civilizations that chose never to travel, but "their number should be small, and could well be zero." The bottom line, if this rationale held, was that "an extensive search for radio messages from other civilizations is probably a waste of time and money."[97]

The prospect of interstellar colonization rather than interstellar communication was stunning, and others soon joined the looming battle of ideas, some on the optimistic side (from the SETI viewpoint), others on the pessimistic side. Suddenly articles appeared with names like "Colonization

to Cameron's, *Interstellar Communication,* he asks, "Where is everybody?," and Menzel had raised the question a decade earlier in the context of his work on UFOs.

[95] On interstellar flight, Hart cites G. Marx, *Nature,* 211 (1966), 22, and articles in Mallove and Forward's 1972 bibliography.

[96] Viewing, "Extra-Terrestrial Technological Communities," 741. Kardashev had raised the same issue in 1969, which he called "The Main Dilemma": "There is a high probability that civilization is a universal phenomenon, and yet there are no currently observed signs of cosmic activity of intelligent creatures," N. S. Kardashev, "The Astrophysical Aspect of the Search for Signals from Extraterrestrial Civilizations," in Kaplan, *Extraterrestrial Civilizations,* 12–58: 14–15.

[97] Hart, "An Explanation," 231.

of the Galaxy," in which it was argued that Hart's and Viewing's ideas needed to be taken seriously, particularly in view of the fact that discussion was currently underway about Gerard O'Neill's space colonies. In two early articles of this type, Eric Jones of Los Alamos argued that unless some unknown barrier to interstellar travel existed, our own species would soon undertake it, and it seemed unreasonable to suppose that societies much more advanced would not have – resulting in the colonization of the Galaxy.[98] Similarly, Tom Kuiper and Mark Morris argued that interstellar travel and colonization were not implausible and discussed their implications for SETI strategy.[99] David Schwartzmann suggested that in light of Hart's criticism we should concentrate on Type II or III civilizations or on searching for Bracewell probes.[100] Boston University astronomer Michael Papagiannis suggested that extraterrestrials might be in the asteroid belt, while David Stephenson preferred the outer solar system.[101]

By 1979 the battle lines were drawn, and the interstellar colonization forces felt confident enough to hold a "Where Are They?" conference centered on the Fermi paradox.[102] Most of the participants undoubtedly agreed with the conclusion that extraterrestrials were not on Earth. The belief that UFOs were of extraterrestrial origin, Robert Schaeffer held, was an extraordinary claim, and "extraordinary claims require extraordinary proof," which was not at hand. [103] But if extraterrestrial civilizations were abundant, why were they not on Earth? The presence of several of the pioneering SETI scientists showed that the question was a serious one. Radio astronomer Bracewell reiterated his belief that probes would certainly have covered the Galaxy by now. Physicist Dyson reviewed the propulsion systems that might be available to advanced civilizations. Perhaps most impressively, Sebastian von Hoerner, who in 1962 had concluded that space travel would be confined to the solar system of

98 Eric M. Jones, "Colonization of the Galaxy," *Icarus,* 28 (July 1976), 421–222, and "Interstellar Colonization," *JBIS,* 31 (1978), 103–107. Gerard O'Neill's ideas of space colonies was given in "The Colonization of Space," *Physics Today,* 27, (September 1974), 32–40, and subsequently in many other places, including his book *The High Frontier* (New York, 1977). A collection of articles on interstellar colonization is Ben Finney and Eric Jones, eds., *Interstellar Migration and the Human Experience* (Berkeley, Calif., 1985).

99 Thomas B. H. Kuiper, "Searching for Extraterrestrial Civilizations," *Science,* 196 (1977), 616, reprinted in Goldsmith, *The Quest,* 170–177.

100 David W. Schwartzman, "The Absence of Extraterrestrials on Earth and the Prospects for CETI," *Icarus,* 32 (1977), 473, reprinted in Goldsmith, *The Quest,* 264.

101 Michael Papagiannis, "Are We Alone, or Could They Be in the Asteroid Belt?," *QJRAS,* 19 (1978), 277, and David Stephenson, "Extraterrestrial Cultures within the Solar System?," *QJRAS,* 20 (1979), 422, both reprinted in Goldsmith, *The Quest,* 243–249.

102 Michael H. Hart and Ben Zuckerman, eds., *Where Are They?* (New York, 1982).

103 Robert Sheaffer, "An Examination of Claims that Extraterrestrial Visitors to Earth Are Being Observed," in Hart and Zuckerman, *Where Are They?,* 20–28.

any given civilization, now found Hart's question a "great puzzle . . . still unsolved."[104] Newer voices also expressed their puzzlement, especially in light of the contemporary debate about O'Neill space colonies. Eric Jones, having run computer simulations of the expansion of a spacefaring civilization in the Galaxy, concluded that on any reasonable assumptions extraterrestrials should be here. That they are not, he suggested, gave credence to the claim of Hart and Viewing that no civilization had arisen in the Galaxy. Other alternatives were possible, including Michael Papagiannis's suggestion that extraterrestrials might be in the asteroid belt or that the evolution to intelligence takes much longer than is usually accepted. But the conference left open the very real possibility that we are alone in the Galaxy. The lack of observational evidence of other solar systems and the indirect nature of all other arguments for them left open "the proposition that planetary systems are extremely rare," warned J. Patrick Harrington. And finally, Michael Hart strengthened his own conclusion that extraterrestrials did not exist by arguing that the probability of life arising spontaneously on any planet was very small. In what amounted to a healthy dose of skepticism, the "Where Are They?" participants abandoned the Drake Equation and what it had come to stand for – the abundance of extraterrestrial civilizations.

Even as the "Where Are They?" conference participants were gathering in College Park, Maryland, mathematical physicist Frank Tipler was putting the finishing touches on an article that would encapsulate the opposition views to SETI, taking the extreme position that the arguments were so compelling that it was a waste of taxpayers' money to undertake a search. Together with the Maryland conference, Tipler's articulate arguments and controversial extreme stance marked a turnabout in the fortunes of SETI, a turnabout given wide circulation in the press. Giving due credit to Fermi, Dyson, Hart, and others who had preceded him, Tipler felt that the force of their arguments had not been appreciated, and he delighted in turning the SETI proponents' own arguments against them. Any civilization that had developed the technology for interstellar communication, he argued, must also have developed the technology for interstellar travel. The rudiments of rocket technology, after all, were developed long before the existence of radio waves was expected. Moreover, any species capable of interstellar communication would also be adept at computer technology and would have developed "a self-replicating universal constructor with intelligence comparable to the hu-

[104] Sebastian von Hoerner, in Hart and Zuckerman, *Where Are They?* In an article the previous year, "Where Is Everybody?" *Naturwissenschaften*, 65 (1978), 553, reprinted in Goldsmith, *The Quest*, 250, von Hoerner found Hart's arguments "puzzling and hard to beat."

man level," something some experts on Earth expected within a century. Such a machine would have explored or colonized the Galaxy within 300 million years, he argued, at a cost less than that of operating a microwave beacon for several hundred years, as SETI advocates postulated alien civilizations might do.[105]

In order to make maximum use of the resources of the other stellar systems being explored, Tipler proposed the universal constructor, "a machine capable of making any device, given the construction materials and a construction program," a so-called von Neumann machine, since the mathematician John von Neumann had first discussed it in theoretical terms in 1966. The most important device it would make – out of the raw materials of asteroids and other debris found in the stellar system – would be copies of itself, which would then be launched to the nearest stars, only to repeat the process until the Galaxy was full of such probes. Even with velocities a few 10/1000ths the speed of light (the speed of the Voyager spacecraft leaving the solar system), each interstellar trip would require less than 100,000 years, and the entire Galaxy would be full of probes in 300 million years.

Furthermore, Tipler argued, if one accepts the observational fact that no traces of extraterrestrial intelligence are evident in our planetary system, this places an "astrophysical constraint" on the evolution of intelligent species. If the Galaxy is about 15 billion years old, as is usually stated, then twice as many stars formed before the Sun did. This is approximately equivalent to 100 billion (or 10^{11}) stars – so that the probability that intelligence will evolve is 1 divided by that number, or 10^{-11}. But since the number of stars born since the Sun was formed is also 10^{11}, the number of civilizations in the Galaxy is one – ours.[106]

[105] Frank Tipler, "Extraterrestrial Intelligent Beings Do Not Exist," *QJRAS*, 21, (September 1981), 267–281: 267–268. Also Tipler, "A Brief History of the Extraterrestrial Intelligence Concept" *QJRAS*, 22 (June 1981), 133–145, and "Additional Remarks on Extraterrestrial Intelligence," *QJRAS*, 22 (September 1981), 279–292. The argument was later summarized in "The Space Travel Argument against the Existence of Extraterrestrial Life," chapter 9 in J. D. Barrow and Frank Tipler, *The Anthropic Cosmological Principle* (Oxford, 1986). Tipler argued the case again in "SETI – A Waste of Time," in Stuart A. Kingsley, ed. (footnote 145), 28–35.

[106] Tipler gives a summary in "Extraterrestrial Intelligent Beings Do Not Exist," *Physics Today*, 34 (April 9, 1981), 9, 70–71. Readers, including Drake, responded in *Physics Today* (March 1982), 26–27, 31–37, with another article by Leonard Ornstein, "A Biologist Looks at the Numbers," 27–31. Drake responded at more length in the June 1982 issue, 9. Other presentations appeared in *Sky and Telescope* (September 1981), 207; *Mercury*, 11 (January 1982), 5; *New Scientist*, 96 (October 7, 1982), 33; *Discovery*, 4 (March 1983), 56; and "Extraterrestrial Intelligence: A Sceptical View of Radio Searches," *Science*, 219 (January 14, 1983), 110–111. See also Tipler, *Physics Today*, 40 (December 1987), 92. For the effect of Tipler's ideas on the NASA SETI program, see Chapter 9.

From Hart to Tipler, the basic assumptions of the SETI community had taken a beating beginning in 1975. Although some had immediately rejected the ideas of Hart, [107] most saw their force, including those in the SETI community. After all, although the Space Age on Earth was still confined to the solar system, at least four probes from Earth – Pioneer 10 and 11 (launched 1972–1973) and Voyagers 1 and 2 (launched 1977) – were on hyperbolic trajectories that would take them out of the solar system. For that very reason, Sagan, Drake, and their colleagues had persuaded NASA to add aluminum message plaques to the Pioneer spacecraft and gold-coated phonograph records to the Voyager spacecraft, to be deciphered by any civilization that might come across them.[108] It was not hard to imagine other civilizations sending similar probes or "manned" spacecraft at much faster speeds.

Among the first to show the effect of the criticisms of SETI was Shklovskii – the father of SETI in the Soviet Union – where the Fermi paradox was referred to as the "Astrosociological Paradox" or the "Cosmic Miracle" problem.[109] According to several Soviet astronomers closely involved in SETI, Shklovskii's change of mind came in 1975 during an "All Union School Seminar on the CETI Problem" held under the auspices of the USSR Academy of Sciences at the Special Astrophysical Observatory. "It was there, in Zelenchukskaya, that I. S. Shklovsky proclaimed for the first time the conception of uniqueness of our terrestrial civilization," they wrote. "In 1976 he published the detailed article in the *Voprosy filosofii* (Problems of Philosophy), substantiating this point of view. This work, reviving the old discussion on the plurality of inhabited worlds, was the reflection of the difficulties, both objective and subjective, met by the SETI problem in the course of its development."[110] In an interview in the *Mos-*

[107] Laurence Cox, "An Explanation for the Absence of Extraterrestrials on Earth," *QJRAS*, 17 (June 1976), 201–208.

[108] The Pioneer plaque is described in Carl Sagan, Linda Salzman Sagan, and Frank Drake, "A Message from Earth," *Science*, 175 (February 25, 1972), 881. The Pioneer plaque and Voyager record are described in detail by the same authors, and by Ann Druyan, Timothy Ferris and Jon Lomberg, in *Murmers of Earth: The Voyager Interstellar Record* (New York, 1978).

[109] L. M. Gindilis and G. M. Ruidnitskii, "On the Astrosociological Paradox in SETI," in Shostak (footnote 129), 403–414. The authors argue that while the paradox has stimulated useful discussions, its significance should not be exaggerated.

[110] L. M. Gindilis, B. A. Dubinskij, and G. M. Rudnitskij, "SETI Investigations in the USSR," unpublished paper delivered at the 39th Congress of the International Astronautical Federation, Bangalore, India (October 1988), p. 27, NASA Ames SETI Office Archives. The seminar was reported on in B. N. Panovkin, "The School-Seminar on the CETI Problem," *Zemlya i Vselennaya*, 4 (1976), 68–71 (in Russian), and the Proceedings of the meeting were published in *The Problem of Search for Extraterrestrial Civilizations* (Moscow, 1981) in Russian. Although Gindilis et al. wrote that "the uniqueness conception had no wide support in the USSR," we shall see in the next section that it had a great effect on SETI observing programs. According to Gindilis, among those

cow News Weekly in 1978, Shklovskii made clear the reasons for the reversal of his optimistic assessment of SETI. The probability of intelligence in the universe was very small, he argued, not only because it requires many factors to evolve, but also because we see no evidence of it in our Galaxy. In a more substantial article written the same year, Shklovskii specifically referred to Hart's concept when he concluded

> an analysis of the facts discussed precludes, with a high degree of probability, the possibility of super-civilizations existing, not only in our Galaxy but also in all the local systems of galaxies (e.g. the Nebula of Andromeda). Since some part of the more primitive civilisations of the terrestrial type, after overcoming numerous crisis situations, must take to the road of unlimited expansion, we must draw the logically inevitable conclusion that the number of civilisations of the terrestrial type in the local system is either insignificant, or, what is most probable, does not exist.[111]

Yet Shklovskii remained fascinated with the problem to the end. In perhaps his last pronouncement on the subject the year before his death in 1985, he made use of the Drake Equation and other arguments to conclude that there were no supercivilizations.[112]

In 1979, 10 weeks before the "Where Are They?" conference, the SETI community addressed Hart's thesis in an open forum at the triennial meeting of the International Astronomical Union in Montreal, a meeting attended by Shklovskii. Amid the usual discussions about strat-

criticizing Shklovskii's position were Kardashev, "On the SETI Strategy," *Voprosy Filosofii,* 12 (1977), 43–54, and P. V. Makovetskij, N. T. Petrovich, and V. S. Troitsky, "The Problem of Extraterrestrial Civilizations: The Search Problem," *Voprosy Filosofii,* 4 (1979), 47–59, both in Russian. Shklovskii and Lem also debated the problem in S. Lem, "Are We Alone in Space?" and in Shklovskii, "A Reply to Lem," *Znanie – sila,* 7 (1977), 40–42, in Russian.

111 Interview with Shklovskii in *Moscow News Weekly,* no. 47 (1978); Josif Shklovskii, "Mind-Endowed Life in the Universe: Can It be Unique?," *Social Sciences,* no. 2, 9 (1978), 199–215. Shklovskii's changed attitude was reported in the Western press in "Soviet Reverses Position," *Astronomy,* 5, no. 1 (January 1977), 56. I am indebted to David Schwartzmann for drawing my attention to these articles. For Shklovskii's response to the idea of probes, see Swift, *SETI Pioneers,* 151.

112 I. S. Shklovskii, "Extraterrestrial Civilizations and Artificial Intelligence," in I. M. Markov, ed., *Cybernetics Today: Achievements, Challenges, Prospects,* translated from the Russian by Felix Palkin and Valerian Palkin (Moscow, 1984). Others, including Kardashev, believed that Shklovskii's pessimism about terrestrial civilization was at least in part responsible for his changed position on the likelihood of extraterrestrial civilizations. "I asked him many times why it changed," Kardashev recalled, "but his position was that it was from his opinion about Earth's civilization, very strong military application and military danger." Steven J. Dick interview with Kardashev, August 9, 1988, p. 14.

egy for the radio search and the search for planets and life in a special session on "Strategies for the Search for Life in the Universe," several participants, including Drake, now felt compelled to discuss the Drake Equation's famous number *N* in light of the question of galactic colonization. Hart himself reiterated the colonization thesis, asserting that the most likely reason extraterrestrials had not arrived on Earth was the very low probability of the origin of life. While agreeing with Hart that the Galaxy would be colonized in 10 million years if interstellar colonization was feasible, Drake argued that interstellar travel and colonization are too expensive; in short, "the workings of biology, the physical laws of energy and the vast interstellar distances conspire to make interstellar colonization unthinkable for all time."[113] The extraterrestrials, he added, "are living comfortably and well in the environs of their own star, thriving in habitats in once uninhabitable climes or in the space surrounding the star." And Boston University astronomer Michael Papagiannis continued to advance his thesis that either extraterrestrials were in the solar system or they did not exist. By 1984 the furor raised by the galactic colonization thesis reached its peak, with another IAU meeting at which the previously expressed opinions were consolidated.[114] One of the simplest but perhaps most powerful arguments against the Fermi–Hart paradox was put forward by Philip Morrison. The logic behind the paradox – the underlying rationale of the question "Where Are They?" – Morrison argued, shared the flaw of Malthus in the concept of the exponential improvement of technology. "In the real world there are no unlimited exponentials. Something limits every growth. . . . With that finiteness the power of the argument fades; it all becomes a discussion over the values of limiting parameters that none of us know."[115] Table 8.2 summarizes the results of two decades of explanations for the apparent absence of interstellar colonization.

The interstellar colonization controversy thus had a sobering, but not fatal, effect on the concept of interstellar communication. By arguing that galactic colonization should already have taken place, it forced greater attention to be given to search alternatives ranging from Bracewell or Tipler probes to spaceships in the asteroid belt. There was a limit, how-

[113] Frank Drake, "N Is Neither Very Small Nor Very Large," in M. D. Papagiannis, ed., *Strategies for the Search for Life in the Universe* (Dordrecht, 1980), 27–34: 34. It is notable that Drake rejected the thesis of Newman and Sagan that the wave of colonization would be much slower than Hart had projected; see W. I. Newman and C. Sagan, "Galactic Civilizations: Population Dynamics and Interstellar Diffusion," *Icarus*, 46 (June 1981), 293–327.

[114] Papagiannis, *The Search*.

[115] Morrison, "Twenty-five Years of the Search for Extraterrestrial Communications," in Papagiannis, *The Search*, 18.

Table 8.2. *Explanations for the apparent absence of extraterrestrials on Earth*

Extraterrestrials exist but: No interstellar travel because									Extraterrestrials do not exist
Too expensive	Physically impossible	Too hazardous	Travel is slow	Interactions slow colonization rate	They are undetected	Limits to growth	Too far	Lack of interest/ persistence	
Drake (1980)			Newman & Sagan (1981)		Ball (1973)				Hart (1975) Viewing (1975)
Drake (1985) Wolfe (1985)		Wolfe (1985)		Turner (1985)	Papagiannis (1978)	Morrison (1985)	Wesson (1990)	Finney (1985) Wolfe (1985)	Hart (1980) Tipler (1980)
					Stephenson (1979)				Jones (1985) Hart (1985)
									Tipler (1986)

Note: Ball (1973), Papagiannis (1978), and Stephenson (1979) are reprinted in Goldsmith (1980), note 1. Hart (1975) and Viewing (1975) are referenced in note 94. For Newman and Sagan, see note 113. The 1980 entries are from the *Proceedings* of the 1979 IAU Montreal conference in Papagiannis (1980), note 113. The 1985 entries are found in the *Proceedings* of the 1984 Boston Conference in Papagiannis (1985), note 33. Tipler is referenced in note 105 and Wesson in note 119.

ever, to what astronomers were willing to follow up. They were clearly unwilling to accept the extraterrestrial hypothesis of UFOs which was one possible explanation of the Fermi–Hart paradox, and most balked at Papagiannis's suggestion of searching for intelligent activity in the solar system. They seemed much more willing to consider Dyson spheres or a search for Kardashev's Type II or III civilizations, and they remained enamored of the now orthodox view of Cocconi and Morrison. Still, they were receptive to alterations in that view, including the proposal in 1979 of University of Washington astronomer Woodruff T. Sullivan III and his colleagues that eavesdropping for radio leakage should be considered in addition to the search for purposeful signals. Sullivan's quantitative study of radio leakage from the Earth convinced some that this was an important, if difficult, idea to follow up.[116] Most of all, SETI advocates did not abandon their conviction that the search should continue, and this is why they reacted so strongly to Tipler.

Nor did the galactic colonization flap kill the Drake Equation. Despite the criticisms, for most optimists and pessimists it continued to be the focus of the debate, although some used it in more radically modified form to take the criticisms into account. Astronomer and science fiction writer David Brin (like Viewing and others before him) called for a modification to the Drake Equation, the "centerpiece of xenological discussion," which he characterized as "now clearly insufficient to encompass the subject" in light of what he called "the Great Silence."[117] Others were more skeptical of the Drake Equation; in particular, Dyson unequivocally disavowed it: "I reject as worthless all attempts to calculate from theoretical principles the frequency of occurrence of intelligent life forms in the universe. Our ignorance of the chemical processes by which life arose on earth makes such calculations meaningless." Most typically, however, the Drake Equation continued to be used, with some skeptical astronomers and physicists such as Robert Rood and James Trefil (playing a role similar to that of skeptical chemist Robert Shapiro, as discussed in the previous chapter) coming up with a number less than 1 for their best estimate.[118]

Increasingly, some were suggesting that we might be alone in the uni-

[116] W. T. Sullivan III, S. Brown, and C. Wetherill, "Eavesdropping: The Radio Signature of the Earth," *Science,* 199 (January 27, 1978), 377–388. See also C. Wetherill and W. T. Sullivan III, "Eavesdropping on the Earth," *Mercury* (March–April 1979), 23–28, and W. T. Sullivan, "Radio Leakage and Eavesdropping," in Papagiannis, *Strategies,* 227–239. Although the problem of leakage vs. beacons from other civilizations had been addressed in Project Cyclops (59ff.) and as early as 1961 by Webb (in Cameron, *Interstellar Communication,* 178), Sullivan's initial paper inspired Carl Sagan, among others, to consider the problem in "Eavesdropping on Galactic Civilizations," *Science,* 202 (October 27, 1978), 374–376, followed by "Sullivan Responds to Sagan," *Science,* 202 (October 27, 1978), 376–377. Sullivan and Stephen Knowles followed up the idea in "Lunar Reflections of Terrestrial Radio Leakage," in Papagiannis, *The Search,* 327–334.

[117] Glen David Brin, "The 'Great Silence': The Controversy concerning Extraterrestrial Intelligent Life," *QJRAS,* 24 (1983), 283–309. Brin, a Hugo and Nebula award winner for his science fiction, is an example of the increasing ties between science and science fiction. Other modifications of the Drake Equation to take colonization into account include Clifford Walters, Raymond Hoover, and R. K. Kotra, "Interstellar Colonization: A New Parameter for the Drake Equation?" *Icarus,* 41 (January 1980), 193–197, and S. G. Wallenhorst, "The Drake Equation Re-examined," *QJRAS,* 22 (December 1981), 380–387.

[118] Dyson, *Disturbing the Universe,* 209; Robert T. Rood and James S. Trefil, *Are We Alone?* (New York, 1981). The entire book is their skeptical reevaluation of the Drake Equation based on new knowledge, summarized in chapter 8, "Summary and Review of the Green Bank Equation." On uncertainties in the Drake Equation see also Peter Sturrock, in Papagiannis, *Strategies,* 59.

verse after all, that life might be so sparse that if intelligence existed at all, it would be at cosmological distances – beyond the range of interstellar communication or colonization.[119] It was, after all, logically possible that the ultimate resolution of the Fermi–Hart paradox was also the simplest: extraterrestrials did not exist in the observable universe. But 30 years after serious discussions were launched at Green Bank, most SETI partisans in both the Soviet Union and the United States were unwilling to accept that conclusion, with its implications for humanity's place in the cosmos. Although the Fermi–Hart paradox would have some impact on observations especially in the Soviet Union, the discussions would go on.[120]

In the end, after all the argument, what was left in logical terms was a true dilemma for science. On the one hand, SETI proponents could well argue that claims about the actions of extraterrestrials were inconclusive at best. On the other hand, based on the terrestrial analogies that SETI proponents were so fond of quoting (at least in the realm of physical law), it was difficult to believe that terrestrial civilization would never spread among the stars, given enough time, or that extraterrestrial civilizations 1 million or 1 billion years older would not have found a method of interstellar travel. Philip Morrison, one of those who had started it all, had an answer to that dilemma: "It is fine to argue about N. After the argument, though, I think there remains one rock hard truth: whatever the theories, there is no easy substitute for a real search out there, among the ray directions and the wavebands, down into the noise. We owe the issue more than mere theorizing."[121] Others agreed, even Dyson, for the history of astronomy gave much support to the primacy of observation over logic. But if the optimists were to pursue the observational goals of SETI, they would have to deal with the problems of equipment, personnel and

[119] J. Richard Gott III, "Cosmology and Life in the Universe," in Hart and Zuckerman, *Where Are They?*, 122, and Paul S. Wesson, "Cosmology, Extraterrestrial Intelligence, and a Resolution of the Fermi-Hart Paradox," *QJRAS*, 31 (1990), 161–170. See also G. F. R. Ellis and G. B. Brundrit, "Life in the Infinite Universe," *QJRAS*, 20 (March 1979), 37.

[120] In order to circumvent the Fermi paradox, Soviet radio astronomer V. S. Troitsky put forward the concept of a one-time and simultaneous origin of life in the universe, implying that all civilizations would be at about the same level of development. See V. S. Troitsky, "A New Approach to the Number *N* of Advanced Civilizations in the Galaxy," in Papagiannis, *Strategies*, 73–76. Among recent books in the Soviet Union dealing with extraterrestrial intelligence were V. V. Rubtsov and A. D. Ursul, *The Problem of Extraterrestrial Civilizations: Philosophical and Methodological Aspects* (Moscow, 1987) [in Russian], and Victor Vizgin, *Plurality of Worlds, Studies in the History of an Idea* (Moscow, 1988) [in Russian].

[121] Morrison, "The Number *N* of Advanced Civilizations in our Galaxy and the Question of Galactic Colonization. An Introduction," in Papagiannis, *Strategies*, 15–18: 18. Dyson, *Disturbing the Universe*, 216: "The question can only be answered by observation."

money – resources that might be used for much surer bets than interstellar communication. In short, whether seeking support from a university or the federal government, on a small or large scale SETI would have to enter the arena of science policy. To the practical considerations of that arena, from the airy heights of rationale, we now turn.

8.4 A STRATEGY FOR SETI: THE DEVELOPMENT OF OBSERVATIONAL PROGRAMS

The widespread acceptance of Philip Morrison's argument for the primacy of observation over theory in science – at least in the case of such a speculative subject as SETI – is evident in the fact that while some 15 observational searches were originated in the 15 years prior to 1975, more than 40 were undertaken on various scales in the following 15 years – even after serious theoretical doubts were raised about the existence of extraterrestrials.[122] In contrast to discussions about the rationale for SETI, as described in the previous section, actual searches required a commitment to specific strategies – of hardware and software; of frequencies, apertures, and bandwidths; of financial commitments and personnel – in short, of all the considerations Drake had to contend with more or less unilaterally during the days of Project Ozma in 1960, now multiplied many times over in cost and complexity. In this section we explore the evolution of representative strategies and their resulting observing programs, with particular emphasis on the most comprehensive program of all – that developed in the United States by NASA. We shall see that while Soviet SETI observing programs were severely impacted by the interstellar colonization crisis, American programs flourished, often inspired or financially aided by NASA. But ironically, in direct proportion to the influx of government funding, SETI became vulnerable to the always unpredictable and sometimes irrational political winds of the U.S. Congress, with ultimately disastrous results for NASA's own flagship SETI program.

Perhaps significantly, the majority of searches prior to 1975 were undertaken in the Soviet Union – an indication of the influence of the discussions inspired by Shklovskii and his radio astronomy group. As we have seen in the previous section, while the United States concentrated on the more immediate problems of its space program, including life in the solar system, discussions on interstellar communication were undertaken in the

[122] All observing programs to 1984 are described in Jill Tarter, "SETI Observations Worldwide," in Papagiannis, *The Search*, 271–290, updated to 1990 in the Appendix to Jill Tarter, "The Search for Life beyond the Solar System," in J. and J. Tran Thanh Van et al., eds., *Frontiers of Life* (Gif-sur-Yvette Cedex, France, 1992), 351–397.

Soviet Union, first with the all-union conference at Byurakan in 1964 and then (with American help) the international meeting at the same location in 1971. Aside from a search by an American astronomer observing one galaxy from Australia, the Russians had a monopoly on searches for interstellar communication until several were carried out at NRAO in Green Bank in the early 1970s.

To judge by their observations, the Soviets were not overly impressed with the strategy of the 21-cm line. Impressed instead with the premise that other civilizations were likely to be millions or billions of years more advanced than us, they turned their attention to the effects such civilizations might produce. Thus, when in 1963 Kardashev and Sholomitskii carried out the first search after Ozma in the wake of Kardashev's paper on Type I, II, and III civilizations, it was at 920 MHz rather than the 1420-MHz equivalent to the 21-cm line. For a few brief weeks, the strategy seemed to have paid off with a spectacular discovery that soon became part of SETI legend. Concentrating on peculiar radio sources with very small angular dimensions, Kardashev and Sholomitski claimed that they had actually detected a possible artificial signal from a Type III civilization associated with a variable radio source designated CTA-102.[123] "There was a great uproar over it," Shklovksii recalled. At a Sternberg Institute press conference, the courtyard

> was crammed with luxurious foreign cars belonging to some 150 of the leading accredited correspondents in Moscow. I led off with a few conservative and skeptical principles. Sholomitsky was extremely restrained. The director of the institute, Dmitry Martynov, basked in the limelight. Unfortunately, it soon became clear that CTA-102 was just an ordinary quasar with a large (but not record-breaking) red-shift.[124]

[123] N. S. Kardashev, "Transmission of Information by Extraterrestrial Civilizations," *Soviet Astronomy – AJ*, 8 (1964), 217, reprinted in Goldsmith, *The Quest*, 136. The Crimea Deep Space Station was used for the observations. See also Sholomitski, *IAU Information Bulletin on Variable Stars* (February 27, 1965) and *NYT* editorial, April 13, 1965, 36. Shklovskii and Sagan, *Intelligent Life*, discuss CTA-102, 394–396. See also Swift, *SETI Pioneers*, 191. There was considerable press reaction; the *New York Times* printed at least five articles, beginning with "Soviet Writers Say Earth Has Received Signals from Space" (March 21, 1964), p. 4.

[124] Shklovskii, *Five Billion Vodka Bottles to the Moon*, 253. For a detailed study of the CTA-102 incident as a lesson for possible future detections, see J. Heidmann, "SETI False Alerts as 'Laboratory' Tests for an International Protocol Formulation," in J. C. Tarter and M. A. Michaud, eds., *SETI Post-Detection Protocol* (Oxford, 1990), 73–80. A few years after the CTA-102 incident, in 1967, a similar uproar ensued when regular pulses were detected by radio telescopes at Cambridge, England. The discoverers considered that the signals might be the products of intelligence, but they were actually the first discovery of rotating neutron stars, or pulsars. For a discussion by one of the

Subsequent Soviet searches headed by the radio astronomers Troitskii, Slysh, and others in the late 1960s and early 1970s also largely shunned the 21-cm strategy, but without the spectacular claims of CTA-102. Troitskii's group in particular felt it would be profitable to use a broadband receiver to look for pulses, eliminating the need for magic frequency arguments.[125]

The first searches in the United States had to wait 10 years after Ozma. Aside from the physical exhaustion caused by Ozma's 12-hour observing schedule, Drake recalled, the effect of always observing noise resulted in another problem: boredom. "After two months of noise, I was quite ready for other conventional astronomical things," he noted. And over the next 10 years, as Drake sensed that telescope time was hard to come by with Struve no longer the director, he did not even ask to undertake more observations.[126] The next searches in the United States, however, clearly followed in the footsteps of Drake, and in fact were also undertaken at the NRAO in Green Bank. The first, performed by G. Verschuur in 1971–1972 (facetiously termed "Ozpa") used the 21-cm line and observed nine stars over 13 hours but made use of the 140-foot and 300-foot telescopes compared to Drake's 85-foot one. The second, undertaken by P. Palmer and B. Zuckerman (Ozma II), was larger in scale and observed some 674 stars over 500 hours, also using the 21-cm line and the Green Bank 140-foot telescope. The third U.S. search initiated before 1975, which has turned out to be the longest-running of all, began under John Kraus and Robert S. Dixon at Ohio State University in 1973 and again used the 21-cm strategy. Utilizing a meridian-transit telescope with a collecting area of 2200 square meters (equivalent to a parabolic dish 175 feet in diameter), the observing program made use largely of student volunteers. Over the course of its program, the Ohio State telescope detected a number of interesting transient signals, most notably one observed in 1977 and known as the "WOW signal" after the exclamation

discoverers see S. Jocelyn Bell Burnell, "Little Green Men, White Dwarfs or Pulsars?" *Cosmic Search* (January 1979), 17–21.

[125] The observations of the Troitskii group, centered at the Radiophysical Research Institute in Gorki, are described in V. S. Troitskii, A. M. Starodubtsev, et al., "Search for Monochromatic 927-MHz Radio Emission from Nearby Stars," *Soviet AJ,* 15 (November–December 1971), 508–510, and by Troitskii, Starodubtsev and L. N. Bondar, "Search for Radio Emissions from Extraterrestrial Civilizations," *Acta Astronautica,* 6 (January–February 1979), 81–94. Slysh made his observations at Nancay, France. Drake later recalled that the problem with the Troitskii observations was that "their sensitivity was one- one hundred millionth of the American sensitivities when it came to detectable signals." Steven J. Dick interview with Drake, May 29–30, 1992, 18.

[126] Steven J. Dick, interview with Drake, May 29–30, 1992, 12–16.

penned on the observing record. But none lasted long enough for positive identification.[127]

Ironically, Drake himself, along with Carl Sagan, was the first in the United States to stray from the single-star 21-cm strategy, if only temporarily. Using the 1000-foot dish at Arecibo, in 1975–1976 they observed four galaxies in their search for Kardashev's Type II civilizations.[128] The year 1975 – during which interstellar colonization began its career in earnest – also marked a shift in actual observing programs in the sense that the United States began its domination of SETI observing programs while the Russians almost completely dropped out. Although the Russians had developed a systematic observing program in 1974 and later proposed to set up an array of 100 one-meter dishes for SETI, neither materialized, at least in part because of the change in heart of Shklovskii regarding the likelihood of life. At the 1991 U.S.–USSR conference on SETI, not a single Soviet SETI radio observing program was reported.[129]

In the United States, on the other hand, Ohio State continued its program; new projects were initiated at the NRAO, Arecibo, and Harvard observatories; and NASA began planning for what would become the flagship SETI program. NRAO continued its tradition by hosting two small programs in 1976–1977, the largest using the 300-foot telescope to observe 200 stars near the 18-cm "OH" end of the waterhole neglected

[127] G. L. Verschuur, " A Search for Narrow Band 21-cm Wavelength Signals from Ten Nearby Stars," *Icarus* (1973), 19, 329; reprinted in Goldsmith, *The Quest,* 142; Patrick Palmer and Ben Zuckerman, "The NRAO Observer," 13, no. 6 (1972), 26, also reported in R. Sheaffer, *Spaceflight*, 19, no. 9 (1977), 307; R. S. Dixon, "A Search Strategy for Finding Extraterrestrial Radio Beacons," *Icarus,* 20 (October 1973), 187–199; R. S. Dixon and D. M. Cole, "A Modest All-Sky Search for Narrowband Radio Radiation Near the 21-cm Hydrogen Line," *Icarus,* 30 (February 1977), 267–273, Dixon describes the first 10 years in "The Ohio SETI Program – The First Decade," in Papagiannis, *The Search,* 305–314, where plans were reported to cover the entire "waterhole" region from 1400 to 1750 Mhz. The WOW signal is described in J. D. Kraus, "We Wait and Wonder," *Cosmic Search,* 1, no. 3 (January 1979), 31.

[128] Carl Sagan and Frank Drake, "The Search for Extraterrestrial Intelligence," *SciAm,* 232 (May 1974), 80. Although 21 cm was one of the observational wavelengths, corresponding to 1420 MHz, they also used the 1667-MHz (OH) and 2380-MHz frequencies.

[129] The Soviet "Research Program on the Problem of Communication with Extraterrestrial Civilizations" was developed by a section of the USSR Academy of Sciences' Council on Radio Astronomy known as "Search for Cosmic Signals of Artificial Origin." Based on the recommendations of the Byurakan meetings of 1964 and 1971, it was approved by the Academy in March 1974 and is described in "The Soviet CETI Program," originally published in *Astronomicheskii Zhurnal,* 51 (September–October, 1974), 1125–1134, English translation in *Soviet Astronomy,* 18 (March–April 1975), 669–675, reprinted in *Icarus,* 26 (November 1975), 377–385. For the Proceedings of the 1991 U.S.–USSR SETI conference see G. Seth Shostak, *Third Decennial US–USSR Conference on SETI* (San Francisco, 1993), where one optical SETI search was reported as part of other experiments in V. Shvartsman, "Results of the MANIA Experiment: An Optical Search for Extraterrestrial Intelligence," 381–392.

by most previous observers. The University of California, Berkeley–based Project SERENDIP (an acronym for Search for Extraterrestrial Emission from Nearby Developed Intelligent Populations) was initiated in the late 1970s in a "parasitic" mode, that is, as a "piggyback" program operated in tandem with other regularly scheduled observing programs.[130] At Arecibo, where Drake served as director, a number of small programs were undertaken beginning in 1977. Notable among these was an attempt at "eavesdropping" by Knowles and Sullivan, a search for Type II and III civilizations, and the more standard 21-cm approach of Paul Horowitz. The last, which made use of prototype equipment built for NASA and further developed by Horowitz, grew into Project Sentinel and the Megachannel Extraterrestrial Assay (META) at Harvard. Funded in part by the Planetary Society, the program was expanded to META II in Argentina in 1990. Horowitz's programs held the record for frequency resolution of .015 Hz and made the assumption that the frequency was corrected at the source to arrive at rest in a heliocentric frame.[131]

By 1985, a quarter century after Drake's first search, NASA's future SETI project scientist Jill Tarter could distinguish three types of SETI strategies in terms of telescope usage: the usual directed searches, in which telescopes could be used for brief SETI observations; shared searches such

[130] On the NRAO programs see J. Tarter, D. Black, J. Cuzzi, and T. Clark, *Icarus*, 42 (April 1980), 136–144. The history of the Serendip Project, the brainchild of Stuart Bowyer, is described in S. Bowyer, G. Zeitlin, J. Tarter, M. Lampton, and W. Welch, "The Berkeley Parasitic SETI Program," *Icarus*, 53 (January 1983), 147–155. An improved observing program is described in Dan Wertheimer, Jill Tarter, and Stuart Bowyer, "The Serendip II Design," in Pagagiannis, *The Search*, 421, and another update is given in S. Bowyer, "The U.C. Berkeley Program," in B. Bova and B. Preiss, eds., *First Contact: The Search for Extraterrestrial Intelligence* (New York, 1990). This and other ongoing observing projects were routinely reported on at SETI meetings.

[131] Results from Knowles and Sullivan were not reported; the search for Type II and III civilizations is described in N. Cohen, M. Malkan, and J. Dickey, "A Passive SETI in Globular Clusters at the Hydroxyl and Water Lines," *Icarus*, 41 (January 1980), 198; and the first of Paul Horowitz's many observations in P. Horowitz, "A Search for Ultra-Narrowband Signals of Extraterrestrial Origin," *Science*, 201 (August 25, 1978), 733–735. The Sentinel program, begun in March 1983, is described in Paul Horowitz and John Forester, "Project Sentinel: Ultra-Narrowband SETI at Harvard/Smithsonian," in Papagiannis, *The Search*, 291; Paul Horowitz, John Forster, and Ivan Linscott, "The 8-Million Channel Narrowband Analyzer," in Papagiannis, *The Search*, 361, describe the system to improve Sentinel by a factor of 200, an 8.4 million channel enhancement of the 128,000 channel Sentinel. Project META is described in P. Horowitz, B. S. Matthews, et al., "Ultranarrowband Searches for Extraterrestrial Intelligence with Dedicated Signal-Processing Hardware," *Icarus*, 67 (September 1986), 525–539, and its first results are given in P. Horowitz and C. Sagan, "Five Years of Project META: An All-Sky Narrow-Band Radio Search for Extraterrestrial Signals," *ApJ*, 415 (September–October 1993), 218–235. The results included some transient signals of unknown origin. On Argentina's interest in SETI, dating from at least 1971, see J. C. Cerosimo, A. A. Cocca, F. R. Colomb, et al., *Inteligencia Extraterrestre* (Cordoba, 1988).

as SERENDIP, which operated in a parasitic mode during other observations or analyzed old signals; and dedicated searches such as the Ohio State effort, in which an instrument was devoted exclusively to SETI over an extended period. These observations had been undertaken with radio telescopes ranging from a few tens of meters in diameter to the giant 305-meter instrument at Arecibo, at a variety of "magic" frequencies based largely on fundamental line radiations from atoms or molecules, with frequency resolutions varying from megahertz to the few hundredths of a Hertz of Horowitz's Sentinel Project at Harvard. Most had taken only a few hours or tens of hours; of the 120,000 SETI hours logged by 1985, 100,000 had been logged at the two dedicated facilities at Ohio State and Harvard. As the summary characteristics of selected observing programs in Table 8.3 show, only a few observations had been undertaken at optical or infrared frequencies; the vast majority were in the radio region originally favored by Cocconi, Morrison, and Drake. Most impressively, Tarter's list demonstrated the accelerating pace of observations; 25 of the 45 entries in the summary of observing programs were searches conducted since 1979, and seven countries were involved instead of three. By the 1990s, SETI had become a global, if still sporadic, effort.[132]

Of all the search programs, NASA's was to be the most comprehensive, and its development illustrates on a large scale many of the problems that all SETI programs faced.[133] The origins of the NASA SETI program may be traced to John Billingham, a physician at NASA's Ames Research Center, where he headed the Biotechnology Division charged with research in space medicine. Inspired by Shklovskii and Sagan's *Intelligent Life in the Universe* (1966), in 1970 Billingham convinced Ames Director Hans Mark that a mini-study of the problem of interstellar communication was within the scope of NASA's mission and should be undertaken. The optimistic results of that study in 1970 led in 1971 to a more ambitious "design study of a system for detecting extraterrestrial intelligent life." Known as "Project Cyclops," the study was undertaken not as a major NASA project, but as part of a summer faculty fellowship program in engineering systems design sponsored by Stanford University, NASA, and Ames.

The key figure in this study, and the author of its influential report, was Bernard M. Oliver, the electrical engineer, vice president of Hewlett-

[132] Tarter, "SETI Observations"; see also Papagiannis's Introduction immediately preceding the Tarter article.

[133] The history of NASA's involvement in SETI is elaborated in detail in Steven J. Dick, "The Search for Extraterrestrial Intelligence and the NASA High Resolution Microwave Survey (HRMS): Historical Perspectives," *Space Science Reviews*, 64 (1993), 93–139, from which parts of the following pages are derived.

Table 8.3. *Characteristics of selected SETI observing programs*

Date	Observer	Site	Instrument size (meters)	Search freq. (MHz)	Frequency resolution (Hertz)	Objects	Total hours	Comments
1960	Drake	NRAO	26	1420-1420.4	100	Two stars	400	Project Ozma, 1 channel
1963	Kardashev Sholomitskii	Crimea DSS	16 (8 antennas)	920	10 MHz	Two quasars	80	CTA 102 - Type III civilization reported
1969-1983	Troitskii Bondar Starodubtsev	Gorky Crimea & others	Dipole	1863, 927, 600		All-Sky	1200/ year	Search for sporadic pulses
1973-	Dixon et al.	Ohio State	53	1420.4	10KHz 1 KHz	All-Sky	C[a]	Longest-running SETI program
1978-	Shvartsman et. al	Zelen-chukskaya	6	Optical		30 radio objects	C	Optical search for short pulses
1975-1976	Drake, Sagan	Arecibo	305	1420, 1667, 2380	1000	Four Galaxies	100	Search for Type II civilizations in galaxies
1976-	Bowyer et al.	Hat Creek and others	26, etc.	1410ff. 1653ff.	2500	All-Sky	C	SERENDIP piggyback system
1983-	Horowitz et al.	Harvard	26	1420.4, etc.	.03	All-Sky	C	Sentinel - "suitcase SETI" followed by META (1985)
1990-	Colomb et al.	Argentina	30	1420.4	.05	All-Sky	C	Southern Hemisphere META II
1990-	Betz	Mount Wilson	1.65	10 microns (infrared)	3.5 MHz	100 stars	C	Infrared search interferometer
1992	NASA	Arecibo	305	1-10 GHz	1-28	Targeted	<1 yr.	HRMS/MCSA[b] terminated by Congress
1992-1993	NASA	JPL/DSN	34	1-3 GHz	30	All-Sky	<1 yr.	HRMS/WBSA[c] terminated by Congress
1995	SETI Institute	Parkes, Australia	64	1-3 GHz	1-28	Targeted		Phoenix HRMS descendant

Source: Adapted from Jill Tarter (footnote 122 in this chapter), by permission of the author and Reidel Publishing Company.

[a] C = Continuing program

[b] High Resolution Microwave Survey/Multichannel Spectrum Analyzer

[c] High Resolution Microwave Survey/Wide Band Spectrum Analyzer

Figure 8.5. Artist's conception of a portion of the Cyclops array, proposed as a result of the NASA/ASEE summer study in 1971, showing antennae and the central control and processing building at the right. Courtesy NASA Ames Research Center.

Packard, and member of the 1961 Green Bank "Order of the Dolphins," whom we encountered earlier in this chapter. Oliver and his colleagues envisioned an "orchard" of perhaps one thousand 100-meter antennas covering a total area some 10 km in diameter; Cyclops was nothing if not ambitious (Figure 8.5). But considered as a phased array of connected antennas, any construction could start out small and sequentially add more antennas in the event that no signals were detected. The study addressed details of such aspects of the project as antenna elements, receiver systems, and signal processing, as well as more general problems about the probability of life in the universe and search strategies. The resulting publication, which emphasized the "waterhole" frequency as the most likely place to search, holds an interesting place in SETI history: although recommendation was made to "establish the search for extraterrestrial intelligent life as an ongoing part of the total NASA space program," the full scope of the Cyclops project itself (requiring some $6 to $10 billion over 10–15 years) was much too ambitious for NASA funding.[134] At the same time, the study not only served as an illustration of what could be done, but also demonstrated the technical feasibility of interstellar communication and provided a benchmark against which smaller programs could be measured. If ever there was a tribute to the dictum "make no small plans," the very idea of Cyclops was it.

Despite the recommendation of the Cyclops report (or perhaps because of its large projected cost), during the 1970s NASA funding for SETI was limited to workshops and conferences, which served the important purpose of keeping interest alive and refining technique. Ironically, it was just as the interstellar colonization crisis of SETI was beginning that NASA began its first major study of a realistic SETI program. Most important was a series of six Workshops on Interstellar Communication held in 1975–1976 under the chairmanship of Philip Morrison.[135] These workshops, and three offshoots dedicated to planet detection and the evolution of technological civilizations, proved collectively to be a landmark in SETI history and were critical in stimulating interest and support in the wider scientific community. Having considered interstellar travel, robot probes, and electromagnetic signals, the report confirmed that radio signals were the optimum method for communication. But it also recognized that the extremely large number of frequencies – analogous to a very extended radio dial – necessitated that the search be limited in direction, in frequency, or both. Although no consensus was reached on search strategy, the report (in a paper by radio astronomer Charles Seeger) was the first public discussion of a possible bimodal method for the search, to include both a targeted search and an all-sky survey. The assumptions behind these strategies would be the subject of much further discussion.

The Morrison report also mentioned for the first time in the SETI context the Deep Space Network and planetary science, signaling the interest of the Jet Propulsion Laboratory (JPL), an agency funded by NASA through Caltech. More particularly, it indicated the support of JPL's prospective new director, Bruce Murray, for the SETI program. Thus, in the case of both JPL and Ames, the high-risk innovative SETI programs stemmed from the personal interest and support of the new directors at each institution. The interest at JPL was a natural development, since JPL ran the Deep Space Network and had expertise in radio astronomy needed for SETI. But the crucial ingredient was Murray, who

[134] B. M. Oliver, and J. Billingham, *Project Cyclops: A Design Study of a System for Detecting Extraterrestrial Intelligence,* NASA CR 114445, (Washington, D.C., 1971, revised 1973). On the events at Ames leading to Cyclops see Steven J. Dick, interview with John Billingham, September 12, 1990.

[135] P. M. Morrison, J. Billingham, and J. Wolfe, *The Search for Extraterrestrial Intelligence,* NASA SP 419 (Washington, D.C., 1977).

as professor of planetary science at Caltech had participated in the April 1975 Morrison Workshop on Interstellar Communication dealing with planet detection. It was Murray who championed the idea of an all-sky survey component for SETI, to the skepticism of some members of the Ames group who pushed the more traditional targeted search. In the fourth workshop in December of that year, Billingham and Seeger presented a paper on "Ames–JPL Plans" for a detector system. By 1977 JPL too had a SETI office, headed by Robert Edelson. In the tradition of agency turf fights, SETI was the subject of conflict as well as cooperation, but what emerged was an Ames–JPL partnership that would become a major feature of NASA's formal SETI program.[136]

With the impetus of the Morrison workshops, NASA's attention turned to an actual program that might be funded. SETI would be a program significantly unlike most NASA endeavors. No spacecraft would be built, no launch risks encountered, no possibility of equipment failure in space. SETI was to be a ground-based program, and political and economic realities dictated that it would be no Cyclops, with a vast array of new equipment; the embryonic program would use existing radio telescopes to which would be attached specialized detectors, and it was the detectors that would be the main object of funding. The proposed total cost of the SETI program, including 5 years of research and development and 10 years of the operational phase, would be about $100 million, some 10 percent of the billion-dollar Viking project but roughly equal to the cost of Viking's biological experiments.

With the possibility of significant NASA funding on the horizon, in June 1979 NASA sponsored a major conference at Ames on "Life in the Universe." With the impetus provided by the Morrison workshops, NASA by this time had formally adopted a strategy for search – the bimodal strategy that not only made sense scientifically but (not incidentally) also satisfied the desire of both JPL and Ames to work on the project. Scientists at both Ames and JPL therefore authored the paper given at the 1979 conference, the first to lay out the NASA program in detail .[137] Terming it "a modest but wide ranging exploratory program,"

[136] The JPL program is described in R. E. Edelson, "At the Technological Frontier: The JPL Search for Extraterrestrial Intelligence," *Mercury*, 6, no. 4 (July–August 1977), 8–12; B. Murray, S. Gulkis and R. E. Edelson, "Extraterrestrial Intelligence: An Observational Approach," *Science*, 199 (February 3, 1978), 485–492. An overview from the Ames perspective is given in D. Black, J. Tarter, J. N. Cuzzi, M. Conners, and T. A. Clark, "Searching for Extraterrestrial Intelligence: The Ultimate Explanation," *Mercury*, 6, no. 4 (July–August 1977), 3–7.

[137] J. H. Wolfe, R. E. Edelson, J. Billingham, et al. "SETI – The Search for Extraterrestrial Intelligence: Plans and Rationale," in Billingham, *Life in the Universe*, NASA CP 2156 (Washington, D.C., 1981), also published by MIT Press (1981).

the authors described a 10-year effort "using existing radio telescopes and advanced electronic systems with the objective of trying to *detect* the presence of just one signal generated by another intelligent species, if such exists." (The emphasis on detection was significant; probably for political reasons, NASA was not prepared to communicate.) JPL would undertake Murray's all-sky survey, at wavelengths ranging from 1.2 to 10 GHz, while Ames would concentrate with more sensitivity on the "targeted search" among some 700 stars within 25 parsecs. This effort, according to the group, was made possible by a maturing radio technology, recent digital solid-state advances for the detectors, and "a minimum number of ad hoc assumptions." The authors characterized the concept of intelligent life as a "widely held hypothesis" in the scientific community, following from two further hypotheses: that life is a natural consequence of physical law acting in appropriate environments and that once a physical process has been found (as on Earth), it can be found elsewhere. A further assumption – which the authors called a matter of practicality – was that any intelligence would be "providing an electromagnetic signal we can recognize." They pointed out that numerous past searches were undertaken with comparatively primitive data processing equipment, and that more sophisticated and sensitive data processing systems were now needed. They contemplated a 10 millionfold increase in capability over the sum of all previous searches, and the equipment to accomplish this would be the focus of the R&D effort. Known as the "Multi-Channel Spectrum Analyzer (MCSA)," this detector – along with its software algorithms – was the heart of the system, the means by which the "cosmic haystack" could be searched for its hidden needle, a favorite metaphor employed by NASA. A graphical representation of the cosmic haystack in this article first dramatically depicted the magnitude of the task (Figure 8.6). By 1979, then, NASA had a detailed idea for a coherent SETI program but not much money to carry it out.

The story of NASA's trials and tribulations in obtaining this funding will be told in the next chapter as part of the question of discipline emergence. But the result was that beginning in the early 1980s, NASA's Ames and JPL centers embarked on an intensive program, known initially as "Microwave Observing Program (MOP)" and since 1992 as "High Resolution Microwave Survey (HRMS)" to build the instrumentation necessary for a systematic search for intelligent life. A 5-year R&D program costing about $1.5 million per year was carried out from 1983 to 1987. Following a period of uncertainty and minimal funding, in 1990 SETI took on the status of an approved NASA project, no longer in the R&D phase, and entered a 10-year phase of final development and opera-

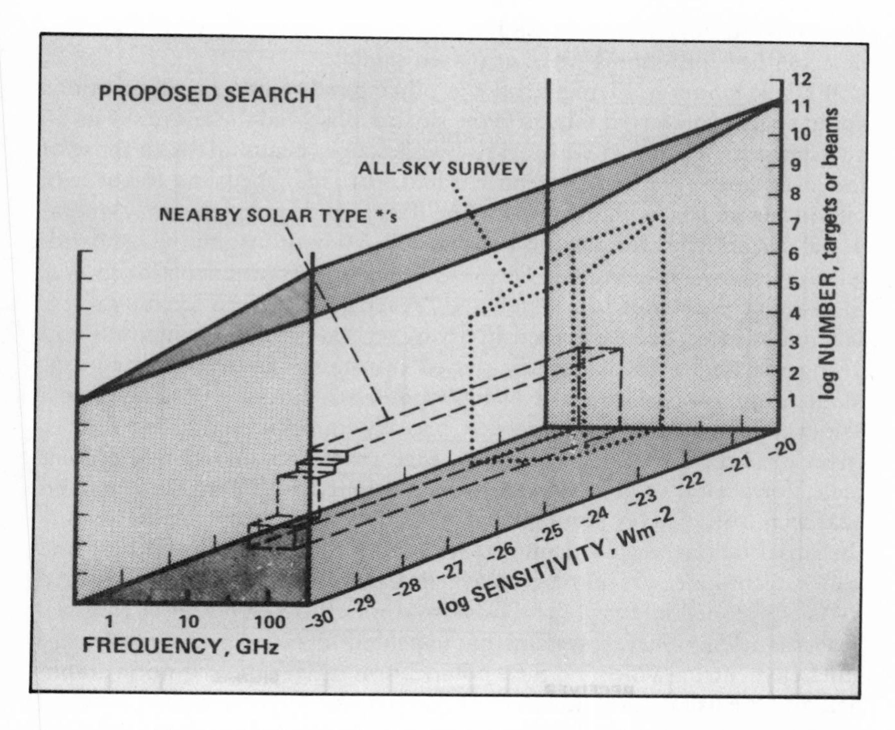

Figure 8.6. The cosmic haystack, showing the search space to be covered by the NASA/JPL Sky Survey and the NASA Ames Targeted Search. The Targeted Search was designed to have greater sensitivity, while the Sky Survey would observe in more directions and over a broader frequency range. Both were terminated in 1993, with the Targeted Search continued by Project Phoenix.

tions, to be completed by the new millennium at a total cost of about $108 million.

The final program that emerged when inaugurated in 1992 was quite similar to that envisioned in 1979 and well illustrates the complexity of detail required for a systematic SETI program. The Ames Targeted Search Element of the NASA SETI program was to search for 800–1000 solar-type stars within about 100 light years. Beginning with Arecibo, it would use the largest radio telescopes possible, observe each star for 300–1000 seconds, and focus on the 2 billion Hz in the 1- to 3-GHz region of the microwave spectrum. Because of practical limitations, it would process 20 MHz of bandwith at one time, necessitating that each star be observed 100 times to cover the entire 2 GHz. The six simultaneous channel resolutions would range from 1 to 28 Hz. The system would have the ability to detect either continuous-wave or pulsed signals.

JPL's Sky Survey Element, on the other hand, made no assumptions about specific preferred targets in the sky but observed the entire sky at 1–10 GHz with smaller (34-meter) radio telescopes beginning with those of the Deep Space Network. Because it had a broader spectrum to cover (9 billion Hz versus 2 billion Hz for the Targeted Search), the fully operational system was designed to process 320 MHz of instantaneous bandwidth, with 20 Hz channels. The Sky Survey's observational strategy was to examine each spot in a tessellated "racetrack" pattern for only a few seconds at most, decreasing sensitivity by several orders of magnitude and losing the ability to detect any pulsed transmissions over time periods longer than its observation at a single spot. Each mosaic would build up a "sky frame," and approximately 25,000 sky frames would be required to cover all directions and frequencies, each taking about 2 hours to complete, for a total of about 7 years for the complete survey. The Targeted Search and Sky Survey strategies were in many ways complementary; only the observations would demonstrate which assumptions were best and which technique was most effective in terms of successful detection.[138]

As envisioned in 1979, the hardware components of both the Targeted Search and Sky Survey systems implemented in 1992 consisted of three chief elements: a wideband dual polarization receiver, a low-noise amplifier, and the usual "downconverter" to intermediate frequencies; a digital spectrum analyzer to break the signal down into many channels; and a signal processor to search for the intelligent signals. The heart of the system and the key to its success was the digital spectrum analyzer. In 1979, it was envisioned that the spectrum analyzer would be constructed of modules that could be configured for each of the two search strategies.[139] In fact, as events developed, Ames and JPL developed separate spectrum analyzers, the MCSA at Ames and the Wide Band Spectrum

[138] The Ames Targeted Search system would achieve minimum sensitivities of 7×10^{24} m²/watt (and a maximum of 2×10^{26} m²/watt observing with the Arecibo telescope for 1000 seconds). For the JPL Sky Survey the sensitivity was 10^{23} m²/watt. For the latter system four 80-MHz modules were needed for each polarization (right and left) to achieve the 320-MHz bandwidth. This operational system was not planned for completion until 1996; the prototype system inaugurated in 1992 was capable of processing 20 MHz for each polarization. The technical details of the two systems implemented in 1992 have been described by the participants elsewhere; see, for example, Gary R. Coulter, Michael J. Klein, Peter R. Backus, and John D. Rummel, "Searching for Intelligent Life in the Universe: NASA's High Resolution Microwave Survey," *Advances in Space Biology and Medicine*, 4 (1994), 189–224.

[139] S. Gulkis, E. Olsen, and J. Tarter, "A Bimodal Search Strategy for SETI," in Papagiannis, *Strategies*, 93–105; J. Wolfe, R. E. Edelson, J. Billingham et al., in J. Billingham, *Life in the Universe* (Cambridge, Mass., 1981), 391–417.

Analyzer (WBSA) at JPL, each suited to the particular needs of its observing program.

Radio astronomy had never before attempted multichannel spectrometers at the scale needed for the SETI search. Standard spectrometers had been developed for a wide range of requirements, from 200-Hz resolution over a band of 40 KHz (for studies of the OH radical emission) or 20-kHz resolution over a band of 3 MHz (for extragalactic 21-cm studies), but nothing approaching the resolution and millions of channels needed for SETI. The key to the new spectrometer was the advance in digital technology. Work on the design of a 74,000-channel prototype MCSA with 0.5 Hz resolution had begun in 1977 at the Engineering College Laboratories at Stanford University headed by Allen Peterson and was built under the immediate supervision of Ivan Linscott.[140] Faced with the necessity to scale up this spectrum analyzer by more than 100-fold to produce more than 14 million 1-Hz channels, MCSA 2.0 replaced the wire-wrap technology by a customized very large integrated circuit (VLSI) chip built under contract to NASA Ames by the Silicon Engines Company.[141] It was the upgraded version of MCSA 2.0 that became operational at Arecibo on October 12, 1992.

Another crucial component of the SETI system was the method for extracting an extraterrestrial signal coming through the spectrum analyzer. While detection of signals from noisy data was a standard problem in communications, SETI was a particular challenge because nothing was known with certainty about the nature of an artificial extraterrestrial signal. The signal detection team at Ames, assuming that the signal would consist of narrowband carriers, single pulses, or pulse trains, designed the signal detection algorithms accordingly. Aside from detecting a continuous wave, the software algorithms were designed to search for pulses over the range of 45 milliseconds to 1.5 seconds . In addition, because the system had to to reject any terrestrial radio frequency interference (RFI), the RFI problem was studied extensively and the result implemented in the software algorithms. Finally, because millions of channels were to be analyzed in real time, great demands were placed

140 On previous spectrometers in radio astronomy see, "Spectrometers," in P. F. Goldsmith, *Instrumentation and Techniques for Radio Astronomy* (New York, 1988), and the accompanying papers. On the work at Stanford of Peterson and colleagues see A. M. Peterson, K. S. Chen, and I. Linscott, "The Multichannel Spectrum Analyzer," in Papagiannis, *The Search*, 373–383.

141 The basic task of MCSA 2.0 was to perform Fourier transforms extremely fast, providing six simultaneous frequency resolutions ranging from 1 to 32 Hz. See I. Linscott, J. Duluk, J. Burr, and A. Peterson, in G. Marx, ed., *Bioastronomy: The Next Steps* (Dordrecht, 1987), 319–335.

on the data acquisition system, which had to be specially designed for the project.[142]

As these events unfolded at Ames, parallel events took place at JPL. There Michael Klein (who had taken over from Edelson as head of the JPL SETI project in 1981) forged a collaboration with the Telecommunications and Data Acquisition Technology Development Office to use part of the Deep Space Network and to design and build an engineering development model (EDM) of their system, including the Wide Band Spectrum Analyzer (WBSA), the equivalent of Ames's MCSA. SETI drove the design of the spectrum analyzer, but the multimission users of the Deep Space Network would share in its use. The purpose of the WBSA was in general the same as that of the MCSA, but its architecture was tailored to the needs of the Sky Survey. Like the Targeted Search element, the Sky Survey has its own signal-processing and data acquisition problems to address.[143]

Administratively, SETI had gone from a few people within a division at Ames in 1976 to two project offices in two centers with a combined staff and subcontractors of about 65 in 1992. Its annual budget had risen from a few hundred thousand dollars in the early 1970s to over 10 million in the 1990s. Conceptually, its strategy had been honed and reduced to politically realistic proportions since the visionary Cyclops days. While at NASA headquarters the SETI program had spent most of its lifetime in the Life Sciences Division, in 1992 it was renamed by Congress the "High Resolution Microwave Survey (HRMS)," and it became the first element in the Solar System Exploration Division's Toward Other Planetary Systems (TOPS) program designed to detect other planetary systems.

On October 12, 1992, symbolically the quincentennial of Columbus's landfall in the New World, the NASA HRMS was inaugurated amid considerable fanfare. On that date, the 305-meter radio telescope at Arecibo began the Ames Targeted Search, while the 34-meter antenna at

[142] On pulse detection see D. K. Cullers in Papagiannis, *The Search,* 385–390, and Marx, *Bioastronomy,* 371–376. For studies of the RFI problem see E. T. Olsen, E. B. Jackson, and S. Gulkis, "RFI Surveys at Selected Radio Observatories," in J. Heidmann and M. Klein, *Bioastronomy: The Search for Extraterrestrial Life* (Berlin, 1991), 210–216, and J. W. Dreher, J. Tarter, and D. Wertheimer, "A 1 Hz Resolution RFI Survey: Preliminary Results," ibid., 217–227. On the data acquisition system see P. Backus, J. Jordan, and D. Harper, *Acta Astronautica,* 26 (1992), 169–192.

[143] The prototype system used in 1992 consisted of a pipelined fast Fourier transform architecture that transformed 40 MHz of bandwith into 20-Hz channels, for a total of 2 million channels. It could also be configured to analyze 1 million channels on each of two polarizations. On the WBSA see M. P. Quirk, M. F. Garyantes, H. C. Wilck, and M. J. Grimm, *IEEE Transactions on Acoustical Speech Signal Processing,* 36 (1988), 1854–1861. On JPL signal processing and data acquisition see E. Olsen in Papagiannis, *The Search,* 405–410.

the Venus station of the Deep Space Communications Complex at Goldstone in the Mohave Desert began the JPL All-Sky Survey. Although these observations were intended to mark only the beginning of an extended enterprise, hopes were soon dashed by Congress. After more than 15 years of planning, $60 million dollars of R&D, and less than 1 year of observing, the program was terminated as part of congressional budget cuts. SETI, a relatively small program by NASA standards, had often been singled out for scrutiny in the past by Congress, and this time Senator Richard Bryan of Nevada successfully led the opposition to what he called "The Great Martian Chase." "The Great Martian Chase may finally come to an end," Bryan said. "As of today, millions have been spent and we have yet to bag a single little green fellow. Not a single martian has said take me to your leader, and not a single flying saucer has applied for FAA approval," leaving the discerning reader to figure out that SETI had nothing to do with either Martians or flying saucers.

Almost exactly a quarter century after Billingham had stirred NASA's interest in SETI, it was unceremoniously excised from the U.S. government, cutting off not only the federal government's financial and intellectual sponsorship of the country's flagship SETI effort, but also denying the support NASA had given to other projects. A significant number of the project's personnel, however, joined the nonprofit SETI Institute, which had some success in raising private funding to carry on at least a scaled-back version of the original Targeted Search.[144] And the long-running programs sponsored by the Planetary Society, Ohio State University, and the University of California, Berkeley, not only continued, with even greater importance, but were upgraded and sporadically joined by new observations. Thus, although the demise of the NASA program was a severe blow to SETI advocates, and although the political hazards of government funding for SETI had been amply demonstrated, in the last decade of the century SETI remained well entrenched both as a topic for interdisciplinary scientific discussion and as an observational activity. Once again, SETI proponents demonstrated, their subject was too important to be derailed by mere politics.

[144] Richard Bryan, "Bryan Amendment Passes to Cut Expensive Search for 'Martians': Great Martian Chase to End?" Bryan Press Release, September 22, 1993; John Noble Wilford, "Ear to the Universe Is Plugged by Budget Cutters," *NYT,* October 7, 1993, p. B12; "Giggle Factor Helps Kill Project to Contact Aliens," *Washington Times,* October 10, 1993, p. D8; "Team Scrambles to Find New Funds for HRMS," *Space News,* 4 (October 18–24, 1993), 27. As Columbus Day 1993 approached, one *New York Times* columnist wrote of the SETI cancellation, "It was as though the Great Navigator, having barely sailed beyond the Canary Islands, was yanked home by Queen Isabella, who decided that, on second thought, she'd rather keep her jewels." George Johnson, *NYT,* October 10, 1993, p. E2.

It is always difficult to assess the importance of any endeavor at close range, let alone one as far-reaching as SETI, whose importance will depend to a large extent on its success. But comparison with the first half of the century may be instructive in placing modern SETI in historical perspective. One of the striking characteristics of the era of interplanetary communication is the speed with which scientists sought to use their embryonic technologies to resolve an age-old problem. Hertz had barely discovered radio waves before Tesla embraced interplanetary communication. Those who followed, including Marconi, Todd, and Menzel, were all pioneers in their fields, unafraid of venting controversial ideas. All were well grounded in technical interests and employed those interests in attempting to confirm an idea long the subject of mere discussion. With time the technical concerns increased in sophistication, beginning with unexplained signals and ending with concerns about optimal wavelengths, the effects of the Earth's atmosphere, and power requirements. Moreover, prospective interplanetary communicators anticipated not only technical problems, but also philosophical, linguistic, and cultural factors. Although they faced ridicule from colleagues, they were driven by a faith that communication with extraterrestrial intelligence was not only possible but desirable, holding the potential for bringing new knowledge and ancient wisdom.

All of these traits were mirrored to some extent in the era of interstellar communication. Radio telescope technology was barely able to detect artificial signals from the nearest stars similar in strength to those generated on Earth when Cocconi and Morrison proposed interstellar communication and Drake first attempted it. Modern SETI pioneers were well-trained scientists grappling with the opening of the electromagnetic spectrum, with all the possibilities that implied. They too devised observing strategies, used every enhancement in technology to further their goal, were driven by the potential of their discovery, and had a strong interest in the implications of success.

Unlike their interplanetary predecessors, however, modern SETI pioneers formed networks; held meetings at the national and international levels; sought consensus on observing strategy; and, above all, secured funds, telescope time, and personnel on scales large and small in order to carry out a variety of observing programs. Pushed forward by the success of their rapidly developing radio telescope technology, they achieved heights never reached in the interplanetary era, including recognition from their scientific peers in the form of journal publications and support from the government in the form of funding. With entry into the arena of science policy, however, they also faced political ridicule unknown in the earlier era and acute disappointment when funds were withdrawn for reasons that seemed to them frivolous.

Modern SETI research continued, and even thrived, despite a serious crisis in rationale that undermined the very basis of its observing programs. The challenge to SETI optimists around 1975 marked an important cautionary turning point in its fortunes, which for a time seemed boundless. It was then that the interstellar colonization question implicit in the Fermi paradox began to take root, and it was then that the shift took place from USSR- to U.S.-dominated SETI searches, even as NASA began serious deliberations about its program. There is no doubt that concerns about the Fermi paradox, which certainly affected Shklovskii, dampened enthusiasm in the Soviet Union for further SETI searches. In the end, however, the Fermi paradox and all the talk about alternative approaches had little effect on the strategy adopted. NASA's program and other ongoing programs would not search for Type II or III Kardashev or Dyson civilizations, or search for laser communication with civilizations more similar to ours, or (although more directly related to NASA's mission) institute any plans for interstellar travel. In a neat closure to the events of Cocconi, Morrison, and Drake three decades before, NASA and most other groups would undertake a search for Type I civilizations in an expanded region of the electromagnetic spectrum that included the 21-cm line, similar in principle to the earliest thoughts about the opening of the spectrum for SETI, even if now much more powerful. Although there was renewed debate about the merits of optical SETI toward the end of the century, the 21-cm choice "still appeals to most of the radio astronomers who have attended to the problem," Morrison wrote in 1985. "Plainly this is no objective matter; one can only argue plausibility, and then depend upon hope."[145]

Although NASA and other groups hedged their bets by including a much broader part of the spectrum for their SETI programs, only observation would demonstrate the success of this strategy. Success was by no means guaranteed, even as, in the wake of the downfall of the NASA program, scientific curiosity continued to carry forward this most modern attempt to answer an age-old question. The potential for misinterpretation and failure remained high. For all their imaginative insight, it is well to remember that Tesla, Marconi, and others were mistaken in their interpretation of radio signals as artificial and extraterrestrial, a pattern

[145] Morrison, in Papagiannis, *The Search,* 16. Morrison points out that even the so-called 21-cm line is 1 MHz wide as a consequence of the Doppler shifts of the hydrogen gases moving in space. This could be narrowed to 100 kHz by using as the rest frame the thermal background radiation. Optical SETI arguments were reviewed in Stuart A. Kingsley, ed., *The Search for Extraterrestrial Intelligence (SETI) in the Optical Spectrum* (Bellingham, Wash., 1993), especially in Kingsley's "The Search for Extraterrestrial Intelligence (SETI) in the Optical Spectrum: A Review," 75–113 and references therein.

that would-be interstellar communicators hoped not to repeat. The false alarms of Ozma, CTA 102, and pulsars injected a cautionary note into the SETI culture, even as Ohio State's WOW signal and the unidentified transient signals of META and other programs lured investigators on.

The verdict on modern SETI therefore remained uncertain at century's end. If we consider the era of interplanetary communication to be bracketed by Tesla in 1901 and Menzel in 1937 – encompassing the transition from belief in the reality and possibility of such communication to its discussion in only hypothetical terms – its total life span was about 35 years. The era of interstellar communication, begun with Cocconi and Morrison in 1959, reached just that age before the flagship NASA SETI program was terminated in 1993. Yet along with the remnants of NASA SETI under new sponsors, other programs remained, with the entire universe as the search domain rather than our own parochial solar system. Whether lack of detections, or lack of political will, or an increasingly narrow focus on practical projects with immediate benefits, or flagging human curiosity would result in a limited life span for modern SETI, and whether a century from now it would be seen as only a curious episode in the history of science like interplanetary communication, only the future will tell. Either way, both eras will take their place as the twentieth century's effort in the venerable tradition of seeking humanity's place in the universe.

On one thing everyone agreed: if, by judicious rationale and strategy or by good fortune, SETI programs defied all the odds and resulted in the discovery of extraterrestrial intelligence, the effect on humanity would be profound. The human implications of that discovery had been a favorite subject of science fiction for decades, and the contemplation of the problem quickly led to philosophical and religious questions about purpose in the universe. It is those issues that we shall address in Chapter 10.

9

THE CONVERGENCE OF DISCIPLINES: BIRTH OF A NEW SCIENCE

The last two decades have witnessed a synthesis of ideas and discoveries from previously distinct disciplines. Out of this synthesis have grown new fields of research. Thus we now have exobiology, which represents a synthesis of discoveries in astronomy, atmospheric physics, geophysics, geochemistry, chemical evolution, and biochemistry.

Bernard M. Oliver (1973)[1]

Let us face the fact that this is a gamble at the most adverse odds in history. Then if we want to go on gambling, we will at least recognize that what we are doing resembles a wild spree more than a sober scientific program.

George Gaylord Simpson (1964)[2]

The Great Martian Chase may finally come to an end. As of today, millions have been spent and we have yet to bag a single little green fellow.

Senator Richard Bryan (1993)[3]

Thus far in this study, we have treated the major threads of the extraterrestrial life debate as more or less separate entities. This is not without justification, since for much of the century planetary science, planetary systems science, origin of life studies, and SETI remained largely separate research programs, undertaken by different groups of scientists, even if in the second half of the century SETI played a somewhat unifying role via its keen interest in the results from the other three fields. In terms of technique, research programs, and even goals, aside from the shared general culture of astronomy, the planetary spectroscopy of Kuiper and Sinton had little in common with Van de Kamp's astrometric studies of stellar motions or Drake's radio astronomy, while all three (with the rare exception of a polymath like Sagan) were far removed from the biochemists in their laboratories studying the origins of life. And certainly most members of all these groups disavowed the extraterrestrial hypothesis of UFOs, which if anything served to retard

1 Bernard M. Oliver, *Project Cyclops* (Moffett Field, Calif., 1973), 3.

2 George Gaylord Simpson, "The Non-Prevalence of Humanoids," *Science,* 143 (February 21, 1964), 769, reprinted in D. Goldsmith, *The Quest for Extraterrestrial Life* (Mill Valley, Calif., 1980), 214–221, and in Simpson's *This View of Life* (New York, 1964), 253–271: 270.

3 Press release, September 22, 1993, after the vote led by Bryan in the U.S. Senate to terminate the NASA SETI project; *Space News,* October 18–24, p. 27.

more mainstream scientific study of extraterrestrial life because of guilt by association.

All of these studies, however, clearly related to the larger question of life beyond the Earth. Setting UFOs aside (as scientists in these other fields deliberately did when carrying out their own work), if we now survey the results of our four "hard science" chapters (3, 4, 7, and 8), we are at once led to the conclusion that during the decade 1953–1963, all four areas were profoundly transformed in relation to the extraterrestrial life debate. During this period, not only did planetary science produce the ground-based claims of Sinton regarding vegetation on Mars (later proven spurious) but, more important, the space program made practical the biological exploration of the planets. Planetary systems science remained an infant study but produced its first modern claim of extrasolar planets in 1963. Origin of life studies began their modern era with the Urey–Miller experiments in 1953. And Morrison and Cocconi made their theoretical suggestions for SETI in 1959, quickly followed by Drake's observations in 1960.

All of these studies followed their own trajectories but, at the same time, became inextricably intertwined as parts of a larger scientific problem. Extraterrestrial life studies – termed "exobiology" by most biologists and "bioastronomy" by most astronomers – arose from the convergence of these four disciplines, each with its own history in terms of scientific content and discipline formation and each still quite young as a discipline. Though planetary science is the best developed of the four, planetary systems science, for example, has coalesced only in the last 15 years: astronomer Tom Gehrels wrote in the first volume of *Protostars and Planets* in 1978 that "with this book we hope to stimulate a new discipline" involving star and planet formation. By the second meeting of scientists interested in planetary systems in 1984, David Black declared that "the elements of Gehrels' 'new discipline' are now beginning to emerge and their interrelationships are being defined," and he named the new field "planetary systems science."[4] Because the study of extraterrestrial life is both a subset of each of these research areas and a combination of them, its formation as a discipline in its own right must be understood, at least in part, as a convergence of other disciplines.[5]

[4] Tom Gehrels, ed., *Protostars and Planets: Studies of Star Formation and of the Origin of the Solar System* (Tucson, Ariz., 1978), 3; David C. Black and Mildred Shapley Matthews, eds., *Protostars and Planets II* (Tucson, Ariz., 1985), Preface, xviii. Aspects of the growth of planetary science have recently been examined by Ronald E. Doel, *Science on the Periphery: Solar System Astronomy in America, 1920–1960* (New York, 1996), and J. N. Tatarewitz, *Space Technology and Planetary Astronomy* (Bloomington, Ind., 1990).

[5] See also E. Ezell and L. Ezell, "The Rise of Exobiology as a Discipline," in *On Mars* (Washington, D.C., 1984), 54–74. In my view, it is counterproductive to argue at length

In this chapter we examine, from a social and institutional point of view, questions that have remained largely unfocused during our study of the intellectual course of the debate during the twentieth century. Is exobiology (or bioastronomy) a new discipline, and if so, what are its characteristics? How did individuals from disparate fields cross disciplinary boundaries to work together on a problem of common interest? And how did they find patrons to support their work? Such questions are of more than passing interest; many social historians would claim that they are of primary interest. Whatever the answer for the broader history and sociology of science, it is certain that for the extraterrestrial life debate these issues loomed particularly large in the latter half of the century, as scientists were negotiating a new discipline on which their careers and reputations depended. It was one thing for British Astronomer Royal Harold Spencer Jones to write a book on a subject of popular interest in 1940;[6] it was quite another matter for scientists in government agencies to push for public funding of projects typically considered by their fellow scientists to belong in the realm of science fiction. Yet the latter is exactly what happened in the second half of the twentieth century. In this chapter we examine some of the major trends in the emergence of a new interdisciplinary science, laying the groundwork for a very large topic that requires further study.[7]

9.1 PERCEPTIONS OF A NEW DISCIPLINE

Proposed criteria for discipline emergence aside, one point is clear: participants in exobiology themselves frequently and vigorously proclaimed their research efforts to be part of a new discipline. In debating when the

whether extraterrestrial life studies constitute a "research specialty," a "field," a "discipline," a "subdiscipline," or a "science." While there might be some advantage in precision and scale if one could distinguish among these terms, any attempt to do so – or to decide when a "field" becomes a "discipline," for example – inevitably runs into grave problems and artificially imposed boundaries. The point being examined here is that over the last three decades, a qualitative change has occurred in studies dealing with life beyond the Earth, whether one applies any of these terms – or calls this activity "astrobiology," "cosmobiology," "exobiology," or "bioastronomy."

[6] *Life on Other Worlds* (New York, 1940).

[7] The birth of scientific disciplines has been a subject of considerable interest to historians and sociologists of science. In astronomy, David Edge and Michael Mulkay's book *Astronomy Transformed* (New York, 1976) examines the emergence of radio astronomy in Great Britain. See also Robert E. Kohler, *From Medical Chemistry to Biochemistry* (Cambridge, 1982), and John W. Servos, *Physical Chemistry from Ostwald to Pauling: The Making of a Science in America* (Princeton, N.J., 1990), among others. While a comparative study of discipline emergence is beyond the scope of this chapter, clearly many similarities and differences exist. Our study differs qualitatatively from most others in that, in the case of exobiology, some scientists questioned whether the subject matter was science, much less a scientific discipline.

new field of study emerged, one can quickly narrow the possibilities, if not pinpoint the exact date. No one would claim that a field of extraterrestrial life studies existed in the first half of the twentieth century. Even the Lowellian uproar was relatively restricted in terms of the numbers of scientists doing original research on the problem; moreover, the techniques applicable to the problem of the canals of Mars were very limited. By 1955, when Otto Struve pondered the use of the word "astrobiology" to apply to the broad study of life beyond the Earth, the concept of a new discipline could not be dismissed as completely out of hand. In the end, however, he decided against such a status because "the time is probably not yet ripe to recognize such a completely new discipline within the framework of astronomy. The basic facts of the origin of life on earth are still vague and uncertain; and our knowledge of the physical conditions on Venus and Mars is insufficient to give us a reliable background for answering the question: Does life exist on these planets, or has it existed in the distant past?"[8]

The dawn of the Space Age rapidly changed attitudes toward the problem. The imminent or actual birth of a new science was soon proclaimed, at first by its practitioners and eventually – sometimes grudgingly – by larger numbers of their peers. The imminent birth of "exobiology" was palpable in 1960, when Joshua Lederberg coined the term and set forth an ambitious but practical agenda.[9] Reviewing the questions raised by the possibility of life beyond the Earth, Gregg Mamikunian wrote in the Preface to the Proceedings of the 1963 JPL conference on *Current Aspects of Exobiology,* "The answers to all of these questions must await the development of exobiology or cosmobiology to the level of a scientific discipline." He emphasized that exobiology was already creating "specific methods of investigation which are fundamentally different from routine laboratory or field biology methods," leaving little doubt that exobiology, in his view, was well on its way to discipline status in 1963.[10] Nor was this attitude exclusively an American phenomenon. At the opening of the first Soviet conference on extraterrestrial civilizations, held at the Byurakan Astrophysical Observatory the following year, the astrophysicist Josif Shklovskii declared, "we are witnessing the inception of a new science, which occupies a boundary position between astrophysics, biology, engineering, and even sociology. This new science remains un-

[8] Otto Struve, "Life on Other Worlds," *S&T,* 14 (February 1955), 137–146.

[9] Joshua Lederberg, "Exobiology: Experimental Approaches to Life beyond the Earth," in *Science in Space,* eds. Lloyd V. Berkner and Hugh Odishaw (New York, 1961), 407–425. The term "exobiology" was distinguished from "esobiology," or "Earth's own" biology, in Joshua Lederberg, "Signs of Life: The Criterion System of Exobiology," in Pittendrigh et al. (footnote 24), 127–140.

[10] G. Mamikunian and M. H. Briggs, *Current Aspects of Exobiology* (Oxford, 1965), ix.

named for the time being, but it has already attracted the keenest attention of both specialists and laymen."[11]

One sees such statements repeated again and again by participants during the 1960s and 1970s, including Oliver's statement at the beginning of this chapter. By 1979, NASA's SETI chief, John Billingham – with a certain amount of vested interest, to be sure – wrote that "over the past twenty years, there has emerged a new direction in science, that of the study of life outside the Earth, or exobiology. Stimulated by the advent of space programs, this fledgling science has now evolved to a stage of reasonable maturity and respectability."[12] Since $100 million had been spent on the Viking biology experiments, experiments that failed to demonstrate unambiguously any life on Mars, "maturity and respectability" for the field were a precondition for further support and funding from NASA. As another sign of respectability, in the same year events were set in motion leading to the the formation of a commission on "bioastronomy" in the International Astronomical Union, a status usually reserved for fields accepted as of legitimate astronomical interest to a large number of members.[13]

There is no doubt, therefore, that practitioners of extraterrestrial life studies often saw themselves as participants in a new science. There is equally no doubt that these claims were challenged for a variety of reasons. Evolutionary biologist George Gaylord Simpson's 1964 characterization of exobiology as a "science" that "has yet to demonstrate its subject matter exists" has echoed throughout the century by critics calling exobiology "a science looking for a subject."[14] Physicist Frank Tipler's attack on the assumptions behind SETI was accompanied by harsh statements about the scientific status of the subject and its institutionalization. Arguing that "no experiment will ever convince the ETI believers that we are alone," Tipler concluded, "Bioastronomy resembles nothing so much as parapsychology. A century of negative results

[11] I. S. Shklovskii, "Multiplicity of Inhabited Worlds and the Problem of Interstellar Communications," in *Extraterrestrial Civilizations,* Proceedings of the First All-Union Conference on Extraterrestrial Civilizations and Interstellar Communication, Byurakan, May 20–23, 1964 (Erevan, 1965; English translation, Jerusalem, 1967), 5.

[12] J. Billingham, *Life in the Universe* (Cambridge, Mass., 1981), ix.

[13] Astronomer Michael Papagiannis was instrumental in this process. "Commission 51 – Search for Extraterrestrial Life" was approved at the International Astronomical Union's 18th General Assembly in Patras, Greece, in 1982. See Papagiannis, "Search for Extraterrestrial Life – A New Commission of the International Astronomical Union," *JBIS,* 36 (July 1983), 305; and Papagiannis, "Activities and Resolutions of IAU Commission 51," in Papagiannis, ed., *The Search for Extraterrestrial Life: Recent Developments* (Dordrecht, 1985), 553–556. The Commission later included the name "Bioastronomy" in its title.

[14] Simpson, "Non-Prevalence of Humanoids." The quotation marks around "science," of course, indicate that Simpson used the term advisedly in this context.

has not diminished the field: More money is now being spent on ESP research than ever before." He believed that bioastronomy was a "pseudoscience" that should not be admitted into the formal commission structure of the International Astronomical Union.[15] Such attacks, undoubtedly characteristic in some degree of all new disciplines (even if emphasized in this case by philosophical and cultural issues), were firmly rooted in the competition for scarce resources. Simpson, writing as the search for life on Mars was about to go into high gear with large budgets, was explicit on this point. Emphasizing that the prospective discovery of extraterrestrial life was advanced as one of the chief reasons, or "excuses," for the billions of dollars spent on the space program, he called it "a gamble at the most adverse odds in history." We know that we have life on Earth to study, he argued, pleading "that we invest just a bit more of our money and manpower, say one-tenth of that now being gambled on the expanding space program, for this sure profit." Simpson clearly viewed the potentially new and expensive exobiology discipline as placing at risk the health of the terrestrial-bound research of himself and his colleagues.

And yet, it is a matter of history that a significant number of scientists supported the opposite view, represented by geneticist and Nobel laureate Joshua Lederberg's statement that "Exobiology is no more fantastic than is the realization of space travel itself, and we have a grave responsibility to explore its implications for science and for human welfare with our best scientific insights and knowledge."[16] Whether or not they created a discipline may be viewed as a matter of semantics; it may also be viewed as a requirement for significant funding. How exobiology aspired to discipline status through networks and institutions is the subject of the remainder this chapter.

9.2 NETWORKS: FORMATION OF THE SCIENTIFIC COMMUNITY

We have seen in previous chapters that many factors caused individuals to turn the course of their careers toward exobiology – a field that offered potentially high rewards and recognition, but at considerable risk of failure and even derision. Carl Sagan (Figure 9.1), a young scientist just beginning his career, was strongly influenced by science fiction and aided in his entry into the field by the interest and encouragement of well-established mentors, including the geneticist H. J. Muller, the geochemist

[15] Frank J. Tipler, review of M. D. Papagiannis, *The Search for Extraterrestrial Life: Recent Developments*, *Physics Today* (December 1987), 92.

[16] Lederberg, "Exobiology," 407–425: 424.

Figure 9.1. Carl Sagan shown in his earliest days as an Assistant Professor of Astronmy at Harvard, circa 1962. Sagan and Joshua Lederberg (Figure 9.2) are two pioneers in exobiology whose thinking had a synergistic effect on each other.

Harold Urey, and the astronomer G. P. Kuiper. By contrast, Joshua Lederberg (Figure 9.2) was a young but established scientist about to receive the Nobel Prize at the time his interest in exobiology was sparked by the Space Age. When the two met in 1957, a synergism developed that helped spark the growth of exobiology. The origin of SETI shows a similar pattern; Frank Drake could not have carried out Project Ozma without the approval and encouragement of Otto Struve and Lloyd Berkner, elder statesmen of science. In the Soviet Union, too, the well-known radio astronomer I. S. Shklovskii mentored many others interested in the subject. As the Viking program and the NASA SETI programs took shape, many career decisions were made that weighed the risks and rewards for each individual.[17]

Individuals, however, could not themselves constitute a new discipline. As in other fledgling scientific disciplines, those with exobiological inter-

[17] A great deal of biographical material, including oral history interviews, is now available for further study of how individuals entered the field of exobiology.

Figure 9.2. Joshua Lederberg (about 1962) in his laboratory at Stanford.

ests sought a more formal and broader network to share ideas, first on a national and then on an international basis. In practice, this expansion of purpose was carried out in at least three ways: by goal-oriented committees driven by potential or funded programs; by open conferences wherein participants could discuss a broad array of topics unconstrained by programmatic concerns; and by publications which, through their wide distribution, had the potential to provide the strongest influence of all. Without these essential modes of debate and information sharing, no individual – however willing to risk his or her own career – could define a new field of study, and no field of study could aspire to discipline status.

Committees, Conferences, and Commissions

Prior to the Space Age, no network existed for those interested in extraterrestrial life studies, whether in planetary exobiology or SETI. Only with the formation in the late 1950s of the "Panel 2 on Extraterrestrial Life" of the Joint Armed Forces–National Research Council Committee on Bio-Astronautics, and the Space Science Board of the National Academy of Sciences's Panel on Extraterrestrial Life, did small groups of scientists first gather for discussion on planetary exobiology. The first group, chaired by Melvin Calvin at Berkeley and including Carl Sagan, consisted of only eight scientists, who had to contend with military officers and colleagues whose interest was more in biology as applied to spaceflight than in extraterrestrial microbes or intelligence.[18] The Space Science Board panel consisted of West Coast (WESTEX) and East Coast (EASTEX) groups, chaired respectively by Lederberg and Salvador Luria (both appointed by Lloyd Berkner, chairman of the Space Science Board). The membership of the committees, Lederberg recalled, was decided by the chairman on the basis of whoever "would have or should have" had an interest in the subject. Some had to be persuaded that they should be interested, a job made easier by the reputation of the chairmen; Lederberg had just received the Nobel Prize in 1958, and Luria was well known in his field. The WESTEX and EASTEX committees were the first vehicles whereby biologists could relate to NASA. Out of these groups emerged substantial recommendations and some of the leaders in planetary exobiology.[19] As we have seen in Chapter 3, their work was carried on by subsequent committees, including the NASA Bioscience Advisory Committee and the Space Science Board's 1962 Summer Study, both of which endorsed the search for extraterrestrial

[18] S. J. Dick interview with Sagan, January 6, 1993, 11. Panel 2 consisted of Melvin Calvin (chair), Matthew Messelson, Malcolm Ross, Carl Sagan, Wolf Vishniac, Harold F. Weaver, Richard E. Lord, and Henry Linschitz. Panel 2 was dissolved in August 1960 by order of National Academy President D. W. Bronk, and the dissolution was ratified by the Executive Committee of the Joint Armed Forces Committee at its August 10–11 meeting. The reason for the dissolution was so that the National Academy of Sciences would have a single representative in bioastronautics, the Space Science Board. The activities of Panel 2 were taken over by "Committee 14 on Exobiology," chaired by Lederberg. Calvin to R. W. Davies and members of Panel 2, August 12, 1960, Lederberg personal files.

[19] S. J. Dick interview with Lederberg, November 12, 1992, 10; The West Coast group consisted of Melvin Calvin, Richard Davies, Norman Horowitz, Joshua Lederberg, A. G. Marr, Daniel Mazia, Aaron Novick, Carl Sagan, William Sinton, Roger Stanier, Gunther Stent, C. S. van Niel, and Harold F. Weaver. The East Coast group consisted of Dean Cowie, Richard Davies, George Derbyshire, Paul Doty, Thomas Gold, W. R. Sistron, Fred Whipple, H. Keffer Hartline, Martin Kamen, Cyrus Levinthal, Bruno Rossi, S. Luria, E. F. Mac Nichol, Stanley Miller, John W. Townsend, Bruce Billings, Herbert Freeman, and Richard S. Young. EASTEX held its first meeting December 19–20, 1958, at MIT and produced a report circulated as SSB-93; WESTEX met first on February 21, 1959, followed by meetings on March 21, May 2–3, and September 26, and produced a "WESTEX Summary Report" in October 1959. Lederberg Files. These two committees were merged into Committee 14 on Exobiology in August 1960, when Panel 2 was dissolved (see footnote 18).

life, the latter setting it as "the prime goal of space biology."[20] These events, in turn, led to the Viking missions to Mars.

Like planetary exobiology, interstellar communication had its threshold to cross. Drake recalled that before Project Ozma in 1960 no network existed, but with the publication of Morrison and Cocconi's paper and the establishment of Ozma, "correspondence started to flow. People knew who they could write to to find out who was interested."[21] This led to the first meeting on interstellar communication – the legendary Green Bank meeting in 1961, at which the Drake Equation was formulated. Unlike its planetary exobiology counterpart, the Green Bank meeting was not inspired by the advent of spacecraft, but rather by the advancement in radio telescope technology, which happened to coincide with the beginning of the Space Age. "Essentially the whole network," Drake recalled, was formed by those who attended that meeting. "It wasn't limited to eleven because that was the maximum number we had room for; it was merely that that was everybody we knew of who was seriously interested in it."[22]

The counterpart meeting in the Soviet Union, inspired in part by the American developments, was hosted by the distinguished astrophysicist V. A. Ambartsumian at the Byurakan Astrophysical Observatory in Armenia in 1964–1965. Unlike the Americans, the Soviets called for a plan of action, including a scientific agenda, conference proceedings, and further meetings, and constituted a Committee on Interstellar Communications drawn from the Astronomical Council and the Council of Radio Astronomy of the USSR Academy of Sciences.[23] While the Americans conceived of the Green Bank meeting as a somewhat hushed event from which no published Proceedings emerged, the Russians, by recording their discussions and consolidating their network, seemed less inhibited about spreading the word of their new subject and more serious about moving forward in a systematic manner.

[20] See Chapter 3, footnotes 178–180. The Space Science Board's Working Group on Biology consisted of 26 members, including Allan H. Brown (chair), C. S. Pittendrigh (vice-chair), S. W. Fox, N. Horowitz, and 5 NASA representatives (W. Haymaker from Ames, E. Konecci, H. Newell, O. Reynolds, and J. Soffen).

[21] David Swift, *SETI Pioneers: Scientists Talk about Their Search for Extraterrestrial Intelligence* (Tucson, Ariz., 1990), 60.

[22] Ibid., 60. The attendees were Otto Struve (chair), D. W. Atchley, Jr., M. Calvin, G. Cocconi, F. Drake, S. S. Huang, J. C. Lilly, P. M. Morrison, B. M. Oliver, C. Sagan, and J. P. T. Pearman. There was therefore some crossover between the planetary exobiology and SETI groups in the presence of Sagan and Calvin. That this included everyone interested is a slight exaggeration; Bracewell, Dyson, and Miller were also interested but not present.

[23] G. M. Tovmasyan, ed., *Extraterrestrial Civilizations: Proceedings of the First All-Union Conference on Extraterrestrial Civilizations and Interstellar Communication, Byurakan, 20–23 May, 1964* (Erevan, 1965; English translation, Jerusalem, 1967).

As interest in life beyond the Earth grew, so did the meetings, which increased in frequency, size, and scope as the aspiring discipline developed (Table 9.1). Unlike interstellar communication, planetary biology by the early 1960s had the prospect of substantial funding as the space program gave impetus to the subject in a way that it could not for its interstellar companion. These more broadly based discussions on exobiology began at JPL in 1963 and were followed up in numerous meetings, the most seminal of which were the 1964–1965 meetings on "Biology and the Exploration of Mars," sponsored by the Space Science Board of the National Academy of Sciences and chaired by Lederberg and Princeton biologist Colin Pittendrigh. Participants at both meetings contemplated a new discipline; asked 30 years later when the discipline congealed, Lederberg pointed to this meeting and its proceedings:

> I would say by that time there was enough variety of content, calling on information from a variety of places, that anyone writing on the question would have to invoke the body of knowledge that was represented in that book, not necessarily those identical papers, but they would be reflections of it. So at that point there was a common ground of what was agreed upon and what wasn't. I think that constitutes a discipline.[24]

And while pointing to developments in molecular biology and experiments on the synthesis of organic molecules in the 1950s, Chairman Pittendrigh wrote in the Preface of the study that "the real transformation that the subject has undergone stems from the spectacular growth of space technology in the last decade. The possibility of life's origin and occurrence on planets other than ours is no longer limited to idle speculation: it has entered the realm of the testable, of science in the strict sense."[25] The 29 technical papers, 36 members of the working group (representing evolutionary biology, genetics, microbiology, biochemistry, molecular biology, animal physiology, soil chemistry, organic chemistry, planetary astronomy, geochemistry, and theoretical physics), 30 consultants, and a "select" Bibliography of 2000 references lend credence to the claim that exobiology was certainly not an empty set for discussion, even

[24] S. J. Dick, interview with Lederberg, 15. The two volumes were Colin S. Pittendrigh, Wolf Vishniac, and J. P. T. Pearman, eds., *Biology and the Exploration of Mars* (Washington, D.C., 1966), and Elie A. Shneour and E. A. Ottesen (compilers), *Extraterrestrial Life: An Anthology and Bibliography* (Washington, D.C., 1966). It is notable that J. P. T. Pearman played a key role in organizing the meeting, as he had the Green Bank meeting on SETI.

[25] Pittendrigh et al., *Biology,* Preface, vii. For whatever reasons, Pittendrigh, who had been vice-chair of the 1962 Space Science Board Summer Study, is an example of another species of the exobiology community: one who facilitated exobiology through his leadership in such meetings but who produced no original papers on the subject himself.

Table 9.1. *Selected conferences on extraterrestrial life*

Date	Sponsor	Location	Proceedings
November 1961	U. S. National Academy of Sciences	NRAO Green Bank	Cameron, *Interstellar Communication* (1963)
Feb. 26-28 1963	Jet Propulsion Laboratory	Pasadena, Calif.	Mamikunian and Briggs, *Current Aspects of Exobiology* (1965)
May 1964	Armenian Academy of Sciences	Byurakan Astrophysical Observatory	Tovmasyan, *Extraterrestrial Civilizations* (1965, English trans., 1967)
1964-65	U.S. National Academy of Sciences (Space Science Board)	Stanford, Rockefeller Institute	Pittendrigh, Vishniac, and Pearman, *Biology and the Exploration of Mars* (1966)
1970	NASA	Ames Research Center	Ponnamperuma and Cameron, *Interstellar Communication* (1974)
1971	Stanford/NASA	Stanford	*Project Cyclops* (1971)
1971	U.S. and USSR[a] Academy of Sciences	Byurakan	Sagan, *Communication with Extraterrestrial Intelligence* (1973)
1974	American Anthropological Association		Maruyama, *Cultures Beyond the Earth* (1975)
1975-76	NASA/Ames	Ames, Caltech, Arecibo, Goddard	Morrison, Billingham, and Wolfe, *The Search for Extraterrestrial Intelligence* (1977)
1976	NASA/ASEE	Stanford	Black, *Project Orion* (1980)
1975-77	International Academy of Astronautics	Lisbon, California	Billingham and Pesek, *Communication with Extraterrestrial Intelligence* (1979)
June 1979	NASA/Ames	Ames Research Center	Billingham, *Life in the Universe* (1980)
August 1979	International Astronomical Union[b]	Montreal	Papagiannis, *Strategies in the Search for Life in the Universe* (1980)

Table 9.1 (*cont.*)

November 1980		University of Maryland	Zuckerman, *Extraterrestrials: Where Are They?* (1981)
December 1981	Soviet Academy[a]	Tallin, Estonia	None published
June 18-21 1984	International Astronomical Union[b]	Boston, Mass.	Papagiannis, *The Search for Extraterrestrial Life: Recent Developments* (1985)
October 1984	International Academy of Astronautics	Lausanne, Switzerland	
June 22-27 1987	International Astronomical Union[b]	Balaton, Hungary	Marx, *Bioastronomy: The Next Steps* (1988)
June 18-23 1990	Third International Bioastronomy Symposium[b]	Val Cenis, France	Heidman and Klein, *Bioastronomy: The Exploration Broadens* (1991)
August 5-9 1991	University of California[a] SETI Institute	University of California, Santa Cruz	Shostak, *Third Decennial U.S.-USSR Conference on SETI* (1993)
1991-92	NASA/ SETI Institute	Santa Cruz	Billingham et al. *Social Implications of Detecting an Extraterrestrial Civilization* (1994)
August 1993	University of California[b] SETI Institute ISSOL	University of California Santa Cruz	Shostak, *Progress in the Search for Extraterrestrial Life* (1995)

[a] U. S.-U.S.S.R. Decennial SETI Conference series.

[b] Triennial Bioastronomy Conference series, organized by the International Astronomical Union's Commision 51 (Bioastronomy), and others

if its content was still speculative. Even though the search for life on Mars would still see many ups and downs, with this meeting the network was created and the foundations for the subject well established.

Meanwhile, although SETI had no such spectacular growth in its activities during the 1960s or even the 1970s, it grew at a measured pace. Indeed, in 1971 the subject of interstellar communication went international in a way that planetary exobiology never did, with a meeting at Byurakan that Sagan called "a turning point" in acceptance of SETI.[26] It was sponsored

[26] Sagan, in R. Berenzden, ed., *Life beyond Earth and the Mind of Man* (Washington, D.C., 1973), 6. For the Proceedings of the meeting see Carl Sagan, ed., *Communication with Extraterrestrial Intelligence (CETI)* (Cambridge, Mass., 1973). The Soviet organizing committee consisted of Ambartsumian (chair), Kardashev, Shklovsky, and Troitsky, while on the U.S. side were Sagan (chair), Drake, and Morrison.

by the U.S. and USSR Academies of Science, and the 28 Soviet participants, 15 U.S. participants, and 4 participants from other countries, along with a few representatives from the social sciences, made this the most diverse SETI meeting ever held. In a Preface to the Proceedings of the meeting, National Academy of Sciences President Philip Handler went so far as to call the meeting "a punctuation mark in the history of mankind."[27] It had taken a decade, from Green Bank to Byurakan, for the American and Russian pioneers to meet, but it was only the beginning of an extended dialogue; Byurakan proved to be the first in a series of decennial U.S.–USSR SETI meetings continued in 1981 and 1991.[28] Figure 9.3 shows the participants in the 1991 meeting, including many of the SETI pioneers.

Back in the United States, events at NASA Ames demonstrate how discipline building could occur by building networks. The first action of John Billingham, after obtaining support from his superiors for a limited study on interstellar communication, was to begin to build a network from the ground up. In 1970, he and a few colleagues undertook a "mini-study" of the problem, and at his behest Ames simultaneously sponsored a series of lecturers on the subject, including Carl Sagan on interstellar communication, A. G. W. Cameron on planetary systems, Cyril Ponnamperuma on chemical evolution, Ronald Bracewell on interstellar probes, and Frank Drake on search strategy.[29] As we have seen, the optimistic results of the mini-study led to the decision to conduct a full-scale study the following summer as part of a summer faculty fellowship program in engineering systems design sponsored by NASA, Stanford, and the American Society of Engineering Education. This resulted in the landmark Project Cyclops report.

Following the Cyclops study, in late 1972 Billingham created a Committee on Interstellar Communication.[30] By March 1973 it had produced "A Program for Interstellar Communication," Phase A of an Interstellar Communication Feasibility Study, and by March 1974 had developed a

[27] Sagan, *Communication*, Preface, viii. Accounts of the meeting were given by two participants: physicist Freeman Dyson, *The New Yorker* (November 6, 1971), 126, and historian William H. McNeill, *Chicago Magazine* (May–June, 1972).

[28] The Proceedings of the 1981 meeting in Tallinn, Estonia, were never published. The meeting is described in Woodruff T. Sullivan, "SETI Conference at Tallinn," *S&T*, (April 1982), 350–353, and Jon Lomberg, "Soviets Host International SETI Conference," *The Planetary Report*, 2 (May–June 1982), 12–13. The Proceedings of the 1991 meeting, held in Santa Cruz, California, are G. Seth Shostak, ed., *Third Decennial U.S.–USSR Conference on SETI* (San Francisco, 1993). The last meeting included 60 participants.

[29] C. Ponnamperuma and A. G. W. Cameron, eds., *Interstellar Communication: Scientific Perspectives* (Boston, 1974).

[30] The original members included Billingham as chief, J. Wolfe as deputy chief, and D. Black, E. Duckworth, R. Eddy, M. Hansen, H. Hornby, R. Johnson, and D. Lumb as members.

Figure 9.3. Participants in the third Decennial U.S.–USSR meeting on SETI at Santa Cruz, California, in 1991. SETI pioneers include (front row standing, left to right) Michael Klein (JPL), Frank Drake (holding N EQLS L, a reference to the Drake Equation), Barney Oliver of Project Cyclops fame, and Jill Tarter, John Billingham, and Carl Sagan. Kneeling below Billingham is Soviet SETI pioneer Nikolai Kardashev, with Amahl Drake to the left of Kardashev.

more comprehensive "Proposal for an Interstellar Communication Feasibility Study." These documents remained unpublished, but briefings to NASA Administrator James Fletcher and NASA's Office of Aeronautics and Space Technology (OAST) led to funding of $140,000 from the latter in August 1974.

As the Viking project reached its culmination in 1976, Billingham was still agitating; his call for further studies in interstellar communication was aimed at increasing the network still further, following up on the groundwork laid by Project Cyclops. At the beginning of 1975 he had established an Interstellar Communications Study Group,[31] which concluded that the OAST funding should be used primarily for a series of six SETI science workshops chaired by Philip Morrison (Figure 9.4), two further workshops on extrasolar planet detection, and one workshop on cultural evolution. As a consensus-building exercise, the "Morrison workshops" of 1975–1976 may be viewed as analogous in importance to the 1964–1965 Space Science Board recommendations for planetary exobiology. Explicitly to build support for an observational program, the participants not only concluded that "it is both timely and feasible to begin a serious search for extraterrestrial intelligence," but also stated that

> it is particularly appropriate for NASA to take the lead in the early activities of a SETI program. SETI is an exploration of the Cosmos, clearly within the intent of legislation that established NASA in 1958. SETI overlaps and is synergistic with long term NASA programs in space astronomy, exobiology, deep space communications and planetary science. NASA is qualified technically, administratively and practically to develop a national SETI strategy based on thoughtful interaction with both the scientific community and beyond to broader constituencies.[32]

This, of course, was just the endorsement that NASA sought. Accordingly, in 1976 a small SETI Program Office was established at NASA Ames, headed by Billingham. This was the first institutionalization of SETI within NASA.

Billingham also played a key role in the 1976 *Outlook for Space,* a study prepared by the NASA centers to guide NASA's thinking for the

[31] The Interstellar Communications Study Group consisted of Billingham, astronomers Charles Seeger and Mark Stull, and Vera Buescher. Others, including Oliver, Black, and Wolfe, remained closely associated with the group.

[32] P. Morrison, J. Billingham, and J. Wolfe, *The Search for Extraterrestrial Intelligence: SETI* (Washington, D.C., 1977), 34. Consensus was not always achieved, as in the JPL push for a Sky Survey.

Figure 9.4. Philip Morrison, who opened the modern era of SETI with his paper co-authored by Cocconi. Morrison is seen here in 1975 at about the time of the SETI science workshops that he chaired.

next 25 years. That study viewed the origins and existence of life, whether microbial or intelligent, as an important part of NASA's space objectives through the end of the century.[33] By 1976 – the year of the Viking landers and the bicentennial of the United States – SETI had become respectable in NASA. Billingham's goal was to make it respectable enough for government funding, and toward that end he and others succeeded in gaining the endorsement of the National Academy of Sci-

[33] National Aeronautics and Space Administration, *Outlook for Space: Report to the NASA Administrator by the Outlook for Space Study Group* (Washington, D.C., 1976), 38, 145–149.

ences via its decadal reviews of astronomy. These reviews were important because they set priorities for government funding in astronomy, and once a putative program gained the support of the review, that endorsement was a substantial argument to those who controlled funding.

Billingham's efforts at network building culminated in the 1970s with the meeting at the Ames Research Center on "Life in the Universe." Appropriately, it was here, where the Viking biology effort was centered and where NASA's SETI program originated, that planetary exobiology and interstellar communication were first combined in a single meeting to review the entire scope of extraterrestrial life studies. It was here that Billingham declared that the fledgling science of exobiology had evolved to a stage of "reasonable maturity and respectability" and that a team of NASA scientists proposed a comprehensive SETI program. This being true, the next step was to secure funding. At this stage, networking took a back seat to science policy and institutional considerations, and we shall see in the next section how NASA fared in this final step of discipline building.

Meanwhile, although interest in planetary exobiology declined in the wake of Viking, international interest in SETI outside the U.S–USSR axis was increasing. The International Academy of Astronautics (IAA), under the leadership of Rudolf Pešek, had set up a study group on SETI as early as 1965, and beginning in 1972 it held annual reviews of developments in interstellar communication at the meetings of the International Astronautical Federation, aimed at those involved in space engineering and science.[34] Only 2 months after the Ames meeting, Boston University astronomer Michael Papagiannis organized a session on "strategies for the search for life in the universe" at the International Astronomical Union meeting in Montreal. This meeting was almost entirely devoted to SETI and planet searches, but by 1982 Papagiannis had formally organized a Bioastronomy Commission in the IAU, and the series of meetings held triennially since 1984 (see Table 9.1) have been the prime venue for discussing developments in the broad field of life in the universe, including both SETI and exobiology. In addition, the International Society for the Study of the Origin of Life (ISSOL) sponsored sessions on exobiology at its triennial meetings (see Table 7.1), where attendance had reached 350 by 1993. The prime interaction between astronomers and biologists, however, continued at the triennial meetings centered on the IAU Bioastronomy Commission.

At these meetings, representatives of the four disciplines of solar system

[34] R. Pesek, "Activities of the IAA CETI Committee from 1965–1976 and CETI Outlook," in *Communication with Extraterrestrial Intelligence*, eds. J. Billingham and R. Pesek (Oxford, 1979), 3–9.

Table 9.2. *Selected extraterrestrial life conference papers by discipline (number of papers/percentage of papers at each conference)*

Conference	Solar System Exobiology	Origin/ Evolution Life	Planetary Systems	SETI	Miscellaneous/ Interdisciplinary	Total Papers
National Academy of Sciences/Space Science Board (1964-1965)	29 (100%)	—	—	—	—	29
Armenian Academy of Sciences (1964)	—	—	—	12 (100%)		12
NASA Ames (1979)	—	18 (69%)	3 (12%)	5 (19%)	—	26
IAU Montreal (1979)	—	—	5 (31%)	11 (69%)	—	16
IAU Boston (1984)	1 (2%)	15 (27%)	9 (16%)	28 (50%)	3 (5%)	56
IAU Hungary (1987)	6 (9%)	27 (40%)	12 (18%)	18 (26%)	5 (7%)	68
IAU France (1990)	—	25 (29%)	17 (20%)	28 (33%)	15 (18%)	85
IAU California (1993)	1 (1%)	15 (21%)	14 (19%)	33 (45%)	10 (14%)	73
Average since 1984:[a]	3%	30%	18%	39%	11%	70

[a] After the formative Montreal meeting in 1979, the 1984 Boston meeting was the first of the triennial Bioastronomy conference series, organized by the International Astronomical Union's Commission 51 (Bioastronomy) and others. Percentages in this row therefore represent the average content by discipline of the papers presented in this series.

exobiology, origins of life, planetary systems science, and SETI debated issues of common interest. Table 9.2 shows the convergence of the four disciplines, as seen by the number of papers given in each, and highlights the fact that over three decades of interaction, the relative importance of each of these disciplines has changed, especially as the Viking project came and went. In the 1960s, solar system exobiology was highlighted as enthusiasm built toward the Voyager and Viking projects. The planetary exobiology and SETI advocates, however, tended to meet separately since their goals were quite different. With the NASA conference in 1979, however, they were for the first time conjoined in a substantial way via the origin of life issue, which applied equally to solar system exobiology and SETI. Origins of life contributions remained relatively steady throughout the triennial bioastronomy meetings, as did planetary systems science papers at a lower level, reflecting the smaller number of workers in the latter field. Radio astronomy SETI papers remained the strongest component through 1993. Finally, the increase in the miscellaneous category reflects an upward trend in papers on the implications of contact with extraterrestrial intelligence and an injection of social science experts into the discussion.

Literature

While committees, commissions, and conferences played a crucial role in network building, another factor of prime importance was published literature. For many years, a few books dominated the field. When exobiology was in its earliest stages of formation, Sir Harold Spencer Jones's *Life on Other Worlds* (1940) was the primary general text on the subject in both the popular and the scientific sense. Shklovskii and Sagan's *Intelligent Life in the Universe* (1966) rapidly took precedence after the mid-1960s and remained the standard exobiology source for many years, superseded only by the publication of Donald Goldsmith and Tobias Owen's *The Search for Life in the Universe* (1980), followed by a second edition in 1992. These works served the traditional crucial role of texts in other fields: they imparted to students, professionals, and the public alike the "standard" body of knowledge that comprised the discipline.

Published journal literature, however, is a better gauge of how the field has developed. We have seen in Chapter 3 (Figure 3.1) how the literature on Mars increased dramatically and in a cyclical fashion in rhythm with the Martian close approaches to Earth. Figure 9.5 shows the increase in literature on interstellar communication (exclusive of the Mars tradition) since 1969.[35] It also demonstrates how the interstellar communication literature has been dominated by conference Proceedings, which account in almost all cases for the high points on the graph. Thus the peak at 1980 reflects the IAU Montreal session, the peak at 1984 the Boston meeting (whose Proceedings were edited by Papagiannis), and the peak at 1988 the IAU meeting at Hungary. These peaks reflect an important characteristic of the discipline, graphically demonstrating just how dependent on conferences (and the institutions that sponsor them) the bioastronomy network has been.

[35] Based on entries in *Astronomy and Astrophysics Abstracts*. It is germane to this chapter that "Extraterrestrial Life" became a category in these volumes only in 1979, and even then in the section on "Miscellaneous Papers (Philosophical Aspects, Extraterrestrial Civilizations," etc.).

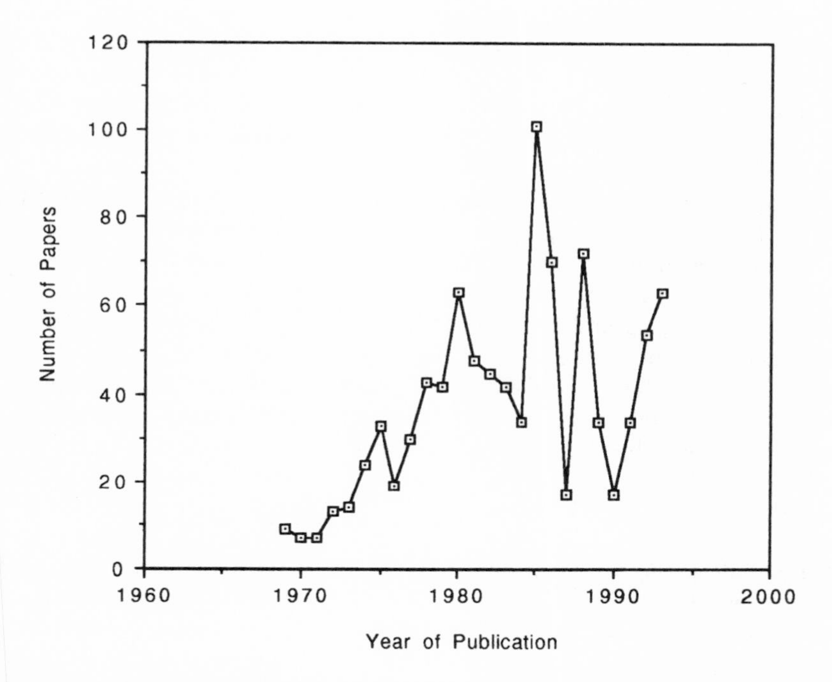

Figure 9.5. Published literature in bioastronomy, excluding Mars. Peaks represent papers from conferences, showing how important conferences were to the discipline. Source of data: *Astronomy and Astrophysics Abstracts,* Section 15, Miscellaneous Papers, including Extraterrestrial Life.

Before, during, and after the Viking mission, the published literature in planetary exobiology was scattered in journals ranging from special issues of *Science* to narrower disciplinary journals including *Icarus* and the *Journal of Geophysical Research*. This highlights an interesting phenomenon: despite the proclamations of a new discipline, more than 30 years after Cocconi and Morrison fired the opening shot in interstellar communication, and despite the outpouring of research centered on the Viking project, there still existed no journal that covered the broad scope of bioastronomy itself. Moreover, some of the disciplines that bioastronomy drew on also did not have their own journals. Although for planetary science *Icarus* was founded in 1962, and for the field of origins of life *Origins of Life* was founded in 1974 (derived from *Space Life Sciences,* published since 1968), neither planetary systems science nor SETI founded its own journal, much less one for the general discipline of bioastronomy itself. The communities of researchers in those fields perhaps had not yet reached the critical mass necessary to support a new journal; in addition, researchers worried about isolating themselves from the peer review and acceptance brought about by publishing in broader-based journals. In any case, the literature in a broad array of journals served the important purpose of increasing the reach of the network.

Finally, a massive popular literature was generated on the subject. By 1978, a NASA *Bibliography on the Search for Extraterrestrial Intelligence* contained almost 1500 entries. Along with the scientific articles, a substantial number represented coverage in *Time, Newsweek,* the *New York Times,* and popular science magazines. These were only the tip of the iceberg for the period covered and only the vanguard of the continued coverage since that time. To the extent that magazine and newspaper coverage measures the pulse of popular interest, this indicates the pervasiveness of the subject in the modern world.[36]

9.3 INSTITUTIONS: PROGRAMS AND FUNDING

Individual career decisions and network formation were indispensable steps toward discipline status, but they were not sufficient in themselves to define a discipline. As is evident from Table 9.1, many institutions, from the National Academies of Sciences of the United States and the USSR to NASA and the International Astronomical Union, spurred the formation of the fledgling extraterrestrial life community by sponsoring meetings. But meetings were one thing; programs were quite another level of commitment. Whatever else may characterize them, a prime characteristic of new fields and disciplines is the ability to secure funding for research programs. Without institutions, individuals may cogitate and groups may discuss, but the observational process that is the hallmark of science (unless one is a theoretician) cannot be undertaken on any large scale. In this section, we outline the primary role that NASA played in building the discipline of extraterrestrial life studies beyond the stage of networks.[37]

It is clear from an institutional perspective that exobiology did not obtain discipline status, and could not have, in the first half of the twentieth century. Even the private financing of one man to search for canals on

[36] Eugene F. Mallove, Mary M. Connors, Robert L. Forward, and Zbigniew Paprotny, *A Bibliography on the Search for Extraterrestrial Intelligence,* NASA Reference Publication 1021 (Washington, D.C., March, 1978).

[37] Institutions also act as homes to the science teams. For the various life detection teams eliminated in the Viking biology competition and the biology team finally selected, some idea of internal dynamics and external competition may be gained from Ezell and Ezell, *On Mars,* and Henry S. F. Cooper, *The Search for Life on Mars* (New York, 1980).

Mars did not approach the necessary elements of a discipline. There were no shared goals, no programmatic consensus, and, most of all, no technique to provide a definitive answer to the question of life beyond the Earth. Individual observations provided tantalizing hints of planetary conditions and eventually eliminated the Lowellian theory of Martian canals, but could not confirm the existence of life in any form. Although some professional scientists were interested in the subject, the first half of the century at best saw only a fragmented community with disparate ideas about extraterrestrial life, no consensus on how to approach the problem, and no institution to fund a consensus program in any case.

As we have emphasized repeatedly, only with the Space Age did it become possible to answer with any degree of certainty the question of life in the solar system. It was "the opportunity for experimental design that focused attention to what were the problematics of extraterrestrial life," Lederberg recalled. "In order to think about designing missions to look for life, you had to ask what were the signatures of life. How narrow or parochial a view do you have of it? Might you have completely different evolutionary patterns from the ones we know on Earth?"[38] It was clear from the beginning that in the United States NASA was the primary agency to undertake planetary exobiology, if it was to be undertaken at all, and we have indeed seen how at first individuals, then advisory groups, and eventually the agency itself made this one of its prime goals.[39]

Aside from relatively small grants for life detection devices, the selling of the exobiology program began in earnest with the enormous amounts of money needed for the spacecraft itself. This selling was, of course, done in part via the general argument for planetary exploration; after all, Venus was also a spacecraft target bereft of any real chance for life detection. There is no doubt, however, that for Mars life detection was a large part of the argument, and NASA built on this public fascination with the problem in proposing in 1965 a $2.5 billion Voyager spacecraft to be launched by the giant Apollo Saturn 5 vehicles still under development. This strategy involved a considerable risk, for some believed that the chances of finding life were very slim. As the Viking historians wrote:

> At the end of 1965, the scientists who believed that looking for life on Mars was a respectable enterprise faced those who were equally devoted to the proposition that such an exercise was foolishness of the gravest order. Voyager, with its goal of

[38] S. J. Dick interview with Lederberg, 6–7. The subsequent planetary contamination events are described in Section 3.4 of this volume.

[39] In the Soviet Union, by contrast, Mars was a primary target for the "Mars" probes beginning in 1960, but without the complex life detection equipment. See Chapter 3, Section 3.4.

> placing automated biology laboratories on Mars, would become the focus of the two groups' debate. Voyager would be scrutinized because of costs and general disenchantment with the space program, but the central issue would continue to be the validity of searching for life on the Red Planet.[40]

The cancellation of the Voyager program in 1967, however, was not primarily a referendum on the question of life on Mars, but rather reflected general budgetary constraints in the Vietnam era.[41] That Congress and the public were willing to spend a great deal of money on the search for life on Mars is evident in the resurrection of Voyager's goals in the smaller Viking project that placed two orbiters around Mars and two landers on its surface in 1976. The Viking project spent about $1 billion, including funding of the Centaur launch vehicles, and of that amount, the biology package and molecular analysis experiments cost about $100 million. If anything, biology was even more central for Viking than it had been for Voyager, and from its inception in late 1967, NASA Administrator Webb looked for – and found – support for the Viking biology goals from the Space Science Board and from its own Lunar and Planetary Missions Board.[42] Some were uncomfortable with the idea of the Mars mission's being propelled primarily by the search for life. Planetary scientist Bruce Murray, who would become director of JPL just months before Viking touched down on Mars, argued that "the extraordinarily hostile environment revealed by the Mariner flybys made life there so unlikely that public expectations should not be raised." Sagan accused Murray of pessimism; Murray accused Sagan of optimism. Murray, like H. P. Klein and Soffen, argued that biological questions should be pursued only after more knowledge had been gained about the Martian environment. For this reason, Murray was a critic of Viking.[43] He also pointed out the legacy of failure to detect life: the intense public disappointment in its absence made large parts of the public lose interest.[44] Nevertheless, it is also true in hindsight that, had a more cautious approach been taken with multiple spacecraft, the biology experiments would never have been funded in the era of budgetary constraints that followed.

[40] Ezell and Ezell, *On Mars,* 82.
[41] Ibid., 115.
[42] Ibid., 142.
[43] Bruce Murray, *Journey into Space* (New York, 1989), 61, 68–69. "Postulating the existence of Martian life publicly as a means of developing support for Viking made me uncomfortable," Murray wrote.
[44] Ibid., 74, 77. By contrast, we recall that Murray supported SETI and was instrumental in the early stages of development of the JPL Sky Survey, for which planning was begun shortly after he became director at JPL.

By contrast to planetary exobiology, support for SETI did not begin at NASA, a reflection of SETI's ground-based technique. Rather, we have seen that SETI in its modern form emerged in the university, where institutions provided moral support and an environment in which controversial ideas could be explored without worrying about congressional and taxpayer reaction. Cornell University is a prime example; Swift has noted the Cornell connection of many SETI pioneers.[45] And when it came to observations, it was still not NASA, but rather the federally funded National Radio Astronomy Observatory in Green Bank, that provided meager funds, encouragement, and telescope time for Project Ozma. This was in no way a science policy decision, but rather, as we have seen, a result of the personal interest of Lloyd Berkner, Otto Struve, and Frank Drake, an interest that could be satisfied at low cost without much disruption to ongoing programs.

That NASA should have become involved in SETI at all was not a foregone conclusion. In contrast to the Viking project, the search for life outside the solar system was not the clear-cut province of national space agencies such as NASA, even if activities in space had the general effect of turning the attention of humanity toward the stars. The 1961 Green Bank meeting had not been attended by any NASA representative, although a collection of papers from the meeting was edited a few years later by NASA astronomer A. G. W Cameron. As we have seen, NASA was understandably more interested in the immediate prospects offered by space exploration.

Following the activity of John Billingham and his Ames colleagues during the 1970s, however, NASA decided to undertake a program of interstellar communication, and with that decision SETI followed its planetary exobiology cousin in entering science policy in a big way. The space agency's entry into the SETI field, the consensus building within the agency required to reach a critical mass of intellectual support, and the search for the considerable funding needed for a powerful and comprehensive search provide a case study of the crucial role of an institution in discipline building.

If the 1970s may be seen as a time of serious study of the SETI problem by NASA, the 1980s may be viewed as a time for implementation of the study's recommendations – two very different processes. Following the Cyclops report (1971), the Morrison workshops (1975–1976), and the Conference on "Life in the Universe" (1979), SETI proponents felt it was time for action. Studies and refinements would continue, of course – notably in a series of SETI Science Working Group

[45] Swift, *SETI Pioneers*, 393–394.

(SSWG) meetings in 1980–1981, headed by Frank Drake.[46] But to convert concepts and discussion into hardware and software required funding, and it was undoubtedly with some trepidation (considering the tradition in popular culture) that SETI advocates in NASA now entered the arena of space policy and space politics – always a hazardous and uncertain endeavor even for the tamest of subjects. Nevertheless, enter that arena they did, and with surprising initial success. Detailed hearings were held on the subject in September 1978, with the result that the Subcommittee on Space Science and Applications of the House Committee on Science and Technology supported NASA's proposal to initiate a SETI program.[47] The House and Senate both authorized $2 million for this purpose in fiscal year 1979 (FY 79), but the House and Senate Appropriations Committees elected not to provide any funds after the program received Senator William Proxmire's infamous Golden Fleece Award for unnecessary expenditures by the federal government. NASA nevertheless continued to fund SETI at a subsistence level until thwarted again by Senator Proxmire, who through the amendment process killed all funding for SETI beginning in FY 82.[48] Still, NASA decided to return to Congress for full funding in FY 83, emboldened by a National Academy of Sciences report that recommended SETI as one of seven moderate programs that NASA should implement; the external consensus building of Billingham and others was paying off.[49] After maneuvers that included a friendly chat between astronomer Carl Sagan and Senator Proxmire and an international petition, SETI funding was again restored for FY 83, at a level of $1.5 million.[50] The hazard and uncertainty of funding even for a relatively small project had been amply demonstrated, but this was only the beginning of SETI's checkered political history.

Internally, SETI was also buffeted by varying opinions regarding strategies, and had to suffer through the same bureaucratic and programmatic

[46] F. Drake, J. Wolfe, and C. Seeger, *SETI Science Working Group Report,* NASA Technical Paper 2244 (Washington, D.C. 1983).

[47] U. S. Congress, *Extraterrestrial Intelligence Research: Hearings before the Subcommittee on Space Science and Applications of the Committee on Science and Technology,* U.S. House of Representatives, 95th Congress, second session, September 19–20, 1978 (Washington, D.C., 1979).

[48] On Proxmire's actions and excerpts from the *Congressional Record* – and his citation of Tipler's arguments – see "Senator Proxmire and the Killing of SETI," *S&T,* 63 (April 1982), 353.

[49] National Research Council, *Astronomy and Astrophysics for the 1980's* (Washington, D.C., 1982). Also known as the "Field Report" after its chairman, George Field.

[50] Carl Sagan et al., "Extraterrestrial Intelligence: An International Petition," *Science,* 218 (October 29, 1982), 426. See also F. Drake, "Will the Real SETI Please Stand Up?," *Physics Today,* 35, no. 6 (June 1982), 9 and 70–71.

battles as did much larger NASA projects. In 1985 Ames and JPL signed a memorandum of understanding delineating the responsibilities of each group. The project underwent definition reviews in 1986 and 1987, the formal Program Plan was adopted in March 1987, and the Project Initiation Agreement was signed by NASA headquarters in 1988. Finally, with funding in 1990, SETI took on the status of an approved NASA project beyond the R&D phase and began "Final Development and Operations," marked by observations begun on October 12, 1992, to be completed by the year 2000 at a total cost of $108 million.

Almost every year, however, SETI was in mortal danger not only of budget cuts but also of termination at the congressional level. Moreover, SETI had an uneasy home at NASA headquarters, where its advocates not only had to wage internal battles for its share of the funding, but also had to worry about where it best fit. The program had spent most of its lifetime in the Life Sciences Division, but one could argue that what was primarily a radio astronomy project really did not belong in there. In the midst of ever-fiercer budget battles, in 1992 the NASA SETI program (previously known as the "Microwave Observing Program," inelegantly shortened to "MOP") was renamed by Congress the "High Resolution Microwave Survey (HRMS)." At the same time, it became the first element in the Solar System Exploration Division's "Toward Other Planetary Systems (TOPS)" program designed to detect other planetary systems. This again was not an entirely judicious match; although many planet hunters were sympathetic to SETI's goals, they were battling to obtain their own NASA funding and understandably feared that SETI's dubious reputation in Congress would make them guilty by association.

All of these political problems caught up with NASA SETI when, in September 1993, almost exactly 1 year after it began observations, the program was terminated by Congress. The termination demonstrated the risks of entering the science policy arena and placed bioastronomy on shaky ground. Although some attempted to struggle on with private funding through the appropriately renamed "Project Phoenix" sponsored by the SETI Institute, the prospect of continued observations, much less the prospect for success, remained uncertain. Clearly, career risks and successful networking were not enough, nor was the desire of a major agency to fund the program; in the end, the funding institution itself was buffeted by larger events in the U.S. Congress beyond its control. The contrast with planetary exobiology is quite striking. While those involved in the Viking project at least had the satisfaction of seeing two spacecraft successfully land on Mars and undertake their experiments as designed, the SETI branch of exobiology – at least at NASA – was ended before the experiment was systematically implemented, and therefore before any semblance of success could be proclaimed.

At century's end, then, the last word was not yet in on the status of bioastronomy as a discipline, any more than on the long-range significance of SETI and exobiology themselves. As astronomer Tom Gehrels wrote in reference to the large attendance at the meetings on planetary systems science, "A new discipline cannot be launched by definition or declaration, but the active participation of [such] a large number of the best scientists is a sign of success."[51] By that standard, extraterrestrial life studies have surely achieved some measure of success in their claims of discipline status. But what institutions – especially government institutions – give they may also take away. The U.S. Congress's termination of the NASA SETI program deprived SETI of the majority of its scientific staff and the field of bioastronomy of a significant fraction of its members (perhaps one-half of the 40 percent listed in Table 9.2, to judge by representation at scientific meetings). Nevertheless, NASA continued to fund broader aspects of exobiology even as smaller SETI programs sponsored by universities and the Planetary Society continued. And most were willing to bet that NASA SETI would be back, financed privately, if not by the fickle U.S. government.[52] To its advocates, the problem at century's end still seemed too important to be permanently derailed by mundane concerns like funding.

The misfortunes of SETI in NASA should not obscure the central role that NASA played in the field of exobiology broadly defined. Not only had NASA held a leading institutional role in SETI, as it had in planetary exobiology, it had also sponsored studies on the search for extrasolar planets, research on biogenic compounds, the origins of life, the conditions in outer space that might affect the evolution of life, and even exobiology in the Earth's orbit.[53] Moreover, elements of its exobiology

[51] Gehrels, *Protostars and Planets,* 9.

[52] SETI differed from planetary exobiology in one significant way: because it was a ground-based activity, patrons other than NASA could afford to support it. In the United States at least three such patrons did: Ohio State University, with its ongoing program running since 1973; the Planetary Society, with its project META, carried out at Harvard and in Argentina; and the Berkeley group, with its SERENDIP parasitic project. Although all three received support or funding from NASA, all planned to continue in the wake of the NASA termination. Linda Billings, "SETI on Campus," in Ben Bova and Bryon Preiss, eds., *First Contact: The Search for Extraterrestrial Intelligence* (New York, 1990), 197–220.

[53] For some of the NASA-sponsored projects in exobiology, see D. C. Black, *Project Orion: A Design Study of a System for Detecting Extrasolar Planets,* (Washington, D.C., 1980); D. Black and W. C. Brunk, eds., *An Assessment of Ground-Based Techniques for Detecting Other Planetary Systems,* NASA CP-2124 (Washington, D.C., 1986); DeFrees, D. Brownlee, J. Tartar, D. Usher, W. Irvine, and H. Klein, *Exobiology in Earth Orbit* (Washington, D.C., 1989); H. Hartman, J. G. Lawless, and P. Morrison, eds., *Search for Universal Ancestors,* NASA SP 477 (Washington, D.C., 1985); D. Milne, D. Raup, J.

program continued to perform vigorous research in the 1990s, despite the expectations of some in the aftermath of Viking. As H. P. Klein wrote 10 years after the Mars landings:

> After the Viking mission, when it became apparent that prospects were dim for the discovery of extraterrestrial life within our solar system, many people concluded that this new field of endeavor would soon expire. Quite the contrary, over the past decade, the field has broadened considerably into a multidisciplinary approach to understanding the circumstances that led to the origin of life and the interplay between the evolution of this planet and its biota.

As chairman of the Space Studies Board's Committee on Planetary Biology and Chemical Evolution, Klein helped lay the groundwork for an ambitious plan for exobiology's future. "Exobiology today includes some of the best work in this country, or anywhere, on how life arose on *this* planet," Klein empahsized. "Some of the key individuals working on RNA replication, some of the key work is being funded by NASA, not by the NIH or the NSF."[54] Although continued funding was not assured, the success of exobiology – and its status as a discipline – were not yet a matter of history. As the century entered its last decade, exobiology's practitioners, supported by NASA, believed that the potential for their aspiring discipline remained strong and its greatest achievements still in the future.

Billingham, K. Niklaus, and K. Padian, eds., *The Evolution of Complex and Higher Organisms,* NASA SP 478 (Washington, D.C., 1985); J. A. Wood and S. Chang, eds., *The Cosmic History of the Biogenic Elements and Compounds,* NASA SP-476 (Washington, D.C., 1985).

54 H. P. Klein, "Exobiology Revisited," *Advances in Space Research,* 6, no. 12 (1986), 187–192; Space Studies Board, Committee on Planetary Biology and Chemical Evolution, H. P. Klein, chair, *The Search for Life's Origins: Progress and Future Directions in Planetary Biology and Chemical Evolution* (Washington, D.C., 1990); Steven J. Dick, interview with H. P. Klein, September 15, 1992, p. 26.

10

THE MEANING OF LIFE: IMPLICATIONS OF EXTRATERRESTRIAL INTELLIGENCE

And now at last the highest truth on this subject remains unsaid; probably cannot be said. . . .

Emerson[1]

We are in deep waters here, in a sea of great mysteries.

E. A. Milne[2]

O, be prepared, my soul!
To read the inconceivable, to scan
The million forms of God those stars unroll
When, in our turn, we show to them a Man.

Alice Meynell[3]

Never before have so many been so enthusiastic about being so trivial.

Rabbi Norman Lamm[4]

At the end of our history, we return to the question with which we began: the meaning of extraterrestrial life for humanity. We have seen in a general sense (Chapter 2) that belief in extraterrestrial life is a worldview in itself, a "biophysical cosmology," that – once accepted – forever changes the perception of our place in the universe. We have also seen (Chapter 1) that the discussion of the mere *possibility* of extraterrestrial life has in prior centuries given rise to a great diversity of opinion on the possible meanings for humanity, in particular in the religious context.

In this final chapter, we examine how twentieth-century thinkers from diverse backgrounds have historically viewed the implications of life beyond the Earth for human culture, theology, and philosophy. In exploring humanity's relation to other intelligent species, these thinkers have, however tentatively, extended the bounds of philosophy and theol-

[1] Ralph Waldo Emerson, "Essay on Self-Reliance" (1841), in *Essays and Lectures* (New York, 1983), 271.

[2] E. A. Milne, "The Second Law of Thermodynamics: Evolution," in *Modern Cosmology and the Christian Idea of God* (Oxford, 1952), 153, speaking of the problem of Incarnation on other planets.

[3] Alice Meynell, "Christ in the Universe," *Poems* (New York, 1923), 92.

[4] N. Lamm, "The Religious Implication of Extraterrestrial Life," in A. Cannell and C. Domb, eds., *Challenge: Torah Views on Science and Its Problems* (2nd edition, New York, 1978), 361.

ogy beyond their usual anthropocentric perspectives, themselves becoming pioneers whose ideas remain for the most part undeveloped. Unlike the scientific issues in the debate, the question of implications has received very little systematic study, for the simple reason that the humanities and social sciences have not yet taken to heart the potential implications of the biological universe, having had quite enough to contend with in studying the sometimes inexplicable activities of the local intelligent species. It is appropriate that we should treat the question in our final chapter, for it remains a subject for the future, one that may draw on the sporadic and widely divergent results that the history of life on Earth has offered thus far.

10.1 PERCEPTIONS OF CULTURAL IMPACT

Though the subject had occasionally been addressed earlier, the Space Age gave immediacy to the problem of the impact of a discovery of extraterrestrial life. In 1961, a NASA-sponsored study undertaken by the Brookings Institution in compliance with the National Aeronautics and Space Act to identify long-range goals of the U.S. space program and their effect on American society offered a discussion of "the implications of a discovery of extraterrestrial life."[5] The broad-based report touched only briefly on life in space, but in doing so raised a variety of important questions. The social science authors viewed the recently completed Project Ozma as having popularized and legitimized speculations about the impact of such a discovery on human values; it also led them to agree with their astronomical colleagues that radio signals seemed the most likely mode of contact over the next 20 years.[6]

The NASA–Brookings authors emphasized that both individual and government reactions to radio contact with intelligence would likely depend on religious, cultural, and social backgrounds, as well as on any information communicated. The mere knowledge of life in the universe, they speculated, could lead to greater unity on Earth based on the uniqueness of humanity or its unified reaction to something alien. Because of the

[5] *Proposed Studies on the Implications of Peaceful Space Activities for Human Affairs. Prepared for the National Aeronautics and Space Administration by the Brookings Institute.* Report of the Committee on Science and Astronautics, U.S. House of Representatives, 87th Congress, 1st session, March 24, 1961 (Washington, D.C., 1961), 215–216 and notes on 225–226. The report was prepared under the direction of D. N. Michael, a social psychologist "primarily responsible for the interpretations, conclusions, and recommendations in and the final drafting of this report" (viii).

[6] Ibid., 215. It is notable that, in those days when all things seemed possible, the authors did not rule out the possibility that a growing space program could find artifacts left by life forms on the Moon, Mars, or Venus.

difficulties in communication, a discovery could become "one of the facts of life" not requiring any action. On the other hand, in a statement often cited since, the authors also warned that substantial contact could trigger a more foreboding effect:

> Anthropological files contain many examples of societies, sure of their place in the universe, which have disintegrated when they had to associate with previously unfamiliar societies espousing different ideas and different life ways; others that survived such an experience usually did so by paying the price of changes in values and attitudes and behavior.[7]

Scientists, they suggested, might be the most devastated of all by the discovery of superior intelligence, since "an advanced understanding of nature might vitiate all our theories at the very least, if not also require a culture and perhaps a brain inaccessible to earth scientists." Finally, the authors pointed out that philosophical problems could be raised, in particular in deciding whether aliens were to be treated morally and ethically as human beings.[8]

The NASA–Brookings study set forth in broad outline the alternative impacts of extraterrestrial intelligence: good or bad, and possibly worst of all for scientists. As for the discovery of plant life or "subhuman intelligence" on Mars or Venus – the planets soon to be explored by the space program – they proposed that after the initial novelty had worn off, large parts of the American public might not be affected any more than they were by the discovery of the coelacanth or the panda.

Members of the Space Science Board of the National Academy of Sciences, however, in reviewing (again at NASA's request) the scope of space biology while contemplating the first Mars missions in 1962, were not so sanguine about the philosophical import of exobiology, even when dealing with low forms of life. They wrote:

> It is not since Darwin – and before him Copernicus – that science has had the opportunity for so great an impact on man's understanding of man. The scientific question at stake in exobiology is, in the opinion of many, the most exciting, challenging, and profound issue, not only of this century but of the whole naturalistic movement that has characterized the his-

[7] Ibid., 215. The passage is reminiscent of Jung's statement in relation to the UFO debate (Chapter 6, footnote 42) that the "reins would be torn from our hands and we would, as a tearful old medicine man once said to me, find ourselves 'without dreams,' that is, we would find our intellectual and spiritual aspirations so outmoded as to leave us completely paralyzed."

[8] Ibid., 226.

> tory of western thought for three hundred years. What is at stake is the chance to gain a new perspective on man's place in nature, a new level of discussion on the meaning and nature of life.[9]

In making this statement, the Space Science Board was not merely expressing an academic opinion: its members not only foresaw significant implications of a discovery of extraterrestrial life, they also used their positive conception of these implications to argue for an exobiology program of considerable expense to the public.

The ability to understand a more generalized biology, however, was quite different from a concern about the implications of the discovery of intelligence. It was the latter that continued to intrigue most academics drawn to the subject, and that was the focus of a 1-day symposium on "Life beyond Earth and the Mind of Man" sponsored jointly by NASA and Boston University in 1972. Billed by its moderator, Boston University astronomy professor Richard Berenzden, as the first meeting in which a distinguished panel from diverse fields discussed in an open forum the ramifications of the discovery of extraterrestrial intelligence, the symposium produced some provocative (but hardly systematic) ideas, and is perhaps most notable for its diversity of opinion than for any consensus. Nobel Prize–winning Harvard biologist George Wald, allowing that the universe was probably full of life and that long-distance communication was more likely than physical contact, declared that he could "conceive of no nightmare as terrifying as establishing such communication with a so-called superior (or if you wish, advanced) technology in outer space." Though some gleefully prophesy great benefits from such communication, Wald continued – conjuring the scenario already broached by the NASA–Brookings report – "the thought that we might attach, as by an umbilical cord, to some more advanced civilization, with its more advanced science and technology, in outer space does not thrill me, but just the opposite." For humanity itself to discover a cure for cancer, or to control nuclear fusion, was one thing, he said. "But just to get such information passively from outer space through that transmission is altogether different. One could fold the whole human enterprise – the arts, literature, science, the dignity, the worth, the meaning of man – and we

[9] National Academy of Sciences–National Research Council, *A Review of Space Research: The Report of the Summer Study Conducted under the Auspices of the Space Science Board of the National Academy of Sciences at the State University of Iowa, Iowa City, Iowa, June 17–August 10, 1962* (Washington, D.C., 1962), 9-2, 9-3. NASA also funded the report. In their opinion, the Space Science Board was opposed by G. G. Simpson, "The Non-Prevalence of Humanoids," *Science,* 143 (February 21, 1964), 769–775.

would just be attached as by an umbilical cord to that 'thing out there.' "[10]

Wald's fears found little sympathy from other participants. Theologian Krister Stendahl, dean of the Harvard School of Divinity, viewed the growing awareness of extraterrestrial intelligence as not at all threatening. Rather, he believed such a discovery would raise cosmic consciousness, make God's universe even better, and leave humanity "relatively unique" rather than "absolutely unique" in the universe. Notably, however, Stendahl raised no discussion of dogma. The three astronomers present – Philip Morrison, Carl Sagan, and Berenzden – all found Wald's fears groundless, as did anthropologist Ashley Montagu.

In a more general sense, Philip Morrison argued that the discovery of intelligence and its subsequent impact would not be a quick event; it would resemble more the development of agriculture than the discovery of America. Reaction to even the latter discovery, he pointed out, was slow. He believed the message would not be simple; the signal would be immediately recognizable, but the message's content would be only slowly deciphered. "It will be more of a subtle, long lasting, complex, debatable effect than a sudden revelation of truth, like letters written in fire in the sky," he said. Morrison was "neither fearful nor terribly expectant. I am anxious for that first acquisition, to make sure that we are not alone. But once that is gained – it might be gained in my lifetime – then I think we can rest with some patience to see what complexities have turned up on other planets."[11]

Also in a positive vein, Carl Sagan argued that discovering extraterrestrial intelligence would reestablish a context for humanity that had been long lost:

> The old secure sense of where we are in the universe has eroded. . . . The kind of exploratory ventures we are talking about seem to be to be precisely the kind that are needed to reestablish a cosmic context for mankind. By finding out what the other planets are like – by finding out whether there are civilizations on planets of other stars – we reestablish a context for ourselves.[12]

Rejecting the likelihood of invasion, exploitation, or subversion of the human race by extraterrestrials, Berenzden provided the most encompass-

[10] R. Berenzden, ed., *Life beyond Earth and the Mind of Man,* NASA SP-328 (Washington, D.C., 1973), 17–19.

[11] Ibid., 44–45. Morrison also noted the importance of historical models in envisioning the impact.

[12] Ibid., 56–57.

ing counterpoint to Wald's fear: the discovery of intelligence beyond the Earth, he argued,

> would perhaps provide us with the opportunity of joining what has been called the 'galactic heritage'. . . . It might also lead us to better social forms, possibly to ways to solve our environmental crises, and even improve our own social institutions. . . . The benefits could come not merely in terms of technology and science, but also in the arts, literature, and humanities. And beyond that – and probably the most significant of all – contact would end our social and cultural isolation.[13]

The cultural shock from radio contact, Berenzden believed, was problematic. The first contact would likely be only a confirmation of existence, followed "many hundreds of years later" by information trickling in, and then by tens or hundreds of years of decipherment.

Neither Berenzden nor the other participants could begin to predict the impact of extraterrestrial contact on humanity. But NASA Administrator James Fletcher, noting that technically advanced civilizations in other planetary systems were not only possible but likely, urged that the provocative ideas of the six panelists were "worthy of consideration by thoughtful persons everywhere." In calling for more study, Fletcher echoed the NASA–Brookings report, which also recommended continued studies of the attitudes toward the possibility of extraterrestrial intelligence and its discovery, as well as historical and empirical studies of the behavior of humans when confronted with dramatic and unfamiliar events or social pressures. The NASA–Brookings authors believed such studies would provide guidance on how the news of discovery should be presented to or withheld from the public, and the role of scientists and government in releasing this information.[14]

That NASA had a hand in all three of these studies – the Brookings report, the Space Science Board, and the Boston discussion panel – is significant; the broader intellectual community was still reluctant to raise the subject without some prodding. One exception was philosopher Lewis W. Beck, who echoed the call for further study and offered two conjectures of his own after reviewing the history of the debate. First, he argued, popular science and science fiction have so prepared the public for the idea of signals from space that such a discovery will be forgotten after a few weeks, just as the details of the Moon landing were forgotten

13 Ibid., 49–50.

14 The authors suggested studies of reactions to UFOs, hoaxes, and psychic manifestations, using as a guide Hadley Cantril's study *Invasion from Mars* (Princeton, N.J., 1940).

by the majority of people, who had more immediate concerns in their daily lives. But he also conjectured that in a larger sense, the discovery of intelligence beyond the Earth would never be forgotten:

> For what is important is not a single discovery, but the beginning of an endless series of discoveries which will change everything in unforeseeable ways. [If further discoveries are made,] "there is no limit to what in coming centuries we might learn about other creatures and, more portentously, about ourselves. Compared to such advances in knowledge, the Copernican and Darwinian Revolutions and the discovery of the New World would have been but minor preludes.[15]

By the early 1970s, then, there were many calls for further study but no consensus among the few groups that had discussed, all under NASA sponsorship and on a very small scale, the potential impact of contact. Nor, Beck aside, was there any inclination among social scientists or philosophers to address the problem of the implications of extraterrestrial life independently and no answer to the call for more study during the 1970s.

On the other hand, both inclination and near-consensus did exist among the members of one group – the SETI participants themselves. These began with neutral declarations of significance: Cocconi and Morrison, in their pioneering article in 1959, wrote that "Few will deny the profound importance, practical and philosophical, which the detection of interstellar communications would have," and in the year of Project Ozma (1960), Drake closed one of his articles on the subject with the words "the scientific and philosophical implications of such a discovery will be extremely great."[16] But these neutral statements quickly took on an air of advocacy. The Introduction to the earliest anthology of papers on "interstellar communication" – many of them based on the 1961 Green Bank meeting – commented that in admitting that millions of advanced societies existed in our Galaxy, we were completing the Copernican revolution. Moreover, "If we now take the next step and communicate with some of these societies, then we can expect to obtain an enormous enrichment of all phases of our sciences and arts. Perhaps we shall also receive valuable lessons in the techniques of stable world

[15] Lewis White Beck, "Extraterrestrial Intelligent Life," in Edward Regis, Jr., ed., *Extraterrestrials: Science and Alien Intelligence* (Cambridge, Mass., 1985), 3–18: 15.

[16] G. Cocconi and P. Morrison, "Searching for Interstellar Communication, *Nature,* (1959), reprinted in D. Goldsmith, ed., *The Quest for Extraterrestrial Life* (Mill Valley, Calif., 1980), 102–104: 104; F. Drake, "How Can We Detect Radio Transmissions from Distant Planetary Systems?" *S&T,* 19 (January 1960), 140–43, reprinted in Goldsmith, *The Quest,* 114–117.

government,"[17] editor A. G. W. Cameron commented at the height of the cold war.

This brief comment reflected the attitude of most other SETI researchers, an attitude that by the time of the Boston symposium had become widespread. The Project Cyclops report, whose influence in this area, as in its technical treatment of SETI, cannot be underestimated, was particularly important in this regard. In this report, Bernard Oliver had spoken of "our galactic heritage" and "the salvation of the human race" in the context of extraterrestrials while downplaying possible hazards such as invasion, exploitation, subversion, and culture shock. Galactic civilizations are likely to have first formed 4 or 5 billion years ago, Oliver pointed out, and thus interstellar communication may have been going on for a long time.

> One of the consequences of such extensive heavenly discourse would be the accumulation by all participants of an enormous body of knowledge handed down from race to race from the beginning of the communicative phase. Included in this galactic heritage we might expect to find the totality of the natural and social histories of countless planets and the species that evolved: a sort of cosmic archaeological record of our Galaxy.[18]

This knowledge, Oliver argued, might lead to our self-preservation, to new branches of science, to the end of the cultural isolation of the human race, and to a reorientation of our philosophy and mores. While not totally lacking risk, Oliver argued, the culture shock exhibited repeatedly by culture contact on Earth was unlikely to be repeated via radio contact. Rather, the "long delays and initially slow information rate" would allow us to adapt to the new situation.

Philip Morrison espoused a similar view almost simultaneously at the joint U.S.–USSR Byurakan meeting in 1971. The complex signal arriving at our radio telescope, he argued, "is the object of intense socially required study for a long period of time. I regard it as much more like the enterprise of history of science than like the enterprise of reading an ordinary message. . . . The data rate will for a long time exceed our ability to interpret it."[19] This scenario of gradual assimilation, he asserted, directly illuminates the nature of the impact on humanity:

17 A. G. W. Cameron, ed., *Interstellar Communication* (New York, 1963), 1.
18 Bernard Oliver and J. Billingham, eds., *Project Cyclops*, (Moffett Field, Calif., 1971), p. 31.
19 Philip Morrison, "The Consequences of Contact," in *Communication with Extraterrestrial Intelligence*, ed. Carl Sagan (Cambridge, Mass., 1973), 333–349: 336–337.

> The recognition of the signal is the great event, but the interpretation of the signal will be a social task comparable to that of a very large discipline, or branch of learning. In that light, I think that a sober study will show that a message channel cannot open us to the sort of impact which we have often seen in history once contact is opened between two societies at very different levels of advance. . . . We could imagine the signal to have great impact – but slowly and soberly mediated.

Frank Drake was more effusive about the immediate impact. Contact, especially with the "immortals," could give us a shortcut to wisdom, he believed. Not only would we gain scientific and technical information, we might also learn about "ultimate social systems," art forms, and other aspects of life as yet unimagined. If this is so, he concluded, "let us know it now. We need not be afraid of interstellar contact, for unlike the primitive civilizations on earth which came in contact with more advanced technological societies, we would not be forced to obey – we would only receive information."[20] Both Drake and Sagan believed such a discovery "would inevitably enrich mankind beyond measure."[21]

These attitudes were in the background of the Morrison workshops of 1975–1976, which, while most notable as a technical landmark in converting SETI from discussions into a realistic program, also offered some thoughts on the impact of SETI. A single extraterrestrial signal, the participants noted, would lead immediately to one great truth: "that it is possible for a civilization to maintain an advanced technological state and *not* destroy itself." Pointing out that humanity was under no obligation to reply to a radio message and could even ignore one that seemed offensive, the Morrison report rejected negative impacts and argued that a signal would "pose few dangers to mankind; instead it holds promise of philosophical and perhaps practical benefits for all of humanity."[22]

In contemplating an actual SETI search, the report also broached another issue that had heretofore received little attention: that early claims of signal detection may turn out to be mistaken; that the press and the public must use caution to avoid raising hopes and fears and, thus, "the importance of a skeptical stance and the need for verification." Beyond the acquisition of the first genuine signal, the report echoed Morrison's previously stated belief that study of any information

[20] F. Drake, "On Hands and Knees in Search of Elysium," *Technology Review*, 78 (June 1976), 22–29.

[21] Sagan and Drake, "The Search for Extraterrestrial Intelligence," *SciAm*, 232 (May 1975), 80–90.

[22] P. Morrison, J. Billingham, and J. Wolfe, eds., *The Search for Extraterrestrial Intelligence*, NASA, SP-419, (Washington, D.C., 1977), 7–8.

content would occupy generations. The commonality of radiophysics and mathematics, it stressed, might allow a signal to be decoded, but perhaps only "in a slow and halting manner." Still, the report urged that the question of impact "deserves rather the serious and prolonged attention of many professionals from a wide range of disciplines – anthropologists, artists, lawyers, politicians, philosophers, theologians – even more than that, the concern of all thoughtful persons, whether specialists or not." One of those professionals, Theodore Hesburgh, the president of Notre Dame University, declared in the Foreword to the report that SETI and theology were not incompatible. To the contrary, Hesburgh held, echoing eighteenth-century natural theology arguments, "As as theologian, I would say that this proposed search for extraterrestrial intelligence (SETI) is also a search of knowing and understanding God through His works – especially those works that most reflect Him. Finding others than ourselves would mean knowing Him better."

Not everyone agreed with the rosy view of the impact offered by the Morrison report. In particular, British Astronomer Royal and Nobelist Sir Martin Ryle fell in the Wald camp. After reading of Drake's attempt to signal (as opposed to listen to) extraterrestrial intelligence as part of the dedication of the newly resurfaced Arecibo dish, Ryle addressed an appeal to the International Astronomical Union urging that no attempts be made to signal other civilizations.[23] *The New York Times,* in an editorial entitled "Should Mankind Hide?" opted for the optimistic side, concluding that "there is no reason to assume that alien intelligence among the stars must be hostile or predatory."[24] Nor, of course, was there any reason to assume that it would be benevolent, nor did they seem worried by Wald's thesis of umbilical subservience.

Even a few SETI proponents urged caution. In discussing strategies for SETI, astronomers T. B. H. Kuiper and Mark Morris argued that the implications for humanity of a discovery of intelligence in space would depend on the relative stages of development of the two civilizations: "Before a certain threshold is reached, complete contact with a superior civilization (in which their store of knowledge is made available to us) would abort [our] further development through a 'culture shock' effect. If we were contacted before we reached this threshold, instead of enriching the galactic store of knowledge we would merely absorb it."[25] A terres-

23 Walter Sullivan, "Astronomer Fears Hostile Attack; Would Keep Life on Earth a Secret," *NYT,* November 4, 1976, p. 46, reprinted in Goldsmith, *The Quest,* 267–268.

24 "Should Mankind Hide?" *NYT,* November 22, 1976, p. A 24, reprinted in Goldsmith, *The Quest,* 269. Note the contrast to the *Times* opinion in 1919 in connection with Marconi's claim of signals possibly from Mars (Chapter 8, footnote 7).

25 T. B. H. Kuiper and Mark Morris, "Searching for Extraterrestrial Civilizations," *Science,* 196 (1977), 616, reprinted in Goldsmith, *The Quest,* 170–177: 175.

trial researcher, they argued, would lose motivation to pursue new ideas, "as the best human minds could be occupied for generations digesting the technology and cultural experiences of a society advanced far beyond our own." Realizing they were transcending the proper bounds of astronomy and physics, the authors did not elaborate, but called for further study.

During the 1980s no further consensus was reached on the implications of contact with extraterrestrials, nor, perhaps, could this have been expected. SETI enthusiasts such as Sagan took their message of hope to an ever-larger population. Sagan's *Cosmos* (1980) spread the message that discovering another civilization would be "a profoundly hopeful sign. . . . It means that someone has learned to live with high technology; that it is possible to survive technological adolescence." Evolutionary paleontologist Stephen Jay Gould argued, without committing to beneficial or catastrophic implications, that "A positive result [of SETI] would be the most cataclysmic event in our entire intellectual history." But others, notably philosopher Edward Regis, Jr., argued against profound consequences. If a signal was intercepted, he claimed, ""it is probably as likely to be wholly unintelligible as it is that it will bring us the answer to life, the universe, and everything." The resistance to decryption of even some earthly texts, in his view, did not inspire confidence that a truly alien message would be decipherable.[26]

The 1980s and 1990s did see the beginnings of systematic discussion, if not resolution, of some of the problems raised in the previous two decades. More tractable than the long-term impact, however, and increasingly important as search programs such as NASA's neared reality, was the problem of the receipt, verification, and announcement of a signal, which came to be subsumed under the heading of cultural impact, since the manner in which the announcement was made could well determine its impact. Considerable energy was spent on what came to be known as the "postdetection protocols," a discussion in which NASA and the International Academy of Astronautics (IAA) again took the lead. Building on work begun a few years earlier, in 1986 and 1987 the IAA and the International Institute of Space Law cosponsored a series of meetings during the annual Congress of the International Astronautical Federation (IAF), where SETI had been a topic of more technical discussions for two decades. The hope of these meetings was "to uncover particular princi-

[26] Sagan, *Cosmos* (New York, 1980); Stephen Jay Gould, "SETI and the Wisdom of Casey Stengel," in *The Flamingo's Smile: Reflections in Natural History* (New York, 1985), 413; E. Regis, Jr., "SETI Debunked," in E. Regis, Jr., ed., *Extraterrestrials: Science and Alien Intelligence* (Cambridge, 1985), 231–244. Two popular works also took up the subject: James Christian, ed., *Extraterrestrial Intelligence: The First Encounter* (Buffalo, N.Y., 1976), and B. Bova and Byron Preiss, *First Contact, The Search For Extraterrestrial Intelligence* (New York, 1990).

ples or guidelines that should be applied to the conduct of SETI programs and the communication of their extraordinarily important results." The discussions also went beyond that goal to discuss diplomatic and legal issues relating to contact and communication concerns raised in 1970 by the distinguished Austrian jurist Ernst Fasan. In the end, these efforts succeeded in drafting postdetection protocols in the aftermath of a message but did not address the more difficult question of response.[27] And only 2 of the 16 papers at the IAF meetings dealt with the thornier problem of the impact after a message was received.[28]

Whereas the IAF meetings were inspired by a desire to discuss the issues of near-term receipt, verification, and announcement of a signal, in the 1990s more systematic studies began to examine the entire scope of short- and long-term implications. In particular, under the direction of John Billingham and on the eve of the launching of its own observations, in 1991–1992 NASA sponsored a series of three workshops on Cultural Aspects of SETI (CASETI) to discuss the cultural, social, and political consequences of a successful detection of extraterrestrial intelligence. Toward this end, a variety of about 25 social and physical science specialists addressed the relevance of four broad areas to such a detection: history, behavioral science, national and international policy, and education. Dividing the historical viewpoints into "millenarian" and "catastrophist" categories, the CASETI historians found that history could offer useful, though not definitive, analogs of the discovery of intelligence. In this respect they recommended that further study be given not to physical culture contacts on Earth, but to intellectual contacts between cultures, as for example when Greek science and learning was passed via the Arab culture to the Latin West in the thirteenth century. Offering their own assessment of impact, the historians wrote that "it takes a very dour catastrophist to believe that mere knowledge of the existence of ETI could extinguish the manifold creative energies that have enabled mankind to overcome many sorts of menaces over the centuries."[29] The behav-

[27] J. C. Tarter and M. A. Michaud, eds., *SETI Post Detection Protocol*, Special Issue of *Acta Astronautica*, 21 (February 1990), 69–154. For an account by one of the principal participants in the history leading to these efforts, see M. A. Michaud, "A Unique Moment in Human History," in Bova and Preiss, eds., *First Contact*, 256–261. The postdetection protocols are published in "Declaration of Principles Concerning Activities Following the Detection of Extraterrestrial Intelligence," in J. C. Tarter and M. A. Michaud, *SETI Post-Detection Protocol*, 153–154. Fasan's book is *Relations with Alien Intelligences* (Berlin, 1970).

[28] Ben Finney, "The Impact of Contact," in Tarter and Michaud, *SETI Post-Detection Protocol*, 117–121; Roberto Pinotti, "Contact: Releasing the News," ibid., 109–115.

[29] J. Billingham, R. Heyns, D. Milne, et al., *Social Implications of Detecting an Extraterrestrial Civilization: A Report of the Workshop on the Cultural Aspects of SETI* (SETI Institute, 1994). The first report on this workshop was given by Stephen E. Doyle,

ioral scientists, however, emphasized the extremely diverse reactions likely to take place among many social and religious groups and cautioned that predicting any reaction was an extremely complex endeavor, even if mass education prior to detection attempted to precondition responses. This did not preclude the recommendations of the education specialists, who urged that announcement be made in the broadest possible manner. And the policy specialists, looking beyond the work of the IAA, encouraged broader institutional policies beyond the postdetection protocols already envisaged.

It seemed likely that by the end of the century the discussions of impact, so tentatively begun with the dawn of the Space Age, would rise to a new level of sophistication. The concrete achievement of the postdetection protocols seemed likely to be expanded and the ideas of longer-term impact debated. While consensus on the latter seemed unlikely, many participants undoubtedly shared the opinion of anthropologist Ashley Montagu, offered two decades earlier at the NASA symposium on "Life beyond Earth and Mind of Man." While counseling that that we should learn to communicate better among ourselves, Montagu nevertheless also declared the importance of preparation for extraterrestrial contact, no matter how unlikely the prospect of detection or communication: "I do not think we should wait until the encounter occurs," he urged. "We should do all in our power to prepare ourselves for it. The manner in which we first meet may determine the character of all our subsequent relations."[30] As a cautionary statement this was difficult to argue, and the proposed course of action seemed likely to carry the day at the end of the century.

10.2 ASTROTHEOLOGY

The question of the impact of extraterrestrial life on religion and theology has very deep roots, at least in the Western tradition. The problem was perceived in the fifteenth century in relation to the reconciliation of Chris-

"Social Implications of NASA's High Resolution Microwave Survey," IAF Congress, Graz, Austria, October 19, 1993, just 3 weeks after the U.S. Congress canceled the HRMS project.

[30] Berenzden, *Life beyond Earth,* 25. Another sign of increased interest in the problem of impact was the increasing number of papers on the subject at the international SETI meetings. In particular, the 1993 Bioastronomy Symposium in Santa Cruz, California, included papers on the subject, with I. Almar and S. J. Dick arguing the usefulness of historical analogues of terrestrial intellectual contact, as well as the value of studying the historical reception of scientific worldviews. I. Almar, "The Consequences of a Discovery: Different Scenarios," and S. J. Dick, "Consequences of Success in SETI: Lessons from the History of Science," in G. Seth Shostak, ed., *Progress in the Search for Extraterrestrial Life* (San Francisco, 1995), 499–505, 521–532.

tianity with the Aristotelian doctrine opposing a plurality of worlds. Most theologians by that time agreed that God could create other worlds. But if so, they wondered, "whether Christ by dying on this earth could redeem the inhabitants of another world." The standard answer was that he could, because Christ could not die again in another world.[31] Very early in the Protestant tradition, Martin Luther's supporter, Philip Melanchthon, not only objected to such a speculative idea but also used it as an argument against the Copernican theory. "It must not be imagined that there are many worlds, because it must not be imagined that Christ died and was resurrected more often, nor must it be thought that in any other world without the knowledge of the Son of God, that men would be restored to eternal life."[32] For Copernicans of any religious persuasion, the problem was a thorny one that extended beyond specific religious doctrine. Kepler stated the conundrum in the early seventeenth century in more general terms that might equally apply to other religions of the world: "If there are globes in the heavens similar to our earth, do we vie with them over who occupies a better portion of the universe? For if their globes are nobler, we are not the noblest of rational creatures. Then how can all things be for man's sake? How can we be the masters of God's handiwork?"[33]

These provocative Keplerian questions were still alive at the end of the nineteenth century, when H. G. Wells quoted them as the prelude to his novel *War of the Worlds*. By that time, as we have seen in Chapter 1, Christianity had explored these implications quite thoroughly. Despite Scriptural objections raised during the seventeenth century, by the early eighteenth century the Anglican priest and Royal Society Fellow William Derham reflected accepted theological opinion when he incorporated extraterrestrial life into natural theology; it is in the sense of inhabited worlds reflecting the magnificence of God's universe that Derham wrote his book *Astro-Theology*.[34] The matter did not rest there, however. Thomas Paine bluntly stated in his *Age of Reason* (1793) that extraterrestrials and Chris-

[31] Steven J. Dick, *Plurality of Worlds* (Cambridge, 1982), 43, 88, 198 (note 48); Grant McColley and H. W. Miller, "Saint Bonaventure, Francis Mayron, William Vorilong, and the Doctrine of a Plurality of Worlds," *Speculum,* 12 (1937), 388.

[32] Dick, *Plurality of Worlds,* 88–89, translating from Melanchthon's *Initia doctrinae physicae* (Wittenberg, 1550), fol. 43.

[33] *Kepler's Conversation with Galileo's Sidereal Messenger,* trans. Edward Rosen (New York, 1975), 43. For more on religious attitudes toward plurality of worlds prior to the mid-eighteenth century see Dick, *Plurality of Worlds,* and Karl S. Guthke, *The Last Frontier: Imagining Other Worlds from the Copernican Revolution to Modern Science Fiction* (Ithaca, N.Y., 1990).

[34] William Derham, *Astro-Theology: Or a Demonstration of the Being and Attributes of God from a Survey of the Heavens* (London, 1714). On the plurality of worlds concept in natural theology see Dick, *Plurality of Worlds,* chapter 6.

tianity did not mix and that "he who thinks that he believes in both has thought but little of either."[35] In a history that would repay study by those interested in the theological implications of an actual discovery of extraterrestrial intelligence, during the nineteenth century some writers rejected Christianity, others rejected plurality of worlds, and still others found ways to reconcile the two.

The twentieth century thus inherited a considerable discussion of the theological implications of extraterrestrial life, mostly within the Christian tradition, inspired by the mere *possibility* of intelligence beyond the Earth. Although the relation between theology and plurality of worlds occasionally reached the level of sustained debate in the eighteenth and nineteenth centuries, by the mid-twentieth century this controversy echoed only faintly in the background as scientists began to contemplate the possibility of a search for extraterrestrial intelligence. In the twentieth century, Derham's "astrotheology" assumed new meaning in light of efforts to detect signals from extraterrestrial intelligence, efforts that, if successful, would surely affect traditional theology, with its emphasis on the relation between God and humanity. Rather than focusing on *confirming* evidence of the glory of God in the best tradition of natural theology, astrotheology in the twentieth century came to describe the considerable *modifications* to theology and religion that might develop in the wake of the discovery of intelligence in the heavens.[36] It is this latter tradition that we trace in this section.

During the twentieth century, theological explorations of contact with extraterrestrial intelligence have been sporadic, peripheral, and never raised to the same level of substantial debate achieved in the previous two centuries.[37] Indeed, in the absence of a more explicit discussion of the problem by practitioners, the attitudes of many religions have had to be deduced by others from the general doctrines of those religions. The major exception has been Christianity, where the doctrine of Incarnation has

35 *Basic Writings of Thomas Paine* (New York, 1942), 67–78; M. Crowe, *The Extraterrestrial Life Debate* (Cambridge, 1986), 163. For details on religious reactions to plurality of worlds during the eighteenth and nineteenth centuries see Crowe, *The Extraterrestrial Life Debate,* and Guthke, *The Last Frontier.*

36 The term "astrotheology" was used in this sense in Kenneth Delano, *Many Worlds, One God* (New York, 1977), chapter 7, "Astrotheology." The term "exotheology" has also been used (e.g., in Lamm, "Religious Implication"), mirroring the uncertainty in terminology seen in the scientific aspects of the debate.

37 Twentieth-century religious responses have been explored at varying levels of detail in Guthke, *The Last Frontier;* W. Sullivan, *We are Not Alone* (New York, 1964), chapter 22, "What if We Succeed?"; and Michael Ashkenazi, "Not the Sons of Adam: Religious Responses to ETI," *Space Policy* 8, (1992), 341–350. See also W. Hamilton, "The Discovery of Extraterrestrial Intelligence – A Religious Response," in Christian, *Extraterrestrial Intelligence,* 99–114.

been a central focus of discussion, and where the consensus has been that a discovery of intelligence beyond the Earth would not prove fatal to the religion or its theology. In general, for Christianity as well as for other religions, theologians see little problem, while those external to religion proclaim the fatal impact of extraterrestrials on Earth-bound theologies.[38]

While most theologians would undoubtedly have preferred to remain silent on the subject, the issue was pushed into the public and theological consciousness by the approach of the Space Age. As Arthur C. Clarke, one of the prophets of the new era, remarked in his popular book *The Exploration of Space* (1951), some people

> are afraid that the crossing of space, and above all contact with intelligent but nonhuman races, may destroy the foundations of their religious faith. They may be right, but in any event their attitude is one which does not bear logical examination – for a faith which cannot survive collision with the truth is not worth many regrets.[39]

Religion could not for long avoid such a commonsense challenge, whose force could only increase as rocketry neared reality.

Christian flexibility is evident in the earliest twentieth-century discussions of the subject, centered at Oxford University, where (not coincidentally) C. S. Lewis had written his novels of Christian apologetics in a science fiction setting.[40] The first Oxford contribution came from a cosmologist's attempt to reconcile the modern view of the universe with Christianity. In one of a series of essays written just days before his sudden death in 1950, Oxford professor of mathematics Edward A. Milne took up the problem in connection with the second law of thermodynamics. Milne concluded that that the universe would not end in a "heat death" as entropy reached a maximum; rather, it would continue to exist with an infinite variety of forms. What, then, was the fate of this

[38] For a few religions, especially Mormonism, the concepts of other worlds and extraterrestrials were actually part of church doctrine. Erich Robert Paul, *Science, Religion and Mormon Cosmology* (Urbana, Ill., 1992), especially chapter 9, "Extraterrestrial Intelligence and Mormon Cosmology."

[39] Arthur C. Clarke, *The Exploration of Space* (New York, 1951), 191.

[40] Lewis, a fellow of Magdalen College, Oxford, focused on Christian doctrinal issues in his novels *Out of the Silent Planet* (1938), *Perelandra* (1943), and *That Hideous Strength* (1945). See our Chapter 5. The British philosopher Olaf Stapledon placed humanity in the cosmic context in *Star Maker* (1937). In the United States, Ray Bradbury had space travelers follow Christ on an interplanetary mission of salvation in "The Man" (1949), while James Blish, in his classic *A Case of Conscience* (1953), explored how a priest might deal with an inhabited planet lacking original sin. These were only the vanguard of an explosion of stories after World War II exploring theological issues of aliens via science fiction. See "Religion" in John Clute and Peter Nichols, *The Encyclopedia of Science Fiction* (New York, 1993), 1000–1003.

infinite variety? Recalling Eddington's statement that nature might require millions of acorns to grow a single oak, and his implication that an infinity of galaxies might be needed to produce a single planetary system, and an infinity of planetary systems to produce one with life, Milne found another possibility more attractive. "The infinity of galaxies," he wrote, could be considered "as an infinite number of scenes of experiment in biological evolution," a view Milne found more appropriate to an omnipotent God. Not only did Milne's God endow the universe "with the only law of inorganic nature consistent with its content," he also "tended his creation in guiding its subsequent organic evolution on an infinite number of occasions in an infinite number of spatial regions. That is of the essence of Christianity, that God actually intervenes in History."[41]

But, Milne noted, this raised a difficulty "which many Christians have felt" with regard to the most notable case of God's intervention in history – the Incarnation:

> Was this a unique event, or has it been re-enacted on each of a countless number of planets? The Christian would recoil in horror from such a conclusion. We cannot imagine the Son of God suffering vicariously on each of a myriad of planets. The Christian would avoid this conclusion by the definite supposition that our planet is in fact unique.

Milne himself was not satisfied with this supposition, and seeking a way to reconcile Incarnation and life in the universe, he found an answer in interstellar communication, one of the earliest references to the subject. Pointing to the new science of radio astronomy and the reception of radio signals from sources in the Milky Way announced two decades earlier, Milne suggested, "It is not outside the bounds of possibility that these are genuine signals from intelligent beings on other 'planets', and that in principle, in the unending future vistas of time, communication may be set up with these distant beings." In that case, Milne continued, even groups of galaxies would "by inter-communication become one system." Therefore, "there would be no difficulty in the uniqueness of the histori-

[41] E. A. Milne, "The Second Law of Thermodynamics: Evolution," in *Modern Cosmology and the Christian Idea of God* (Oxford, 1952), 151–153. Milne has been called "one of the foremost pioneers of theoretical astrophysics and modern cosmology" by his colleague G. J. Whitrow, *Dictionary of Scientific Biography* vol. 9 (New York, 1974), 404–406. Milne's father was headmaster of St. Mary's Church of England School, and Milne's studies at Trinity College, Cambridge, were directed by E. W. Barnes, later the Anglican Bishop of Birmingham, who had already expressed his belief in the possibility of interstellar radio communication in the 1930s (see Section 8.1 of this volume). More on Milne may be found in W. H. McCrea, "Edward Arthur Milne," *MNRAS*, 111 (1951), 160–170.

cal event of the Incarnation. For knowledge of it would be capable of being transmitted by signals to other planets and the re-enactment of the tragedy of the crucifixion in other planets would be unnecessary."[42] Were Milne alive today, therefore, he would undoubtedly champion SETI as a means of interplanetary salvation.

The second Oxford volley came only 4 years later, when Milne's clergical contemporary at Oxford, E. L. Mascall, agreed that Christianity could be flexible on the issue but disagreed with Milne's solution. As part of a broad-ranging discussion of the relation of science and theology, Mascall, an Anglican priest and lecturer in philosophy of religion, elaborated Milne's discussion as an example of how modern science could stimulate and expand the scope of theological discussion. Mere knowledge of terrestrial Incarnation brought about by interstellar radio communication was, in Mascall's view, not enough for universal salvation, as conceived in classical Christian thought. The latter, he believed, required a "hypostatic union" between Redeemer and redeemed, which Mascall viewed as possible on other planets, no less than on Earth:

> There are no conclusive *theological* reasons for rejecting the notion that, if there are, in some other part or parts of the universe than our own, rational corporeal beings who have sinned and are in need of redemption, for those beings and for their salvation the Son of God has united (or one day will unite) to his divine Person their nature, as he has united to it ours. . . ."[43]

Mascall went on to more theological technical points – of interest, one would imagine, only to theologians – and probably an indication of the kind of discussion that would follow if SETI were successful. Though he disagreed with Milne on the details, the important point was theological flexibility, and in concluding, Mascall emphasized "how wide is the liberty that Christian orthodoxy leaves to intellectual speculation."

The Oxford excursions into speculative theology, brief though they were, would not be entirely forgotten in subsequent decades. Nor, how-

[42] Milne, "Second Law," 153–154. It is notable that Milne remarks that many Christians have felt this difficulty, indicating that some earlier tradition, or perhaps C. S. Lewis's novels, were generating some discussion on the subject.

[43] E. L. Mascall, *Christian Theology and Natural Science* (London, 1956), 36–45. Mascall, ordained an Anglican priest in 1932, cites C. S. Lewis's *Perelandra* theme, wherein humanity is the only creature in the solar system who has fallen from grace. Mascall notes in his memoirs his interaction with C. S. Lewis via the Oxford Socratic Club, *Saraband: The Memoirs of E. L. Mascall* (Herefordshire, 1992), 253. Mascall also was familiar with E. W. Barnes, and cites his *Scientific Theory and Religion* (New York, 1933) as raising the question of extraterrestrial intelligence, though Barnes's view of Christ, he remarks disapprovingly, was far from the orthodox Christian point of view. Lewis, Milne, and Mascall, therefore, form a core group on the subject, with links to Barnes.

ever, were theologians especially anxious to encourage this kind of speculation. In offering his own view that Incarnation "is unique for the special group in which it happens, but it is not unique in the sense that other singular incarnations for other unique worlds are excluded," Protestant theologian Paul Tillich characterized the question as one that "has been carefully avoided by many traditional theologians, even though it is consciously or unconsciously alive for most contemporary people."[44]

Nevertheless, the actual dawning of the Space Age pressed reluctant theologians even further. By 1958 C. S. Lewis, now willing to transport his ideas from fiction to nonfiction, downplayed the effect of extraterrestrials on Christianity, holding that such a discovery would do no more to disprove its principles than did the ideas of Copernicus, Darwin, or Freud.[45] The problem of the Incarnation, he believed, "could become formidable," but only if a variety of conditions were met. Not only would intelligence have to be discovered beyond the Earth, it would also have to possess "rational souls," and it would have to have fallen from grace. If all of this were known to be true, then – and only then – do we need to worry about the mode of Redemption – whether it could take place on other planets or whether Christ, by dying on Earth, had already saved extraterrestrials. Lewis, professing to be skeptical about the possibility of intelligence on other worlds, concluded that "a Christian is sitting pretty if his faith never encounters more formidable difficulties than these conjectural phantoms."

Other theologians, however, felt the need to tackle the problem more squarely, perhaps in response to popular interest. The Catholic version of Christianity, like the Protestant, was remarkably open-minded on the subject.[46] Father Daniel C. Raible was typical of this open-mindedness when he wrote in the wake of Project Ozma:

44 P. Tillich, *Systematic Theology*, vol. 2 (Chicago, 1957), 95–96.

45 C. S. Lewis, *Christian Herald*, April 1958, reported in "Other-Worldly Faith," *Newsweek*, March 24, 1958, p. 64, and "Faith and Outer Space," *Time*, March 31, 1958, p. 37. The article, "Will We Lose God in Outer Space?" was reprinted for wide distribution in pamphlet form, a copy of which may be found in the Georgetown University Special Collections. Guthke, *The Last Frontier*, 201, traces back to the eighteenth century Lewis's view that planetary inhabitants might be free from sin. Lewis returned to this theme in one of the last articles of his life, "Onward Christian Spacemen!" *Catholic Digest* (August 1963), 90–95: 94–95.

46 Positive opinions are given by the Jesuits Angelo Perego, "Origine degli esseri razionali extra-terreni," *Divus Thomas* (Piacenza, 1958), 3–24, summarized in "Rational Life beyond the Earth?" *Theology Digest*, 7 (Fall 1959), 177–178; and Domenico Grasso, theology professor at the Gregorian University in Rome, "Missionaries to Space," *Newsweek*, February 15, 1960, p. 90. J. Edgar Bruns, a priest at St. John's University in New York City, also argued that Christian theology could be expanded to include extraterrestrials in "Cosmolatry," *The Catholic World* (August 1960), 283–287, as did T. J. Zubek, O.F.M., "Theological Questions on Space Creatures," *The American Ecclesiastical Review*, 145 (December 1961), 393–399.

> Yes, it would be possible for the Second Person of the Blessed Trinity to become a member of more than one human race. There is nothing at all repugnant in the idea of the same Divine Person taking on the nature of many human races. Conceivably, we may learn in heaven that there have been not one incarnation of God's son but many.[47]

The Catholic Church also had an eye on history; quoting a cardinal that "one Galileo case is quite enough in the history of the Church," an editorial in one Catholic journal suggested that "today's theologians would welcome the implications that such a discovery might open – a vision of cosmic piety and the Noosphere even beyond that of a Teilhard de Chardin."[48]

The most substantial theological discussion of the subject, and the closest the Roman Catholic Church came to an official position, was given by the priest Kenneth Delano in his book *Many Worlds, One God* (1977). Complete with the official "nihil obstat" and "imprimatur" sanctions, the author's position was that any person with a religious faith including "an adequate idea of the greatness of God's creative ability, of humanity's humble position in the universe, and of the limitless love and care God has for all His intelligent creatures" should not be afraid to examine the implications of intelligence in the universe.[49] Delano characterized the fears of some in the religious community with regard to extraterrestrials as analogous to early Church skepticism that any humans could live in the terrestrial "antipodes" because none of Adam's descendants could have reached the Southern Hemisphere. Reacting to an early-twentieth-century writer who claimed that "If he [man] is not the greatest, the grandest, the most important of created things, the one to whom all else is made to contribute, then the Bible writers have misrepresented entirely man's relation to God and the universe," Delano pointed out that God was not obliged to reveal

[47] "Daniel C. Raible, "Rational Life in Outer Space?" *America: National Catholic Weekly Review,* 103 (August 13, 1960), 532–535, condensed in *Catholic Digest* (December 1960), 104–108.

[48] "Messages from Space," *America,* vol. 111 (December 12, 1964), 770. In his major work, *The Phenomenon of Man* (in French, Paris, 1938–1940; in English, New York, 1959), the Jesuit priest and paleontologist Pierre Teilhard de Chardin (1881–1955) put forth the concept that life on Earth would evolve toward a super-sapient "noosphere." Though in 1953 Teilhard extended the idea to other planets, for most of his career he considered extraterrestrial intelligence an idea "probability too remote to be worth dwelling on" (New York, 1975 ed., 286). For the extension of the noosphere concept to other planets see de Chardin, *Christianity and Evolution* (New York, 1971), 229. See also Barrow and Tipler (footnote 68), 195–204. Teilhard's concept was also elaborated in connection with ETI in Father J. Donceel, "A Pangalactic Christ," *Continuum* (Spring 1968).

[49] Kenneth Delano, *Many Worlds, One God* (Hicksville, N.Y., 1977).

extraterrestrials in the Bible, when it would have served no moral purpose.[50] The Space Age requires a theology that is neither geocentric nor anthropomorphic, he argued, and it follows that the Earth may not be the only planet that has seen an incarnation: "Any one or all three Divine Persons of the Holy Trinity may have chosen to become incarnated on one or more of the other inhabited worlds in the universe." This he considered much more likely than a theory of the "cosmic Adam," in which the single redemptive act by Christ on Earth is applicable to the entire universe. On the other hand, humanity's "mission" could be to spread the Gospel among the inhabited planets while refraining from any form of religious imperialism.[51] The Church, while spreading the story of terrestrial redemption, might also encourage fallen races to seek salvation. In Delano's view, Alice Meynell's poetic vision of the universe was edifying:

> . . . in the eternities
> Doubtless we shall compare together, hear
> A million alien Gospels; in what guise
> He trod the Pleiades, the Lyre, the Bear.

Although Delano made it clear that Catholic opinion was not unanimous,[52] he certainly reinforced the prevalent idea of flexibility toward a discovery of extraterrestrials in the Church doctrines.

The same flexibility was expressed in a study of religious implications of the problem for Jewish thought, where the primary concern was, of course, not the Incarnation, but the uniqueness of humanity and its relationship to God. Cautioning that extraterrestrial intelligence was far from proven, Rabbi Norman Lamm nevertheless pointed to precedents in medieval Jewish thought and declared that in the spirit of open-mindedness toward new knowledge, it was prudent to explore "a Jewish exotheology, an authentic Jewish view of God and man in a universe in which man is not the only intelligent resident, and perhaps inferior to many other races."[53] Medieval

50 Ibid., xv, 9. The writer cited is L. T. Townsend, *The Stars Not Inhabited* (New York, 1914).

51 Ibid., 110–120. On the cosmic Adam theory, Delano, (ibid.,) cites Richard J. Pendergast, "Terrestrial and Cosmic Polygenism," in *Science and Faith in the 21st Century* (1968).

52 See the debate in J. A. Breig and L. C. McHugh, "Other Worlds – For Man," *America*, 104 (November 26, 1960), 294–296, and the agonizing discussion in Charles Davis, "The Place of Christ," *The Clergy Review*, new series, 45 (December 1960), 706–718. The latter concluded that while "no absolute impossibility excludes other incarnations," there was no reason to suppose them, and other incarnations would reduce the Christian scheme "to a merely local significance in this universe." See also L. C. McHugh, "Life in Outer Space," *The Sign* (December 1961), 26–29. For the Meynell poem, see footnote 3.

53 N. Lamm, "The Religious Implication of Extraterrestrial Life," in A. Cannell and C. Domb, eds., *Challenge: Torah Views on Science and Its Problems* (2nd edition, New

Jewish philosophy had rejected the uniqueness of humanity, Lamm pointed out, but the nonsingularity of humanity did not mean insignificance. Harvard astronomer Harlow Shapley and others, he argued, were "profoundly mistaken" in assuming that the number of intelligent species had any relation to the significance of humanity, and even more mistaken in holding that a peripheral position in the Galaxy implied metaphysical marginality and irrelevance. That "geography determines metaphysics" he called a "medieval bias" that should have disappeared with the collapse of geocentrism. Judaism, therefore, "could very well accept a scientific finding that man in not the only intelligent and bio-spiritual resident in God's world" as long as the insignificance of man was not an accompanying conclusion. Man could still be considered unique in "spiritual dignity," and the existence of innumerable intelligences does not lessen God's attention to man. "A God who can exercise providence over one billion earthmen," Lamm concluded, "can do so for ten billion times that number of creatures throughout the universe. He is not troubled, one ought grant, by problems in communications, engineering, or the complexities of cosmic cybernetics. God is infinite, and He has an infinite amount of love and concern to extend to each and every one of his creatures."[54]

Within various religions, therefore, the consensus was that terrestrial religions would adjust to extraterrestrials, an opinion echoed in a recent study of religious attitudes.[55] And, as the same study also pointed out, if the "Adamist religions" of Judaism, Christianity, and Islam – those that share a view of the creation of humanity that links it directly to the godhead – can survive extraterrestrials, non-Adamist religions such as Buddhism, Hinduism, Confucianism, or Taoism should have no trouble. In fact, there is already some indication that this would be the case.[56]

York, 1978), 354–398: 371, first published in Lamm, *Faith and Doubt – Studies in Traditional Jewish Thought* (New York, 1971), chapter 5. Among Jewish thinkers, Lamm noted, "the 11th century halachic authority Rabbi Judah ben Barzilai of Barcelona, in his commentary to *Sefer Yetzirah,* speculated concerning the probable existence of intelligent beings on the 18,000 'other worlds' postulated in the Talmud. In the 14th century the great halachist and philosopher Rabbi Hasdai Crescas, who was the mentor of Rabbi Joseph Albo, author of the *Sefer Ha-ikkarim,* also considered at length the possibility of life in other worlds from the point of view of Jewish theology," 359.

54 Ibid., 394.

55 Michael Ashkenazi, "Not the Sons of Adam: Religious Response to ETI," *Space Policy,* 8 (1992), 341–350. In interviews with 21 religious authorities from a variety of religions, the author found that none believed ETI created a theological or religious problem, not even the 17 who were virtually certain ETI existed.

56 Guthke, *The Last Frontier,* 17, cites Wolfgang Bauer, *China und did Hoffnung auf Gluck* (Munich, 1971); Joseph Needham, *Science and Civilization in China,* vol. 2 (Cambridge, 1956), 219–221, 419, 438–442; and Edward H. Schafer, *Pacing the Void: T'ang Approaches to the Stars* (Berkeley and Los Angeles, 1977). See also Ronald Huntington, "Mything the Point: ETIs in a Hindu/Buddhist Context," in Christian, *Extraterrestrial Intelligence,* 115–127.

This flexible attitude of the world's religions toward extraterrestrials, however, was not shared by some external to those religions. In particular, this was the view taken in perhaps the most detailed discussion of the impact of extraterrestrials on theology by a nontheologian, philosopher Roland Puccetti. Writing in 1968 in the broad philosophical context of the concepts of "persons" and "moral agents in the universe," Puccetti argued that the world's religions would not be able to survive the implications of an inhabited cosmos. He believed monotheistic religions in particular, involving a personal relationship between a creator and its creatures, would be in trouble. It is not only that "the prospect of extraterrestrial intelligence, concerning which the principal sacred writings of Christianity, Judaism, and Islam are absolutely silent, generates a profound suspicion that these terrestrial faiths are no more than that."[57] An even graver problem, in Puccetti's view, was the "particularism" of these religion, especially when it comes to doctrines like the Incarnation. Referring to the Milne–Mascall positions regarding that doctrine, he found neither satisfying, Milne's because the size of the universe precluded spreading the news by radio communication to all potential planets and Mascall's because his proposed "hypostatic union" led to multiple divine beings in the universe.[58]

For the Oriental faiths, Pucetti believed that such a discovery would be a far less serious matter from the point of view of doctrine, since "in their 'higher' forms at least they teach salvation through individual enlightenment and conceive the supreme Reality in strictly impersonal terms." Even they, however, suffer from a characteristic "that may be said to affect all terrestrial religions without distinction." This is what Puccetti terms "particularism," those traits that make each religion specific not only to the Earth, but even to particular cultures on the Earth. For example, Confucianism and Taoism are deeply immersed in Chinese culture, the former bound up with social, political, and familial duties and the latter with Chinese superstitions. Hinduism has its caste system and regard for sacred cows; Shintoism holds that the first divine creation was the islands of Japan; Judaism is intimately tied to a particular part of the human race, the "chosen people"; Islam has its Mecca and its feasts tied to the Moon; and Christianity, of course, has the greatest

[57] Roland Puccetti, *Persons: A Study of Possible Moral Agents in the Universe* (London, 1968), chapter 5, "Divine Persons," 121–145: 125–126.

[58] Ibid., 135ff. Puccetti's approach raises a related problem: the moral question of how to behave toward extraterrestrials when they could be very different forms of intelligence. On this see Jan Narveson, "Martians and Morals: How to Treat an Alien," in Regis, *Extraterrestrials,* 245–265; and Michael Tooley, "Would ETIs Be Persons?" in Christian, *Extraterrestrial Intelligence*. 129–145.

particularism in its emphasis on Jesus Christ and the Incarnation. Particularism, Puccetti argued, has two problems. First, because of this trait, no one religion can claim to be universal. Second, if there are multiple independent religions on each potentially inhabited planet, then there are billions of independent claims to truth, surely undermining any concept of truth one would wish to offer. And if one decides to take the easy way out and deny the possible existence of extraterrestrials, Puccetti argued, this makes terrestrial faiths falsifiable, something we surely wish to avoid as well. From these considerations, Puccetti concluded, the existence of extraterrestrials causes us to abandon all particularistic religions, and thus all terrestrial religions, since none of them may be considered universal.

The dichotomy between internal and external attitudes of the impact on religion and theology is highlighted by the response to Puccetii. Notre Dame University philosopher Ernan McMullin rejected Puccetti's claim that the Christian doctrine of Incarnation made no sense in an inhabited universe. His response confirmed the now characteristic attitude of flexibility adopted by many religions toward extraterrestrials: "There is an odd, ungenerous fundamentalism at work here, a refusal to allow for the expansion of concept, the development of doctrine, that is after all characteristic of both science and theology."[59] The most important thing about Puccetti's book, McMullin concluded, was that it drew attention to the need for more study of theology and philosophy in the context of new astronomical discoveries, including the possibility of extraterrestrial intelligence.

In retrospect, the internal religious response to extraterrestrial intelligence is perhaps not surprising, considering that the alternative to adaptation is destruction. While maintaining flexibility, these religions did not underestimate the challenge. As Delano said, "To build up an immunity to the possibility of a shock that could put an end to, or at least cripple, the practice of religion, we should all do well to exercise the virtue of humility in evaluating our standing in God's creation." This sentiment was echoed by McMullin, who noted that "a religion which is unable to find a place for extraterrestrial persons in its view of the relations of God and the universe might find it difficult to command terrestrial assent in days to come."[60] Still, McMullin admitted in 1980, theologians have

[59] Ernan McMullin, "Persons in the Universe," *Zygon,* 15, no. 1 (March 1980), 69–89: 88. McMullin's review (incorporating his review of the same volume 10 years earlier) dealt with the logical fallacies of the SETI arguments.

[60] Delano, *Many Worlds,* 11; McMullin, "Persons in the Universe," 69–89.

been mostly silent on this question, "no doubt feeling that the problems of earth are more than enough to occupy them."[61]

No true astrotheology was developed in the twentieth century in the sense that new theological principles were created, or existing ones formally modified, to embrace other moral agents in the universe. Although the mere possibility of extraterrestrial intelligence generated sporadic attempts at a universal theology, systematic astrotheology will probably be developed only when – and if – intelligence is discovered beyond the Earth. In the meantime, merely posing the problem demonstrates the anthropocentricity of our current conceptions of religion and theology and suggests that they should be expanded beyond their parochial terrestrial bounds. Though theologians have gone some way toward addressing Clarke's challenge, even the theological legacy of the Space Age in a broader sense is still unfulfilled. And as C. S. Lewis suggested, if extraterrestrials are actually discovered, the problem will become much more urgent.

In the end, the effect on theology and religion may be quite different from any impact on the narrow religious doctrines that have been discussed during the twentieth century. It may be that in learning of alien religions, of alien ways of relating to superior beings, the scope of terrestrial religion will be greatly expanded in ways that we cannot foresee. It may even be that, as a search for superior beings, the quest for extraterrestrial intelligence is itself a kind of religion. Even Puccetti argued that if extraterrestrials doomed religions, the religious attitude – the striving for "otherness," for something beyond the individual that offers understanding and love – might survive. "It is in that direction, and not in the direction of sophisticated machines or theological abstractions, that a plausible 'otherness' lies. Thus the religious attitude, or an important element of it, may yet survive the death of all determinate religious beliefs." It may be that religion in a universal sense is defined as the never-ending search of each civilization for others more superior than itself. If this is true, then SETI may be science in search of religion, and astrotheology may be the ultimate reconciliation of science with religion.

[61] It should be emphasized that a few theologians stoutly resist embracing extraterrestrials. One Catholic priest has warned that extraterrestrial intelligence (ETI) may undermine morality: "The cavalier approach of ETI champions to the question of human intellect is acting in the cultural context in a most harmful way. The ETI program is a systematic devaluation of the dignity of human intellect. The ETI program is doing now on a far vaster scale . . . what had been done earlier by the popularization of mechanistic science." He even warned that "ETI historians who take lightly the moral relativism latent in the ETI program are so many unwitting travel guides into the land of anarchy." Stanley L. Jaki, "Extraterrestrial Intelligence and Scientific Progress," paper presented at the 1985 History of Science Society meeting, Bloomington, Indiana.

10.3 LIFE AND PURPOSE IN THE UNIVERSE: THE ANTHROPIC PRINCIPLE

In the last half of the twentieth century, as the search for extraterrestrial life reached new heights, another approach to the meaning of life came from an unexpected quarter: cosmology itself. This approach, known as the "Anthropic Principle," was introduced in 1961 (though not under that name), was elaborated during the 1970s, and by the last decade of the century was a topic of considerable discussion among cosmologists and philosophers. In some ways a counterpoint to the "theistic principle" that God designed the universe, the anthropic principle found significance in the fact that the universe in its deepest structure and most fundamental properties seemed tailor-made for humanity; in order to have produced life and intelligence, those fundamental properties could not have been much different from those we now observe. A link between the physical and biological universes, between mind and matter, the anthropic principle may be seen as a secular search for the meaning of life based on physical principles rather than theological dogma. Transcending the usual critical but more parochial concerns about local conditions for life, it gave the extraterrestrial life debate a cosmological component of the broadest possible scope.

Though the suitability of the cosmos for life is clearly an argument that applies to life anywhere in the universe, it is revealing that in its early formulations at least, the idea was applied only to humanity (thus the term "anthropic"). Contemplating the work of Arthur S. Eddington, Paul Dirac, and P. Jordan on certain coincidences in the relations between fundamental physical constants,[62] in 1961 Princeton physicist Robert Dicke pointed out that life could not exist at any random time in the history of the universe, but only after the universe had reached a certain age. Thus, the physical constant known as the "Hubble age" of the universe "is not permitted to take on an enormous range of values, but is somewhat limited by the biological requirements to be met during the epoch of man." Among those requirements was that the universe must be old enough for carbon to have been formed inside stars and then distributed, since physicists and their fellow humans are carbon based. And the universe must be young enough to provide a hospitable home in the form of a planet circling a stable star, yet not so old that the inevitable stellar

[62] Dicke cites A. S. Eddington, *Theory of Protons and Electrons* (Cambridge, 1936); P. A. M. Dirac, *Proceedings Royal Society A*, 165 (1938), 199; and P. Jordan, *Schwerkraft und Weltall* (Braunschweig, 1955). On the history of "large number coincidences" see Barrow and Tipler (footnote 68), "The Rediscovery of the Anthropic Principle," 219–287. For the context of Dirac's work see Helge Kragh, *Dirac: A Scientific Biography* (Cambridge, 1990), especially chapter 11, "Adventures in Cosmology."

death precludes life. Thus, Dicke concluded (arguing against Dirac's idea that the fundamental "constants" might vary with time), the age of the universe as observed by us is not random, "but is limited by the criteria for the existence of physicists."[63]

This early formulation of the idea that our very existence tells us something about the fundamental constants, though broached only 1 year after Drake's Project Ozma had raised consciousness about extraterrestrial intelligence, still spoke only in terms of how the "epoch of humanity" constrained the age of the universe. In fact, Dicke need not have generalized, since nothing was known about the existence of extraterrestrials, and the argument derived its force from the existence of the one set of intelligent observers known with certainty to exist. But Dirac's reply in the same issue of *Nature* indicated that he did not agree with Dicke's argument precisely because it placed limits on life in the universe. On Dicke's assumption of nonvarying constants, he wrote, "habitable planets could exist only for a limited period of time. With my assumption they could exist indefinitely in the future and life need never end. There is no decisive argument for deciding between these assumptions. I prefer the one that allows the possibility of endless life."[64]

Despite Dirac's objection, however, Dicke's concept was revived more than a decade later, and in such anthropocentric terms that it was now called the "Anthropic Principle." The occasion was a 1973 symposium on "Confrontation of Cosmological Theory and Observational Data," in which, at the suggestion of quantum physics pioneer John Wheeler, Cambridge University physicist Brandon Carter expanded on Dicke's idea. Influenced by his reading of Hermann Bondi's *Cosmology* (1959), Carter pointed out that the anthropic principle could have been used to predict certain "large number coincidences" in cosmology that Bondi described, following in the footsteps of Eddington and Dirac. In its simplest form, dubbed the "Weak Anthropic Principle," the idea was that "what we can

[63] R. H. Dicke, "Dirac's Cosmology and Mach's Principle," *Nature,* 192 (November 4, 1961), 440–441. Dicke's argument had stemmed from Dirac's observation that three of the fundamental constants – the mass of the universe, the gravitational constant G, and the Hubble age of the universe – seemed to be related, since the first was 10^{80} and the latter two were, respectively, 10^{-40} and 10^{40}. Dirac believed this was too much of a coincidence; that there was some fundamental relation between cosmology and atomic theory; and that since the Hubble age obviously changed with time, so might the other so-called constants. Dicke argued that because the mass of the universe stayed the same, the mass of the universe and the gravitational constant *G* must not vary, since they were linked by Mach's principle that an object's inertia is due to forces from all the matter in the universe. And if this was true, then only the Hubble age varied, and we lived in a special time when life could develop and survive.

[64] Dirac, reply to Dicke, "Dirac's Cosmology," 441. This is an example – presumably after the fact – of an aesthetic consideration influencing a theory. As Kragh notes (*Dirac,* 236), it was Dirac's first public announcement on cosmology in 22 years.

expect to observe must be restricted by the conditions necessary for our presence as observers." Thus, Carter cited Dicke's example that the Hubble age (or expansion rate) of the universe is constrained by the fact that we exist. Moreover, he went on to expound a "Strong Anthropic Principle" based on other constants, namely, that "the Universe (and hence the fundamental parameters on which it depends) must be such as to admit the creation of observers within it at some stage," a much more problematic claim. And he pointed out that the gravitational constant *G* was also critical to the existence of life, for if it were slightly weaker, only red stars would form, and if it were slightly stronger, only hot blue stars would form, neither being able to harbor life-supporting planets. In an ensemble of universes with all conceivable combinations of fundamental constants, Carter wrote, the existence of an observer will be possible only for certain restricted combinations of parameters. Although such reasoning turned the deductive method on its head, Carter argued that it might be considered a kind of explanation for why our universe is the way it is.[65]

Without addressing the question of extraterrestrial life, Carter viewed the Anthropic Principle as "a reaction against exaggerated subservience to the 'Copernican Principle.' " This, he pointed out, in its most extreme form had resulted in the "perfect cosmological principle" at the foundation of the steady-state theory, a theory since disproven. Carter also cautioned that just because we do not occupy a privileged *central* position in the universe, this does not mean that our position cannot be privileged in any sense. Even if there were other life forms in the universe, our position in the universe as observers was a privileged one – at least as privileged as theirs. Thus, even if there were extraterrestrials (a point that Carter did not discuss), the Anthropic Principle was conceived from the beginning as a step away from Copernicanism and toward anthropocentrism.

Going one step further, and drawing on his pioneering role in quantum physics, John Wheeler proposed a Participatory Anthropic Principle in which the observer has an intimate relationship with the origin of the universe. "Has the universe had to adapt itself from its earliest days to the future requirements for life and mind?" Wheeler asked, accustomed from quantum physics to statements that made no common sense. "Until we understand which way truth lies in this domain, we can very well agree

[65] Brandon Carter, "Large Number Coincidences and the Anthropic Principle in Cosmology," in M. S. Longair, ed., *Confrontation of Cosmological Theories with Observational Data* (Dordrecht, 1974), 291–298. Fifteen years later, Carter indicated that the term "anthropic" was regrettable and that "observational biassing principle" was better. See Carter (footnote 77), 185–187.

that we do not know the first thing about the universe."[66] It is possible, Wheeler suggested, that in the same way that quantum physics found the observer and the observed to have a close and totally unexpected relationship, the study of the physical world "will lead back in some now-hidden way to man himself, to conscious mind, tied unexpectedly through the very acts of observation and participation to partnership in the foundation of the universe." In its preoccupation with the relationship between observer and observed, Wheeler viewed the Anthropic Principle as analogous to something like the uncertainty principle in quantum mechanics.

While some began to wonder whether the Anthropic Principle was really saying anything at all, cosmologists especially could not resist developing the idea further. In 1979 B. J. Carr and and Martin Rees showed how the Anthropic Principle could be used to determine most of the fundamental constants of physics; in other words, life was remarkably sensitive to cosmologically related numerical values. But at the same time, they realized that from a conventional physical point of view, the anthropic explanation of these coincidences was unsatisfactory, in part because it was based on an unduly anthropocentric concept of the observer.

> The arguments invoked here assume that life requires elements heavier than hydrogen and helium, water, galaxies, and special types of stars and planets. It is conceivable that some form of intelligence could exist without all of these features – thermodynamic disequilibrium is perhaps the only prerequisite that we can demand with real conviction.[67]

Still, in these terms, the existence of humanity outweighed the uncertainty about exotic extraterrestrials, thus focusing on humanity as the only known center of the biological universe.

The idea that life was very sensitive to the values of the fundamental physical constants need not have been anthropocentric in the sense that it implied, particularly in its stronger forms, that life was somehow essential to the cosmos. To the extent that the only life we know of *is* terrestrial, and that we consequently know nothing about the properties of any potential extraterrestrial life, the focus on humanity was perhaps justified. But anthropocentrism could be carried to greater extremes in this context, and was by physicists J. D. Barrow and Frank Tipler, whose

[66] J. A. Wheeler, in *The Nature of Scientific Discovery*, Owen Gingerich, ed., reprinted in "The Universe as Home for Man," *American Scientist* (November–December 1974), 683–691: 688–689.

[67] B. J. Carr and M. J. Rees, "The Anthropic Principle and the Structure of the Physical World," *Nature*, 278 (April 12, 1979), 605–612. They also pointed out that the principle was post hoc and that it did not explain the exact values of the constants, only their order of magnitude.

massive volume *The Anthropic Cosmological Principle* (1986) provided by far the most detailed treatment of the subject. In this volume, the concept became strongly anthropocentric in conjunction with arguments that extraterrestrial life did not exist. Placing the idea of an Anthropic Principle in the context of a long history of design arguments in theology, Barrow and Tipler argued – contrary to the opinion of the vast majority of contemporary scientists – that teleology (the search for purpose) may have a role in modern science, especially in biology. Having laid the foundations for their Anthropic Principle, they detailed that teleological role in physics and astrophysics, cosmology, quantum mechanics, and biochemistry.[68] And they pointed out that the principle they were championing was a much more general version of the "fitness of the environment" argument proposed by Harvard biochemist Lawrence J. Henderson seven decades before.[69]

This general point of view would seem to favor life in the universe, since it meant that life could arise anywhere that both the fitness of the environment and the fitness of the universe were favorable. But in fact, the Anthropic Principle had more ambiguous implications, and was used by Barrow and Tipler to argue against extraterrestrial life. Carter had first emphasized that the time for the evolution of intelligence on the Earth (5 billion years) was within a factor of 2 of the entire time available before life around a main sequence star would become impossible (10 billion years). In other words, if the average time for the evolution of intelligence in the universe exceeded by only a factor of 2 that evolutionary time on Earth, intelligence could not evolve elsewhere in the universe.[70] Barrow and Tipler accepted this as an argument against extraterrestrial intelligence, though they emphasized that a testable prediction of Carter's claim was the uniqueness of terrestrial intelligence in the Galaxy, and that his argument would collapse if any SETI program made a successful detection.[71] They also pointed out that while SETI proponents

68 J. D. Barrow and Frank Tipler, *The Anthropic Cosmological Principle* (New York, 1986). See also Barrow, "Anthropic Definitions," *QJRAS*, 24 (1983), 146–153.

69 L. J. Henderson, *The Fitness of the Environment* (Cambridge, Mass., 1913), reprinted with an Introduction by George Wald (Gloucester, Mass., 1970), and *The Order of Nature* (Cambridge, Mass., 1917) give the anthropic argument as applied to biochemistry. On Henderson see Barrow and Tipler, *Anthropic Cosmological Principle*, 143–148.

70 Carter, "Large Number Coincidences" and "The Anthropic Principle and Its Implications for Biological Evolution," *Philosophical Transactions of the Royal Society A*, (1983), 347; also in *The Constants of Nature*, ed. W. H. McCrea and J. J. Rees (London, 1983).

71 Barrow and Tipler, *Anthropic Cosmological Principle*, 556–575. The space travel argument, entirely independent of the anthropic principle, is elaborated in chapter 9. The great majority of Barrow and Tipler's book deals with life on Earth; only in these few instances does extraterrestrial life enter in a significant way – all negative.

believed that intelligence was an inevitable outcome of the evolutionary process, many evolutionary biologists did not agree with this application of teleology. Thus, they held that intelligence on Earth was not foreordained and that it may not have occurred anywhere else in the Galaxy. The presence of intelligence on Earth, they argued, can be understood only by the Weak Anthropic Principle, namely, that "only on that unique planet on which it occurs is it possible to wonder about the likelihood of intelligent life."[72] In conjunction with the "space travel" (Where are they?) argument against extraterrestrial intelligence, of which we have seen in Chapter 8 that Tipler was a prime advocate, the Anthropic Principle for these physicists became truly an anthropocentric view of the universe. Ruling out extraterrestrials left nothing but humanity as the measure of the universe. Indeed, anthropocentrism may be seen as the main argument of the Barrow and Tipler book as a whole, since the space travel argument was otherwise irrelevant to the Anthropic Principle. In the quarter century from Dicke's first enunciation of the concept in 1961 to Barrow and Tipler's volume, the Anthropic Principle was transformed from an incidentally anthropocentric to a purely anthropocentric concept. That this occurred as proponents of extraterrestrial life were becoming more vociferous is perhaps no coincidence.[73]

Although not everyone agreed with an extremely anthropocentric form of the Anthropic Principle, by the 1980s it was a well-established if still controversial subject that found a place in cosmology texts as well as in popular literature.[74] Commenting on the unusual post hoc mode of reasoning, philosopher George Gale found the utility, and even the legitimacy, of the principle still questioned by both philosophers and cosmologist as of 1981. Philosopher Kenneth Winkler believed anthropic reasoning was

[72] Ibid., 124. Yet another argument that might be taken as a negative one for extraterrestrial life is Wheeler's idea that the universe must be as large as it is for intelligent life to exist. Otherwise, it would not be old enough for life to have developed. See ibid., 385, and 451, note 30. Although this does not prove that there is only one case of intelligence in the universe, it *is* an answer to those who argued teleologically that surely life must exist elsewhere because otherwise the entire universe would be empty.

[73] Applying teleological reasoning to Barrow and Tipler, one might wonder if they themselves had a purpose in arguing that the universe was made for humanity. See, for example, Tipler's book, *The Physics of Immortality: Modern Cosmology, God and the Resurrection of the Dead* (New York, 1995), in which he argues his Omega Point Theory for resurrection of the dead, the existence of God, and a Judeo-Christian heaven.

[74] E. R. Harrison, *Cosmology: The Science of the Universe* (Cambridge, 1981), 111–112. Paul Davies, *Other Worlds* (New York, 1980), chapter 8, "The Anthropic Principle," 142–161, summarized in "The Anthropic Principle and the Early Universe," *Mercury* (May–June 1981), 66–77; see also *The Accidental Universe* (Cambridge, 1982). Among other popularizations see Dietrick Thomsen, "A Knowing Universe Seeking to Be Known," *Science News,* 123 (February 19, 1983), 124, and Ann Finkbeiner, "A Universe in Our Own Image," *S&T* (August 1984), 107–111.

based on concealed anthropocentric assumptions, and wondered further why we should even broach the idea that life could bring about an adjustment of physical constants if we had no idea how this could be done. Others rendered even harsher judgments. Heinz Pagels found it

> a farfetched explanation for those features of the universe which physicists cannot yet explain. Physicists and cosmologists who appeal to anthropic reasoning seemed to me to be gratuitously abandoning the successful program of conventional physical science of understanding the quantitative properties of our universe on the basis of universal physical laws.

Pagels found the interminable debate about the status of the principle to be symptomatic of its chief weakness – that there was no way to test it. He found it to be "needless clutter in the conceptual repertoire of science" and claimed that its influence had been sterile. Life is not a principle acting on the laws of nature, he wrote, but rather a consequence of them.[75] Contrasting the Anthropic Principle with the Theistic Principle, Pagels noted that because many scientists were unwilling to accept God, the human-centered principle was "the closest some atheists can get to God."

Finally, astronomer Fred Hoyle undermined the whole anthropic program when he concluded that although the occurrence of life is the greatest of all problems, "It is not so much that the Universe must be consistent with us as that we must be consistent with the Universe. The anthropic principle has the position inverted."[76] Instead of cosmology telling us about life, he suggested, we should let life (which we know more about) tell us about cosmology. Like the UFO debate and other controversies we have seen repeatedly in the history of the extraterrestrial life debate, the controversy over the Anthropic Principle is yet another case demonstrating the diversity of scientific cultures, particularly when it comes to scientific method.

Despite the criticism, the Anthropic Principle continued to be discussed by both philosophers and cosmologists as the century entered its last dozen years. Reviewing Barrow and Tipler's book, cosmologist G. F. R. Ellis questioned many of their claims but concluded that they had raised

[75] Heinz Pagels, *Perfect Symmetry* (New York, 1985), 358–360; George Gale, "The Anthropic Principle," *SciAm*, 245, 12 (December 1981), 154–171; Kenneth Winkler, "A Philosopher's Viewpoint," *S&T*, 68 (August, 1984), 110.

[76] Pagels, "A Cozy Cosmology," in *The Sciences* (New York, March–April 1985), 34–38; Fred Hoyle, "Some Remarks on Cosmology and Biology," in *Memorie* (footnote 78), 513–518. Pagels also gives Dicke's latest opinion here, 38. For other criticisms see Martin Gardner, review of Barrow and Tipler, *The Anthropic Cosmological Principle*, in *The New York Review of Books*, 33. no. 8 (May 8, 1986), 22–25.

important and stimulating questions. Foremost among them was the importance of life and mind in the universe, which he now labeled as "of central concern in any complete understanding of cosmology." Carter himself elaborated his opinion by arguing that the Anthropic Principle was most powerful in the biological, and particularly the evolutionary, realm, where he used it to argue that intelligent life would be rare even on planets with favorable environments. Among the philosophers, Patrick Wilson argued that the principle was not anthropic at all. In its emphasis on humanity as the focus of the universe, he found it more of a religious principle than a scientific one, "either a rather trivial truth or an impotent teleological principle." John Leslie, in a volume culminating two decades of interest in the philosophical status of the Anthropic Principle, concluded that the evidence that the universe was fine tuned was "impressively strong," that the "God hypothesis" (although perhaps not with its traditional meaning) was a strong one for explaining this fine tuning, and that unless life is a fluke, "God is real and/or there exist vastly many, very varied universes."[77]

Despite claims that it was tautological, illogical, or irrelevant, the Anthropic Principle showed no signs of disappearing from scientific or philosophical discourse by the end of the century. To the contrary, it was a major point of discussion in the Venice Conference on Cosmology and Philosophy, held in 1988, where at least one participant presented it as an argument in favor of extraterrestrial life: "If moderately uniform conditions exist and life can exist *somewhere* (at suitable times), it can almost certainly exist *almost everywhere;* and this implies infinite repetition of Life forms in an infinite universe."[78] Inevitably, the Anthropic Principle

[77] G. F. R. Ellis, Essay Review of Barrow and Tipler, *The Anthopic Cosmological Principle,* in *General Relativity and Gravitation,* 20, no. 5, (1988) 497–511; Brandon Carter, "The Anthropic Principle: Self-Selection as an Adjunct to Natural Selection," in S. K. Biswas, D. C. V. Mallik, and D. C. V. Vishveshwara, eds., *Cosmic Perspectives* (Cambridge, 1989), 185–206, an extension of his article "The Anthropic Principle and Its Implications for Biological Evolution," 347. In this article (196, 203), Carter states his belief that life is not common in planetary systems. Patrick Wilson, "The Anthropic Principle," in *Cosmology: Historical, Literary, Philosophical, Religious and Scientific Perspectives,* Norriss S. Hetherington, ed. (New York, 1993), 505–514. John Leslie, *Universes* (New York, 1989), especially 165–204; see Leslie's references in this volume for his previous publications on the subject.

[78] F. Bertola and U. Curi, eds., *The Anthropic Principle: The Conditions for the Existence of Mankind in the Universe* (Cambridge, 1989), summarized in Marek Abramowicz and George Ellis, "The Elusive Anthropic Principle," *Nature,* 337 (February 2, 1989), 411–412. George Ellis,"Major Themes in the Relation between Philosophy and Cosmology," *Memorie della Societa Astronomica Italiana* (Journal of the Italian Astronomical Society), 62, no. 3 (1991), 553–606. This was part of the 1989 Venice Conference on Cosmology and Philosophy on the "Origin and Evolution of the Universe." See also G. F. R. Ellis and G. B. Brundrit, "Life in the Infinite Universe," *QJRAS,* 20, 37 (1979), and J. Heidmann, "The Anthropic Principle" in *Life in the Universe* (New York, 1992).

was also used for theological purposes by those who saw no contradiction in the evidence of design in the universe; for them it was a more scientific and sophisticated version of the traditional "design argument" for the existence of God.[79] In conjunction with Barrow and Tipler's use of the Anthropic Principle, at the end of the century one could therefore choose from the full spectrum of possibilities in the context of the extraterrestrial life debate: a positive argument, a negative argument, and the extraterrestrially neutral argument from design. But it is remarkable that just when anthropocentrism seemed irretrievably banished from the repertoire of reputable worldviews, it returned in a more sophisticated but remarkably similar form to that of A. R. Wallace, who in arguing against the plurality of worlds at the beginning of the century concluded that "the supreme end and purpose of this vast universe was the production and development of the living soul in the perishable body of man."[80]

Although the Anthropic Principle bears on the relationship between mind and the cosmos, unless one takes seriously Carter's argument that the time for the evolution of intelligence *may* on average be longer than the time scale when life will be possible (a very big "may"), it says nothing about the abundance of life in that cosmos. In this sense, it adds nothing new to the extraterrestrial life debate. But the Anthropic Principle does remind us that, in the broadest possible sense, conditions are such in our universe that the fundamental constants give rise to stars, planets, and life, and that this state of affairs has not been, and may not always be, true for the entire history of the universe. Thus, even while keeping veiled the question of life's abundance, it gives added meaning to the importance of life and mind in the universe. At the same time, the Anthropic Principle tells us that there can be no extraterrestrial intelligence fundamentally different from humanity in our universe, in the sense of evolving outside the constraints imposed by the physical constants. This, indeed, still gives scope for a tremendous diversity of life and intelligence.

Finally, the elegantly and tellingly misnamed Anthropic Principle focuses on the question of purpose in the universe, and thus on the issue of life and mind. Although most scientists shy away from questions of pur-

[79] For example, Hugh Ross, *The Creator and the Cosmos: How the Greatest Scientific Discovery of the Century Reveals the Existence of God* (Colorado Springs, Colo., 1993); Michael A. Corey, *God and the New Cosmology: The Anthropic Design Argument* (Lanham, Md., 1993); and Stanley L. Jaki, *The Purpose of It All* (Washington, D.C., 1990). Among the continuing popular treatments were Reinhard Breuer, *The Anthropic Principle: Man as the Focal Point of Nature* (Boston, 1991), and John Gribbin, *Cosmic Coincidences: Dark Matter, Mankind and Anthropic Cosmology* (New York, 1989).

[80] For a comparison of the anthropic principle to Wallace's view, see Stephen Jay Gould, "Mind and Supermind," *Natural History* (May 1983), 34–38, reprinted in Gould, *The Flamingo's Smile: Reflections on Natural History* (New York, 1985), 392–402.

pose, most are also willing to concede that we do not yet know everything, and that the science of the twenty-first or the thirtieth century may be vastly different in both method and content. It was the problem of purpose that Nobel physicist Steven Weinberg broached when he wrote in a widely quoted pessimistic passage "The more the universe seems comprehensible, the more it also seems pointless."[81] That view did not take into account the possibilities inherent in the biological universe, where intelligence, whatever else its characteristics, is likely to be purposeful by definition. What that purpose may be we have not yet the slightest inkling, but if there is a meaning to life on Earth, it is undoubtedly linked ultimately to intelligence in the universe, if this exists. Whether or not life is found beyond the Earth, having posed the question of purpose in a cosmic context is itself a significant contribution of the extraterrestrial life debate in the twentieth century.

[81] Steven Weinberg, *The First Three Minutes* (New York, 1977), 154–155.

11

SUMMARY AND CONCLUSION: THE BIOLOGICAL UNIVERSE AND THE LIMITS OF SCIENCE

The discussions in which we are engaged belong to the very boundary regions of science, to the frontier where knowledge, at least astronomical knowledge, ends, and ignorance begins.

William Whewell (1853)[1]

The properties of matter and the course of cosmic evolution are now seen to be intimately related to the structure of the living being and to its activities; they become, therefore, far more important in biology than has been previously suspected. For the whole evolutionary process, both cosmic and organic, is one, and the biologist may now rightly regard the universe in its very essence as biocentric.

Lawrence J. Henderson (1913)[2]

Unlike most of science, this topic extends beyond the test of a well-framed hypothesis; here we try to test an entire view of the world, incomplete and vulnerable in a thousand ways. That has a proud name in the history of thought as well; it is called exploration. We are scientists and engineers from a dozen backgrounds joined in the early ingenuous stages of a daring exploration, become real only during recent years.

Philip Morrison (1985)[3]

When the twentieth century began, the idea of a universe filled with life was widely accepted, completely unproven, and heavily burdened with a long and checkered history that finally held the promise of more successful scientific scrutiny. By the end of the century, scientists and the public embraced the biological universe even more enthusiastically, now in the context of an enormously larger, dynamically expanding universe. But nowhere among the innumerable stars and the fleeing galaxies had life been found. In the interim, as documented in this study, the best attempts of science to answer this age-old question yielded negative results in the solar system, tantalizing hints among the stars, and a revealing picture of the scientific enterprise and popular hopes and fears on Earth. In this Conclusion we summarize the nature and significance of a debate that

[1] William Whewell, *Of a Plurality of Worlds: An Essay* (London, 1853), 115.

[2] L. J. Henderson, *The Fitness of the Environment* (Cambridge, Mass., 1913), reprinted with an Introduction by George Wald (Gloucester, Mass., 1970), p. 312.

[3] Philip Morrison, "Twenty-Five Years of the Search for Extraterrestrial Communications," in M. D. Papagiannis, ed., *The Search for Extraterrestrial Life: Recent Developments* (Dordrecht, 1985), 19.

proponents and critics alike have viewed as crucial to humanity's self-image while remaining at the outer limits of the capabilities of science.

11.1 THE TRIUMPH OF COSMIC EVOLUTION

Despite the lack of a final answer to the question "Is there life in the universe?" the twentieth-century extraterrestrial life debate above all else demonstrates, as no other scientific controversy does, how completely the idea of cosmic evolution has triumphed, notwithstanding the continued controversy over terrestrial biological evolution led by those with vested religious interests. It is, indeed, this idea that unites all the seemingly disparate elements of the debate. Percival Lowell understood at the beginning of the century that the solar system had evolved and was evolving; his compelling picture of a dying Mars whose inhabitants were desperately trying to distribute their water resources was the epitome of a solar system constantly subject to change. The Space Age failure to detect vegetation and even organic molecules on Mars, while a temporary setback to the cause of cosmic biological evolution, was after all a conclusion limited to a single planet. That innumerable planets might exist was an implication of the nebular hypothesis, whereby the birth of planetary systems was a natural by-product of star formation. Despite a temporary eclipse and Jeans's accompanying claims of planetary rarity during the 1920s and 1930s, by the 1940s the nebular hypothesis was revived, increasingly elaborated, and widely accepted throughout the remainder of the century, even as astronomers sought – with limited success – observational evidence of planetary systems. On such planets throughout the universe, it was postulated, chemical evolution and the origins and evolution of life had taken their own course, perhaps similar in a general way to that on Earth but subject, according to the principles of natural selection, to the environmental conditions on each planet. Beginning in 1953, Miller–Urey-type experiments demonstrated that the building blocks of life, if not life itself, could be produced under presumed primitive Earth conditions. By 1975 it was confirmed that Nature itself had produced some of these building blocks in a variety of outer-space environments ranging from meteorites to interstellar molecules. Though by the end of the century the Oparin–Haldane model of the origin of life was under fire in details ranging from the nature of the first living organism to the constituents of the primitive Earth atmosphere, its essential premise that biological evolution on Earth followed chemical evolution as night follows day was not only widely accepted but had been extended to the universe at large.

The Viking missions failed in their local test to detect microbial life, but the goal of SETI was to test for the ultimate product of this hypothesis of cosmic evolution – extraterrestrial intelligence. The importance of the Drake Equation was not that it gave any definitive indication of the number of communicative technological civilizations in the Galaxy – a role even its users denied – or that it was a tool on which discussion could focus. Rather, it was the very embodiment of the concept of cosmic evolution. Each of its disparate elements – individually uncertain and collectively almost devoid of meaning in any observational sense – represented one of a series of steps in astronomical, biological, and social evolution. Despite the uncertainties in astronomical and biological evolution, the fact that the number of technological civilizations hinged on L – the average lifetime of a technological civilization – emphasized that social evolution was the most crucial parameter of all. It was also the most unknown. The Drake Equation was not a puzzle to be solved for the century but a problem for the centuries to follow; its chief contribution for our age was to codify how cosmic evolution had become the paradigm of modern times.

Cosmic evolution became a paradigm not just in some vague and abstract way but as a real research program, with NASA as its flagship patron and a community of researchers and enthusiasts around the world. It did so at first in piecemeal fashion, beginning with concern about planetary contamination and back contamination, then more directly with the search for life on Mars, leading to broadly conceived research on the origins of life. When Viking failed to detect life on Mars, the cosmic evolution program expanded beyond the solar system to the search for planetary systems, and finally to the search for radio signals as part of the SETI program. SETI, more than any other single activity, embodied all the elements of cosmic evolution in a unified research program explicitly enunciated by NASA, as graphically shown in Figure 11.1. That program was viewed by many researchers around the world as the cosmic context of their work, whether in planetary science, planetary system science, or origin of life studies. Aside from a sporadic interest mainly in the Soviet Union (which by the 1990s itself dramatically demonstrated social evolution on Earth), only toward the end of the century was there a slowly awakening interest in social evolution in a cosmic context.

The concept of cosmic evolution was not a product exclusively of the Space Age, but it was the Space Age that brought it – and all its consequences for humanity's status in the universe – to the forefront as a research problem susceptible to science. Although Harvard biochemist Lawrence J. Henderson, for one, clearly grasped the idea (and even employed

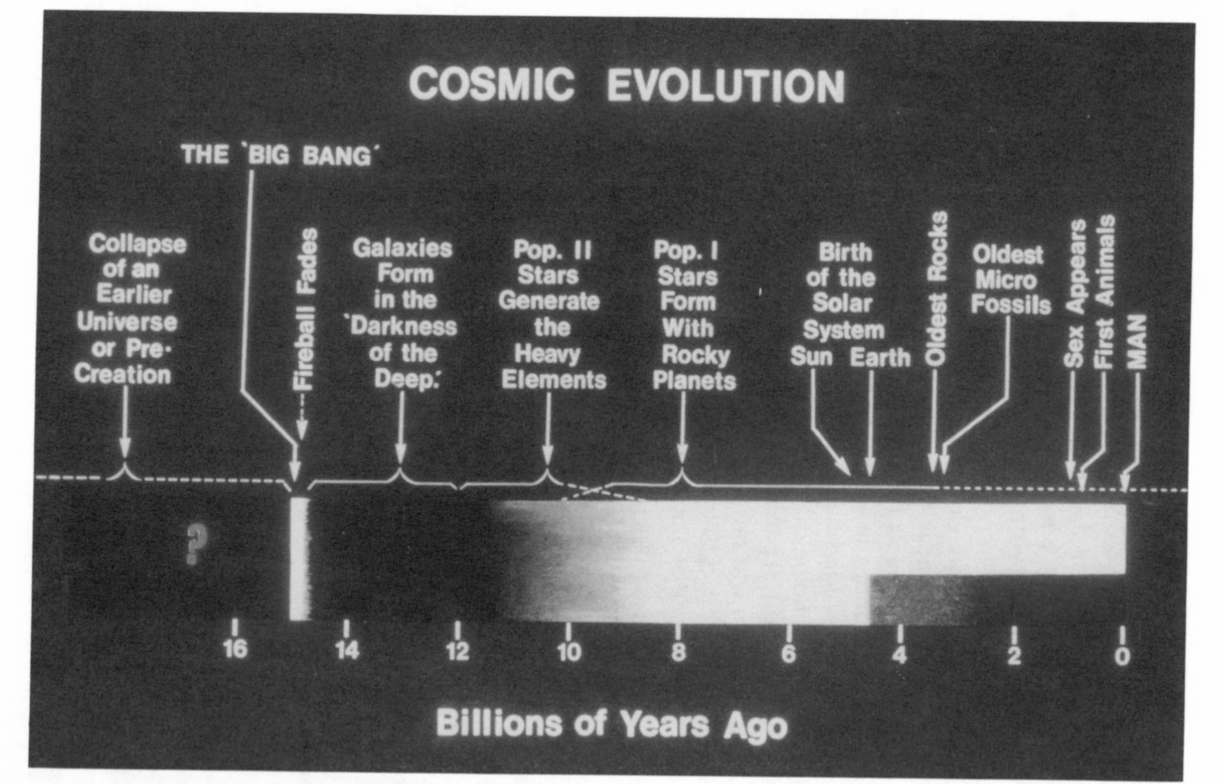

Figure 11.1. Cosmic evolution, as depicted in this NASA drawing, encompasses events from the Big Bang to the development of intelligence. NASA was the chief patron of research in cosmic evolution. Courtesy NASA.

the terminology) of cosmic evolution at the beginning of the century, his sentiment, as stated in the passage at the beginning of this chapter, was not taken to heart in the first half of the century. Spurred by the Space Age, only during the last half of the century was cosmic evolution embraced by an increasing number of scientists from many disciplines. But whether we may yet "regard the universe in its very essence as biocentric" remained the essence of the conundrum at the end of the century, whether in the scientific sense of the evolution of life, the philosophical sense of the Anthropic Principle, or the religious sense of the meaning and purpose of life. While the twentieth century proposed, elaborated, and tested the concept of cosmic evolution, the question of biocentricity is likely to occupy the twenty-first century as one of its chief scientific and philosophical problems.

11.2 THE BIOLOGICAL UNIVERSE AS COSMOLOGICAL WORLDVIEW

Based on the concept of cosmic evolution, those championing the biological universe held that planetary systems were common, that whenever and wherever conditions were favorable life would originate and evolve, and that this evolution would culminate in intelligence, even if not necessarily similar to that on Earth. SETI enthusiasts believed further that intelligence would develop the same facility with radio technology as terrestrials, and that the lifetime of technological civilizations was sufficiently long for a large number of civilizations to coexist in the universe – large enough for Earthlings to mount significant efforts to listen for any radio messages from outer space. In short, supporters of the biological universe favored optimistic estimates for all the parameters in the Drake Equation. The biological universe was more than an idea, more than another theory or hypothesis; it was sufficiently comprehensive to qualify as a worldview. Since according to SETI advocates it was testable, it was a *scientific* worldview perhaps best described as a cosmology. And because it combined the biological universe with the characteristics of the physical universe, it may be most accurately termed the "biophysical cosmology."

That the idea of extraterrestrial life and all its accompanying assumptions should be viewed in cosmological terms is not a framework that we need impose on history; several participants in the debate implicitly or explicitly recognized it themselves. Among the first was former Harvard Observatory Director Harlow Shapley, who already in 1958 described extraterrestrial life as the "Fourth Adjustment" in humanity's view of itself in the universe since the time of ancient Greece. Astronomer Otto

Struve – crucial to our story for his advocacy of abundant planetary systems and for his support of Project Ozma – agreed, viewing the question as a revolution comparable to the Copernican cosmology in the sixteeenth century and the "galactocentric" cosmology in the early twentieth century. Others gave it a similar status, including Project Cyclops author Bernard Oliver, who spoke of "Biocosmology" in his influential study of 1971.

Like other cosmologies, the biophysical cosmology made claims about the large-scale nature of the universe, most comprehensively that life was one of its basic properties. Adherents believed that life was not merely an accidental or incidental property but an essential one; whether this meant that it might eventually change our perception of the physical universe via some anthropic or (in Henderson's words) "biocentric" principle, in the way that Einstein's theory of relativity changed our concept of the physical universe, was still uncertain. Like all cosmologies, the biophysical cosmology redefined the place of humanity in the universe, now precluding the last hope for anthropocentrism, unless one used a strict (and not very plausible) interpretation of the Anthropic Principle that the universe was made entirely for humanity. And as with other cosmologies, its advocates believed the existence of widespread intelligence was testable, resulting in SETI programs supported by groups ranging from grass-roots space advocates (via the Planetary Society) to universities and government.

Viewed in these terms, the study of the extraterrestrial life debate would undoubtedly benefit from a historical comparison with those other great shifts in cosmological worldview of the past. Valid parallels might be drawn with other cosmologies with respect to the development and evolution of the idea, the arguments and strategies of its critics, and its widespread acceptance in the face of inconclusive evidence. Moreover, such a comparison might provide insights into the implications of the biophysical cosmology, for one of the hallmarks of cosmologies is that they affect the broader culture, including philosophy, literature, and religion. Although many aspects of culture have already been affected to some extent by the mere possibility of extraterrestrial intelligence, they are likely to be affected even more if and when the biophysical cosmology is confirmed. Nor need the parallels be confined to physical worldviews; the Darwinian worldview, with its consequences for humanity's status and broad social implications, might serve as an even closer parallel. Indeed, the biophysical cosmology raises Darwinism to a cosmic context in which humanity is part of a much larger chain of being rather than simply at the apex of the apes.

11.3 THE PROBLEMATIC NATURE OF EVIDENCE AND INFERENCE

In attempting to understand why the question of extraterrestrial life has not been resolved in the twentieth century, the factor that looms largest is the problematic nature of evidence and inference. This is true whether dealing with the nearest planets of the solar system, the search for more distant planetary systems, or theories and experiments about the origin and evolution of life relevant to both. It is true whether using ground-based telescopes or spacecraft that actually land and undertake in situ observations. And it applies equally to the visual, photographic, and spectroscopic techniques of astronomy, as well as to the variety of techniques employed in studies of the origin and evolution of life.

The most notorious example of the difficulties of evidence and inference is Lowell's claim that he had detected canals on Mars. But these difficulties neither began nor ended with Lowell, nor with the controversy he fueled, nor were they endemic only to the extraterrestrial life debate. Some scientists believed that continuous canals were illusory, caused by the mind conjoining rows of spots or detecting the boundaries of shadings; others believed they were real and continuous but represented natural cracks or strips of vegetation; still others (it needs to be remembered, lest Lowell seem an aberration) agreed with Lowell that the canals were real and artificial. The apparent and much applauded triumph of Antoniadi, who believed he had resolved the canals into dark patches by 1909, was in the end no triumph at all, for we now know that most of the claimed canals correspond to no dark features on Mars, so that even Antoniadi had overturned Lowell's claims only by accident. Even once the canal controversy was past its peak, astronomer R. J. Trumpler and others believed they had detected vegetation, and the so-called wave of darkening and other factors led Kuiper to proclaim at midcentury that a lichenlike vegetation was the most likely cause of some of the Martian dark areas. This seemed to be confirmed by Sinton's sophisticated spectroscopic observations beginning in 1957, setting the stage for the search for life on Mars as one of the prime goals of the Space Age. Even with the Viking landers, the evidence was ambiguous at first, and although a consensus was reached relatively quickly that the results were not biological in origin, a few still dispute the result; even more would like to return to Mars to search for past or present life in other locations.

The search for planetary systems chronicles similar difficulties with observational evidence – now complicated by increased interaction with theory. The claims of Reuyl, Holmberg, and Strand for planetary systems during the 1940s helped to revive the nebular hypothesis, with its implication for abundant planetary systems. Yet, so long-term were the observations needed for these delicate conclusions that they could not be disproven for decades, by which time van de Kamp and his colleagues had new candidates for planetary systems, equally difficult to confirm. Our study of van de Kamp – and the numerous claims for "first planet detection" outside the solar system – show how great was the potential for error when dealing with the limits of detection, as so often was perversely the case during the extraterrestrial life debate. The manner in which Struve's interpretation of stellar rotation data, and Kuiper's binary star separation data, were delayed as arguments for abundant planetary systems until the nebular theory indicated that abundance was plausible clearly demonstrates the effect of theory on observation. Even determining the most basic prerequisite for life – the existence of the planets themselves – turned out to be most difficult, and it was only the first step in a long chain of observation and theory needed to reach a conclusion about extraterrestrial life.

In the case of the origin and evolution of life, the difficulties were of a different nature. At stake was not so much the difficulty of observations as the inability to make them, at least directly, resulting in heavy reliance on inference drawn from experiment. No one could observe the primitive Earth atmosphere or the origin of life. Experiments had to be performed under conditions of a presumed primitive atmosphere, and even if one accepted that atmosphere, the products of prebiotic experiments were only suggestive, not definitive, in showing how life might have originated on Earth. Experiments were no more definitive in determining whether the origin of life was a common process, and what exactly this might tell us about the abundance of life on other planets. The pioneering and startling observations of complex organic molecules in space, and even the triumph of unambiguously demonstrating the presence of amino acids in carbonaceous meteorites, were, in the end, still only suggestive. Neither experiments on Earth nor Nature's experiments in interplanetary and interstellar space produced the equivalent to life, only the prebiotic building blocks of life. In the end, all arguments about the origin of life on Earth and its abundance in the universe remained presumptions.

In both the specific arguments represented by each element in the Drake Equation and the grand argument represented by the elements in the Drake Equation taken together, evidence and inference were the culprits preventing definitive solution. This problem was only compounded when indirect evidence was brought to bear. Physicists like Coblentz and Lampland could use state-of-the-art techniques to determine planetary temperatures (subject to their own experimental error), but physical factors alone could not clinch the case for life on Mars. One had to consider spectroscopic evidence for atmospheric pressure and atmospheric constituents; surface observations such as the changing size of the polar cap, the wave of darkening, and the relation between the two; and many other factors, each again with its own experimental error and each subject to errors in inference. Attempts to quantify observational data were no assurance of success, as shown by Very's claim for water vapor on Mars. And in moving beyond the solar system, with the exception of artificial radio signals that were the holy grail of SETI programs, all observations became indirect. In the search for planetary systems (which, of course, did not prove the existence of life), the astrometric method of measuring extremely small perturbations was a technique taking decades to consummate, during which time changes in the telescope, the observers, and the photographic plate measurments had to be taken into account. When the spectroscopic technique of radial velocities held out hope for a quicker solution, worries arose about whether the data were really showing a planetary tug or were an indication of astrophysical effects, such as pulsation or surface activity on the star itself. Even the first announcements of the unexpected discovery of pulsar planets by the radio timing technique had to be retracted, although evidence of other pulsar planets seemed to be confirmed by the early 1990s. And if it were true, the implications of pulsar planets for extraterrestrial life advocates were also subject to inference. Advocates argued that if planets could form around pulsars, they could form anywhere; skeptics argued that pulsar planets were irrelevant to normal planet formation. Similarly, advocates argued that observations of circumstellar material implied that planets would eventually form; skeptics replied that such debris might never form into planets. None of these questions had easy answers, and when the problems were compounded by the many factors in the Drake Equation, one can begin to see why no solution has yet been reached.

Finally, evidence and inference were inevitably grounded in certain methodological assumptions, and scientists, philosophers, and historians have pointed out that at least some of the assumptions made in the pursuit of exobiology and SETI may well be false.[4] This is undoubtedly true, yet at the same time, advocates emphasized that the assumptions could not be proven false. In short, they were the assumptions of those who adopted the biophysical cosmology as their worldview, and sought to test it by whatever means possible. While exobiological assumptions might be more grandiose than those found in most sciences, they were not qualitatively differ-

[4] Ernan McMullin, "Persons in the Universe," *Zygon*, 15 (March 1980), 69–89, points out some methodological fallacies in SETI, as does Michael Crowe, *The Extraterrestrial Life Debate, 1750–1900* (Cambridge, 1986), 552–555.

ent, and those who termed exobiology a science without a subject would seem either to have mistaken the nature of science or to have had an agenda of their own. No forefront science could be sure if its subject existed until the observations were made or the experiments undertaken. From gravitational wave astronomy to the magnetic monopole and the search for the gene as a unit of heredity, over whatever period one chooses in the history of science, substantial progress was most often made by making bold assumptions and following leads to their sometimes dead-end conclusions. Objections to the contrary are interesting exemplars of the rhetoric of science, no more evident than in the case of evolutionist George Gaylord Simpson, who coined the phrase "science without a subject" in connection with exobiology while arguing that the money would be better spent on down-to-Earth research – such as his own.

11.4 THE LIMITS OF SCIENCE

The lack of a definitive resolution to the question of life in the universe, due in part to the problematic nature of evidence and scientific inference, leads us to the conclusion that science has limits in its ability to resolve certain questions. The phrase "limits of science" has been used in so many ways that it is important to define how it applies to the extraterrestrial life debate. Nobelist Sir Peter Medawar, among others, has pointed out the inability of science to answer certain ultimate questions – such as "What are we all here for?" and "What is the point of living?" – that "neither arise out of nor require validation by empirical evidence." In the terminology of philosopher of science Nicholas Rescher, such questions are external to the domain of science. Though we may agree that science cannot answer questions such as these, it is not in this sense that we apply the term "limits of science" to the the question of life in the universe, because at least certain parts of it are subject to empirical validation. Such validation was the central purpose of the Viking mission and remains the purpose of all SETI programs. Although the criticism remains that to search the entire universe is a rather large research program, extraterrestrial life is verifiable in principle, and for relatively nearby regions of the universe, using certain assumptions and techniques, it is also verifiable in practice. In other words, although there are limits to what can be done at present in the search for extraterrestrial life, all but the harshest critics would agree that it falls in principle within the realm of science.

Another question that might be asked is whether science *should* tackle certain questions – in other words, whether limits should be placed on scientific inquiry into subjects such as recombinant DNA, cloning, and mind control. Although we have seen this most common usage of the term applied to the extraterrestrial life debate by evolutionist George Gaylord Simpson, physicist Frank Tipler, astronomer Sir Martin Ryle and certain theologians, among others, it is not this moral question that primarily concerns us here. Rather, we wish to emphasize that the extraterrestrial life debate demonstrates the limits of both scientific reasoning and scientific practice; in Rescher's terminology, it is doable in theory and partly doable in practice, but the task has not yet been achieved except for the most local cases. In short, the search for extraterrestrial life is a subject that during the twentieth century has remained at the very limits of science in the sense that the required techniques have been barely sufficient, leading to problematic evidence and stretching the normal principles of scientific inference.[5]

In this sense, then, our history may be seen as a case study of how science functions at its limits, and we may summarize the characteristics of science in the realm of its "boundary regions," as the philosophically astute Whewell put it in connection with his discussion of a plurality of worlds. The first point to be made is sociological: it is a significant fact that scientists tackled the question of extraterrestrial life at all. Whewell and his successors could have thrown up their hands and pronounced the case hopeless; to the contrary, recognizing that they were acting in the boundary regions, they nevertheless proceeded as best they could. A variety of motivations, ranging from religion and a desire to understand humanity's place in the universe, to reaping profit from a subject known to be of intense public interest, drove scientists forward into the boundary regions, where they were forced to make the best argument possible. Boundary science (to adopt Whewell's term) was not for everyone. We should not forget that, for a variety of reasons related to career risk and other factors, the vast majority of scientists never entered the debate, and among those who did, most had self-imposed limits. While scientists in the second half of the twentieth century increasingly came to accept the goals of exobiology, few would enter the UFO controversy. And among those who entered the UFO controversy, even fewer would venture into its nether regions such as alien abduction. In part this was because of discipline limitations (astronomers had less to say about alien abductions

[5] Peter Medawar, *The Limits of Science* (Oxford, 1984); Nicholas Rescher, *The Limits of Science* (Berkeley, Calif., 1984); David Faust, *The Limits of Scientific Reasoning* (Minneapolis, 1984). For essays on the question of whether limits should be placed on scientific inquiry see Gerald Holton and Robert S. Morison, eds., *Limits of Scientific Inquiry* (New York, 1979); for examples of science at its limits see Richard Morris, *The Edges of Science: Crossing the Boundary from Physics to Metaphysics* (New York, 1990), where issues such as dark matter, superstrings, and the origin of the universe are discussed. Morris distinguishes (in order of being increasingly far out) frontiers, boundaries, and edges of science. He places the Anthropic Principle at the edges of science.

than psychiatrists and did not consider hypnosis part of their scientific method), but in part there was a line that even most boundary scientists would not cross.

Second, once in the boundary region, scientists preferred observation over theory, but they did not hesitate to use theory in the absence of observation to further their argument. In the arena of planetary systems, in the absence of much empirical data, scientists were forced to lean heavily on theory and even more heavily on such shreds of evidence as were available to tell them which theory was correct. Jeans had little empirical evidence to validate the close encounter theory for the origin of solar systems, and when only two independent observations seemed to show planetary systems around two different stars, that evidence, combined with difficulties in the theory itself, was enough to cause the return to the nebular hypothesis during the 1940s and 1950s. On another level, we have seen that theory could affect the interpretation of observations, as in Struve's interpretation of stellar rotation data as evidence for the abundance of planetary systems in the face of widespread acceptance of the nebular hypothesis. But given a choice, scientists always preferred observation. Few were convinced, for example, by the "Where are they?" Fermi paradox that SETI programs should not be undertaken.

Third, we have seen that where evidence was available, scientists took a variety of approaches to the same data. Faced with claims of rectilinear features on Mars, some scientists chose the simplest explanation, which, they were quick to emphasize, was not to be found in Martian engineers. Yet, faced with the wave of darkening and other Martian phenomena, other scientists did not hesitate to propose vegetation as the explanation, even though simpler explanations (including wind-borne dust, now known to be the cause) were available. By the time of the Viking probes, most scientists had learned their lesson, at least in the Martian context: Norman Horowitz led the way among biologists in arguing that the simplest explanation of the Viking data was nonbiological, and this proved to be the consensus view.

Similarly, in planetary systems science, some astronomers saw patterns in their data that they chose to interpret as evidence of planetary systems, a claim that others would not have made using the same data; conversely, some astronomers cautiously referred to their own discoveries as "millisuns," while others emphasized that this meant large planets. The line between interesting brown dwarfs and spectacular large planets was sometimes creatively construed. The same may be said for origins of life studies; to mention only one case, some saw much more significance than others for extraterrestrial life in the continuing discoveries of complex organic molecules in space. As a corollary to different approaches, diverse interpretations of the data, as just discussed, were more likely than in normal science.

Fourth, to return to a sociological point following these methodological points, the distinction between optimists and pessimists in the search for life was very sharply drawn. How scientists assumed their roles as optimists and pessimists is a subject for further study, but once they took these roles, the debate between those functioning at the limits of science and their colleagues doing more normal science was often vigorous, encompassing not only technical arguments but also broad issues of methodology and philosophy. Whether this is due to the emotional nature of this particular issue, or whether other practitioners functioning at the limits of science in other subjects face similar rancorous encounters, remains for further study, as indeed does the generality of all of the characteristics of frontier science found in the extraterrestrial life debate.

A last point to be made in this connection is the charge of critics that exobiology had been marked by lack of progress. Compared to the beginning of the century, we now know that no intelligence or vegetation exists on either Mars or Venus; that not even organic molecules are to be found on Mars; that material exists around stars that may (or may not) form into planets; that interstellar space contains complex organic molecules that could serve as the basic building blocks for life; and that numerous civilizations do not seem to be beaming messages toward us at the 21-cm frequency. There is still no definitive answer, and although the biological universe is not verifiable globally, it is verifiable light year by light year as observations advance outward in an expanding sphere of search space for particular frequencies and other sets of assumptions. In the sense that it may need to search the entire universe, SETI is perhaps the largest research program ever proposed. Although the answer is bound by the limits of scientific reasoning and technique, SETI is limited only by the size of the universe.

Finally, Rescher has described another sense in which science has limits, namely, that it "is limited by the very fact of being *our* science," rather than the science of an alien civilization.[6] Rescher argues that the idea that alien civilizations may be more advanced than ours is unlikely. Since they have been conditioned by their own worlds, he reasons, there is no common intellectual journey, and thus no comparisons between science or knowledge in general. One need not adopt the extreme view of incompatibility to agree that our science may well be limited compared to that of other civilizations. If comparison is possible, one wonders whether the

[6] "Extraterrestrial Science," in Rescher, *Limits of Science*, reprinted in Edward Regis, Jr., ed., *Extraterrestrials: Science and Alien Intelligence* (Cambridge, 1985), 83–116.

long-sought "objective knowledge" might be found at last by gleaning the common elements remaining after processing by many sensory and mind systems independently evolved throughout the universe. If so, many of the problems of interpretation and inference might be resolved, or at least reach a new level.

11.5 THE CULTURES OF SCIENCE

The different approaches to the problem of life in the universe, the diverse interpretations of data, and the vigorous nature of the debate between protagonists and antagonists in the debate lead us to a conclusion about the nature of scientists, as distinct from the science they produce. In his classic work *The Two Cultures* (1959), C. P. Snow contrasted the sciences and humanities as two cultures, "literary intellectuals at one pole – at the other scientists . . . between the two a gulf of mutual incomprehension."[7] Our study of the extraterrestrial life debate demonstrates more clearly than most other areas of science what is evident to anyone with a broad overview of scientific ideas and practice but often remains unspoken, especially by scientists themselves: that whether in terms of problem choice, of method, of inference, or whatever other characteristic one wishes to choose, science itself has many cultures. Again and again we have seen this throughout our study: in Lowell's interpretation of the canals compared to that of others (Table 3.1); in the Martian vegetation controversy; in claims of planetary systems; in the unwillingness of some to take the UFO phenomenon seriously and in their diverse interpretations of the same data (Table 6.2); in the meaning of prebiotic experiments for the origin of life; and in the diverse estimates for success in SETI (Table 8.1).

These cultures of science undoubtedly operate at many levels, from the routine to the frontier. They may have been glimpsed by those who have studied "styles" of science, whether geographic, conceptual, or methodological.[8] But the idea of "culture" is at once more personal and runs deeper than "style"; it may involve not only methodology, but also worldview; not only the problems that scientists actually take up, but also the ones they are willing to consider taking up; and not only day-to-day research, but what are viewed as the ultimate goals of science. As Snow's concept served a purpose for its time, the time may be ripe for a

[7] C. P. Snow, *The Two Cultures and a Second Look* (Cambridge, 1969), 4.

[8] Evidence for scientific styles has been found independently in several disciplines. For a recent review concentrating on biology see Marga Vicedo, "Scientific Styles: Toward Some Common Ground in the History, Philosophy and Sociology of Science," *Perspectives on Science*, 3 (Summer 1995), 231–254.

study of the many cultures of science. In particular, the field of science studies, which has often emphasized the multiplicity of the material, social, and conceptual elements of "scientific culture," could use the extraterrestrial life debate to study more precisely the many cultures that compose science.[9] A study of how these cultures internal to science originate, dynamically interact, and affect the production of scientific knowledge promises to shed new light on the nature of the scientific enterprise.

11.6 EXOBIOLOGY AS PROTOSCIENCE

Those who did take up the quest for extraterrestrial life functioned in a world often seen by their peers as of dubious scientific status and, at least in its early stages, perhaps not as science at all. Real people risked real careers to study a problem that critics ridiculed as without content. The question of whether or not they succeeded in forming a discipline may be academic (unless one is among those participants searching for funding), but what is clear is that exobiology was at least an emerging, if not yet emergent, science. To put it another way, exobiology (taken in its broadest sense to include SETI and the entire program of cosmic evolution, sometimes also referred to as "bioastronomy") was a protoscience in the most literal sense of the word. Using an astronomical metaphor, like protostars and protoplanets, exobiology was a system (in this case of concepts, techniques, and researchers) still in some state of chaos. No one could quite be sure whether, like stars and planets, the new discipline would coalesce or whether, after a promising start, centripetal forces would cause it to dissipate before a critical mass was reached.

The forces of dissipation for protoscience may be great, having to do chiefly with funding, and therefore with politics, and ultimately with a society's self-image. The demise of government funding for the NASA SETI program demonstrated the fragility of the protoscience of exobiology, showing at once how the irrationality of the political moment could affect any program, no matter how carefully planned, even as society itself debated the larger issue of practical research versus curiosity-driven research. If the adherents of practicality won out, the entire program of exobiology (and indeed much else in science) would be seen as irrelevant, and exobiology, along with other aspiring disciplines, would dissipate into the void. How much would be lost in the process was anyone's guess. Proponents of curiosity-driven research would point to the mold and penicillin, among many other examples, waxing eloquent about curiosity

[9] See Ian Hacking, "The Self-Vindication of the Laboratory Sciences," in Andrew Pickering, ed., *Science as Practice and Culture* (Chicago and London, 1992), 29–64. This volume is a good entree to the vigorous field of science studies.

as the essence of humanity; critics would emphasize the dire needs of the world. In this sense, the fate of exobiology was part of a much larger debate at the end of the century, whose outcome was unpredictable and indeterminate. What seems clear is that, unlike established sciences like physics, biology, and chemistry and many of their subdisciplines, exobiology had little chance of enunciating a rationale of practical application. Nothing, advocates would say, except for gaining the knowledge and wisdom of the universe, long a theme in alien literature (Figure 11.2) and a hope of SETI practitioners.

11.7 CULTURAL SIGNIFICANCE OF THE DEBATE

Illuminating as the extraterrestrial life debate is for understanding the triumph of cosmic evolution and the nature of science and its many cultures, it is clear from our study that its significance goes well beyond the relatively parochial boundaries of science. Although the idea of extraterrestrials clearly fascinated a wide audience as early as the time of Bernard le Bovier de Fontenelle's phenomenal best-seller *Entretriens sur la pluralité des mondes* (Conversations on the Plurality of Worlds) in 1686, no one could have foreseen the extent to which the idea would pervade popular culture by the twentieth century. The contrasting alien approaches of H. G. Wells's *War of the Worlds* (1897) and Kurd Lasswitz's *Auf zwei Planeten* (On Two Planets, 1897) were only the vanguard of an enormous number of diverse treatments of the alien theme in science fiction. And in the real world, one might have predicted that the first wave of reports of UFOs in 1897 would be a relatively minor flap and no concern of the sophisticated twentieth century. But the revival of the UFO phenomenon in 1947, its peak in the 1960s with attention from government agencies, and the continued bizarre reports of alien abductions as the century ended showed that the interest in aliens was far from limited to fiction. That interest was confirmed when films such as *ET*, *Close Encounters of the Third Kind*, and the *Alien* series became among the most popular in cinematic history. First in science fiction literature, then in the UFO debate, and finally through the emotionally powerful medium of film, the idea of extraterrestrials insistently appealed to the popular mind and became an integral part of popular culture.

More than this, the idea of extraterrestrials for many people constituted a worldview no less influential and appealing than the geocentric spheres and choirs of angels of the medieval worldview. The move from the physical world to the biological universe, ending with what we have called in scientific terms the biophysical cosmology, was one of the most distinctive traits of the twentieth century, constituting what Harvard Pro-

Figure 11.2. Cover by Leo Morey for the April 1930 issue of *Amazing Stories*, showing an alien imparting knowledge of the Galaxy. Copyright 1930 by Experimenter Publishing Company. The search for extraterrestrial knowledge and wisdom is one of the major themes of the debate.

fessor Karl Guthke has called not only *a* myth, but perhaps *the* myth, of modern times. Although we normally associate mythology with something untrue, Joseph Campbell and others have shown that mythology is in fact a reflection of the deepest beliefs and characteristics of a culture. Nor does this at all contradict the assertion that the biological universe constitutes a cosmology, a broad view of the world which itself may or may not be proven true. Whether the particular myth of our study is true or not we do not yet know, but that it has been claimed through different approaches to have the status of both myth and cosmology is some measure of its impact on modern culture.

Finally, the extraterrestrial life debate causes us to reexamine the very foundations of our knowledge and belief in a wide variety of subjects. Whether one chooses mathematics, religion, philosophy, science, the arts, or humanities, viewed from an extraterrestrial perspective they are all only local examples of what may perhaps be a much more generalized knowledge and belief. All take on new meaning if life is abundant in the universe – and if it is not. Even if extraterrestrials are not discovered, an extraterrestrially motivated reexamination of the knowledge and belief that we all too often take for granted is a considerable legacy for one of the twentieth century's most persistent debates.

SELECT BIBLIOGRAPHICAL ESSAY

PRIMARY SOURCES

As is evident from the references in this volume, extraterrestrial life is the subject of an enormous literature in the twentieth century, much of it accessible through three bibliographies. *Extraterrestrial Life: A Bibliography*, published as *Part I: Report Literature, 1952–64* (NASA SP 7015, 1964) and *Part II: Published Literature, 1900–1964* (NASA SP 7015, 1965), includes detailed abstracts, as well as author and subject indexes. Eugene F. Mallove, Mary M. Connors, Robert L. Forward, and Zbigniew Paprotny, *A Bibliography on the Search for Extraterrestrial Intelligence* (NASA Reference Publication 1021, March 1978), is very useful for its categorization by topic and source, and carries the entries forward another dozen years. Beyond that, *Astronomy and Astrophysics Abstracts*, the standard index for all astronomical subjects, lists general technical literature on extraterrestrial life under Section 15 (Miscellanea) in its annual volumes beginning in 1969 and more specific related entries under the relevant sections, such as "Mars." For publications prior to that year, one must search in its predecessor publication, the *Astronomische Jahresbericht*.

Among general works on extraterrestrial life, a succession of books has codified the knowledge of each generation in almost textbooklike fashion, beginning with Sir Harold Spencer Jones, *Life on Other Worlds* (New York, 1940), followed by I. S. Shklovskii and Carl Sagan, *Intelligent Life in the Universe* (San Francisco, 1966), and Donald Goldsmith and Tobias Owen, *The Search for Life in the Universe* (Menlo Park, Calif., 1980; 2nd edition, 1992). Primary readings and a bibliography related chiefly to the origin of life and life on Mars prior to 1965 are found in Elie Shneour and Eric A. Ottesen, *Extraterrestrial Life: An Anthology and Bibliography* (Washington, D.C., 1966), while the very early papers on what later became known as the "Search for Extraterrestrial Intelligence (SETI)" are reprinted in A. G. W. Cameron, *Interstellar Communication* (New York, 1963). Reprinted articles covering all these topics through the 1970s are found in Donald Goldsmith, *The Quest for Extraterrestrial Life: A Book of Readings* (Mill Valley, Calif., 1980).

Because no journal exists covering the broad range of extraterrestrial life studies, the largest single concentration of literature will be found in conference Proceedings, as listed in Table 9.1. The 1978 NASA bibliography cited earlier, with its categorization by publication, shows the extent to which the remaining literature is dispersed throughout an enormous number of popular and technical magazines and journals.

SECONDARY SOURCES

On the history of the debate, the Cambridge University Press trilogy, of which this is the final volume, began with Steven J. Dick, *Plurality of Worlds: The Origins of the Extraterrestrial Life Debate from Democritus to Kant* (1982), followed by Michael J. Crowe, *The Extraterrestrial Life Debate, 1750–1900: The Idea of a Plurality of Worlds from Kant to Lowell* (1986). Karl S. Guthke, *The Last Frontier: Imagining Other Worlds from the Copernican Revolution to Modern Science Fiction* (Ithaca, N.Y., 1990) is also a major contribution to that history, especially from the literary point of view. Also very useful is the popularized account by the distinguished *New York Times* writer Walter S. Sullivan, *We Are Not Alone* (New York, 1964; 2nd edition, 1993). All of these sources list numerous further references.

Issues of anthropocentrism and plurality of worlds are dealt with in Paolo Rossi, "Nobility of Man and Plurality of Worlds," in *Science, Medicine and Society in the Renaissance*, ed. Allen G. Debus (New York, 1972), 157. More general issues of the status of humanity following the Copernican revolution are treated in Hans Blumenberg, *The Genesis of the Copernican World* (Cambridge, Mass., 1987). How we came to our modern view of the universe is described historically in Robert Smith, *The Expanding Universe: Astronomy's Great Debate, 1900–31* (Cambridge, 1982), and in Richard Berenzden, Richard Hart, and Daniel Seeley, *Man Discovers the Galaxies* (New York, 1976). More popular treatments are C. A. Whitney, *The Discovery of our Galaxy* (New York, 1972), and Timothy Ferris, *The Red Limit* (New York, 1977; revised edition, New York, 1983).

The history of solar system astronomy, which serves as background to Chapter 3, is given in Ronald E. Doel, *Solar System Astronomy in America: Communities, Patronage, and Interdisciplinary Research, 1920–1960* (Cambridge, 1996), and J. N. Tatarewitz, *Space Technology and Planetary Astronomy* (Bloomington, Ind., 1990). General problems of planetary observation are discussed in William Sheehan, *Planets and Perception: Telescopic Views and Interpretations, 1609–1909* (Tucson, Ariz., 1988). The controversy over the canals of Mars is best covered in William G. Hoyt, *Lowell and Mars* (Tucson, Ariz., 1976), in chapter 10 of Professor Crowe's *The Extraterrestrial Life Debate*, and in a variety of more specialized studies cited in those books and in the present volume. More about both Lowell and his observatory will be learned in David Strauss's forthcoming biography of Lowell. The crucial story of Campbell's work on the Martian atmosphere is examined in David H. DeVorkin, "W. W. Campbell's Spectroscopic Study of the Martian Atmosphere," *QJRAS* (1977), 18, 37–53. The 1909 expedition is discussed in more detail in Donald E. Osterbrock, "To Climb the Highest Mountain: W. W. Campbell's 1909 Mars Expedition to Mount Whitney," *JHA*, 20 (June 1989), 77–97.

Antoniadi's *La planéte Mars* (Paris, 1930) is the apex of the visual tradition of Martian studies. Gerard DeVaucouleurs, *Physics of the Planet Mars: An Introduction to Areophysics* (London, 1954), holds the same place in the astrophysical tradition of Martian studies that Antoniadi's 1930 volume and Flammarion's *La planéte Mars et ses conditions d'habitabilite* (Paris, 1892, 1901) hold for the visual tradition, each discussing all relevant observations to the time of publication.

The official history of the Viking mission is Edward C. Ezell and Linda Neuman Ezell, *On Mars: Exploration of the Red Planet, 1958–1978*. The results of the mission are given in papers as referenced in Table 3.5. The view of one of the principal investigators of the biology experiments is given in Norman Horowitz, *To Utopia and Back: The Search for Life in the Solar System* (New York, 1986). The final word on the canals of Mars in light of modern research is Carl Sagan and P. Fox, "The Canals of Mars: An Assessment after Mariner 9," *Icarus*, 25 (August 1975), 602–612. The latest research on Mars is described in massive detail in H. H. Kieffer et al., eds., *Mars* (Tucson, Ariz., 1992). A similar volume for Venus is D. M. Hunten, L. Colin, T. M. Donahue, and V. I. Moroz, eds., *Venus* (Tucson, Ariz., 1983).

On the history of theories of the origin of the solar system, see especially Stephen G. Brush, "From Bump to Clump: Theories of the Origin of the Solar System 1900–1960," in *Space Science Comes of Age: Perspectives in the History of the Space Sciences,* ed. Paul Hanle and Von del Chamberlain (Washington, D.C., 1981), 78–100; Brush, "Theories of the Origin of the Solar System 1956–85," *Reviews of Modern Physics,* 62 (January 1990), 43–112; Aleksey E. Levin and Stephen G. Brush, eds., *The Origin of the Solar System: Soviet Research, 1925–1991* (New York, 1995); and Stanley L. Jaki, *Planets and Planetarians: A History of Theories of the Origin of Planetary Systems* (Edinburgh, 1978). Although the last is a polemic that often ridicules its subjects rather than informs, it remains a useful source. Professor Brush analyzes a specific instance of solar system cosmogony in "A Geologist among Astronomers: The Rise and Fall of the Chamberlin–Moulton Cosmogony," *Journal for the History of Astronomy*, 9 (February and June 1978), 1–41, 77–104.

As this volume went to press a Jupiter-size planet was reported and confirmed circling the Sun-like star 51 Pegasi, tilting the argument further in favor of abundant planetary systems. The results of the discovery, made by Michel Mayor and Didier Queloz, and the confirmation by Geoffrey Marcy and R. Paul Butler, were not yet published, but were reported in *Science News*, 148 (October 21, 1995), 260, and in newspapers around the world.

On the history of science fiction, I have found indispensable the volume edited by John Clute and Peter Nicholls, *The Encyclopedia of Science Fiction* (New York, 1993). Another volume that I have used extensively is Neil Barron, ed., *Anatomy of Wonder: A Critical Guide to Science Fiction* (New York, 1987), a complete guide to science fiction and its historical and critical studies. Other useful sources on the history of science fiction include Brian W. Aldiss, *Billion Year Spree* (New York, 1973), and its revision Brian W. Aldiss and David Wingrove, *Trillion Year Spree: The History of Science Fiction* (New York, 1986); James E. Gunn, *Alternate Worlds: The Illustrated History of Science Fiction* (New York, 1975); and Lester del Rey, *The World of Science Fiction: The History of a Subculture* (New York, 1979). Encyclopedic treatments include Brian Ash, ed., *The Visual Encyclopedia of Science Fiction* (New York, 1977).

Bibliographies of UFO literature include Lynn Catoe, *UFOs and Related Subjects: An Annotated Bibliography* (Washington, 1969), compiled by the Library of Congress for the U.S. Air Force in support of the Condon study, and Richard Michael Rasmussen, *The UFO Literature: A Comprehensive Annotated Bibliography of Works in English* (Jefferson, N.C., and London, 1985). Histories of the UFO debate, largely limited to the United States, are found in David M. Jacobs, *The UFO Controversy in America* (Bloomington, Ind., 1975), and Curtis Peebles, *Watch the Skies! Chronicle of the Flying Saucer Myth* (Washington, D.C., and London, 1994). The reader may keep up with both sides of the controversy in *The Skeptical Inquirer*, the *Journal of Scientific Exploration*, and the *Journal of UFO Studies*.

The two chief historical studies on the origin of life are John Farley, *The Spontaneous Generation Controversy from Descartes to Oparin* (Baltimore and London, 1977), and Harmke Kamminga, "Studies in the History of Ideas on the Origin of Life from 1860," Ph.D. thesis, University of London (November 1980). Robert Shapiro's *Origins: A Skeptic's Guide to the Creation of Life on Earth* (Toronto and New York, 1987) is a very readable guide to present theories of the origin of life, while current issues may be followed in the various conference Proceedings listed in Table 7.1 and in the journal *Origins of the Life and Evolution of the Biosphere*. As this book went to press, an excellent volume appeared by Nobelist Christian de Duve, *Vital Dust: Life as a Cosmic Imperative* (New York, 1995). For the latest on self-organization in the context of origin of life see Stuart Kauffman, *At Home in the Universe: The Search for the Laws of Self-Organization and Complexity* (New York and Oxford, 1995). Many of the pioneering papers in the origin of life are reprinted in David W. Deamer and Gail R. Fleischaker, *Origins of Life: The Central Concepts* (Boston, 1994).

SETI is still a young enough science that its history has never before been written. Much source material will be found in David W. Swift, *SETI Pioneers: Scientists Talk about Their Search for Extraterrestrial Intelligence* (Tucson, Ariz., 1990), 49–53. The history of NASA's involvement in SETI is elaborated in detail in Steven J. Dick, "The Search for Extraterrestrial Intelligence and the NASA High Resolution Microwave Survey (HRMS): Historical Perspectives," *Space Science Reviews*, 64 (1993), 93–139, on which Chapter 8 is based in part. A fascinating inside account from the point of view of one of the SETI pioneers is F. Drake and D. Sobel, *Is Anyone Out There? The Scientific Search for Extraterrestrial Intelligence* (New York, 1992). A prime example of how a substantial interest in SETI has spread to other countries is Guillermo A. Lemarchand, *El Llamado de las Estrellas: Busqueda de Inteligencia Extraterrestre* (The Call of the Stars: Searching for Extraterrestrial Intelligence) (Buenos Aires, 1992). A psychologist critiques SETI's assumptions in John C. Baird, *The Inner Limits of Outer Space* (Hanover, N.H., and London, 1987), an unusual contribution because its author is a social scientist.

Very little scholarly literature exists dealing specifically with the subject of the last two chapters. Although we have dealt in this volume only sparingly with issues of career entry into exobiology and SETI, much of interest will be found in the contemporary interviews by Shirley Thomas in *Men of Space: Profiles of Scientists Who Search for Life in Space* (Philadelphia, 1963), as well as in David Swift's interviews cited earlier, and those undertaken by myself. More generally, however, the literature on discipline emergence is considerable, and several studies in other areas are worthy of emulation. These include David Edge and Michael Mulkay's *Astronomy Transformed* (New York, 1976); Robert E. Kohler, *From Medical Chemistry to Biochemistry* (Cambridge, 1982), and John W. Servos, *Physical Chemistry from Ostwald to Pauling: The Making of a Science in America* (Princeton, N.J., 1990).

Aside from science fiction treatments, a systematic study of the implications of discovering extraterrestrial intelligence remains a subject for the future. Popular treatments have appeared in James Christian, ed., *Extraterrestrial Intelligence: The First Encounter* (Buffalo, N.Y., 1976), and B. Bova and Byron Preiss, *First Contact: The Search For Extraterrestrial Intelligence* (New York, 1990). More serious discussions touching on the subject are found in Edward Regis, Jr., ed., *Extraterrestrials: Science and Alien Intelligence* (Cambridge, 1985), and in the NASA-commissioned study by John Billingham, Roger Heyns, David Milne, et al., *Social Implications of Detecting an Extraterrestrial Civilization: A Report of the Workshop on the Cultural Aspects of SETI* (SETI Institute, 1994). As this book went to press, a volume appeared by Paul Davies, *Are We Alone? Philosophical Implications of the Discovery of Extraterrestrial Life* (New York, 1995).

The question of the limits of science is approached from several directions in Peter Medawar, *The Limits of Science* (Oxford, 1984); Nicholas Rescher, *The Limits of Science* (Berkeley, 1984); David Faust, *The Limits of Scientific Reasoning* (Minneapolis, 1984); Gerald Holton and Robert S. Morison, eds., *Limits of Scientific Inquiry* (New York, 1979); and Richard Morris, *The Edges of Science: Crossing the Boundary from Physics to Metaphysics* (New York, 1990),

INDEX

Abbot, Charles G., 94; and Martian vegetation, 106; and Venus, 130–2, 133; and wireless communication with Venusians, 132, 403
Abelson, Philip H., 143
Academy of Sciences, USSR, 338, 435, 438–9, 482, 486, 494
Adams, Walter S., 94; on life on Mars, 106; oxygen on Mars, 114, 115, 116; water vapor on Mars, 114; CO_2 on Venus, 133
Adamski, George, 284, 317
Adel, Art, 118, 133
Air Force, U. S., and UFOs: 4, 268–9, 272–82, 288–307
Aitken, R. G., 95; and planetary systems, 180; and rotation rate of Venus, 129
Algol, 163
alien: development of, 238–66; invention of, 223–38; *see also* film; life, extraterrestrial; science fiction; television; unidentified flying objects
Alvarez, Luis, 282
Ambartsumian, V. A., 436, 439, 482
American Association for the Advancement of Science, 290, 298, 302, 306–7, 308
American Astronomical Society, 182, 188, 201
American Institute of Aeronautics and Astronautics: and UFOs, 304–5, 307
Ames Research Center, *see* National Aeronautics and Space Administration
amino acids, 3, 348, 360; and comets, 373; enzyme production from, 383; heating of, 362; and interstellar molecules, 374; left-handedness, 372 n95; and life, 387, 389; in meteorites, 351, 369, 372, 375; in Miller–Urey experiment, 345–6; and proteins, 333 n24, 334; synthesis of, 345 n53, 353, 360, 361, 381 n112, 387; *see also* organic synthesis, experiments in
analogy, 30; and planetary systems, 165; Whewell argues against overuse of, 25–6
Anders, Edward, 371, 375
Anderson, Poul, 254 n61
Andromeda Galaxy (M31), 56f
anthropic principle, 57–8, 527–36, 542; and Copernicanism, 529; and SETI, 531–2
anthropocentrism, 36–58, 502–26; and Clerke, 51–2; decline of, 50–7, 542; and Flammarion, 50; and Hubble, 53; Lowell opposes, 74–5; and Maunder, 50; and planetary systems, 162; and Newcomb, 50–1; and plurality of worlds, 27–8, 32, 36–58; and Russell, 55; and Shapley, 52–3, 55; and Sun's position in space, 50–6; in Wallace, 38–49; *see also* anthropic principle; mediocrity, assumption of
Antoniadi, E. M., 64, 79, 91–9, 120; and 33-inch Meudon telescope, 92; life of, 91–3, 93; Martian canals resolved, 93, 94–5, 543; Martian life probable, 99; *La planète Mars* (1930), 95–9; and Strughold, 137
Arecibo Observatory: message sent, 440; SETI observations, 457, 458, 459, 460t, 465, 467, 468
Aristotle: belief in a single world, 13–14; *De caelo*, 13; geocentric cosmology, 13; and kosmos, 13; and life, 45; natural motion and natural place, doctrine of, 13
Army, U.S., 407, 408, 409f
Arnold, Kenneth, and UFOs: 271–2, 307
Arrhenius, Svante: nonbiological hypothesis for Martian surface features, 120; and panspermia theory, 325–9, 367; and Venus, 129–30
Asimov, Isaac, 254 n61, 261
astrobiology, 137–8, 355, 475 n5, 476; *see also* bioastronomy; cosmobiology; exobiology
astrobotany, 122
astrophysics, and extraterrestrials, 30–1; *see also* spectroscopy
astrotheology, *see* theology
Atchley, Dana, 426, 427
atmosphere: oxidizing, 361–2; planetary, 60, 61, 64; reducing, 339, 340, 342,

atmosphere (*cont.*)
343–5, 350, 352–3, 360–2, 363, 387, 388; *see also* Mars; Stoney, G. Johnstone; Venus
atomism, 12
Avery, Oswald, and composition of the gene, 347

Babcock, 94
Bacon, Roger, 14
Bailey, S. I., 87
Baker, Robert M., 298
Bakh, Aleksei N., 338
Ball, Robert S., 33–4
Baly, E. C. C., 342 n44
Barabashev, N. P., 121
Barnard, E. E.: and Barnard's star, 201; and Martian canals, 94 n72, 94 n73, 97 n78, 98, 100; and Newcomb experiments, 87
Barnard's star, possible planets around, 3, 188, 201–6, 207, 209t, 214
Barnes, E. W., 413–14, 518 n41, 519 n43
Barrow, J. D., 530–2
Beck, Lewis W., 507–8
Becquerel, Paul, 328
Belsky, T., 124
Bentley, Richard, 20–1
Berenzden, Richard, 505, 506–7
Bergson, Henri, 331, 337
Berkner, Lloyd: and Green Bank conference, 427; and Project Ozma, 421–2, 479, 497; and Space Science Board, 481; and UFOs, 282–3
Bernal, J. D., 367, 378; and clays, 363; and meteorites, 369, 371; and nature of life, 378–9; and Oparin, 338, 343–4; and panspermia, 368; and probability, 386–7; and Stapledon, 247
Berzelius, J. J., 326, 368
Bessel, Friedrich Wilhelm: stellar parallax, 182, 205; stellar proper motion variation, 162–3
Bethe, Hans, 415
Biemann, Klaus, 147, 153
Billingham, John: and exobiology, 477; and implications of contact, 513; and NASA SETI, 459, 463, 486–90, 497, 498; and Project Cyclops, 439, 459
binary stars, 25; and planetary systems, 192–3; spectroscopic, 163; unsuitable for life according to Wallace, 47
bioastronomy, 325, 474, 475 n5, 477–8, 490, 492–3, 499–500; *see also* exobiology; life, extraterrestrial
biochemistry: at Cambridge University, 342; and extraterrestrial life, 363, 392; International Congress of, 367; International Union of, 349; and organic chemistry, 344; origins of, 333; and origins of life, 337, 339, 340, 350; in the Soviet Union, 338–9
biological universe, acceptance of, 537; and anthropic principle, 527–36; and Aristotle's overthrow, 15; and cosmic evolution, 541; as cosmological worldview, 34, 541–2; implications of, 503–26, 536; as myth of modern times, 3, 5, 552–4; transition from physical world to, 4, 10, 34, 552–3; *see also* cosmology
biology, 321–98; aspirations for universality, 322, 331, 350, 351, 357–67, 377, 378, 379, 382, 387, 388–9, 391, 396, 398; an autonomous science, 322; a unified science, 322; *see also* bioastronomy; exobiology; life, extraterrestrial; space biology
biophysical cosmology, *see* cosmology
Black, David, 160 n4, 474
black dwarf, 207
Blackett, P. M. S., 416–17
Blish, James, 260, 517 n40
Blum, Harold F., *Time's Arrow and Evolution* (1951), 385–6, 392
Bobrovnikoff, Nicholas, 434–5
Bok globules, 193
Bondi, Hermann, 528
Born, Max, 292
Boston Scientific Society, 74
Bracewell, Ronald, 429–30; and Drake Equation, 442; and probes, 260 n75, 432–3, 445, 450, 486
Bradbury, Ray: and aliens, 238, 251–2, 263, 517 n40; *The Martian Chronicles* (1950), 251–2;
Brewster, Sir David, *More Worlds than One* (1854), 26
Brin, David, and science fiction, 260, 261; and Drake Equation, 452
British Association for the Advancement of Science, 337 n29
British Astronomical Association, 92
British Interplanetary Society, 379
Brown, Harold, 291
Brown, Harrison: and planetary systems, 207, 260 n74; and science fiction, 260
brown dwarf, 208, n115, 213, 213 n120, 217–18
Bruno, Giordano, 329; *De l'infinito universo e mondi* (1584), 15–16

Bryan, Senator Richard, 469
Buchner, Eduard, 333
Buddhism, and extraterrestrials, 523
Buhl, David, 374
Buridan, Jean, 14
Burke, Bernard F., 215
Burroughs, Edgar Rice, and aliens, 238, 239–40
Byron, 28
Byurakan Astrophysical Observatory, 436, 438–9, 476, 482, 485–6

Cairns-Smith, A. Graham, 363
Calvin, Melvin, 124; and meteorites, 368–9; and NASA, 357; and organic synthesis, 344–5; and origin of life, 352–5, 380; and Panel on Extraterrestrial Life, 481; and SETI, 354 n65, 427; Space Science Board, 142
Cameron, A. G. W., 497; and Drake equation, 437; and implications of contact, 508–9; and planetary systems, 486
Campbell, Bruce, 211–12
Campbell, John, 241; "Who Goes There?" (1938), 244–5, 262, 265
Campbell, Joseph, 554
Campbell, W. W.: and binary stars, 47; and life on Mars, 90; and Lowell's theory of Martian canals, 79–83, 101; Martian water vapor not detected, 77, 83, 89–90, 99, 100; organic origin of dark areas accepted, 82; and planetary systems, 165–6, 211; and spectroscopic binaries, 165–6; and Trumpler, 95
canals, *see* Mars, canals of
Cantril, Hadley, 507
Carr, B. J., 530
Carter, Brandon, 528–9, 531, 534, 535
Cashman, R. J., 117
cell: composition of, 334–7, 360, 362; and extraterrestrial biochemistry, 392; nucleus of, 347, 394; origin of, 339, 340, 342, 359t, 362–3, 380, 387–8, 394; and origin of life, 336, 337; from space, 376; as unit of life, 334–6, 378; and Wilson, 334–5; zymase isolated from, 333; *see also* life, origin of
Central Intelligence Agency, 282–3
Cerulli, Vincenzo, 85
chain of being, 37, 542
Chalmers, Thomas, 23
Chamberlin, R. T., 339
Chamberlin, T. C.: and organic synthesis, 339 n35; and planetary systems,166–72, 178–9
Chamberlin–Moulton theory, 166–72, 173; and Lowell, 88; *see also* planetary systems; planetesimal hypothesis
chance and necessity, *see* evolution; life, origin of
Chandrasekhar, S., 190
Chardin, Teilhard de, 521
chemical evolution, *see* evolution; life, origin of
Chiu, H. Y., 304
chondrites, carbonaceous, *see* meteorites
Christianity: and plurality of worlds, 14–15, 22–7, 28–9, 34, 514–26; and science fiction, 249–51, 260; *see also* natural theology; theology
CIA (Central Intelligence Agency), 4
civilizations, longevity of, 429, 430, 442, *see also* Drake Equation
Clark, Alvan, 71 n25, 72
Clarke, Arthur C., 254–7, 259, 261; *2001: A Space Odyssey* (1968), 256, 262, 264, 265; *2010: Odyssey Two* (1982), 264; *2061* (1988), 264; *Against the Fall of Night* (1953), 254; and alien encounter, 254–6, 260, 265; and Asimov, 261; and Burroughs, 239; *Childhood's End* (1953), 255–6; *The City and the Stars* (1956), 254–5, 257; and extraterrestrial civilizations, 255; and Hoyle, 257; and imagination; 404 n12; and implications of contact, 517; and Kubrick, 264; and Stapledon, 249, 253, 254, 260; and UFOs, 256 n66; and Wells, 231
Claus, George, 370, 376
Clemence, Gerald M., 301–2
Clement, Hal, 253, 256–7
Clerke, Agnes: planetary systems unlikely, 51, 162, 170; solar system central, 51, and spectroscopic binaries, 162; and Wallace, 40, 40 n13, 44, 51
coacervates, 340, 363
Coblentz, W. W.: life on Mars, 106, 109–13, 544–5; temperature of Mars, 109–14, 279
Cocconi, Giuseppe: and Green Bank conference, 427, 431; and interstellar communication, 415–19, 421, 422, 425, 431, 432, 433, 434, 436, 451, 459, 470, 471, 472, 474, 508
colonization, interstellar, *see* interstellar colonization
comet, *see* molecules, organic
Commoner, Barry, 143
communication, interplanetary and interstellar, *see* SETI

Condon, Edward U., 292–307, 309; *see also* Condon Report
Condon Report, 268, 290, 292–307, 308; and AIAA, 304–5; choice of Condon for, 292; conclusions of, 300; and Congress, 298; and extraterrestrial hypothesis, 300–1, 309, 318; and Hynek, 309; and Kuiper, 303; and MacDonald, 303, 304; and Menzel, 292; methodology of, 297; and Morrison, 302; and Murray, 310; National Academy of Sciences review, 301–2, 311; origins of, 292; reaction to, 301–6; staff of, 297–8; and Sturrock, 310–11; *see also* unidentified flying objects
Confucianism, and extraterrestrials, 523, 524
Congress, U. S., 4; and exobiology, 496; and UFOs, 289, 290, 291, 298–9, 306, 307; and SETI, 454, 469–70, 498–500
contamination, *see* planets, contamination of
contingency, *see* evolution; life, origin of
Cook, Stuart, 297
Copernicanism: and anthropic principle, 529; and anthropocentrism's demise, 40, 53; exobiology's implications compared to, 141, 504, 508, 542; extraterrestrials used to argue against, 515; and planetary science, 59; and planetary systems, 199, 220, 221; and plurality of worlds, 15–18, 19, 26; as revolution, 10; and theology, 520
cosmic evolution, *see* evolution, cosmic
cosmobiology, 379, 475 n5, 476; *see also* astrobiology; bioastronomy; exobiology
cosmochemistry, 372; birth of, 368
cosmology: Aristotelian, 5, 13–15; atomist, 5, 12–13; biophysical, 11, 35, 502, 541–2, 545, 552–4; Cartesian, 5, 17–19; Copernican, 5, 15–17; modern, 52–3, 55; Newtonian, 19–22; and plurality of worlds, 11–22
COSPAR (Council on Space Research), 139
Crick, Francis: and genetic code, 347; and panspermia, 377; and RNA first concept, 362
Crowe, Michael, 2, 5, 22 29, 401; height of Martian canals controversy, 64; on Lowell, 62; reviews of Wallace, 50 n26; on Whewell, 23
Curtis, H. D., 165–6, 211
Cyclops, *see* SETI
61 Cygni, and planetary companion, 186, 188, 202, 203, 207, 214
Däniken, Erich von, 316
Darrach, H. B., Jr, 28
Darwin, Charles, 29, 141, and biology, 322; and Darwinism in cosmic context, 31–5, 37, 238, 353–4, 386, 391, 542; and evolution, 31, 44, 324, 391; Flammarion and, 32–3, 238; Hoyle and, 257; and Huxley, 36, 231, 332; and implications of exobiology, 141, 504, 508, 520; Lowell and, 88; and origin of life, 323, 326, 332; and *Origin of Species* (1859), 322, 323; Schiaparelli and, 79; and Wallace, 38, 44, 389; Wells and, 231, 238; *see also* evolution
Darwin, Sir George, 172
Dauvillier, A., 344
Dawkins, Richard, 386
Deep Space Network, 462, 466, 468–9
Deich, A. N., 188
Delano, Kenneth, 521
Democritus, 10, 379
Derham, William, 515
Descartes, René, 17–18
determinism, *see* evolution; life, origin of
de Vaucouleurs, Gerard: and astrophysics applied to planets, 107; and life on Mars, 125
Dick, Thomas, 23
Dicke, Robert, 527–8
Dirac, Paul, 527, 528
Dixon, Robert S., 456–7
DNA, *see* nucleic acids
Dobzhansky, Theodosius, 393
Doel, Ronald, 115
Dollfus, A., and Martian vegetation, 137
dolphins, and intelligence, 396–7, 397 n136, 427, 430
double stars, *see* binary stars
Douglass, A. E.: at Arequippa, Peru, 68; on Lowell and W. H. Pickering, 72; Mars diameter, 77; Martian canals, 82, 84 n46, 94; and Schiaparelli, 74; site testing for Lowell Observatory, 72
Drake, Frank, 260, 306, 418–31, 473, 486, 498; and Arecibo message, 440; Arecibo observations, 457, 458; and Fermi paradox, 450; and Green Bank meeting, 427–31, 482; and implications of contact, 508, 510; and interstellar travel, 450; and Kardashev civilizations, 457; and new discipline, 482; and Pioneer and Voyager messages, 448; and Project Ozma, 418–27, 454, 479, 497, 503; *see also* Drake Equation; SETI
Drake Equation, 431–57; abandoned by

some, 446, 452; and cosmic evolution, 539; diffusion of, 437; estimates for factors in, 441t, 443, 541, 544; modifications to, 442, 452; origins of, 428–31; and probability, 439; and Project Cyclops, 439–40; in Soviet Union, 434, 436, 437, 438–9, 449; *see also* SETI
Driesch, Hans, 331
Dunham, Theodore: CO_2 on Venus, 133; methane and ammonia on Jupiter and Saturn, 340; oxygen on Mars, 115, 116
du Noüy, Lecomte, 381, 383, 385, 398
Durant, Frederick C., III, 285
dwarf stars, 182, 188, 196, 202, *see also* black dwarf; brown dwarf
Dyson, Frank, 402,
Dyson, Freeman: and Drake Equation, 452; and infrared Dyson spheres, 432, 434, 435–6, 451, 471; and interstellar travel, 445; and SETI, 438, 453

Earth: conditions for life on, 39–48; position of, 1, 36 n3; *see also* atmosphere
ecosphere, 137
Eddington, Arthur S.: and energy production in stars, 53; and fundamental constants, 527; life in the universe rare, 57, 57 n41, 179, 518; and Martian canals, 79
Edelson, Robert, 463
Edwards, Frank, 289
Eichhorn, Heinrich, 205, 206
Eigen, Manfred, 362
Einstein, Albert, 52, 58; and life on Mars, 403
Eiseley, Loren, 321, 391
Ellis, G. F. R., 533–4
Emerson, Ralph Waldo, 502
enzymes, 333–4, 336, 338, 349, 362, 378, 380, 383
Epicurus, 12
esobiology, 476
Europa, and life, 61
Evans, J. W., 84–5
evolution: biological, 323, 331, 348, 359t, 363, 382, 385, 387, 389; and chance and necessity, 385–6, 389–98; chemical, 329–89, 359t; cosmic, 34, 88, 327, 329, 538–41, 551; and image of man, 37; of intelligence, 389–98, 400, 413, 428, 444, 446; and science fiction, 223–4, 225, 226, 230, 231, 233–4, 235, 238, 248, 249, 266; as scientific foundation for extraterrestrials, 29–35; and Wallace, 38–9, 48–9, 389–91, 392, 394; *see also* Darwin, Charles; life, origin of; Wallace, Alfred Russel
Ewen, H. I., 422, 434
exobiology, 475 n5; and biology, 358; and chance and necessity, 379; as a discipline, 325, 473–501, 551–2; goals of, 322–3, 388; Lederberg coins term, 140, 476; and NASA, 352, 355–6, 364–7, 387; and origins of life, 357–67, 388; progress in, 549; as protoscience, 551–2; science looking for a subject, 5, 323, 325, 392–3, 477; *see also* astrobiology; bioastronomy; cosmobiology; esobiology
extraterrestrial life, *see* life, extraterrestrial; exobiology; science fiction; SETI; unidentified flying objects

Farley, John, 336
Fasan, Ernst, 513
Fermi, Enrico, 415, 443
Fermi paradox, 281, 312–13 n107, 450–1, 453, 471, 548; and Morrison, 450; origins of, 443; in Soviet Union, 444 n96; 448–9, 453, 471; and Tipler, 446–8; and Where are They? conference, 445–6; *see also* Hart, Michael; SETI; Tipler, Frank; Viewing, David
Fesenkov, V., 349
Feuerbach, Ludwig, and anthropocentrism, 27
film, aliens in, 4, 222–3, 253, 259, 262–6, 552; *2001: A Space Odyssey* (1968), 223, 262, 264, 265; *Alien*, 4, 265; *Close Encounters of the Third Kind* (1977), 4, 223, 264; and Communist paranoia, 262; *The Day the Earth Stood Still* (1951), 263; *Earth vs. the Flying Saucers* (1956), 263; *ET: The Extraterrestrial*, 4, 223, 265; *Forbidden Planet* (1956), 263; *Invasion of the Body Snatchers* (1956, 1978), 263; *It Came from Outer Space* (1953), 263; *Solaris* (1971), 259; *Star Trek* (1979–94); *Star Wars* (1977), 4, 264; *The Thing* (1951, 1982), 262–3; *This Island Earth* (1955), 263; *War of the Worlds* (1953), 263; *see also* science fiction; television
Findlay, John, 419
Fiske, John, 70
Fitch, Frank, 371
Flammarion, Camille: influence on Lowell, 72; Mars habitable, 68–9; and Maunder, 84; and panspermia, 326; *La planète Mars* (1892), 68–9, 107 n91,

Flammarion, Camille (*cont.*) 228; *La pluralité des mondes habités* (1862), 32–3, 34, 224; and science fiction, 224–5, 226, 228, 232, 238, 243; and Wallace, 50
Fletcher, James, 488, 507
Fontenelle, Bernard le Bovier de: and Cartesian cosmology, 18; *Entretriens sur la pluralité des mondes*, 18, 552; and the plurality of solar systems, 18
Ford, Gerald, 291
Fort, Charles, 271 n8, 272
Forward, Robert, 260
Fournet, Dewey, 282
Fox, Philip, 87
Fox, Sydney, 362–3, 364
Frost, E. B., on life on Mars, 106
Fuller, John, 296

Gale, George, 532
Galilei, Galileo: *Dialogue on the Two Chief World Systems* (1632), 17; Moon similar to Earth, 16; *Siderius nuncius* (1610), 16
Gallup poll, 3
Galton, Francis, 236 n29
gamma rays, 415–16
Gamov, George, 190
Gatewood, George, 205, 206
Gehrels, Tom, 160 n4, 474, 500
gene: as unit of life, 334, 336; composed of DNA, 347
Geneva Observatory, 213
Gernsback, Hugo, 273f, 411 n24
Gindilis, L. M., 436
Ginna, Robert, 278
Godfrey of Fontaine, 14
Goldsmith, Donald, 492
Gore, J. E., 34, 162, 232
Goudschmidt, Samuel A., 282
Gould, Stephen Jay, 394–6, 512
Green, Nathaniel, 83, 93
Greenstein, Jesse, 411
Griffin, Roger, 211
Gunn, James, 260
Guthke, Karl, 2–3, 5, 29, 227 n13, 231, 554

Haldane, J. B. S.: and definition of life, 379; and extraterrestrial life, 348, 350; and Oparin, 341–2, 364, 386–7; and Oparin–Haldane theory, 341–5, 350, 360–1, 362, 363, 367–8, 378, 386, 538; and probability of life, 386; and Stapledon, 249
Hale, George Elery, and canals of Mars, 93–4
Hall, Asaph, 66
Hall, Robert L., 298
Handler, Philip, 486
Handlin, Oscar, 296
Harder, James, 298
Harkness, William, 66
Harrington, Patrick, 446
Harrington, Robert S., 208
Hart, Michael: and Fermi paradox, 443–54; and origin of life, 446
Hartmann, William, 303, 306
Harvard-Smithsonian Center for Astrophysics, 213
HD 114762, 213
Heeschen, David, 419, 420
Hegel, G. W. F, and anthropocentrism, 27
Helmholtz, Hermann von, 326, 339
Henderson, Lawrence J.: and biocentric principle, 542; and cosmic evolution, 537, 539–41; and fitness of the environment, 531
Hendry, Allan, 311
Hennessy, Douglass, 369
Henry of Ghent, 14
Herelle, Felix d', 334, 337, 342 n43
Herschel, John, 23, and Sun's position, 39
Herschel, William: central location of solar system, 37; temperature on Mars, 112
Hershey, John, 205
Hertz, Heinrich, 401, 470
Hertzsprung, Ejnar: photographic techniques for double stars, 182; teacher of Kuiper, 117; teacher of van de Kamp, 183
Hesburgh, Theodore, 426, 511
Hess, V. F., 407 n16
Hey, J. S., 414
Hill, Betty and Barney, 295–6, 317
Hinduism, and extraterrestrials, 523, 524
Hoerner, Sebastian von: and Drake Equation, 429–30; and Fermi paradox, 445–6; and interstellar travel, 433–4; and Ozma, 426
Holmberg, Erik, 186, 187, 202, 543
Hopkins, Budd, 317
Hopkins, Frederick, 342
Horowitz, Norman H.: and astrobiology, 355 n68; and NASA, 357; and National Academy study, 357 n74; and Oparin, 343; and origin of life, 381–2; and reducing atmosphere, 361–2; and Space Studies Board, 356; and Viking pyrolytic release experiment, 150–8, 548

Horowitz, Paul, 458–9
Housden, C. E., 129
Hoyle, Fred: and anthropic principle, 533; and organics, 373, 375–7; and planetary systems, 197 n95, 198, 354; and science fiction, 257–8, 260
Hoyt, William S., 62, 64
Huang, Su-Shu: and Green Bank conference, 427; and planetary systems, 197, 433
Hubble, Edwin P.: and expanding universe, 53, 58, 171; and external galaxies, 53; life in the universe likely, 57 n41
Hubble Space Telescope, and planetary systems, 211, 215, 218 n128
Huggins, Sir William: and astrophysics, 30; belief in extraterrestrial life, 30; and gaseous nebulae, 30
Hulburt, E. O., 409–10
humanism, Renaissance, 37
Huxley, Julian, 337, 342 n44; 391
Huxley, Thomas H., 337, 343, 378, 379; and man's place in nature, 36; *Man's Place in Nature* (1863), 39; and protoplasm, 45, 332, 334–5, 336; and Wells, 231
Huygens, Christiaan: and Cartesian vortices, 18–19; *Kosmotheoros* (1698), 18; and plurality of solar systems, 18
Hynek, J. Allen, 190, 264, 267; and AAAS Symposium on UFOs, 306; as Air Force consultant on UFOs, 276, 279, 290, 292; and astronomers' poll on UFOs, 281, 319; and Condon Report, 309; and Congressional hearings, 291–2, 298; and extraterrestrial hypothesis of UFOs, 276, 296, 309, 311, 314; and MacDonald, 293; and Markowitz, 296; and Menzel, 283, 289; and Michigan UFO sightings, 290–1; and Murray, 310; and New Wave UFO theories, 314–15, 316; panel of scientists proposed to study UFOs, 290; and scientific method, 298, 309–10, 313, 319; turning points on UFOs, 283–4, 291, 296, 314; *The UFO Experience* (1972), 309; as UFO skeptic, 276; UFO studies encouraged, 307, 310, 319; and Vallee, 287, 288, 314, 316

Incarnation, 27, 516, 518–22, 524–5
infrared observations, 109–14, 117, 118–20, 122–5, 210; and Infrared Astronomical Satellite (IRAS), 210; *see also* radiometry; spectroscopy
intelligence: and dolphins, 396–7, 397 n136, 427, 430; evolution of, 389–98, 400, 413, 428, 444, 446; implications of, 472, 502–26; probability of, 428–30, 431, 434–5, 440, 447, 449, 452–3; *see also* Drake Equation; evolution; life, extraterrestrial; Mars, canals of; SETI
International Academy of Astronautics, 490, 512
International Astronautical Federation, 285, 490, 512–13
International Astronomical Union, 440, 449–50, 477, 478, 490, 492–3, 494, 511
International Council of Scientific Unions, 139
International Institute of Space Law, 512
International Society for the Study of the Origins of Life, 490
interstellar colonization, 399–400, 432, 443–54, 462, 471
interstellar communication, *see* SETI
interstellar molecules, *see* molecules, organic
interstellar travel, 438, 453, 471; and Bernal, 249 n51; and Drake, 450; and Dyson, 445; and Fermi paradox, 444, 445; and Hoerner, 433–4, 445–6; and Morrison workshops, 462; and Purcell, 434, 444; and Soviets, 436; and Stapledon, 249 n51; and Tipler, 446; *see also* interstellar colonization
Islam, and extraterrestrials, 523, 524

Jacob, Francois, 382
Jacobi, C., 172
Jacobs, David, and UFOs, 269, 275, 317
Jaki, Stanley, 6
Jansky, Karl, 410–11
Jean of Bassols, 14
Jeans, James, 186; life lacking on Venus, 133; planetary systems common, 55, 180–1, 187–8; planetary systems rare, 53, 172–80, 328, 330, 340, 413, 434, 548; and Stapledon, 246
Jeffreys, Sir Harold, 177–8
Jessup, Morris K., 284
Jet Propulsion Laboratory, 140, 142, 210; and exobiology, 367, 483; and SETI, 462–72
Jodrell Bank, 416, 418
Jones, Eric, and Fermi paradox, 443 n93; and interstellar colonization, 445, 446
Jones, Sir Harold Spencer, *see* Spencer Jones, Sir Harold
Jong, Bungenburg de, 340

Jordan, P., 527
Judaism, and extraterrestrials, 522, 523, 524
Jung, Carl, and UFOs, 286, 299, 504 n7
Jupiter, organic molecules on, 61, 159, 374
Juvisy Observatory, 92

Kallarakal, Varkey, 208
Kamerlingh Onnes, Heike, 328
Kant, Immanuel, 21–2, 224, 229, 230, 238
Kaplan, L. D., 145
Kapteyn, J. C., 185
Kardashev, Nicolai: and Byurakan meetings, 436, 439; and Bruno, 435 n76; and classification of civilizations, 435–6, 445, 451, 455, 457, 458, 471; and CTA-102, 455; and Fermi paradox, 444 n96, 448–9 n110; and SETI observations, 436, 455; and Shklovskii, 449 n110 and n112; and Tikhov, 435, n76
Kauffman, Stuart, 387 n120
Keats, 28
Keeler, James E., 42, 167–8
Keldysh, Mstislaw, 435
Kelvin, William Thomson, Lord: and mathematical physics, 172; meteoric hypothesis used by Wallace, 44; and panspermia, 326–7, 339, 367
Kepler, Johannes: inhabitants on Earth the best, 37, 515; and inhabitants on Moon, 16, 223; Jovians, 16–17; *Somnium* (1634), 16
Kety, Seymour, 355
Keyhoe, Donald, 276–7, 285
Klass, Philip, and UFOs, 295, 296, 298, 307, 312
Klein, Harold P.: and NASA Ames exobiology, 355–6; and origins of life, 388; and Space Studies Board, 501; Viking biology science team head, 147, 152, 153, 154–7, 496, 501
kosmos, as ordered system, 12
Krauss, John, 456
Kubrick, Stanley, 262, 264
Kuettner, Joachim, 305
Kuiper, Gerard P.: and carbon dioxide in Mars atmosphere, 117–20; and double stars, 192–3, 194, 544; infrared spectroscopy, 117–18, 473; and planetary systems, 126, 191–4, 198; and Sagan, 479; and Strughold, 137; and UFOs, 294–5, 298, 303, 318; and vegetation on Mars, 118–21, 122, 123, 543
Kuiper, Tom, 445, 511
Kumar, S. S., 207
Kurtz, Paul, 315
Kvenolden, K. A., 372

Lalande 21185, 188, 202, 203, 206
Lambert, Johann, 21–2
Lamm, Norman, 502, 522–3
Lamont-Hussey Observatory, 122
Lampland, C. O.: photographs Martian canals, 85; temperature of Mars, 109–12, 279, 544–5
Lane, B. W., 84
Langley Research Center, and Viking, 146
Langmuir, Irving, and UFOs, 275, 276 n16
Laplace, 172
Lasswitz, Kurd, 224; *Auf Zwei Planeten* (1897), 227–30, 231, 238, 240, 241, 255, 265, 552
Latham, David, 213
law, biological, 2, 10, 58, 322
law, physical, 1–2, 10
Leander McCormick Observatory, 183
Lear, John: and Project Ozma, 425; and UFOs, 295–6
Lederberg, Joshua: and discipline formation, 483, 495; and exobiology, 140, 355, 358, 391 n124, 427, 478, 479, 483; and planetary contamination, 138–9, 352; pre-Viking prospects, 152; and Sagan, 479; Space Science Board, 142, 427, 481, 483
Leibniz, Gottfried Wilhelm, 27
Leinster, Murray, 260 n73
Lem, Stanislaw, *Solaris* (1961), 258–9
Leslie, John, 534
Leucippus, 12
Levin, Gilbert, 149–58
Lewis, C. S.: and aliens, 249–51, 260, 265, 517; and implications of contact, 519 n43, 520, 526; *Out of the Silent Planet* (1938), 250–1; *Perelandra* (1943), 251; *That Hideous Strength* (1944), 251
Lick Observatory, 47, 75; and Aitken on Venus, 129; and canals of Mars, 68, 79, 82, 92, 94, 95, 98, 100, 108; and Campbell, 77, 79, 89–90; and Coblentz, 109, 113; Keeler and spiral nebulae, 167–8; and Schaeberle, 82; and spectroscopic binaries, 163, 165; and Trumpler, 95, 98, 108; and Wright, 108
life: carbon-based, 151, 363; cell as unit of, 334–6; complexity of, 44; definition of, 45, 331–6, 378; gene as unit of, 334; mineral, 363; probability of, 322, 325, 348, 354, 364, 379–87, 389–98, 417, 428–9, 437, 439, 442, 446, 447, 449,

450, 461; protein enzyme as unit of, 333–4; protoplasm as unit of, 332–3; silicon-based, 232–3, 243, 391; virus as unit of, 334; vitalism vs. mechanism, 331–2, 336–7; *see also* evolution; life, extraterrestrial; life, origin of
life, evolution of, *see* evolution
life, extraterrestrial: biologists vs. astronomers, 321–2, 330–1, 351; communication with, 236 n29; conferences on, 351, 367, 396–7, 476, 480–92, 484t, 491t, 497; cultural significance, 552–3; dissociated from UFOs, 283; and evidence and inference, 543–6; and human knowledge, 552, 554; impact of discovery, 502–26; and metaphysical assumptions, 6; as myth, 3, 5; a persistent theme, 3–4; polls on, 3; progress in the debate, 61; and religion, 514–26; and science fiction, 222–66; separable from planetary science, 59, 60; and the space age, 135–59; *see also* bioastronomy; exobiology; life; life, origin of; Mars; planetary systems; scientific method; SETI
life, origin of: 321–98; and chance and necessity, 325, 330–1, 351, 379–89; chemical theory of, 324, 329–50, 351–89; and complexity, 387; conferences on, 349, 350, 351, 364–7, 365t, 386; evidence and inference problems, 544; mineral theory, 363; panspermia theory, 324, 325–9, 339, 341, 348, 351, 367–77; and self-organization, 387; and SETI, 400; and the Space Age, 350–89; spontaneous generation theory, 323–4, 325, 326, 327, 333, 336, 338–9, 344, 367, 379, 384; *see also* atmosphere; enzymes; evolution; gene; life; life, extraterrestrial; nucleic acids; organic synthesis, experiments in; proteins; virus
Lilly, John C., *Man and Dolphin* (1961), 427, 430
limits of science, *see* science, limits of
Lindsay, David, *A Voyage to Arcturus* (1920), 241–2, 245, 250, 251, 253, 256, 259, 265
Lipp, James E., 274
Lippincott, Sara, 202–03, 205
Locke, John, 27
Lockyer, Sir Norman, and Sun's position, 39
Lorenzen, Coral and Leslie, 285
Lovejoy, Arthur O., and principle of plenitude, 5, 12–13
Lovell, Bernard, 414, 416–17, 417 n34
Low, Robert, 297
Lowell, A. Lawrence, 104
Lowell, Percival, 3, 62–105, 322; and anthropocentrism, 74, 76n 33; and Antoniadi, 94–5; and canals of Mars, 62–105, 277, 543; compared to Proctor and Flammarion, 75; and Douglass, 72, 74, 83, 84; *The Evolution of Worlds* (1909), 88; evolutionary view of universe, 34, 88, 538; and Flammarion, 72; legacy of, 135–6, 146; life of, 70–1; *Mars* (1895), 75; *Mars and Its Canals* (1906), 75, 86, 88; *Mars as the Abode of Life* (1908), 75, 88; and Maunder, 85; nebular hypothesis, 71, 74–5, 88; and Pickering (W. H.), 68, 69, 71–5, 78, 101; rotation rate of Venus, 127; and science fiction, 224, 227, 228, 229–30, 232, 234–5, 236, 238, 239, 240, 246; and Strughold, 137; temperature of Mars, 112; and Schiaparelli, 71–4, 77; and Todd, 406; and UFO phenomenon, 269–71; *see also* Mars; Mars, canals of
Lowell Observatory: astrobiology symposium, 137; and Coblentz, 109–11; and Lowell, 62–105; origins of, 71–5
Lucian, 223
Lucretius, *De rerum natura*, 12
Lumsden, Charles J., 396 n134
Lunan, Duncan, 433
Lunar and Planetary Exploration Colloquium, 138
Luria, Salvador, 481
Luther, Martin, 515
Luyten, W. J., 247 n46
Lwoff, Andre, 344, 382

Macallum, Archibald, 334
McDonald, James, and UFOs, 292, 293–4, 298, 303–4, 305, 306, 307, 309
MacDonald Observatory, 118
Mack, John, 317
Maclaurin, C., 172
McMullin, Ernan, 525–6
Maestlin, Michael, 16
Mamikunian, Gregg, 476
Mantell, Thomas, 274, 276
Marconi, Guglielmo, 236 n29; 402–6, 470, 471
Margulis, Lynn, 377
Mariner spacecraft, 102 n86, 145–6
Mark, Hans, 459
Markowitz, William, 296, 301, 306
Mars, 62–126; 110t, 135–59; altitude of, 66, 67; animals on, 99, 106, 112, 114,

Mars (*cont.*)
116–17, 137; atmosphere, 69, 77, 83, 88–9, 90, 98–9, 100, 103–4, 107–8, 110t, 117–20, 136, 141, 145, 147, 150, 155; carbon dioxide, 117–20, 150; channels, 146, 158; chlorophyll, 118–19, 120 n118, 121; clouds, 68, 74, 75, 83; color changes, 69, 77, 82, 97, 107, 108, 112, 118, 120–1; communication with, 401–14; conditions for life, 46, 88–9, 107, 110t; craters on, 145; diameter of, 66, 67f, 77, 123; distance of, 64–8; life on, 59–126, 135–59, 322, 348, 363, 375, 378, 379, 391, 392, 394; literature about, 62f; oases, 76, 77, 78; oppositions, 64–8; and origin of life, 348, 355, 356, 382, 388; oxygen, 90, 100, 103–4, 114–15, 116–17, 118, 125; photosynthesis, 117, 118–19; polar caps, 69, 77, 78, 89, 95, 98, 107, 118, 121, 136, 158; river beds, 146, 158; in science fiction, 222–66, passim; seas, 69, 77; and SETI, 417; Soviet exploration of, 121–2, 141, 158, 352, 357; temperature, 60, 107, 108, 109–14, 118, 120, 122, 147, 150; and UFO phenomenon, 269–71, 275; vegetation on, 60, 61, 68, 72, 77, 78, 95, 99, 105–26, 136, 137, 143, 145, 159, 330; water on, 59, 64, 69, 76, 77, 83, 89–90, 99, 100, 103–4, 107, 110t, 111, 114, 116, 120, 125, 131, 135–6, 144t, 145, 146, 150, 155, 158; *see also* Mariner spacecraft; Mars, canals of; Viking spacecraft; Voyager spacecraft

Mars, canals of, 38, 62–105, 65t, 82t, 106–7, 146, 330, 406, 476, 495; aftermath of controversy, 99–105; Antoniadi resolves canals, 93, 94–5; and evidence and inference problems, 99–105, 543; genesis of the theory, 66–78; illusionist arguments, 83–105; and imagination, 104; interpretations of, 79–105, 82t; and Lowell, 66–105; and Mariner results, 102 n86, 103; photographed, 85–6, 122; resolution of conflict, 79–99; and Schiaparelli, 66–78; and telescope size, 64, 65t, 66–8, 78, 94, 97, 98, 100; and Trumpler, 95; as vegetation strips, 68, 78, 95; *see also* Mars

Martin, James S., 146

Martynov, Dmitry, 455

Mascall, E. L., 519

maser, 432

Matthew, W. D., 330, 343, 381, 390–1

Matthews, Cliff, 360, 364

Maunder, E. W.: and Wallace, 50; illusion theory of Martian canals, 84–5

Maxim, Hiram Percy, 413, 414

Mayall, Nicholas U., 292

Mayer, C. H., 134

Mayr, Ernst: and evolution of life, 393–4, 395; and origin of life, 386; and SETI, 5, 394

Medawar, Peter, 546

mediocrity, assumption of, 37; replaces assumption of centrality, 53; in Shklovskii and Sagan, *Intelligent Life in the Universe* (1966), 56

Meinschein, Warren, 369

Melanchthon, Philip, 515

Menzel, Donald H.: and interplanetary communication, 411–12, 470; and Keyhoe, 277 n21; and life on Mars, 115, 411; and Martian temperature, 111; and Venus, 133–4; and UFOs, 267, 279–83, 284, 285, 287, 288, 289, 292, 295, 296, 298, 306, 307, 312, 315, 320

Mercury, conditions for life on, 46

meteoric hypothesis, 44

meteorites, 367–72; and amino acids, 351, 369, 372, 375; and Bernal, 369, 371; Berzelius, 326, 368; and Calvin, 368–9; and carbon on early Earth, 340; carbonaceous chondrites, 368–72, 373; and fossil life, 370–2, 387; Ivuna, 370; lunar origin of, 371; and Morrison, 370–1; Murchison, 372; Murray, 369; and Nagy et al., 369–71; and Oparin, 340, 341; and organic compounds, 324, 351, 358, 367–72; organisms in, 341, 341 n40; Orgeuil, 369, 370; and panspermia, 326, 328, 367–72; and the UFO controversy, 267, 307; and Urey, 371

Meynell, Alice, 502, 522

Michael, D. N., 503 n5

Michel, Aime, 287

Michaud, Michael, 513 n27

Michelet, Carl, against teleology, 27–8; and anthropocentrism, 27

microwave window, *see* SETI

Miller, Stanley, 349; and extraterrestrial life, 356, 356 n70; and Miller–Urey experiment, 3, 345–7, 352–3, 358, 360, 375, 376, 388, 474, 538; organic synthesis experiments, 356, 358, 360, 376; and reducing atmosphere, 361, 362 n80; *see also* organic synthesis, experiments in

Miller, William Allen, 30

Millman, Peter, 123

Milne, Edward A., 502, 517–19

INDEX

Milton, John, 28
molecules, organic, 109, 344 n50; and biochemistry, 344 n50; on comets, 351, 367, 372–3, 375, 376, 387; and interplanetary dust, 351, 367; interstellar, 351, 367, 373–4, 538; and life, 344; on Mars, 153–4; on Jupiter, 61, 159, 374; in meteorites, 324, 351, 358, 366–72; on Titan, 159; *see also* amino acids; organic synthesis, experiments in
Monod, Jacques, *Chance and Necessity* (1971), 382–3, 398
Montagu, Ashley, 506, 514
Moon, life on, 223, 225–6, 236
Mora, Peter, 386–7
Mormons, 27
Morowitz, Harold, 384
Morris, Mark, 445, 511
Morrison, Philip: and Fermi paradox, 450, 453; and implications of contact, 506, 508, 509–11; and interstellar communication, 415–19, 421, 422, 425, 431, 432, 433, 434, 436, 451, 459, 470, 471, 472, 474; and meteorites, 370–1; and observation, 453, 454; and SETI workshops, 462, 488, 489f, 497, 510–11; and UFOs, 302–3, 306–7
Moulton, F. R., and Chamberlin–Moulton hypothesis, 167–72; on Martian life, 99
Mt. Wilson Observatory, 54, 93, 114, 115, 145
Mt. Palomar Observatory, 123
Muller, H. J., 249 n51, 334, 391–2, 478
Münch, G., 145, and Martian water vapor
Murray, Bruce, 310, 313; and SETI, 462–3; and Viking, 496

Nagy, Bartholomew, 369–71, 376
National Academy of Sciences, U. S., 4, 494; and implications of extraterrestrial life, 504; and life sciences, 357; and Mars, 322, 483; and origins of life, 388; Panels on Extraterrestrial Life, 139, 481; and planetary contamination issue, 139, 352; and SETI, 427, 438–9, 486, 489–90, 498; Space Science Board, 141, 142, 149, 215–16, 356, 427, 481, 483, 488, 496, 504–5; and UFOs, 301–2, 304, 305, 308; and Venus, 134–5
National Aeronautics and Space Administration (NASA): Ames Research Center, 141, 142, 147, 355–6, 372, 459–69, 486–90; and biological sciences, 351–7, 481; Biosciences Advisory Committee, 140, 355, 481; and cosmic evolution, 539–41; and exobiology, 352, 355–7, 358, 364–7, 377, 387, 477, 479, 481, 494–501; Exobiology Branch, 364; and exploration of Mars, 140–59, 352, 357, 394; and implications of contact with intelligence, 503–14; Office of Life Sciences, 140; and origin of life studies; 351; Planetary Astronomy Committee, 215; and SETI, 394, 439, 454, 457, 459–72, 486–90, 494–501; Space Biology Committee, 355; *see also* Jet Propulsion Laboratory; Viking spacecraft
National Bureau of Standards: and Coblentz, 106, 109; and Condon, 292; and interplanetary communication, 408
National Radio Astronomy Observatory (NRAO), 418–31, 457
natural selection, 30, 38, 382; and extraterrestrials, 386, 389, 390, 391, 393, 396, 397, 398; *see also* evolution
natural theology: and other worlds, 20–2, 24, 26, 511; *see also* Christianity; theology
Naval Observatory, U. S., 66, 109, 186, 208, 296
Naval Research Laboratory, 409, 418
Navy, U. S., 403–4, 408
nebular hypothesis, 22, 34, 162, 172, 195, 538; attack on, 166–7, 179; and Lowell, 71, 74–5, 88; revival of, 180, 189–91, 195, 197, 198, 219, 220; and Shapley, 197; and Wells, 234; *see also* planetary systems; spiral nebulae
Neumann, John von, 447
neutron star, *see* pulsars
Newcomb, Simon, 34; on Martian canals, 86–7; non-committal on life beyond Earth, 51; on planetary systems, 50, 163, 165; on Sun's position, 39; and Wallace, 39, 50–1
Newton, Isaac, and other solar systems, 19–20
Newtonians, and other solar systems, 20–2
Nicholson, Seth B., temperature of Mars, 109, 113–14; and Venus, 131–2
novae, as collisions of stars, 168 n22
nucleic acids: and Avery, 347; and cells, 362; composition of, 347; discovery of, 347; DNA as constituent of genes, 347; and interstellar clouds, 376; and origin of life, 347, 362, 381, 387–8; RNA as first replicative molecule, 362; RNA random replicator, 385; role of, 322; synthesis of 358; and Watson and Crick, 347
nucleoside, synthesis of, 347 n54, 360
nucleotide, 347 n54, 360, 362, 385

Oberth, Hermann, 285
O'Brien, Brian, 290, 292
Ockham, William of, 14
Oliver, Bernard M.: and biophysical cosmology, 542; and Drake Equation, 439–40, 442; and exobiology, 473, 477; and Green Bank conference, 427; and implications of contact, 509; and Project Cyclops, 439, 459–62; and Project Ozma, 426; and SETI, 432
O'Neill, Gerard K., 445, 446
Oparin, A. I., 338–41, 344, 435; and extraterrestrial life, 341, 348–9, 357, 358; and Haldane, 341–2, 364, 386–7; and Oparin–Haldane theory, 341–5, 350, 360–1, 362, 363, 367–8, 378, 538; and organics in comets, 372–3 n96; *Origin of Life* (1938), 340–1
70 Ophiuchi, planetary companions of, 164, 187, 202, 203, 214
Öpik, E. and Mars, 120
Oppenheimer, Robert, 282, 415
Oresme, Nicole, 14–15
organic chemistry, and biochemistry, 344 n50
organic molecule, *see* molecules, organic
organic synthesis, experiments in: 324, 339, 341, 367; in 1828, 344; by Baly (1920s), 338 n30, 342 n44; by Calvin (1951), 344–5, 352–3; as a field of work, 349; implications for extraterrestrial life, 357–8, 373–4; by Miller–Urey (1953), 3, 345–7, 352–3, 358, 360, 375, 376, 388, 474, 538; and necessity, 380, 387; by Oró, 367–8; in oxidizing atmospheres, 361–2; progress in, 356, 358, 359t, 360, 361, 376, 387–8; relevance to origin of life, 376, 387; *see also* life, origin of
Orgel, Leslie, 360, 362, 377
origin of life, *see* life, origin of
Orion, protoplanetary systems in, 209t, 215
Oró, Juan, 367; and organic cosmochemistry, 368, 373, 377 n102
Owen, Tobias, 492
Oyama, Vance, 149–58
Ozma, *see* Drake, Frank

Page, Thornton L.: and planetary systems, 190; and UFOs, 282, 306
Pagels, Heinz, 533
Paine, Thomas, 22–3, 515
Palmer, Patrick, 456
Palmer, Ray, 272
Palomar Observatory, 122
panspermia theory, *see* life, origin of
Papagiannis, Michael: and Bioastronomy Commission of International Astronomical Union, 477 n13, 490; and extraterrestrials in asteroid belt, 445, 446, 450, 477, 490, 492
parallax, stellar, 182, 205
Paris Observatory, 93, 403
Pasteur, Louis, 323, 338
Pearman, J. P. T., 427, 428, 437
51 Pegasi, 557
Percy, Greg, 227 n12
Perrotin, Henri, and Martian canals, 64, 78; rotation rate of Venus, 127
Pešek, Rudolf
Pettit, Edison: canals of Mars photographed, 114, 122; temperature of Mars, 109, 113–14
Pfund, August, 109
philosophy, and plurality of worlds, 11, 22–29
Pickering, E. C., 68; and Lowell, 71; and spectroscopic binaries, 163
Pickering, W. H.: at Arequippa, Peru, 68; intelligence on Mars, 106; and Lowell, 68, 69, 71–5, 78, 101; Martian canals as strips of vegetation, 68, 88; Martian lakes and clouds, 68; and Newcomb experiments, 87; and Strughold, 137; temperature of Mars, 112; and Wallace, 50
Beta Pictoris, 211
Pierce, Benjamin, 70
Pioneer 10 and 11 spacecraft, 448
Pirie, Norman W.: and alien life, 394–5 n132; and Bernal, 344 n48, 378; and nature of life, 336, 378; and Oparin, 343 n46
Pittendrigh, Colin, 483
planetary science, 59–60
Planetary Society, and SETI, 458, 469, 500, 542
planetary systems: 51, 160–221; abundance of, 55, 180–200, 428–9, 430; and angular momentum problem, 178, 196; and astrometric method of detection, 163, 182–9, 201–10, 209t, 213–14, 219, 220–1; and Barnard's star, 3, 188, 201–6, 207, 209t, 214; and Chamberlin–Moulton theory, 166–72, 173; and 61 Cygni, 186, 188, 202, 203, 207, 214; and double stars, 191–4; and evidence and inference problems, 543–4, 548–9; and geochemistry, 194–5; and HD 114762, 213; infrared detection of,

210–11; and Jeans–Jeffreys tidal theory, 172–80; and Lalande 21185, 188, 202, 203, 206; and Morrison workshops, 462, 463; observation of, 161, 185–6, 193, 218 n128; and 70 Ophiuchi, 164, 187, 202, 203, 214; in Orion, 209t, 215; 51 Pegasi, 557; photometric method of detection, 163, 219, 221; and planetesimal hypothesis, 168–72; and Procyon, 163, 180, 186; and pulsars, 214–16; rarity of, 51, 53, 54, 170–80, 181, 187, 328, 330, 340, 413, 434, 446, 518, 538; in science fiction, 247–8, 251, 256; and SETI, 400; and speckle interferometry, 208; and spectroscopic binaries, 163; and spectroscopic method of detection, 163, 165–6, 208, 209t, 211–14, 219, 221; and spiral nebulae, 42, 167–71, 172, 179, 193; and stellar rotation, 195–7; and supernovae, 197, 198; theory important, 161; and VB8B, 208, 211; Wallace argues against, 47; *see also* Drake Equation; nebular hypothesis; protoplanetary systems
planetary systems science, 160, 474
planetesimal hypothesis, 168–72
planets: contamination of, 138–9, 140, 352; definition of, 189, 203, 213; simulated conditions on, 137–8
plenitude, principle of: in Bruno, 16; and Lovejoy, 12–13
plurality of worlds, *see* worlds, plurality of
Poincaré, Henri, 172
Ponnamperuma, Cyril, 356, 380, 486; and exobiology, 358, 364
Pope, Alexander, 28
Princeton Observatory, 94
Proctor, Richard A.: and the evolutionary universe, 31–2; *Other Worlds than Ours* (1870), 31–2; *Our Place Among Infinities* (1875), 32; and planetary systems, 162; *Science Byways* (1875), 32; and teleology, 32; and Venus, 127; and Wells, 224
Procyon, variable proper motion of, 163, 180, 186
proper motion of stars, 163, 185; variations in, 185
proteins: and amino acids, 333 n24, 360; as building block of life, 347f; and coacervates, 340 n38; as constituent of gene, 347; as enzymes, 333–4; and insulin, 382; in interstellar dust clouds, 376; nature of, 333 n24; and nucleic acids, 358; origin of, 349, 362, 381, 382; and origin of life, 336–7, 358, 359t, 387–8; and prebiotic soup, 362; and proteinoids, 362–3; and purposeful activity, 383 n114; role of, 322, 358, 378; synthesis of, 360, 362, 364, 382, 387; as unit of life, 333–4, 336
protoplanetary systems, 210, 211, in Orion, 215
protoplasm, 333, 334; and Huxley, 45, 332, 336, 378; and Oparin, 339–40; and Wallace, 45; and Wilson, 334–5
Proxmire, Senator William, 498
Puccetti, Roland, 524–6
Pulkovo Observatory, 121
pulsars: and artificial radio signals, 455 n124; planetary systems around, 214–16
Purcell, Edward, 422, 434
Putnam, William Lowell, 71

Quimby, Freeman H., 364
Quintanilla, Hector, Jr., 291

radio astronomy, 400, 410–11, 412, 414, 416, 417, 421, 433, 436, 438, 455, 467; *see also* SETI
radiometry, infrared: and Coblentz and Lampland, 109–13; and Pettit and Nicholson, 113–14
Raible, Daniel C., 520–1
Raup, David, 396
Rea, D. G., 124
Reber, Grote, 414
Redemption, 27
Rees, Martin, 530
Regis, Edward, Jr., 512
relativity, theory of, 52
religion, and extraterrestrials, *see* Buddism; Christianity; Confucianism; Hinduism; Islam; Judaism; Mormons; Shintoism; Taoism
Rescher, Nicholas, 546–7, 549
Reuyl, Dirk, 185, 187, 201, 202, 543
Reynolds, Emerson, 233
Richard of Middleton, 14
Richardson, Robert S., 115, 122
Richey, Joseph L., 412–13
Richter, H. E., 326, 339
Riedel, Walther, 279
RNA, *see* nucleic acids
Roach, Franklin, 297, 306
Roberts, Walter Orr, 297, 306
Robertson, H. P., 282–83
Roche, E.A., 172
Roddenberry, Gene, 263
Rood, Robert, 452

Rosny, J. H. (the Elder), 224, 226–7
Ross 614: stellar companion of, 201
Rossi, Paolo, 37
Roush, J. Edward, 298
Royal Greenwich Observatory, 84
Ruppelt, Edward J., 278, 285, 286–7
Russell, Henry Norris: and Algol, 163; American Astronomical Society Symposium on Dwarf Stars and Planet-Like Companions, 188–9; and Coblentz, 111; and Jeans–Jeffreys theory of planetary system formation, 178; and life on Mars, 106, 114, 116–17; on Lowell's theory of Martian canals, 85, 106, 101; Mars as object of observation, 100; and Menzel, 279; and origin of novae, 168 n22; and oxygen on Mars, 116–17; planetary systems common in 1940s, 55, 182, 187; planetary systems rare in 1920s, 53–4; and Venusian atmosphere, 130, 132–3
Ryle, Sir Martin, 440, 511, 547

Sagan, Carl: and Byurakan meeting, 438–9, 485; and *Contact* (1985), 260–1; and *Cosmos* (1980), 512; *Dragons of Eden* (1977), 397; and Drake Equation, 437; and Dyson spheres, 438; and eavesdropping, 452; and evolution of intelligence, 397; and exobiology, 355, 473, 478–9; and Fermi paradox, 443 n94; and Green Bank conference, 427, 431; and implications of contact, 506, 510, 512; *Intelligent Life in the Universe* (1966), 54, 56, 289, 299, 435, 437, 438, 459, 492; and interstellar colonization, 450; and interstellar communication, 486; and interstellar molecules, 374; and Lederberg, 478–9; and Mars, 152, 496; and National Academy Committee, 357 n74; and organics on Jupiter and satellites, 374–5; and origin of life, 380; and Panel on Extraterrestrial Life, 481; and panspermia, 367, 375 n100; and Pioneer and Voyager messages, 448; and Proxmire, 498; and science fiction, 260–1; and SETI observations, 436, 457, 458; SETI petition, 396 n134; and UFOs, 293, 298, 299, 304, 306, 307; and Venus, 134
St. John, Charles E., oxygen on Mars, 114; water vapor on Mars, 114; and Venus, 131
Sakharov, Andrei, 439 n85
Saunders, David R., 297, 303
Schaeberle, J. M., 82
Schaeffer, Robert, 312, 314, 445
Schelling, Friedrich, 28
Schiaparelli, Giovanni: Martian atmosphere, 69; and Martian canals, 64–78, 227; and Martian canals doubled, 67; Martian surface features, 69, 82; and rotation rate of Venus, 127; and water on Mars, 69
Schlesinger, Frank, 182
Schopenhauer, Arthur, 28
Schorn, Ronald, 121 n121, 146 n189
Schrödinger, Erwin, 8
Schwartz, R. N., 432
science: cultures of, 7, 104, 313–14, 533, 550–1; evidence and inference in, 6–7, 87, 99–104, 108, 111, 121, 198, 206, 543–6; limits of, 7, 49, 52, 201, 220–1, 546–50; nature of, 4–7; *see also* unidentified flying objects
science fiction, 7, 222–66; aliens in, 222–66; and exobiology, 478; inspired by Lowell, 224; and SETI, 259–60, 266, 438; and UFOs, 272, 273, 279–80, 287; *see also* evolution; film; scientific method; television; names of individual authors
scientific method: and anthropic principle, 529, 533; astronomers vs. biologists, 330, 393–4; and extraterrestrial life debate, 4–5, 6–7, 11, 26, 29, 34, 330, 545, 548–9; and Lowell, 104; observation vs. theory, 454; Ockham's razor, 155; and probability, 177–8, 442; and Rutherford, 343; and UFOs, 269, 287–8, 297, 298, 301, 306, 308–9, 311–14, 318; *see also* Drake Equation
Scot, Michael, 14
Sedgwick, William, 334–5
See, T. J. J.: and planetary systems, 164–5, 170; and Venus, 128–9
Seeger, Charles, 462, 463
Seeliger, Hugo von, 186
Seitz, Frederick, 301
Serviss, Garrett P., 165
SETI (Search for Extraterrestrial Intelligence), 3, 213, 358, 399–472; and 21-cm line strategy, 417–18, 422, 425, 426, 440, 455, 456, 457, 458, 471; and anthropic principle, 531–2; beacons, 418; and bimodal strategy, 462; and biophysical cosmology, 542; and Bracewell probes, 432–3, 445, 450; and Congress, 454, 469–70, 498–500; Cyclops Report, 439–40, 459–62, 486, 497; as a discipline, 473–501; and Dyson spheres, 432, 434, 436, 438, 451, 471; and eavesdropping, 451–2; and exobiology, 473–

501; implications of success in, 502–26; and International Astronomical Union, 477 n13, 477–8, 490; and interplanetary communication, 401–14; interstellar communication, 414–72; and Kardashev civilizations, 435–6, 445, 451, 455, 457, 458, 471; literature, 492–94; magic frequencies, 418, 422, 440, 456, 459; and META, 458; and META II, 458; microwave window, 412, 423, 423 n43, 424f; and NASA, 459–72, 477, 479, 486–90, 497–9; Ohio State Project, 456–7; Ozma, 418–27, 454, 479, 497; Ozpa, 456; petition, 396 n134; and Phoenix, 499; and Planetary Society, 458, 469, 500; and probability of life, 417, 428–9, 437, 439, 442, 446, 447, 449, 450, 461; and probability of success, 410, 418, 436, 449, 466, 471; programs, 460t; and science fiction, 259–60, 266, 438; and Sentinel, 459; and SERENDIP, 458–9, 500 n52; signal detection, 464, 467; spectrometers for, 464–7; and universities, 497, 500; and UFOs, 313–14 n107 and 108; and waterhole, 422, 440, 457–8, 461; WOW signal, 456–7; *see also* Drake Equation; Fermi paradox; interstellar colonization; NASA

SETI Institute, 469

Seventh-day Adventists, 27

Shane, Donald, 302

Shapiro, Robert, and origin of life, 360–1, 377 n102, 384–5, 386

Shapley, Harlow: and biophysical cosmology, 541–2; and eccentric position of solar system, 52–3, 58, 523; and life on Mars, 106; and life on stars, 247 n46; and Menzel, 281 n31; *Of Stars and Men* (1958), 55, 57, 197–8, 428; views planetary systems as common, 55, 281 n31, 354, 428–9; views planetary systems as rare, 54; worldview compared to Dante, 57

Shelley, Mary, 224

Shelley, Percy Bysshe, 28

Shintoism, and extraterrestrials, 524

Shklovskii, I. S., 434–9, 449, 454, 479; and Byurakan meeting, 439, 476; and CTA-102, 455; and Drake Equation, 434, 439, 442, 449; and Fermi paradox, 448–9, 457, 471; *Intelligent Life in the Universe* (1966), 54, 56, 289, 299, 435, 437, 438, 459, 492; and new discipline, 476–7

Sholomitskii, Evgeny, 455

Simpson, George Gaylord, and exobiology, 5, 352, 392–3, 394, 395, 477, 478, 547

Sinton, William M., 141; and rotation rate of Venus, 127; and spacecraft exploration of Mars, 143–5; and temperature of Mars, 122; and vegetation on Mars, 122–5, 137, 473, 474, 543

Sirius, variable proper motion, 162

Slipher, E. C., Andes expedition, 86, 98; and Mars photographs, 122; and Martian vegetation, 108

Slipher, V. M., life on Mars, 106, 123; and spiral nebulae, 171; water vapor on Mars, 89, 100, 125

Slysh, V. I., 436, 456

Smith, E. E. "Doc," 240 n33

Smithsonian Astrophysical Observatory, 403

Snow, C. P., 7, 550

Soffen, Gerald A.: Viking project scientist, 147, 385 n118; Viking results, 154, 496

solar system: central position of, 33, 39–49, 51; peripheral position of, 37, 52–3, 55, 58, 523; *see also* planets, definition of; planetary systems; and names of individual planets

solar systems, *see* planetary systems

Sommerfeld, Arnold, 292

Southworth, G. C., 412

Soviet Union: and extraterrestrial life, 349, 434–5; and Mars exploration, 141, 352; and origin of life studies, 338, 357; and planetary contamination, 352; and planetary systems, 188 n64, 434; and science fiction, 259 n72, 260 n73; and SETI, 434–7, 438–9, 448–9, 453, 454–7, 482; *see also* Oparin, A. I.

space biology, 136–8, 141–2

spacecraft, exploration of Mars by, 135–59; *see also* Mariner; Viking; Voyager

space medicine, 136–8

Space Science Board, *see* National Academy of Sciences

spectroscopy: and Adams and St. John, 114; and Campbell, 89–90, 99, 100, 131; and Coblentz, 109; and Dunham, 116, 118; extended to orange and red, 89; infrared, 118–20, 122–5; and Kuiper, 118–20; at Lowell Observatory, 89, 100; and Martian CO_2, 118–20; and Martian life, 116–17; and Martian oxygen, 90, 100, 114, 116; and Martian vegetation, 122–5; and Martian water vapor, 69, 83, 89–90, 99, 100, 114, 116, 131, 145; at Palomar, 123; and planetary science, 60, 88–9; and plane-

spectroscopy (*cont.*)
tary systems detection, 163, 211–14; and plurality of worlds debate, 29; and Sinton, 122–5; and Slipher (V.M.), 89, 100, 125; and Spinrad, Münch and Kaplan, 145; and Venus, 128, 131–3; and Vogel, 69; *see also* Mars; Venus
Spencer Jones, Sir Harold: greenhouse effect on Venus, 133; *Life on Other Worlds* (1940), 54, 121 n 121, 133, 137, 161, 200, 342–3, 475, 492; and origin of life, 380; and planetary systems, 161, 200
Sperry, Elmer, 403
Spielberg, Steven, 264
Spinrad, H., and Martian water vapor, 145
spiral nebulae: origin of, 173; as solar systems, 42, 167–71, 172, 179, 193
Spitzer, Lyman, and Jeans–Jeffreys theory, 178
spontaneous generation, *see* life, origin of; Pasteur, Louis
Sproul Observatory, 183, 201
Stapledon, Olaf: and aliens, 239, 245–9; and Bernal, 247, 249 n51; and Clarke, 249; and Flammarion, 225; and Haldane, 245, 249; and Jeans on rarity of planetary systems, 246–7; *Last and First Men* (1930), 245–9; and Lewis, 249; and Lowell's Mars, 246; and Muller, 249 n 51; *Star Maker* (1937), 245, 247–9; 517 n 40; and Toynbee, 249
Steffens, Heinrich, 28
Steinmetz, Charles, 403
Stendahl, Krister, 506
Stephenson, David, 445
Stoney, G. Johnstone: kinetic theory of gases and planetary atmospheres, 34, 47 n22, 51, 83; and Lowell, 77
Strand, Kaj, and 61 Cygni, 186, 188, 194, 202, 543
Strauss, David, 71
Strieber, Whitley, 317
Strong, John, and temperature of Mars, 122
Strugatsky, Arkady and Boris, 259 n72
Strughold, Hubertus, 126, and astrobiology, 137–8; aviation and space medicine pioneer,136–8; *The Green and Red Planet* (1953), 136–7; life on Mars, 136–8
Struve, Otto: and "astrobiology," 476; and biophysical cosmology, 541–2; and Green Bank conference, 427, 430–1; and planetary systems, 194, 195–7, 198, 419, 421, 430, 433; and Project Ozma, 421–2, 425, 479, 497; and SETI, 442; and stellar rotation, 195–7, 434, 544; *Stellar Evolution* (1950), 194, 196; and UFOs, 281–2; and Venus, 134, 476
Sturrock, Peter, 310–11, 319, 313
Sullivan, Walter, *We Are Not Alone* (1964), 437
Sullivan, Woodruff T. III, 451–2
Swedenborgians, 27
Swift, David, 497
Swords, Michael, 315 n111

T Tauri stars, 193
Taoism, and extraterrestrials, 523, 524
Tarter, Jill, 458, 459, 460t
Tatum, E. L., 358
teleology, 27–8, 29; and anthropic principle, 527–36; and Bentley, 20; and Brewster, 26; and Eddington, 179; and Michelet, 27–8; and Proctor, 32; and Wallace, 48; and Whewell, 24; and Wilkins, 17, *see also* anthropic principle
telescope: Antoniadi's 8.5-inch, 93; and controversy over canals of Mars, 64, 65t, 66–8, 78, 94, 97, 98, 100; Flammarion's 9.6-inch, 92; Las Campanas 2.5 meter, 211; Lick Observatory 36-inch refractor, 68, 100; Lowell's 6-inch, 71, 72; Lowell 24-inch, 89; Lowell 40-inch, 109; Meudon 33-inch, 93, 100; Mt. Wilson 60-inch, 93–4, 114; Mt. Wilson 100-inch, 54; Nice Observatory 15-inch, 68; Palomar 200-inch, 122, 123; and planetary science, 59–60; Pulkovo 30-inch refractor, 121; Schiaparelli's 8.5-inch, 78; U. S. Naval Observatory 26-inch, 78; Wyeth 61-inch reflector, 123; Yerkes 40-inch, 100; *see also* radio astronomy
television, aliens in, 61, 262–3; *Buck Rogers* (1939), 262 n83; *Flash Gordon* (1936), 262 n83; *Star Trek* (1966–69, 1987), 263; *see also* film; science fiction
Teller, Edward, 345
Tennyson, 28
Ter Haar, D., 191
Tesla, Nikola, 236 n29, 401–2, 404, 411, 470, 471
Thayer, Gordon, 305
theology, 12; astrotheology, 514–26; and Huggins, 30; and Wilkins, 17
thermocouple, vacuum, 109–14
tholins, 375
Thollon, Louis, 64, 68, 78
Thomas of Strasbourg, 14
Thomson, William, *see* Kelvin, Lord

Tikhov, G. A.: and astrobotany, 122; and Martian vegetation,118, 121–2, 123
Tillich, Paul, 520
Tipler, Frank: and anthropic principle, 530–2; exobiology a psuedoscience, 5, 477–8, 547; and Fermi paradox, 446–8, 450, 451t, 452; and Proxmire, 498
Titan: as an abode of life, 61; organic molecules on, 159, 374–5
Todd, David P.: Andes expedition to photograph Martian canals, 86; and interplanetary communication, 406–8, 470
Tombaugh, Clyde, and UFOs, 279
Townes, C. H., 423, 432
Toynbee, Arnold, 249
Trefil, James, 452
Troitskii, V. S., 436, 439; and SETI observations, 456
Troland, Leonard, 334, 337, 380
Truman, President Harry S., 292
Trumpler, Robert J.: and Martian canals, 95, 98, 107, 108, 114, 543
Tsiolkovsky, K., 442 n40; 35 n74
Turner, H. H., 50
Tyndall, John, 332, 337, 378

UFOs, *see* unidentified flying objects
unidentified flying objects, 3, 267–320, 552; and abductions, 284, 295–6, 308, 317–18; as aircraft, 269; and APRO, 285; and Air Force, 4, 268, 269, 272–82, 288–307; and astronomers, 269, 272, 274–5, 276, 279, 280, 281, 282, 284, 287, 292, 294, 296, 297, 298, 301, 302, 304, 307, 311, 315, 319, 451, 473–4; and Central Intelligence Agency, 282–3; Condon Report, 268; and Congress, 289, 290, 291, 298–9, 306, 307; and cultures of science, 313t; extraterrestrial hypothesis, 269–320; and Fermi paradox, 443, 445; "flying saucer" term coined, 271; and limits of science, 547; and media, 269, 270–1, 272, 275, 278, 291 n54, 307; and meteorite phenomenon, 267, 307; and MUFON, 315 n111; and New Age ideas, 308, 314–16; New Wave theories, 308, 314–16; and NICAP, 285; numbers of, 270t; polls on, 3, 271–2, 281, 319; and pseudoscience, 315–16, 318–19; and science fiction, 272, 273, 279–80, 287; and scientific method, 269, 280, 287, 288, 297, 298, 301, 306, 308–9, 311–14, 317–18 n117, 318; and SETI, 313–14 n107 and n108; "UFO" term coined, 278 n23; wave phenomenon, 269; and Wells, 271 n7
United States, *see* Air Force, U.S.; Central Intelligence Agency; Congress; National Academy of Sciences; National Bureau of Standards; National Radio Astronomy Observatory; Naval Observatory, U.S.; Navy, U.S.
universe, size of: according to Spencer Jones, 54; according to Wallace, 42–3
Urey, Harold: and extraterrestrial life, 356; and Miller–Urey experiment, 345–7, 352–3, 360, 375, 376, 388, 474; and meteorites, 371–2; and organics on Jupiter, 374; and planetary atmospheres, 349, 345, 361 n79; and planetary systems, 194–5; and Sagan, 479; and Space Science Board, 427; and Venus as desert, 133; *see also* organic synthesis, experiments in

Vallee, Jacques F., and UFOs, 287–8, 314–16, 320
van Biesbroeck 8 (VB8), 208, 211
van Biesbroeck 10, 208
van de Hulst, 422 n42
van de Kamp, Peter, 3; and planetary systems, 183–5, 188, 194, 201–6, 320, 473, 544
Vandenberg, Hoyt S., 274
Van Maanen, 205
Vega, 210
Venus, 60–1, 126–34, 128t, 503 n6; and Abbot, 130–2; animals on, 132, 133; atmosphere, 127, 128t, 130, 132, 133, 134–5; carbon dioxide, 132, 133–4; clouds, 129, 131, 132, 135; Coblentz and Lampland work on, 113; conditions for life on, 46, 126–34, 128t; greenhouse effect, 133, 134; and Lowell, 128–9; and origin of life, 348, 356; oxygen on, 60, 131–3, 134; rotation rate, 127, 128; and Schiaparelli, 127; in science fiction, 243, 246, 250, 251; and See, 128–9; sulfuric acid, 135; temperature, 130, 131, 134; and UFO phenomenon, 269, 275; water on, 129, 131–2, 133–4
Verne, Jules, 224, 225–6
Verschuur, G., 456
Very, Frank W.: oxygen on Mars, 90; water vapor on Mars, 89, 90, 125, 545
Viewing, David, 443–6
Viking spacecraft, 146–59, 351, 356, 385 n118, 399, 414, 479, 481, 490, 491, 499; biology experiments, 147–59, 343, 355, 363, 539, 548; cost, 146, 147, 477; gas exchange experiment, 149; labeled

Viking spacecraft (*cont.*)
release experiment, 149–50; molecular analysis experiment, 147, 153, 154; and origins of life, 351; pyrolitic release experiment, 150–1; results, 152–8, 382, 387, 388; and SETI, 463; superoxide interpretation of data, 155
virus, 334
Vishniac, Wolf, 142
vitalism, *see* life
Vivian, Weston, and UFOs, 291
Vogel, Hermann, 69
Voltaire, 27, 223
Voyager 1 and 2 spacecraft, 447; record on, 448
Voyager (Mars) spacecraft, 146, 491, 495–6

Wald, George, 347–8, 354, 380, 396, 505–7, 511
Wallace, Alfred Russel, 34, 38–49, 330, 389–91; anthropocentric world view, 36–49, 535; and binary stars, 47; biological approach to plurality of worlds, 38, 321, 330; compared to Lowell, 38; and complexity of life, 44–6; and conditions for life, 45; and evolution, 38–9, 48–9, 389–91; humanity distinguished from animals, 38–9; improbability of intelligence evolving beyond Earth, 49, 389–90, 392, 394; life of, 38; *Man's Place in the Universe* (1903), 38–49; Martian canals, 79, 88; and natural selection, 38–9; planetary conditions not like Earth, 46–7; planetary systems unlikely, 47, 162; position argument for uniqueness of Earth, 40–8; on protoplasm, 45; reaction to *Man's Place*, 50–8; and Whewell, 38; *The Wonderful Century* (1898), 39
Watson, James, and genetic code, 347
Weinberg, Stephen, 536
Weinbaum, Stanley: and aliens, 241; "A Martian Odyssey" (1934), 242–4
Weizsäcker, Carl Friedrich von, and nebular hypothesis, 189–91, 194, 198, 199, 200, 277, 344
Welles, Orson, and "War of the Worlds," 115, 231, 236, 271
Wells, G. P., 337, 342 n44, 391
Wells, H. G., 231–8; and alien nature, 265, 266, 552; biology text, 337, 342 n44, 391; and Bradbury, 251, 252; and extraterrestrial intelligence, 2, 224; *The First Men in the Moon* (1901), 236; and Gore, 232; and Huxley, 231; and Lasswitz, 227, 267, 515; and Lewis, 250; and Lowell's Mars, 232, 234–5, 236; and Maunder, 93; and planetary contamination, 138; and plurality of worlds tradition, 231–2; and Proctor, 232; and silicon life, 232; and Stapledon, 246; and *The War of the Worlds* (1897), 222, 231–8, 240, 241, 552; and UFOs, 271 n7
Wheeler, John, 528, 529–30
Whewell, William, 23–6; against use of analogy, 25–6; binary stars not suitable for planets, 25; and boundaries of science; 547; and natural theology, 26; rejects pluralism in *Of a Plurality of Worlds* (1853), 24; rejects teleological arguments, 24; stars may not be like Sun, 25; and Wallace, 38
Whipple, Fred L.: and radio astronomy, 411; and solar system, 194, 274–5; and UFOs, 315; and Venus, 133–4
Wickramasinghe, Chandra, 373, 375–7
Wildt, Rupert: greenhouse effect on Venus, 133; methane and ammonia on Jupiter and Saturn, 340
Wilkins, Bishop John, *Discovery of a World in the Moone* (1638), 17
William of Auverne, 14
William of Ware, 14
Wilson, E. B., 334–5, 343
Wilson, E. O., 396 n134
Wilson, Patrick, 534
Wilson, W., 117
Winkler, Kenneth, 532–3
Woese, Carl, 362
world: term redefined in sixteenth century, 15
worlds, plurality of: and anthropocentrism, 27–8, 32, 36–58; and cosmology, 11–22; and God's omnipotence, 14–15; medieval ideas about, 14–15; and philosophy, 22–9; and nineteenth century science, 29–35; and science fiction, 223–5, 228, 231–2, 238; *see also* life, extraterrestrial
Wright, Thomas, 21–2
Wright, W. H., 108
Wright brothers, 269

Yefremov, Ivan, 260 n73
Yerkes Observatory, 92, 98
Young, Richard S., 355, 364

Zuckerman, B., 456